ANALYSIS OF VARIANCE DESIGNS

One of the most fundamental and ubiquitous univariate methodologies employed by psychologists and other behavioral scientists is the analysis of variance (ANOVA). *Analysis of Variance Designs* presents the foundations of this experimental design, including assumptions, statistical significance, strength of effect, and the partitioning of the variance.

Exploring the effects of one or more independent variables on a single dependent variable as well as two-way and three-way mixed designs, this textbook offers an overview of traditionally advanced topics for advanced undergraduates and graduate students in the behavioral and social sciences. Separate chapters are devoted to multiple comparisons (post hoc and planned/weighted), ANCOVA, and advanced topics. Each of the design chapters contains conceptual discussions, hand calculations, and procedures for the omnibus and simple effects analyses in both SPSS and the new point-and-click *SAS Enterprise Guide* interface.

Glenn Gamst is Professor and Chair of the Psychology Department at the University of La Verne, where he teaches the doctoral advanced statistics sequence. He received his doctorate from the University of Arkansas in experimental psychology. His research interests include the effects of multicultural variables, such as client–therapist ethnic match, client acculturation status and ethnic identity, and therapist cultural competence, on clinical outcomes. Additional research interests focus on conversation memory and discourse processing.

Lawrence S. Meyers is Professor of Psychology at California State University, Sacramento, where he teaches undergraduate and graduate courses in research design, data analysis, data interpretation, testing and measurement, and the history and systems of psychology. He received his doctorate from Adelphi University and worked on a National Science Foundation Postdoctoral Fellowship at the University of Texas, Austin and Purdue University.

A. J. Guarino is on the faculty at Auburn University, teaching the statistics sequence of ANOVA, multiple regression, MANOVA, and structural equation modeling (SEM) in the College of Education. He received his bachelor's degree from the University of California, Berkeley. He earned a doctorate in statistics and research methodologies from the University of Southern California through the Department of Educational Psychology.

Analysis of Variance Designs

A Conceptual and Computational
Approach with SPSS and SAS

GLENN GAMST
University of La Verne

LAWRENCE S. MEYERS
California State University, Sacramento

A. J. GUARINO
Auburn University

CAMBRIDGE
UNIVERSITY PRESS

CAMBRIDGE UNIVERSITY PRESS
Cambridge, New York, Melbourne, Madrid, Cape Town, Singapore, São Paulo, Delhi

Cambridge University Press
32 Avenue of the Americas, New York, NY 10013-2473, USA

www.cambridge.org
Information on this title: www.cambridge.org/9780521874816

First published 2008

Printed in the United States of America

A catalog record for this publication is available from the British Library.

Library of Congress Cataloging in Publication Data

Gamst, Glenn.
Analysis of variance designs : a conceptual and computational approach with SPSS and
SAS / Glenn Gamst, Lawrence S. Meyers, A. J. Guarino.
   p.   cm.
Includes bibliographical references and index.
ISBN 978-0-521-87481-6 (hardback)
1. Analysis of variance.   2. SPSS (Computer file)   3. SAS (Computer file)   I. Meyers,
Lawrence S.   II. Guarino, A. J.   III. Title.
QA279.G36   2008
519.5′38 – dc22   2008008948

ISBN   978-0-521-87481-6 hardback

Contents

Preface

TO THE INSTRUCTOR

The present text is an exploration of univariate methodology where the effects of one or more independent variables are assessed on a single dependent variable. Such univariate designs are ubiquitous in the social, behavioral, and biological science literature. We have chosen, in this book, to focus our efforts on analysis of variance (ANOVA). Issues concerning multivariate methodology, including multiple regression analysis, are not covered in the present text as a result of space limitations, but they are addressed in a companion text (see Meyers, Gamst, & Guarino, 2006).

This book owes both a conceptual and computational debt to early ANOVA pioneers, beginning with the seminal work of Fisher (1925, 1935), who focused on solving agricultural problems with experimental methods. Fisher's early work was adapted to other fields, including the social and behavioral sciences, and in doing so moved from childhood to early adolescence with the work of Baxter (1940, 1941), Crutchfield (1938), Garrett and Zubin (1943), Lindquist (1940), Snedecor (1934), and Yates (1937). By the 1950s, ANOVA procedures were well established within most social and behavioral sciences (e.g., Cochran & Cox, 1957; Lindquist, 1953; Scheffé, 1959).

Beginning in the early 1960s, ANOVA procedures were further delineated and popularized by Winer (1962, 1971) and Winer, Brown, and Michels (1991). These works, while sometimes challenging to read, were considered the "gold standard" by many ANOVA practitioners. In the 1970s, and up until the present, Geoffrey Keppel and his associates (Keppel, 1973, 1982, 1991; Keppel & Saufley, 1980; Keppel, Saufley, & Tokunaga, 1992; Keppel & Wickens, 2004) have helped to formalize the teaching of ANOVA through their innovative mnemonic notation system.

Pedagogically, there are at least three ways that an instructor can approach ANOVA; the present text attempts to address each of these course schemas. One approach is to emphasize the drill and practice of computational formulas and procedures for creating sums of squares, mean squares, and so on. Texts that have catered to such courses have run the gamut from purely step-by-step computational procedures (e.g., Bruning & Kintz, 1968; Collyer & Enns, 1987) to conceptual and computational medleys (e.g., Keppel et al., 1992). Instructors of these courses believe that students are best served by teaching them the computational mechanics behind the ANOVA results.

A second approach has been motivated by the proliferation of microcomputer technology and its attendant statistical software (e.g., SPSS and SAS). Statistical software-oriented instructors believe student interest and motivation for conducting statistical analyses may be better enhanced and maintained with a focus on statistical conceptualization accentuated with computer application. Hence, these instructors focus on the interpretation of computer-statistical output and are less inclined to have students "plug away" with a calculator to solve a statistical problem. A number of texts that vary in ANOVA topic coverage have emerged that attempt to address this instructional niche (e.g., Kinnear & Gray, 2006; Page, Braver, & MacKinnon, 2003).

Of course, a third pedagogical approach is to combine the first two approaches. Instructors who operate from this third orientation see the benefit, to the student, of requiring computational drill and practice with a calculator *and* computer application.

First, the present text attempts to bridge the gap between the two primary pedagogical orientations mentioned previously. This is accomplished by providing a thorough, readable, and well-documented conceptual foundation for each of the ANOVA topics covered in the text. Second, computational examples are provided for most of the topics covered. Third, SPSS and SAS screen images and computer output are illustrated for all ANOVA problems covered in the text.

This text consists of seventeen chapters that are segmented into six topical sections.

Section 1, "Research Foundations," consists of two chapters. Chapter 1, "ANOVA and Research Design," explores various types of research strategies, including nonexperimental, quasiexperimental, and experimental research designs. Key concepts of independent and dependent variables, scales of measurement, and between-subjects and within-subjects designs are also introduced. In Chapter 2, "Measurement, Central Tendency, and Variability," the central tendency concepts of the mean, median, and mode, along with the variability concepts of range, variance, and standard deviation, are explained. Numerical examples are also provided.

Section 2, "Foundations of Analysis of Variance," consists of three chapters. In Chapter 3, "Elements of ANOVA," the partitioning of the total variance into between-groups and within-groups variability is explained with a concrete study example. Chapter 4, "The Statistical Significance of F and Effect Strength," includes topics of the F ratio, the sampling distribution of F, statistical significance, and magnitude of treatment effects. In Chapter 5, "ANOVA Assumptions," the three fundamental assumptions underlying ANOVA are covered: independence of errors, normality of errors, and homogeneity of variance. Ways of assessing these violation assumptions with SPSS and SAS are offered.

Section 3, "Between-Subjects Designs," consists of four chapters. Chapters 6, 8, and 9 provide conceptual and computational foundations for one-, two-, and three-way between-subjects designs, respectively. Chapter 7 provides an overview of multiple comparisons procedures.

Section 4, "Within-Subjects Designs," consists of three chapters. Chapters 10, 11, and 12 cover the conceptual and computational details associated with one-, two-, and three-way within-subjects designs.

Section 5, "Mixed Designs," also consists of three chapters. Chapters 13, 14, and 15 cover simple mixed designs, complex mixed designs with two between-subjects factors and one within-subjects factor, and complex mixed designs with one between-subjects factor and two within-subjects factors, respectively.

Section 6, "Advanced Topics," concludes with Chapter 16 on analysis of covariance and Chapter 17 on advanced topics, which provides a brief introduction to a number of advanced ANOVA topics.

We believe that Chapters 1–8, 10, and 13 provide sufficient depth for an advanced undergraduate course that emphasizes ANOVA procedures. Beginning graduate students would benefit from these chapters also, in addition to the chapters dealing with the three-way and mixed designs, as well as the advanced topics chapters.

TO THE STUDENT

This text assumes that you, the student, have had at least one undergraduate course in statistics. Typically, these introductory courses do not cover ANOVA in much depth, if at all. This book will show you when to use these procedures and also how to perform these computations with a calculator and with SPSS and SAS statistical software. In preparing this book, we used SPSS version 16.0 and *SAS Enterprise Guide* 4.0. At the time you are reading this, it is quite likely that newer versions will have become available; however, we anticipate that virtually all of what you find here will be able to be applied to the newer software.

We accomplish these goals by providing you with what we believe to be a readable conceptual overview of all ANOVA topics covered. For many of these topics, we delineate step-by-step computational procedures for calculating various statistics by hand. We follow the hand computations with step-by-step screen images of how to perform these same procedures with SPSS and SAS. We also provide an example results section write-up for each procedure that we cover. Based on the specific task demands of your particular course and instructor, you may want to allocate more or fewer resources to certain parts of each chapter.

We hope you enjoy using this text. We also hope that it helps you in reading the journal articles that make use of ANOVA techniques in their data analysis sections, and that it encourages you to explore the use of ANOVA designs in your own research applications.

ACKNOWLEDGMENTS

We thank the many students each of us has taught over the years (at University of LaVerne; California State University, Sacramento; and Auburn University) for their helpful feedback, thoughtful questions, and enthusiasm for this project. We have tried to keep your varied needs and interests in our "mind's eye" at all times during the writing process.

Special thanks go to Elita Burmas for her invaluable help and always positive spirit in editing and re-editing all of the many figures, tables, and statistical notation used in this book. Her labor made this project attainable, and we owe her a great debt.

Very special thanks go to Dennis R. Bonge (University of Arkansas, Psychology Department, retired) for kindling an embryonic interest (in the first author) for ANOVA procedures more than thirty years ago. Your diligence, rigor, and humor will always be appreciated.

We would also like to thank Lauren Cowles, Senior Editor of Mathematics and Computer Science at Cambridge University Press. Her thoughtful comments and advice on this project have been much appreciated. Mary Paden, our project manager at Aptara Inc., guided us through the process of turning all our words and images into a book. Finally, we are very grateful to Susan Zinninger. Entering the process at the page proofing stage, her copy editing skill and keen eye made a significant contribution to the readability of this book.

Glenn Gamst
Larry Meyers
Tony Guarino

SECTION 1

Research Foundations

CHAPTER ONE

ANOVA and Research Design

1.1 WHAT IS ANALYSIS OF VARIANCE?

Analysis of variance (ANOVA) is a statistical technique used to evaluate the size of the difference between sets of scores. For example, a group of researchers might wish to learn if the room color in which college students are asked to respond to questions assessing their mood can affect their reported mood. Students are randomly assigned to complete a mood inventory in one of two rooms. Random assignment, one of the hallmarks of experimental design, is used in an attempt to assure that there is no bias in who is placed into which group by making it equally likely that any one person could have been assigned to either group. One of the rooms is painted a soft shade of blue that was expected to exert a calming effect on the students; the other room is painted a bright red that was presumed to be more agitating than calming. Higher numerical scores on the mood inventory indicate a more relaxed mood. At the end of the study, we score the mood inventory for all participants.

The research question in this example is whether mood as indexed by the score on the mood inventory was affected by room color. To answer this question, we would want to compare the mood scores of the two groups. If the mood scores obtained in the blue room were higher overall than those obtained in the red room, we might be inclined to believe that room color influenced mood.

One way to start the comparison process is to take an average (a mean) for each group of the responses to the mood questions and visually inspect these two values. But comparing the scores between the groups in order to draw a conclusion about the effect of the color of the room on mood is not always going to be a simple matter. Among the related questions that we face in doing the comparison are:

- How much of a difference is there between the means?
- How much overlap is there in the scores of the members of each group?
- Based on the difference in the means and the overlap in scores, is the mean difference sufficiently large for us to say that room color made a difference?

It is possible to specify a statistical procedure that takes the answers to these questions into account in providing us with a set of results. This procedure is known as the analysis of variance, or ANOVA, and here is its story.

1.2   A BRIEF HISTORY OF ANOVA

The ANOVA began its life in the second decade of the twentieth century with the statistical and experimental design writings of Sir Ronald Aylmer Fisher, who we know today as R. A. Fisher. The ANOVA was born of Fisher's creative mind during the time that he was working as head of the Rothamsted Agricultural Experimental Station. As described by Salsburg (2001), Fisher took this position in 1919 rather than becoming a chief statistician in the Galton Laboratories, which was then under the supervision of Karl Pearson. Rothamsted was the oldest agricultural research institute in the United Kingdom, established in 1837 to study the effects of nutrition and soil types on plant fertility, and probably appealed to the farmer in Fisher. At the same time, it was also the case that he probably did not want to work with Pearson (over his career, Fisher was to have a less than positive relationship with Pearson), making the decision of which job to take a bit easier. He remained at Rothamsted until 1933 when, in an interesting historical development, he replaced Karl Pearson at University College. Fisher moved on to Cambridge in 1943 where he spent the rest of his career until he retired in 1957.

For about ninety years before Fisher arrived, the Rothamsted Station had been experimenting with different kinds of fertilizers by using a single fertilizer product on the entire field during a single year and measuring, together with a variety of other variables such as rainfall and temperature, the crop yield for that year. The institute used a different fertilizer in the next year, a different one the year following, and so forth. It thus attempted to compare fertilizers across years while taking into account differences in temperature, rainfall, and other environmental variables. Fisher (1921a) was able to demonstrate that, despite the elaborate mathematical treatment of the data by those who worked at the station before him, one could not draw any reliable conclusions from all of that work over the course of almost a century (Salsburg, 2001).

What Fisher did was to revolutionize the way in which the agricultural experiments were done (Salsburg, 2001) by comparing the effects of more than one fertilizer within a single year by using all of them simultaneously on different nearby plots. To mostly control for local conditions within the field, he would take a block of plots and randomly assign fertilizers to them. Any differences between the fertilizers in terms of crop yield, aggregated over the entire field of crops, could then be attributed to the product and not to one area receiving more rainfall or having better drainage than another.

Not only did Fisher practically invent a powerful, elegant, and relatively simple experimental procedure, he produced the statistical technique to analyze the data collected through such a procedure. This technique was the ANOVA as well as the analysis of covariance (ANCOVA). He laid the groundwork and documentation for this work as well as the experimental design innovations through a series of what are now considered to be classic publications (Fisher, 1921b, 1925, 1935; Fisher & Eden, 1927; Fisher & Mackenzie, 1923). This analytic technique of ANOVA has now become the foundation of entire curricula in research methods courses in the social and behavioral sciences and is the focus of this book.

1.3   DEPENDENT AND INDEPENDENT VARIABLES

ANOVA is a statistical technique that is applied to a data set. These data may have been collected in either an experimental or nonexperimental manner, but in performing the ANOVA it is essential that the researchers have identified which variables are assigned the roles of the dependent and independent variables.

1.3.1   THE CONCEPT OF A VARIABLE

A variable is anything that can take on different values. According to this definition, almost everything that is a part of our world is a variable. To name just three examples, the speed with which we drive down the street, the intensity of our caring for another person, and the color of our hair are all variables. This is because we can drive down the street at 10, 15, or 25 miles per hour; we can care for another somewhat, a moderate amount, or considerably; and our hair color can be blond, brown, black, or grey.

1.3.2   DEPENDENT VARIABLE

Dependent variables or *dependent measures* reflect the outcome of the study, and a useful way to conceptualize them is as *outcome variables* or *outcome measures*. In the example we used in Section 1.1, the responses to the mood questions were the dependent variable in the study. Participants answered the questions according to their own individual feelings, and so the numerical values we would obtain from them could take on a range of possible values. In other research studies, dependent variables could be the number of patients in a particular category (e.g., females who have contacted a medical provider for a flu vaccination), the number of words correctly recalled, or the relative amount of improvement in mental health that clients exhibited following their psychotherapy.

1.3.3   INDEPENDENT VARIABLE

Independent variables reflect the factors that may be said to influence the dependent variable; they can be conceptualized as *input factors, treatment conditions,* or *treatment effects.* In our example, the color of the room in which participants completed the mood inventory was the independent variable. The room color was either blue or red; these are said to be the *values* or *levels* of the independent variable. Although room color took on only two possible values in this study, such variation is enough to define it as a variable. It is conceived as an independent variable because it would be the presumed cause of differences in mood scores between the groups.

1.3.4 THE RELATIONSHIP OF DEPENDENT AND INDEPENDENT VARIABLES

Dependent and independent variables are intimately related to each other. Dependent variables are presumed to be influenced or affected by independent variables. In the example, the dependent variable of mood was thought to be influenced by the color of the room in which the students answered the mood questions. The form that this influence takes is a function of the nature of the research. In the example, the influence is presumed to be causative; that is, the possibility that room color caused or produced differences in mood was examined.

Another way to express the idea of influence is to predict the dependent variable based on the values of the independent variable. For example, we might want to predict the degree to which individuals hold intrinsically motivated goals such as making contributions to their community and having enriched relationships with other people. This could be measured by a paper-and-pencil inventory. Using a nonexperimental research design, we could administer other inventories such as those measuring self-esteem and spirituality. These latter two variables could then be used to predict the degree of intrinsic motivation reported by participants. In doing so, these predictors would be treated as independent variables in the data analysis.

1.4 THE IMPORTANCE OF VARIATION

The key element to bear in mind when we talk about dependent variables and independent variables is that in order to be a viable variable in our statistical analysis, that variable must demonstrate variation in our study; if it does not vary, it is a constant rather than a variable. Although this may sound obvious, remembering this maxim can be quite helpful in selecting variables for a research design and in designing a research study.

In terms of independent variables, this requirement of variation is really an absolute rule. In our example study, if all of the participants completed the mood inventory in the same color room, then room color would not be able to be used as an independent variable in our study. Yes, room color in the abstract is a variable in that rooms can take on a wide variety of values (colors), but in the specific circumstances of our experiment, in order for room color to be a variable, we must make sure that it varies for the two groups in question.

Other examples of variables can be mentioned to provide you with a sense of what needs to be considered in designing and analyzing the data from a research study:

- If you wish to compare females and males on some performance measure, both genders must be represented in the study.
- If you wish to compare psychodynamic, cognitive-behavioral, and client-centered psychotherapy on improvement in general functioning, then all three types of therapy must be included in the study.
- If you wish to evaluate the differences in standardized math test performance in school districts of different sizes, then different sized school districts must be represented in the study.

The same argument holds for dependent or outcome variables even though we have no control in advance of the study over what values these measures may take. The key is to assure ourselves in advance of the data collection that the values we could have obtained on the dependent variable were capable of being different and that there was nothing about the way we designed the data collection process that may have biased or limited these values in any reasonable way. In the mood study, for example, we would not want to expose the students to stimuli that might make them especially joyous or sad; that could certainly bias their feelings and corrupt any conclusions we might otherwise draw from the results of the experiment.

1.5 EXPLORATORY RESEARCH AND HYPOTHESIS TESTING

Scientists do not collect data at random. Instead, we have a rather specific idea about the particular subject matter on which we intend to focus in a particular study. These ideas can be more specific or less specific. Up to a point, the mood study we have been describing could have resulted from either strategy.

The focus of the research could be somewhat broad and less specific. For example, researchers may wish to study the variety of factors that affect mood, and select the color of the surroundings as one of the many variables they will use. Based on this, they design a study in which students complete a mood inventory in either a blue or red room. This sort of research is considered exploratory, and the researchers may have few expectations but instead intend to simply observe what happens.

The research focus could also be relatively specific. For example, the researchers may know that color of the surroundings can affect mood, and that blue tones create relative calm whereas red tones create more excitement. Now they wish to put their expectations to a specific test. Based on this, they design a study in which students complete a mood inventory in either a blue or red room. In conjunction with this design, they formulate a hypothesis – specific predictions derived from their model or understanding of how external stimuli affect mood – that students in the blue room will report higher scores than those in the red room. They then conduct the study to test their hypothesis.

Hypothesis testing is one of the ideals of social and behavioral research, but much of the research that is published in the psychological literature is more exploratory in nature. Both approaches to research, hypothesis testing and more exploratory studies, are productive in their place, and these approaches are not incompatible with one another. If our models are not sufficiently formulated to frame hypotheses, then it is not often useful to "force" hypotheses to be articulated. Instead, it is appropriate to formulate a more general research question and engage in the exploratory research necessary to answer that question. This more exploratory work will eventually lead to a fuller understanding of the subject matter, which will, in turn, allow us to construct a more complete model that we may be able to then test in a more formal way.

The ANOVA technique is equally applicable to both research strategies. So long as dependent and independent variables are specified, the statistical analysis will compare a group of scores to one or more other groups of scores and supply us with information about the differences between those sets of scores. With that information, researchers can either take the results as is or can evaluate those differences with respect to any hypotheses they have put forward.

Measurement, Central Tendency, and Variability

2.1 SCALES OF MEASUREMENT

In Chapter 1, we indicated that variables can take on different values. Here, we will deal with the process of associating values with variables and how to summarize the sets of values we obtain from that process. This will prove to be important in guiding us toward situations in which it is appropriate to use the ANOVA technique.

2.1.1 THE PROCESS OF MEASUREMENT

Measurement represents a set of rules informing us of how values are assigned to objects or events. A *scale of measurement* describes a specific set of rules. Psychology became intimately familiar with scales of measurement when S. S. Stevens, in 1946, wrote a response to a committee of the British Association for the Advancement of Science. That committee was dealing with the possibility of measuring sensory events, and the members could not agree on whether or not it was possible to perform such measurements. Because they had focused on the work that Stevens had done to measure loudness of sound, he felt that a response to the committee from him was appropriate. But instead of arguing with the committee, Stevens adopted a different strategy, presenting instead a theory of measurement. He elaborated on this theory in 1951 when, as the editor of the *Handbook of Experimental Psychology,* he gave prominent treatment to this topic in his lead chapter.

Stevens identified four scales in his theory: *nominal, ordinal, interval,* and *ratio* scales in that order. Each scale includes an extra feature or rule over those in the one before it. We will add a fifth scale to Stevens's treatment, *summative response scaling,* placing it between the ordinal and the interval scale.

2.1.2 NOMINAL SCALES

A nominal scale of measurement, sometimes called a *categorical scale, qualitative measurement,* or *categorical measurement,* has only one rule underlying its use: different entities receive different values. This is done without implying any quantitative dimension at all; that is, there is no implication that one entity has any more of a property than another. Examples of nominal scales include numbers on football uniforms, types of computer platforms, demographic variables such as gender and ethnicity, and the names of participants or the arbitrary code numbers assigned to them.

Numerical coding of categorical variables is regularly done in the course of data analysis. For example, in specifying the gender of the participants in a study, we would use numerical codes such as 1 and 2. Which gender is associated with which numeric code is arbitrary and is not important; what is important is that the coding scheme be used consistently throughout the data set. Thus, some researchers always code females as 1 and males as 2, whereas others may adopt a different coding strategy for each separate data set.

2.1.3   ORDINAL SCALES

An ordinal scale of measurement uses numbers to convey "less than" and "more than" information. This most commonly translates as rank ordering. Objects may be ranked in the order that they align themselves on some quantitative dimension but it is not possible from the ranking information to determine how far apart they are on the underlying dimension. For example, if we were ranking the height of three people, a person seven feet tall would be ranked 1, someone five feet, one inch tall would be ranked 2, and someone else five feet tall would be ranked 3. From the ranked data, we could not determine that two of the individuals were quite close in height.

2.1.4   SUMMATIVE RESPONSE SCALES

Summative response scales, such as five-point and seven-point scales, require respondents to assign numbers to represent attitudes or judgments. For example, a five-point scale may have an anchor (a definition of a scale point) for the value of 1 as *not very true for me* and an anchor for the value of 5 as *very true for me*. Respondents would be presented with a set of items (e.g., "I feel as though I am a worthwhile person," "I enjoy meeting new people at social gatherings") and their task would be to place the scale value that represented how true the statement was for them by each item. This type of measurement can be traced back to Rensis Likert (pronounced "lick-ert" by the man himself) in the 1930s (Likert, 1932; Likert, Roslow, & Murphy, 1934; Murphy & Likert, 1937). Likert was trying to find a more efficient way to accomplish what Louis Thurstone had done in his pioneering work to develop interval level measurement scales to assess attitudes (Thurstone, 1927, 1928, 1929; Thurstone & Chave, 1929).

The numbers on the scale are based on an underlying continuum defined by the anchors and are ordered, typically in an ascending way, to reflect more of the property being rated. It is called a *summative scale* because it is possible to add (summate) the ratings of several items on an inventory and to divide by a constant (usually in the process of taking a mean) to obtain an individual's score on the inventory.

The key to working with this sort of scale is to appreciate that the average derived from a summative response scale is meaningful. Let's say that we administered a short self-esteem inventory to a class of public policy graduate students and that one item on the inventory read, "I feel that I am a worthwhile person." Assume that items were rated on a five-point scale with higher values indicating more endorsement of the statement. Let's further say that mean for this item based on all of the students in the class was 4.75. Is that value interpretable? Yes, it indicates that the individuals in the sample believed pretty strongly on average that the content of the item was quite true for them, namely, that they were worthwhile people.

2.1.5 INTERVAL SCALES

Interval scales of measurement have all of the properties of nominal, ordinal, and summative response scales but include one more important feature. Fixed distances between the numbers represent equal intervals.

The most common illustrations of an equal interval scale are the Fahrenheit and Celsius temperature scales. According to Stevens (1951, p. 27), "Equal intervals of temperature are scaled off by noting equal volumes of expansion. . . . " Essentially, the difference in temperature between 30 and 40° F is equal to the difference between 70 and 80° F. A less-obvious but important characteristic of interval scales is that they have arbitrary zero points. For example, the term *zero degrees* does not mean the absence of temperature – on the Celsius scale, zero degrees is the temperature at which water freezes.

As was true for summative response scales, it is meaningful to average data collected on an interval scale of measurement. We may therefore say that the average high temperature in our home town last week was 51.4° F. Note, however, that summative response scales are not quite at the interval level. That is because in most cases it is not true that the difference between the scale values of, say, 1 and 2 represent the same psychological distance value as that between, say, 2 and 3.

2.1.6 RATIO SCALES

A ratio scale of measurement has all of the properties of nominal, ordinal, summative response, and interval scales but includes one more important feature. It has an absolute zero point, where zero means absence of the property. Examples of ratio scales are time and measures of distance. Because of this, it is possible to interpret in a meaningful way ratios of the numbers on these scales. We can thus say that four hours is twice as long as two hours or that three miles is half the distance of six miles.

2.1.7 QUALITATIVE VERSUS QUANTITATIVE MEASUREMENT

It is useful for our purposes to identify two general categories into which we can classify subsets of these measurement scales: *qualitative* and *quantitative measurement*. Qualitative measurement is what we obtain from using a nominal scale of measurement. Researchers sometimes call qualitative variables by other names:

- Categorical Variables.
- Nonmetric Variables.
- Dichotomous Variables (when there are only two values or categories).
- Grouped Variables.
- Classification Variables.

It is useful for our purposes to think of quantitative measurement in a somewhat restrictive manner. Although the ordinal scale certainly presumes an underlying quantitative dimension, we would generally propose thinking in terms of those scales for which it is meaningful and informative to compute a mean. With the ability to compute a mean and all that this ability implies, the gateway is open to performing a whole range of statistical procedures such as the ANOVA. Summative response, interval, and ratio scales meet this standard. Researchers sometimes call quantitative variables by other names, such as

- Continuous Variables.
- Metric Variables.
- Ungrouped Variables.

2.2   CENTRAL TENDENCY AND VARIABILITY

In the ANOVA designs that we will cover in this book, the dependent variables will be measured on a quantitative scale of measurement. We are therefore able to generate summary statistics that quickly convey a great deal of information about a set of scores. We will focus on two classes of summary measures: central tendency and variability.

Measures of *central tendency* provide an index (or single-value summary) of the most typical score in a set or distribution of scores. We use a measure of central tendency, the mean, to provide a convenient way to describe the impact of our independent variable on the dependent measure. Thus, we might hypothesize that the mean of one group (e.g., the experimental group) will be greater on the dependent variable than the mean of another group (e.g., the control group).

Variability addresses the issue of how scores within a group or treatment condition vary or deviate from one another. Knowledge of variability (as we will see later in this chapter) helps us gauge whether the manipulation of our independent variable is really producing differences between or among the means of our treatments, or if these observed differences are simply due to random or chance fluctuation. The difference between the means of the experimental and control groups is evaluated with respect to the variability within each of the two groups.

2.3 THE MEAN AS A MEASURE OF CENTRAL TENDENCY

2.3.1 GENERAL CONCEPTION OF THE MEAN

The arithmetic mean, or simply the *mean*, is the most common measure of central tendency used in quantitative research. It is computed by summing all of the scores in a distribution and dividing by the number of scores in the distribution. The mean is generically referred to as an average and is a part of our daily language. Students have encountered means when they receive their grade point average from the Registrar's Office, sports fans read with great anticipation baseball batting averages in newspaper sports sections, and TV weather reporters update their local viewers on average rainfall in inches.

2.3.2 NOTATION FOR THE MEAN

The mean is computed by adding the scores in a group and dividing this value by the number of scores. A verbal account of an arithmetic procedure can often be described more succinctly with a computational formula, and throughout this text we will be providing computational formulas and numerical examples for many of the key statistical processes we have covered at a more conceptual level. For many readers, working out these statistical analyses "by hand" with a calculator is an excellent way of rein-forcing the conceptual lesson already conveyed and may provide a useful comparative bridge to the SPSS and SAS statistical software output.

Before providing you with the computational formula for the mean, we need to introduce some of the basic notational system that we will be using throughout this text. The symbols we use in the formula for the mean are shown in Table 2.1. Generally, the letter Y is used to denote scores on the dependent variable. The subscript i in Y_i is used to represent any score and thus is applicable to every score. A Y with a bar over it ($\overline{Y}$) represents the mean of the scores. Group size is symbolized as a lowercase n; full sample size is symbolized as an uppercase N.

2.3.3 FORMULA FOR THE MEAN

Now we are ready to write our formula for the mean as

$$\overline{Y} = \frac{\Sigma Y}{n}. \tag{2.1}$$

To see this formula in practice, consider the following numerical example in Table 2.2. Suppose we sample a small group of newlyweds ($n = 7$) and ask them as couples the following question: "How happy are you at this time with your marriage?" They are asked to use the following four-point scale:

- *very happy* $= 1$
- *happy* $= 2$
- *unhappy* $= 3$
- *very unhappy* $= 4$.

Table 2.1. Notation symbols

Symbol	How to verbalize	Meaning and comments
Y		This represents a set of scores or observations. When we refer to all of the scores the Y is accompanied by a subscript i or Y_i. When we refer to a particular score the subscript i is changed to a number. Thus, the first participant's score is Y_1 the second Y_2, and so on. Most of the time, we will eliminate subscripts to simplify presentation of a formula.
$\overline{Y}$	Read as "Y-bar."	This is the symbol for the mean of a distribution.
Σ	The Greek uppercase letter sigma.	This is a summation instruction and signals us to add up the scores that follow it.
ΣY	Read as "summation Y."	This instruction indicates we are to sum (add up) the Y scores.
n		This is the number of scores within a group or treatment condition. To simplify computations throughout this book, we will always assume equal sample sizes (equal n) across treatment conditions. The concept of unequal n will be dealt with in Chapter 17. A uppercase N is used to designate the entire sample size.

As we can see from Table 2.2, the couples vary somewhat in their personal perceptions of matrimonial bliss. By summing the seven scores we obtain the sum (ΣY) of 13. By dividing this sum by n (7), we obtain the group mean $\overline{Y}$, which equals 1.86. This mean lies between the anchors *very happy* and *happy* and is much closer to the latter. Based on the mean, we would conclude that as a group these newlyweds appear to be "happy" with their marriages.

When we perform these calculations by hand (with a calculator), we typically report the results to two decimal places (the nearest hundredth) because we rarely obtain whole numbers. As a general heuristic, if the number in the third decimal place is 5 or greater, round up. Conversely, if the number in the third decimal place is less than 5, round down. Thus, in the present case the actual mean was 1.857. Hence, we rounded this value to 1.86. This rounding procedure may occasionally produce small amounts of rounding error, but it is usually not enough to cause concern.

Table 2.2. Raw scores for a two-group example: Newlywed-rated happiness

Y_i	Happiness score
Y_1	3
Y_2	2
Y_3	2
Y_4	2
Y_5	2
Y_6	1
Y_7	1
ΣY	13
$\overline{Y}$	1.86

2.4 THE MEDIAN AS A MEASURE OF CENTRAL TENDENCY

2.4.1 GENERAL CONCEPTION OF THE MEDIAN

A second type of measure of central tendency is called the *median*. The median is the midpoint or middle score when the scores are arranged from lowest to highest; defined in terms of percentiles (the proportion of scores in the distribution lower in value), the median is the fiftieth percentile.

The median can be a particularly useful index of central tendency when a group of scores contains extreme values or *outliers*. Outliers are sometimes found when working with income or age data. For example, suppose in a small graduate seminar with five students we observed the following age distribution: 21, 22, 23, 24, and 62. The mean for the class is $\Sigma Y/n = 152/5 = 30.40$ years of age. However, does this computed mean truly represent the "center" of the ages in the class? The answer is "not really." The one older student, the value of age that we judge to be an outlier, has pulled or affected the mean such that it no longer represents the typical student age. A better indicator of student age is the median, which is 23 in this example. The point to remember is that the median can compensate for high or low outlying scores in a distribution of scores.

2.4.2 CALCULATION OF THE MEDIAN

Calculating the median is straightforward when there is an odd number of scores. You simply arrange the scores in ascending order and the score in the middle is the median. When there is an even number of scores you have, in effect, two middle scores, and the median is the average or mean of these two scores. For example, let's modify our previous example to include one additional student who is twenty-five years old. The group of scores would thus be: 21, 22, 23, 24, 25, and 62. In this case, the mean is $\Sigma Y/n = 177/6 = 29.5$, and the median is $23 + 24 = 47/2 = 23.5$.

2.5 THE MODE AS A MEASURE OF CENTRAL TENDENCY

One last measure of central tendency is the *mode*. The mode is defined as the most common or frequent value in a distribution of scores. No special computation is involved in determining the mode. A simple inspection of the frequency of occurrence of each data value is all that is required. For example, the most frequent or modal happiness score in Table 2.2 is 2. In this particular example, the mode and median are the same; that is, both are 2 with a mean of 1.86.

The mode is particularly useful in describing nominal level data. Are there more Republicans, Democrats, or Independents present at a particular meeting? Who is the most typical or modal mental health service provider for a given community mental health client who is being provided mental health services by a variety of providers (e.g., psychiatrist, psychologist, social worker, marriage and family therapist, case manager)? Because the mode is typically not used when working with quantitative dependent variables, it will not be emphasized in this text.

2.6   RANGE AS A MEASURE OF VARIABILITY

As we have seen, measures of central tendency (the mean, the median, and the mode) provide information of the most typical value in a distribution of scores. Other indicators – the range, variance, and standard deviation – are designed to describe the diversity or variability of scores in a distribution. Perhaps the simplest measure of variability is the *range*; this measure represents what we ordinarily denote by the term in common language. The range is a single value computed as the highest score (the maximum score) minus the lowest score (the minimum score). In our previous student age example, the age range is 41 ($62 - 21 = 41$). The range is a global and relatively imprecise measure of variability for at least two reasons: (a) it takes into account only the two scores at the extremes of the distribution, and (b) a given range, say 41, can be associated with very different minima and maxima (e.g., subtracting ages 62 and 21 and subtracting ages 92 and 51 both yield the same age range). Range is used relatively infrequently in the behavioral and social sciences.

2.7   VARIANCE AS A MEASURE OF VARIABILITY

2.7.1   GENERAL CONCEPTION OF THE VARIANCE

A very useful index of the dispersion of scores within a distribution is called the "variance" (symbolized as s^2) and is crucial to all the subsequent computational work we do in this text. The variance tells us how dispersed the scores are with respect to the mean. Larger variances represent greater spreads or variability of the scores from the mean. In practice, we will see that the variance is an average of the squared deviations from the mean.

We will show you how to compute the variance with two different formulas that arrive at the same answer. The first is known as the *defining* or *deviational formula* and the second is the *computational formula*. The first formula is intuitive, making clear exactly how the variance can be conceptualized. The latter formula may be computationally more convenient with larger data sets but makes it somewhat harder to "see" what is represented conceptually by the variance.

In both computational strategies, the variance is calculated in a two-step process. The first step is to calculate what is called the *sum of squares*, or *SS*, which becomes the numerator of the variance formula. The second step is to create a denominator that adjusts the sum of squares, called the *degrees of freedom*, or *df*.

Table 2.3. Defining or deviational formula calculation steps for the variance

Y_i	Score	Deviation from mean	Squared deviation
Y_1	8	-3	9
Y_2	9	-2	4
Y_3	10	-1	1
Y_4	13	2	4
Y_5	15	4	16
$n = 5$ $\quad$ $\Sigma Y = 55$		$\Sigma(Y - \overline{Y}) = 0$	$SS = \Sigma(Y - \overline{Y})^2 = 34$
$\overline{Y} = \dfrac{\Sigma Y}{n} = \dfrac{55}{5} = 11$			$s^2 = \dfrac{\Sigma(Y - \overline{Y})^2}{n-1} = \dfrac{34}{4} = 8.5$

2.7.2 DEFINING OR DEVIATIONAL FORMULA

The defining formula for the variance is as follows:

$$\text{variance} = s^2 = \frac{\Sigma(Y_i - \overline{Y})^2}{n-1}. \tag{2.2}$$

This formula indicates that the variance is a function of the sum of the squared deviations from the mean divided by the sample size minus one. This formula and definition are easier to understand within the context of some actual data, as in Table 2.3.

Note that in Table 2.3 we begin with a distribution of five scores whose sum (ΣY) is equal to 55. We divide this sum by the number of scores to obtain the treatment mean of 11. For example,

$$\overline{Y} = \frac{\Sigma Y}{n} = \frac{55}{5} = 11.$$

Recall that the goal or purpose of measures of variability such as the variance is to describe how the scores in a distribution deviate or vary on or about the mean. One way to approach this goal is to take each Y_i score and subtract the treatment mean from it $(Y_i - \overline{Y})$. We have made this calculation in the middle of Table 2.3 under the heading "Deviation from mean." For example, the first score Y_1 has a value of 8 and we then subtract the mean of 11 to obtain a value of -3. Symbolically and computationally, we have $(Y_1 - \overline{Y}) = (8 - 13) = -3$. This is done for each of the five scores.

An interesting feature of these deviations $(\Sigma Y - \overline{Y})$ is that their sum equals zero. This will always be true regardless of how much or how little a set of scores is dispersed about the mean. This is happening because the negative values are balancing out the positive values.

The fact that the sum of the deviations from the mean always equals zero poses a minor quandary for statisticians. However, we can sidestep this dilemma by eliminating the negative valences on our deviation scores. This can be achieved by squaring all the deviation scores (i.e., multiplying each deviation by itself). These calculations are provided on the far right side of Table 2.3. For example, the first deviation score, -3, is multiplied by itself $(-3)^2$ and equals 9. When we sum these squared deviations $\Sigma(Y - \overline{Y})^2$ we create a value known as the *sum of squares*. In the present case, $SS = \Sigma(Y - \overline{Y})^2 = 34$. By dividing the sum of squares by the sample size minus one or $n - 1$, we determine the variance, which is 8.5 in the present example. In Chapter 3, we will describe this denominator or $n - 1$ value as the *degrees of freedom* and discuss it in more detail. For now, think of the degrees of freedom as a way of adjusting the sum of squares. Thus, in practice the variance is equal to the sum of squares divided by the degrees of freedom, or symbolically, $s^2 = SS/df$.

2.7.3 COMPUTATIONAL FORMULA FOR THE VARIANCE

The computational formula for the variance takes the following form:

$$s^2 = \frac{\Sigma Y^2 - \dfrac{(\Sigma Y)^2}{n}}{n - 1}. \tag{2.3}$$

This formula is based upon calculating two separate sums:

ΣY^2 (read "summation Y squared"), the sum of the squared scores.

$(\Sigma Y)^2$ (read "summation Y quantity squared"), the square of the total Y-score sum.

The calculations for this computational formula can be more readily understood by examining the numerical example of Table 2.4, which uses the same raw scores used with the defining formula example (shown in Table 2.3).

Table 2.4. Computational formula calculation steps for the variance

Y_i	Score	Y_i^2 (or Score^2)
Y_1	8	64
Y_2	9	81
Y_3	10	100
Y_4	13	169
Y_5	15	225

$n = 5$

$\Sigma Y = 55$

$\Sigma Y^2 = 639$

$(\Sigma Y)^2 = 3{,}025$

$SS = \Sigma Y^2 - \dfrac{(\Sigma Y)^2}{n} = 639 - \dfrac{3{,}025}{5} = 639 - 605 = 34$

$\overline{Y} = \dfrac{\Sigma Y}{n} = \dfrac{55}{5} = 11$

$s^2 = \dfrac{\Sigma Y^2 - \left[\dfrac{(\Sigma Y)^2}{n}\right]}{n - 1} = \dfrac{639 - \left[\dfrac{3{,}025}{5}\right]}{5 - 1} = \dfrac{639 - 605}{4} = \dfrac{34}{4} = 8.5$

To obtain our sum of squares with the computational formula, we must calculate two quantities. The first, ΣY^2 or the sum of the squared scores, is calculated by squaring each Y_i score and then summing these squared scores. These values are found on the right side of Table 2.4. Their sum, ΣY^2, is equal to 639.

The second quantity we need to calculate is $(\Sigma Y)^2/n$, or the squared total sum of the Y scores divided by the number of scores. In Table 2.4, we see that the sum of the Y scores is equal to 55 or $\Sigma Y = 55$. When we square this total (55×55), we arrive at $(\Sigma Y)^2 = 3,025$. We now have the constituents for computing the sum of squares.

$$SS = \Sigma Y^2 - \frac{(\Sigma Y)^2}{n}$$

$$SS = 639 - \frac{3025}{5}$$

$$SS = 639 - 605 = 34. \tag{2.4}$$

As we noted previously, by dividing the sum of squares by the degrees of freedom, SS/df, we obtain the variance. Hence, $34/4 = 8.5$. This is the same value that we obtained with the defining formula.

2.8 STANDARD DEVIATION AS A MEASURE OF VARIABILITY

You will recall that in order to remove the problem of our deviations from the mean always summing to zero, we squared the deviations to eliminate the negative values. While this numerical operation sidesteps the zero sum problem, it makes it difficult to "intuitively" interpret the value of the variance. For example, what exactly does a variance of 8.5 really mean? This value reflects the squaring operation and, technically, informs us that the five scores deviate from the mean by 8.5 "squared units."

This description is not particularly helpful, and so most researchers compute an additional measure of variability known as the *standard deviation*, symbolized as either *s* or *SD*. The formula for the standard deviation is simply the square root of the variance.

$$SD = \sqrt{\text{variance.}} \tag{2.5}$$

Thus in the present example, $\sqrt{8.5} = SD = 2.92$. By computing the square root of the variance, we are literally "unsquaring" the variance, which allows us to interpret the variability of the scores in the original units of measurement. For example, if the original (raw) scores in Tables 2.3 and 2.4 were five hourly wages, we can now say that the average wage was $11.00 per hour ($\overline{Y} = 11$) and that on the average these wages deviated or varied above or below the mean by $2.92. This is more informative than to say the wages deviated by 8.5 squared dollars and cents.

Most social and behavioral scientists prefer to work with means and standard deviations. A useful heuristic to remember is that when you report a mean from a continuous variable, you should also report its standard deviation. This information provides the reader with the necessary context to interpret the variability within your study.

CHAPTER 2 EXERCISES

2.1. Students were asked to rate the quality of the university lunch menu on the following Likert-type scale: 4 = *excellent*, 3 = *good*, 2 = *fair*, and 1 = *poor*.

Ratings
3
3
2
1
4
3
2
2

Compute the mean, median, mode, range, variance, and standard deviation for this distribution of scores.

2.2. A second sample of cafeteria lunch evaluators produced the following data:

Ratings
4
4
3
2
4
3
3
4
2
3

Compute the mean, median, mode, range, variance, and standard deviation for this distribution of scores.

SECTION 2

Foundations of Analysis of Variance

$x_1 x_1 x_1^2 1 \ 3 \ 3 \ 3 \ 4$

CHAPTER THREE

Elements of ANOVA

3.1 PARTITIONING OF THE VARIANCE

The term *analysis of variance* is very descriptive of the process we use in the statistical treatment of the data. In a general sense, to analyze something is to examine the individual elements of a whole. In ANOVA, we start with a totality and break it apart – partition it – into portions that we then examine separately.

The totality that we break apart in ANOVA, as you might suspect from the name of the procedure, is the variance of the scores on the dependent variable. As you recall from Chapter 2, variance is equal to the sum of squares – the sum of the squared deviations from the mean – divided by the degrees of freedom. In an ANOVA design, it is the variance of the total set of scores that is being partitioned. The various ANOVA designs are different precisely because they allow this total variance to be partitioned in different ways.

ANOVA is a statistical procedure that allows us to *partition* (divide) the *total variance* measured in the study into its sources or component parts. This total measured variance is the variance of the scores that we have obtained when participants were measured on the dependent variable. Therefore, whenever we talk about the variance associated with a given partition or whenever we talk about the total variance, we are always referring to the variance of the dependent variable.

3.2 A SIMPLE EXAMPLE STUDY

3.2.1 A DESCRIPTION OF THE STUDY

To keep this chapter's discussion somewhat concrete, we will flesh out the example introduced in Chapter 1, a hypothetical two-group design. Assume that a set of researchers wished to learn if the room color in which college students are asked to respond to questions assessing their mood can affect their reported mood. A total of fourteen participants were randomly assigned to one of two groups. The students were asked to complete an inventory that assessed their current mood. One group of seven students answered the questions in a room painted a soft shade of blue (Blue Room group); this color was expected to exert a calming effect on the participants. The other group of seven students answered the questions in a room painted a bright red (Red Room group); this color was expected to be more agitating than calming. The dependent variable was the overall score on the mood questionnaire with higher scores indicating a more relaxed mood. It was hypothesized that students answering the questions in the blue room would produce higher scores than those answering the questions in the red room (i.e., students in the blue room would indicate more calmness or relaxation than those in the red room).

3.2.2   THE DESIGN OF THE STUDY

ANOVA is a general statistical technique that we use to compare the scores in one condition to those of one or more other conditions. It is useful to distinguish among three classes of ANOVA designs: between-subjects designs, within-subjects or repeated-measures designs, and mixed designs. These will be described in detail in Chapters 6–15. For now it is sufficient for our purposes to note that this example study falls into the class of between-subjects designs. The defining element of a *between-subjects design* is that each participant contributes just one score to the analysis. Thus, the number of data points we enter into the analysis is equal to the number of cases in the study.

In the room color study we have fourteen participants, seven in each group; consequently, we have fourteen pieces of data, seven for each condition. It follows from this arrangement that the levels of the independent variable are represented by different cases or participants; in this example some of the students are assigned to the blue room and others are assigned to the red room.

We can further specify the particular design of our example study. The between-subjects design that we are using as our example is a one-way design. In a *one-way between-subjects design* there is only one independent variable (the *one* in *one-way* signifies this). In the present instance the independent variable is room color and it has two levels: the room color is either blue or it is red.

In the ANOVA designs that we cover in this book, all participants in a single design are measured on the same dependent variable. In the current example, we are assessing calmness of mood via scores on a paper-and-pencil inventory.

3.2.3   THE DATA COLLECTED IN THE STUDY

The data from this hypothetical study are shown in Table 3.1. As can be seen in the table, participants in the Red Room condition had scores ranging from 11 to 17 and those in the Blue Room condition had scores ranging from 16 to 22. There is some overlap between the two distributions of scores but the means of the two groups are in the direction of the experimental hypothesis: The mean of participants in the Blue Room condition is higher than that for the Red Room condition. Thus, qualitatively

Table 3.1. Mood scores for students answering a questionnaire in rooms painted different colors

Red room	Blue room
11	16
12	17
13	18
14	19
15	20
16	21
17	22
$\overline{Y} = 14$	$\overline{Y} = 19$

at least, it appears that answering the mood questions in the blue room might have produced greater calmness than answering the questions in the red room. The results of our ANOVA will inform us whether we can say that these two means are significantly different.

3.2.4   THE STATISTICAL RESULTS

We have performed the ANOVA of the data already, but this chapter is not about teaching you the details of calculating the numbers we are presenting – Chapter 6 will take you through a worked ANOVA example. Rather, our goals here are to talk you through the logic of the variance partitioning process and to orient you to some of the computational strategies that are used in ANOVA. To accomplish these goals, it is useful to have the results of the analysis presented to you.

In the days when only hand calculations were done, it was common practice to place the outcome of ANOVA computations in one neatly organized place rather than have the numbers scattered about over several pages of calculations. This organized place was known by the very appropriate name of a *summary table*. So useful was this practice of placing the information in a summary table that the statistical software applications perpetuated it. Thus, even today SPSS and SAS show a summary table as part of their ANOVA output.

3.2.5   THE SUMMARY TABLE

Table 3.2 presents the summary table for the room color study. Summary tables have a relatively standard structure. The rows represent the partitioning of the variance and will vary with each type of experimental design that was used. The columns depict statistical information relevant to the analysis; because we compute the same type of statistical information in all of the ANOVA designs, these columns are relatively fixed in the designs that we cover in this book. The elements in this table, and the concepts underlying them, are central to understanding the ANOVA technique. We will discuss the entries in the columns for sources of variance,

Table 3.2. Summary table for the analysis of the data contained in Table 3.1

Sources of variance	Between-subjects effects					
	SS	df	MS	F ratio	Probability	Eta squared (η^2)
Between groups (Factor A)	87.50	1	87.50	18.75	.001	.610
Within groups (Error or S/A)	56.00	12	4.67			
Total variance	143.50	13				

sum of squares, degrees of freedom, and mean square (MS) in the remaining portion of this chapter. In the following chapter, we will address the F ratio, the probability associated with the F ratio, statistical significance, the eta squared (η^2) statistic, and statistical power.

3.3   SOURCES OF VARIANCE

The first column of the summary table is labeled "Sources of variance." A "source" of variance represents a portion or partition of the total variance. As shown in Table 3.2, the *total* variance occupying the last row is partitioned into *between-groups* variance and *within-groups* variance. The details concerning these sources of variance will be explicated when we discuss the sum of squares column in Section 3.4. For the moment, here is an overview of each of these sources of variance:

- The between-groups portion of the variance deals with the differences between the group means. In the notation system that we will be using, independent variables, often called *effects*, *factors*, or *treatments*, are designated by uppercase letters in alphabetic order starting with A. With only one independent variable in the study, its effect is noted as Factor A. If there were two independent variables, the effect of one (arbitrarily determined) would be called "Factor A" and the effect of the other would be called "Factor B."
- The within-groups portion of the variance deals with the variation of the scores within each of the experimental groups. It is referred to as error variance for reasons we will discuss shortly. In our notation system we designate it as S/A; this notation stands for the expression "subjects within the levels of Factor A."
- The total variance of the dependent variable deals with the variation of all the scores taken together (regardless of group membership).

3.4   SUMS OF SQUARES

3.4.1   TOTAL SUM OF SQUARES

The total variability of the scores can be seen in pictorial form in Figure 3.1. The data in the circle are the values of the dependent variable taken from Table 3.1. There are fourteen scores in the set – each score in the study is considered individually when we focus on the total sum of squares. That is, when considering the total sum of squares, we ignore the fact that there are multiple groups involved in the study and simply pool all of the data together.

As you may recall from Chapter 2 (Section 2.7.2), the numerator of the variance computation is the sum of squares, which calls for summing the squared deviations of scores from a mean. In the case of the total variance, the scores in the sum of squares computation are those shown in Figure 3.1.

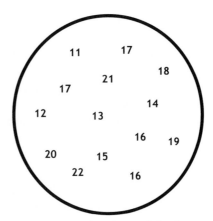

Figure 3.1 Total variance of the dependent variable.

The mean from which these scores deviate is the average of all these scores. Such a mean, based on all of the individual scores, is known as the *grand mean.*

It is this grand mean that becomes the reference point for the deviations we speak of concerning the total sum of squares. For the total sum of squares we deal with differences (variability) of each individual score from the grand mean. Thus,

$$\text{total sum of squares} = \sum(\text{individual score} - \text{grand mean})^2.$$

Our notation system calls for using uppercase *Ys* to represent the scores on the dependent variable. The formula for computing the total sum of squares is therefore written as

$$SS_{\text{Total}} = \sum(Y_i - \overline{Y}_T)^2, \tag{3.1}$$

where SS_{Total} is the total sum of squares, Y_i is the score for a given case, and $\overline{Y}_T$ is the grand mean.

3.4.2 BETWEEN-GROUPS SUM OF SQUARES

The focus of the between-groups sum of squares is on the group means, and it is therefore the between-groups source of variance that represents the effect of the independent variable. As we saw from Table 3.1, the scores for the Red Room condition are generally lower than those for the Blue Room condition. These scores generate respective group means that are quite different from each other – a mean of 14 for the Red Room group and a mean of 19 for the Blue Room group. These means summarize the scores by providing a single value that captures (denotes, represents) the center of their respective distributions and serve as proxies for the sets of scores that gave rise to them.

We often speak about the *effect of the independent variable.* This form of expression comes from the experimental setting in which we intend to draw a causal inference by asserting that the presence of the treatment – the presumed cause – produced or caused the changes in behavior – the effect – that we measured. The way we quickly summarize the validity of this presumed causal relationship is by examining the means of the groups in the study. If the treatment was "effective," the mean of the one group should be significantly different from the mean of the other group.

When we compute any sum of squares, a mean must be subtracted from a "score," and that must be true in computing the between-groups sum of squares. We therefore deal with the mean differences in what may seem to be an indirect manner, although it does get the job done: The group means become the scores from which a mean is subtracted. The mean that is subtracted is the grand mean. Thus,

between-groups sum of squares

$$= \sum(\text{group mean} - \text{grand mean})^2.$$

In our notation system the formula for computing the between-groups sum of squares is therefore written as

$$SS_A = \sum(\overline{Y_j} - \overline{Y_T})^2, \tag{3.2}$$

where SS_A is the between-groups sum of squares, $\overline{Y_j}$ is the mean for a given group, and $\overline{Y_T}$ is the grand mean.

The between-groups source of variance is one of the portions or partitions of the total variance. Here is one way to understand what this means. In Figure 3.1 the total variability of the dependent variable is depicted. When we examine the between-groups source of variance, we are asking if any variation we see in the total set of scores is related to or associated with the group to which the participant belongs. This "association," should any be observed, is treated as variance that is "explained" by the independent variable. In our example, the lower values are generally associated with participants who took the mood survey in the red room, and the higher values are generally associated with participants who took the mood survey in the blue room.

The idea that the independent variable is associated with or that it explains some of the total variability in the dependent variable can be understood from the standpoint of prediction. Suppose we are attempting to predict the scores on the dependent variable. Simply treating all of the participants as a single large group, and knowing nothing about them, our best prediction in the long run of any one participant's score will be the overall or grand mean, which gives us little in the way of precise prediction (it's better than selecting any number at random). If the independent variable is shown to explain some of the variance of the dependent variable, then considering it in the prediction effort should increase our precision over using only the grand mean. Specifically, if the independent variable explains some of the variance, that is, if the means of the groups are significantly different, and if you know the group to which a given participant belongs, then using the group mean rather than the grand mean should improve our prediction precision. For example, knowing that a given score was associated with a participant who took the test in the blue room, we would predict on a better than chance basis that the score was relatively higher than those of the participants who took the test in the red room. This is treated as statistical explanation.

The prediction based on knowing the participants' group membership, however, will not be perfect. This is because, as we can see from viewing Table 3.1, the scores of the participants in each group differ from each other to a certain extent. Knowing whether a student was associated with the Blue Room or Red Room group can significantly increase the preciseness of our prediction, but in the long run there will still be some error in predicting the value of each student's score in this study. How much of the total variability can be explained, as we will see later, can be indexed by the proportion of the between-groups sum of squares with respect to the total sum of squares.

3.4.3 WITHIN-GROUPS SUM OF SQUARES

As can be seen from the summary table in Table 3.2, the remaining source of variance is associated with within-groups variance. The term *within groups* provides a useful cue of what this portion of the total variance represents: variability within each of the two groups. Consider the Red Room group, whose mean on the dependent variable is 14. If room color completely (100 percent) determined the score on the mood survey, then each of the participants should have scored 14. Of course, they did not. Therefore, factors other than room color affected their performance; perhaps the participants were of different temperaments or different ages, had different emotional histories, different genders, and so on. But which combination of the wide range of possible factors was operating here is unknown. Because of this, none of these variables can be statistically analyzed to determine whether they are associated with the dependent measure; to the extent that they are important, these other variables contribute measurement variability to the study, and that unaccounted for variability within each group is defined as *error variance.*

Note that these other "unknown" factors may be rather important contributors to the differences in the scores that we observe. The point is that we can statistically analyze the effects only of known (measured) variables. Thus, if some important other variables were not assessed in the study, they contribute to measurement error. A one-way between-subjects design is limited to a single independent variable, and thus the design limits the number of "known" effects to just one. Some of the more complex designs that we will cover in this book allow us to incorporate additional variables enabling us to explore complex interrelationships among potentially important variables.

The within-groups sum of squares represents error variability – variability within each group that cannot be attributed to the level of the independent variable those participants experienced. In the context of sum of squares, the scores within each group are the scores from which a mean is subtracted. The mean that is subtracted is the mean of the group containing the score. Thus,

$$\text{within-groups sum of squares} = \sum(\text{score} - \text{group mean})^2.$$

In our notation system the formula for computing the within-groups sum of squares is therefore written as

$$SS_{S/A} = \sum(Y_i - \overline{Y_j})^2, \tag{3.3}$$

where $SS_{S/A}$ is the within-groups sum of squares, Y_i is the score in a given group, and $\overline{Y_j}$ is the mean of the particular group. The heart of this computation is the residual values when we subtract the group mean from each of the scores. Because some of these residuals would be positive (the score would be higher than the mean) and others would be negative, such that their sum would be zero (the sum of deviations around the mean must add to zero), it is necessary to square these residuals or deviations to obtain values that, when added, will not necessarily produce a zero value.

3.4.4 SUMS OF SQUARES ARE ADDITIVE

One feature of the sum of squares worth noting at this point is its additive nature: The between-groups sum of squares added to the within-groups sum of squares is equal to the total sum of squares. This allows us to take the proportion of the total sum of squares accounted for by the between-groups sum of squares. We will use such a proportion to quantify the "strength of effect" of the independent variable indexed by eta squared in Chapter 4.

3.5 DEGREES OF FREEDOM (*df*)

The degrees of freedom, commonly abbreviated as *df*, that are associated with each source of variance are contained in the third column of the summary table in Table 3.2. These are, very roughly, computed by subtracting the value of 1 from the number of scores that are being processed in a computation.

3.5.1 A BRIEF EXPLANATION OF DEGREES OF FREEDOM

Consider a set of scores with these two constraints:

- There are a total of three scores (negative values are allowed).
- Their sum is 11.

How many of the scores are you free to fill in with any values of your choosing before the others are completely determined? The answer is that we are free to fill in two of the slots before the third one is now determined. For example, if we selected 2 and 4 for our free choices, the third number has to be 5 in order to meet the constraint that the total is 11. We therefore have 2 *df* (two free slots) when there are three numbers in the set.

Consider another set of scores with these constraints:

- There are a total of four scores (negative values are allowed).
- Their sum is 20.

How many of the scores are you free to fill in with any values of your choosing before the others are completely determined? The answer is that we are free to fill in three of the slots before the fourth one is determined. For example, if we selected 3, 6, and 8 for our free choices, the fourth value has to be 3 in order to meet the constraint that the total is 20. We therefore have 3 *df* in this example.

The general rule that these two examples illustrates is that you can freely fill in all but one of the slots before the last value is determined. Calculating the degrees of freedom for the three sources of variance that we have, although more complex, involves a similar logic.

3.5.2   DEGREES OF FREEDOM FOR TOTAL VARIANCE

The degrees of freedom for the total variance is equal to the total number of observations minus 1. Expressed as a formula where our group sizes are equal as in the current example with 7 cases in each of two groups, our computation is as follows:

$$df_{Total} = (a)(n) - 1 = (2)(7) - 1 = 13, \tag{3.4}$$

where a is the number of groups and n is the number of cases contained in each group.

3.5.3   DEGREES OF FREEDOM FOR BETWEEN-GROUPS VARIANCE

The degrees of freedom for the between-groups effect (Factor A) are equal to the number of levels of the independent variable minus one. Expressed as a formula, our computation is as follows:

$$df_A = a - 1 = 2 - 1 = 1, \tag{3.5}$$

where a is the number of groups.

3.5.4   DEGREES OF FREEDOM FOR THE WITHIN-GROUPS (ERROR) VARIANCE

The degrees of freedom for the error variance is equal to the sum of the degrees of freedom for each of the groups. For the Red Room group, we have 7 scores, which translate to 6 *df*. For the Blue Room group, we also have 7 scores and 6 *df*. Adding these two values together results in 12 *df* associated with the within-subjects source of variance. Assuming equal sample size, this computation can be written as the following formula:

$$df_{S/A} = (a)(n-1) = (2)(7-1) = 12. \tag{3.6}$$

Note that we could have figured out the value of the degrees of freedom for the unexplained (error) variance based on what we already had computed, since, as was true for the sum of squares, the degrees of freedom for the partitions of the total variance are additive. Given that there are 13 df in total and that one of them is associated with the between-groups effect, the rest had to be associated with the error variance. We could have thus computed this value as the residual, and this is a common way for researchers to obtain certain degrees of freedom values in the summary table when they are doing the calculations by hand. Of course, by performing the analyses in SPSS or SAS, all of these values are provided to you.

3.6   MEAN SQUARE (MS)

The mean square column in the summary table (Table 3.2) shows the variance of each respective source of variance. As a variance statistic, it is computed by dividing the sum of squares by the corresponding degrees of freedom.

3.6.1   MEAN SQUARE TOTAL

The mean square corresponding to the total variance is nothing more than the variance of the entire set of scores in the study. Its value by tradition is not entered in the summary table. If we did enter it, it would correspond to the total variance of the sample and would take on a value of 11.038 (total sum of squares ÷ total degrees of freedom = $143.50 \div 13 = 11.038$). It could also be computed by calculating the value of the variance of the scores based on the sample as a whole.

3.6.2   MEAN SQUARE BETWEEN GROUPS

The value for the mean square between groups (MS_A) is the variance of the group means; it computes to 87.50 (between-groups sum of squares ÷ between-groups degrees of freedom = $SS_A \div df_A = 87.50 \div 1 = 87.50$).

3.6.3   MEAN SQUARE WITHIN GROUPS

The value for the mean square within groups ($MS_{S/A}$) is the average variance of the groups; in this example, each group has a variance of 4.67, and so they average to 4.67. It is computed by dividing within-groups sum of squares by within-groups degrees of freedom ($SS_{S/A} \div df_{S/A} = 56.00 \div 12 = 4.67$). It could also be computed by calculating the value of the variance for each group and taking the mean of those values assuming equal group size.

3.6.4   MEAN SQUARES ARE NOT ADDITIVE

Note that, unlike the sum of squares and the degrees of freedom, the mean squares are not additive. A mean square can be roughly thought of as an average sum of squares. The averages of between-groups and within-groups sources of variance are not directly additive because the degrees of freedom (the denominators of the calculation) are based on very different kinds of counts – the between-groups sum of squares is divided by the one less than the number of groups, whereas the within-groups sum of squares is divided by the sum of the degrees of freedom of each group.

3.7   WHERE ALL THIS LEADS

With the mean square values together with their associated degrees of freedom in hand for each source of variance, we are ready to compute the F ratio, determine its probability of occurrence based on its sampling distribution, determine whether the mean difference was statistically significant, and compute an eta squared value representing the strength of the effect of the independent variable. These portions of the summary table, and some of the issues surrounding them, are discussed in Chapter 4.

CHAPTER FOUR

The Statistical Significance of F and Effect Strength

We continue here with our discussion of the elements of the summary table provided in Table 3.2. We have discussed the sources of variance, the sum of squares, the degrees of freedom, and the mean squares in Chapter 3; we are now ready to deal with the F ratio, the probability of the F value occurring by chance based on its sampling distribution (so that we can determine whether or not it is statistically significant), and the eta squared value (η^2) representing the strength of the effect of the independent variable. We begin with the F ratio.

4.1 THE F RATIO

4.1.1 THE F RATIO AND THE NULL HYPOTHESIS

As Snedecor (1946) stated so well, ANOVA tests the null hypothesis, which states that the observed means for the various conditions in the study represent "... random samples from the same normal population" (p. 219). Expressed in more conversational language, if the null hypothesis is true, then the means of the conditions in a study are not, statistically, significantly different. An extremely convenient way to test the null hypothesis is to compute the ratio of the between-groups variance (the mean square value associated with the between-subjects source of variance) to the within-groups variance (the mean square value associated with the within-subjects source of variance). This may be written as follows:

$$F = \frac{\text{between-groups } MS}{\text{within-groups } MS} = \frac{MS_A}{MS_{S/A}}. \tag{4.1}$$

In our example study (see Table 3.2), the F ratio is 18.75. To evaluate this ratio, it is necessary to know how the ratio is distributed based on chance. And that brings us back to R. A. Fisher.

4.1.2 THE F RATIO AND R. A. FISHER

How the ratio of mean square between groups (MS_A) to mean square within groups ($MS_{S/A}$) is distributed based on chance was worked out and presented to an international mathematics conference by R. A. Fisher in 1924 (Kirk, 1995). Because this ratio plays such a pivotal role in ANOVA, it was bound to take on a name early in its history. The name that stuck was provided by George W. Snedecor. Snedecor's *Statistical Methods,* first published in 1937, was probably the most influential statistics textbook of the times. Snedecor taught at Iowa State University from 1913 to 1958, and for much of that time he served as the director of its statistics laboratory. Under his leadership, Iowa State became one of the leading statistics centers in the world. R. A. Fisher was a somewhat regular summer visitor to Iowa State during the 1930s (Salsburg, 2001). Snedecor held Fisher in high regard, and Snedecor's prestige within the statistical community led to the acceptance of the name he suggested in 1934 for the ratio (see Snedecor, 1934).

As for the name of this ratio, Snedecor (1946, p. 219) tells us that "Fisher and Yates [1938–1943] designate *F* as the *variance ratio,* while Mahalanobis [1932], who first calculated it, called it *x.*" But the more complete story as told by Snedecor (1946, p. 219) is as follows:

One null hypothesis that can be tested is that all of the [sets of scores] are random samples from the same normal population.... It is necessary to next learn the magnitude of variation ordinarily encountered in such ratios. The ratio of the two estimates of variance ... has a distribution discovered by Fisher. I named it *F* in his honor [Snedecor, 1934].

4.2 THE SAMPLING DISTRIBUTION OF *F*

4.2.1 OBTAINING THE SAMPLING DISTRIBUTION OF *F*

The *F* ratio is computed as the ratio of two variance estimates. Let's build a hypothetical illustration to show the sampling distribution of this ratio, not worrying, for the moment, about distinguishing between-groups variance from within-groups variance. Instead, using the null hypothesis as our base, we can decide on a sample size and randomly draw two samples of values from the same population. We then compute the variances of each sample and divide the variance of the first sample by the variance of the second sample. We repeat this process an infinite number of times. We then plot these ratios as a frequency distribution. This plot from our hypothetical procedure is the *sampling distribution* of the *F* ratio for samples of this given size. Although the distribution is slightly different for different sample sizes, the plot shown in Figure 4.1 is representative of what the general shape looks like.

4.2.2 THE SHAPE OF THE SAMPLING DISTRIBUTION OF *F*

As shown in Figure 4.1, the sampling distribution of *F* is positively skewed – the bulk of scores are relatively lower values that are represented toward the left portion of the *x* axis with the tail "pointing" to the right. Here is why it is positively skewed. If the larger variance is in the denominator, then the value that results from the division will be less than one. Although there may be a theoretically infinite number of such values, with increasingly

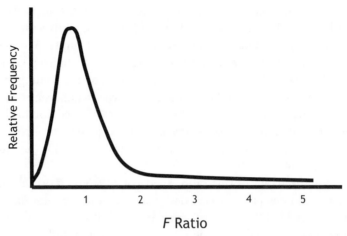

Figure 4.1 Sampling distribution of F.

larger denominators, we approach a barrier of zero. Thus, the distribution "compresses" or "bunches up" as it approaches zero.

If the larger variance is in the numerator, many of the resulting ratios will be large numbers well in excess of one. The curve actually extends to infinity, never quite touching the x axis. Although we cannot carry the curve out all the way to infinity, we can extend the x axis to values in the range of F values of 5, 6, or 7 because the distribution is noticeably above the x axis in this range of F.

If the null hypothesis is true, then most of the values of the variance ratio turn out to be, approximately, between 0.5 and 2.0. Here is why. Because the two samples were drawn from the same population and are composed of equal numbers of scores, we would expect that their variances would be of roughly the same magnitude; thus, their ratio should be someplace in the general region of 1.0 much of the time.

4.2.3 THE EXPECTED VALUE OF F

Most researchers may not need to do so very often, but it is possible to calculate the value of the mean of a given sampling distribution. We present this for those who might be interested.

The mean is the *expected value* of F based on drawing an infinite number of samples. It is a function of the within-groups degrees of freedom (Hays, 1981, p. 314–15). When there are more than 2 df for this source of variance, as there will invariably be in the research that readers will conduct, the expected value of F (the mean of the sampling distribution of F) is computed as follows:

$$\text{expected } F \text{ value} = \frac{df_{\text{Within Groups}}}{df_{\text{Within Groups}} - 2}. \tag{4.2}$$

In our example and in Table 3.1, we see that there are 12 df associated with the within-groups source of variance; thus, the sampling distribution of

F applicable to our data set would have a mean of 1.20, that is, $12 \div (12 - 2) = 1.20$. With increasingly greater degrees of freedom the expected value – the mean of the sampling distribution – approaches one.

4.3 THE AREA UNDER THE SAMPLING DISTRIBUTION

Although this may seem rather obvious, we can state for the record that 100 percent of the area is subsumed by, or contained within, the *F* distribution. And thanks to the work of many statisticians, it is possible to figure out the percentage of the area contained in a designated portion of the curve. For our purposes, we want to know the *F* value at which 5 percent of the area lies beyond.

The *F* distribution is based on an infinite number of variance ratios, but in a single research study, we actually obtain just a single *F* ratio, and we must make a determination about the viability of group differences based on that piece of information alone. Because *any* mean difference between two groups *can* occur by chance, how can we possibly evaluate "true" differences between the groups in a study by evaluating the single *F* ratio that we have obtained?

The answer is that we can make an inference about the differences being "probably true" or "probably reliably different" provided that we are willing to accept some chance of being wrong. That is, we can interpret a group difference that is "large enough" with respect to the error variance as "probably true, reliable, or valid" even though we know that large differences do occur by chance and that we may be looking at one now.

So researchers have learned to live with a certain amount of uncertainty. How much uncertainty? The answer to that question relates precisely to dividing the area of the *F* distribution into two parts: one containing 95 percent of the area and the other containing 5 percent of the area. And because we are interested only in *F* ratios greater than one (where the between-groups variance is larger than the within-groups variance), we make our dividing point toward the right side of the distribution.

4.4 STATISTICAL SIGNIFICANCE

4.4.1 THE TRADITIONAL 5 PERCENT BENCHMARK

The scientific community has agreed that *F* ratios falling into the 95 percent region are unremarkable in the sense that they occur fairly frequently. On the other hand, the scientific community has also agreed that, all else being equal, *F* ratios falling in the 5 percent area will be treated as "remarkable" or noteworthy even though it is recognized that these do occur by chance when the null hypothesis is true. The probability level at the boundary separating remarkable from unremarkable regions is called an *alpha level* or *statistical significance level*. Those *F* ratios falling into this 5 percent region of the distribution – those that are said to be remarkable – are said to be *statistically significant*. When we obtain such a result in our research,

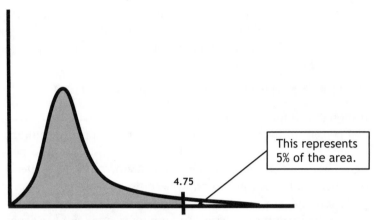

Figure 4.2   *F* distribution showing the 95 percent and 5 percent areas for 1 *df* and 12 *df*.

we can assert that the mean difference between the groups is statistically significant at a .05 alpha level.

In Figure 4.2, we have marked the 5 percent boundary for the *F* distribution applicable to our room color study. That boundary occurs at an *F* value of 4.75 given that we have 1 *df* and 12 *df* (1 *df* for between-groups variance and 12 *df* for within-groups variance). The area under the curve to the left of the *F* value of 4.75 in Figure 4.2 is shown by a shaded region and represents 95 percent of the area. The region to the right represents 5 percent of the area. *F* ratios based on the given degrees of freedom in our example whose values are equal to or greater than 4.75 fall in that 5 percent area and thus may be said to represent a statistically significant amount of between-groups variance in the study, that is, to represent a statistically significant difference among the means of the groups in the study.

4.4.2   PROBABILITY OF *F*

Before the days of computer-based statistical analysis, researchers would look up the value of the *F* ratio they calculated in a table similar to the Table of Critical *F* Values presented in Appendix C. Such tables display *F* values that correspond to particular benchmarks (probability breaks) in the *F* distribution (e.g., .05, .01).

To read the table in Appendix C, we first recall that in computing the *F* ratio, the between-groups mean square (MS_A) was in the numerator and the within-groups mean square ($MS_{S/A}$) was in the denominator. As can be seen from the summary table (Table 3.1) for our worked example, we have 1 *df* associated with between groups (df_A) and 12 *df* associated with within groups ($df_{S/A}$). In the Table of Critical *F* Values, the columns indicate the *F* ratio numerator degrees of freedom and the rows indicate the *F* ratio denominator degrees of freedom. We therefore focus on the first column (for 1 *df*) and the twelfth row (12 *df*). The intersection of these two degrees of freedom within the Critical Values of the *F* Distribution table reveals the critical value of *F* at the .05 level to be 4.75 (this is the top entry; the value of 9.33, which appears below it, gives the .01 breaking point). Since our observed (computed) F of 18.75 exceeds the value of 4.75, we know that it lies to the right of the 5 percent benchmark in the 5 percent area under the curve, and we can conclude that our obtained F value is statistically significant at the .05 alpha level. We would therefore reject the null hypothesis that the two groups of scores were drawn from the same population, and instead interpret the mean difference as a valid difference between the Blue Room and Red Room groups.

Technological advances now enable users of major statistical software packages to bypass the Critical Values table. SPSS and SAS compute the exact probability of obtaining the F ratio by chance alone with the given degrees of freedom on the presumption that the null hypothesis is true. The column in the summary table (Table 3.2) labeled "Probability" represents this exact probability truncated to a value of .001 because that is the number of decimal places SPSS is designed to report (SAS provides probability values to four decimal places). Thus, the exact probability is not actually zero but some very small value that cannot be displayed with only the three decimal places that are permitted in the output. Our obtained F value would therefore ordinarily occur less than once in every thousand occasions if the null hypothesis is true. Students should not report the probablity as .000 but should instead evaluate and report it with respect to the alpha level they are using (e.g., $p < .05$).

4.4.3 DIFFERENCES OF GROUP MEANS

In our room color study, we evaluated the obtained probability value against our alpha level, .05, and determined that we have a statistically significant effect of the independent variable. Because there are only two group means compared, we know automatically that they are significantly different. In looking at the actual values of the means of the two groups, we can conclude that the students completing the mood survey in the blue room reported significantly more relaxation than those completing the survey in the red room, thus confirming the experimental hypothesis.

Had there been more than two group means, that is, if there were three or more groups in the design, we would no longer know which means were significantly different from which others given a statistically significant F ratio. It is possible that only some of the means would differ significantly from others; in fact, although it is not common, it is possible that no two means, when taken individually, differ significantly from each other. As a result of this ambiguity, when there are three or more means to compare, it is necessary to perform a follow-up, post-ANOVA analysis to examine the difference between means. We will elaborate on this in Chapter 7.

4.4.4 ALPHA LEVEL CONSIDERATIONS

Most of the statistical analyses that we perform will be run under an alpha level of .05, a standard across the scientific disciplines. But there are a few times when circumstances will either suggest or demand that you modify the alpha level that you use for a particular statistical analysis. For example, in a very preliminary and exploratory phase of a research program, where you are not sure which variables are important and which are not, you may not want to miss an effect just because you have not controlled for all of the effects that you should. Here you might want to modify your alpha level to .10 or greater to be able to recognize a potential effect that is worth pursuing in the future.

On the other hand, if you are performing a large number of comparisons of means, or performing several related analyses, you will presumably find 5 percent of them to be significant even if the null hypothesis is true. Thus, you may not be able to distinguish which effects are obtained by chance and which are due to some valid effect of the independent variable. Under these situations, it is common practice to avoid such *alpha inflation* and change your alpha level to a more stringent value. There are a variety of rubrics that are documented in a number of sources (e.g., Keppel & Wickens, 2004; Maxwell & Delaney, 2000) that can be used to help ensure that the alpha level of .05 holds across the range of your analyses. For example, a *Bonferroni correction* would involve dividing the alpha level of .05 by the number of comparisons we were making. Thus, if we were making five comparisons, our Bonferroni-corrected alpha level would be .01 ($.05/5 = .01$). We discuss this topic in more detail in Chapter 7 in Section 7.2.

4.4.5   TYPE I AND TYPE II ERRORS

In Equation 4.1, we obtained an F ratio of 18.75 and asserted that we were looking at a significant mean difference. The result is statistically significant in the sense that a variance ratio of 18.75 would ordinarily be obtained less than 5 percent of the time; that is, the probability of that value occurring by chance alone if the null hypothesis is true is less than .05 ($p < .05$). Of course, in the long run we will be wrong 5 percent of the time (because F ratios that large actually are observed in the sampling distribution even though such occurrences are infrequent), but we are willing to risk committing this error because we will never be able to be absolutely certain of anything.

The error just described is called a *Type I error*. It occurs when we are wrong in rejecting the null hypothesis. In such a situation, the means are not "truly" different (their difference is essentially zero), because the means are derived from the same population, but we make the judgment that the means were significantly different. A Type I error is therefore a *false positive* judgment concerning the validity of the mean difference obtained. The chances of making a Type I error correspond to our alpha level. In this case the probability of committing a Type I error is .05.

A *Type II error* is the other side of the coin. The reality is that the means did come from different populations, and we should have properly rejected the null hypothesis. However, here when we compute the ratio of between-groups to within-groups variance, the *F* ratio is not large enough to fall into the 5 percent area under the curve. Thus, we fail to reject the null hypothesis and claim instead that the means are not significantly different. A Type II error is therefore a *false negative* judgment concerning the validation of the mean difference obtained. There are several possible reasons that may account for why a Type II error was made. Among these reasons is that we had insufficient statistical power in the study to detect a true difference between or among the means; this topic is discussed in Section 4.7.

4.5 EXPLAINING VARIANCE: STRENGTH OF EFFECT

4.5.1 OVERVIEW

We infer from a statistically significant *F* ratio that the difference between or among the means of the levels of the independent variable is reliably different. We can therefore assert (with 95 percent confidence) that the independent variable did distinguish our groups from each other on the dependent measure. Using our room color study as an example, we concluded that room color seemed to affect the mood of the students. But the *F* ratio does not inform us of the strength (magnitude, potency) of the effect of the independent variable.

One way to conceptualize what is meant by "strength of effect" is to recognize that if the means are significantly different, then we can predict or differentiate with better-than-chance likelihood the scores of the participants on the dependent variable given the group to which they belong. But one can defeat the chance odds by a little, a moderate amount, or by a large margin. The degree to which we can perform such a prediction or differentiation is indexed by the degree to which the scores on the dependent variable are associated with the particular levels of the independent variable.

Strength (or magnitude) of effect measures are indexes of how strongly the scores and the levels of the independent variable are related to each other. The two most commonly used indexes evaluating the strength of the effect are omega squared and eta squared indexes (Keppel, 1991; Keppel & Wickens, 2004).

4.5.2 OMEGA SQUARED

Omega squared represents the strength of the effect of the independent variable in the population. Because we rarely know the population parameters needed to perform that calculation, we ordinarily estimate omega squared based on the results of the ANOVA. The formula for this calculation can be found in Keppel's (1991) text. Whereas omega squared varies between zero and one, the estimate of omega squared can yield a negative value when the *F* ratio is less than one.

4.5.3   ETA SQUARED

Eta squared is also known as R^2 (Kirk, 1995) and as the *correlation ratio* (Guilford & Fruchter, 1978). This statistic is descriptive of the data in the sample rather than being an estimate of some population parameter. Eta is a correlation coefficient and therefore varies between zero and one. Because it is based on a correlation coefficient, eta squared (η^2) can be directly interpreted as the proportion of total variance of the dependent variable that is accounted for or explained by (or associated with) the independent variable in the sample data set. Eta squared will typically yield a value higher than the estimated omega squared value by about .08 or so (Meyers et al., 2006); it is the strength of the effect index that we will use in this book.

Most professional organizations, such as the American Psychological Association (APA), strongly encourage researchers to report strength of effect indexes in addition to statistical significance information, and many journals are now requiring such reporting in the manuscripts that they review for publication (APA, 2001; Wilkinson et al., 1999). You are therefore well advised to always supply this information when reporting the F ratio.

The interpretation of whether the eta squared value is "high" or not is a relative matter, depending on the context of the research. Kirk (1995, p. 178), citing work done by Jacob Cohen, suggested that, in the absence of other criteria, omega squared values of .01, .06, and .14 or greater could be described in the behavioral sciences as small, medium, and large, respectively. That translates to approximate eta squared values of .09, .14, and .22 or greater. Kirk (1996) also talked about a concept of *practical significance* and Thompson (2002) added to that a notion of *clinical significance*. The point that Kirk and Thompson emphasized was that we should take into account the context within which we will use the information that the independent variable produces a particular effect rather than exclusively focusing on the statistical significance of the result. In such a light, it is possible that researchers will be thrilled in one context with an eta squared value of .10 or less while being disappointed in another context with an eta squared value of .20. The "potency" of an effect is a judgment made by researchers; it is therefore a somewhat subjective evaluation informed by and relative to the state of theory and research within the particular topic area representing the research study.

Eta squared is the proportion of total variance attributable to the independent variable. It is computed as follows:

$$\eta^2 = \frac{SS_A}{SS_{\text{Total}}}. \tag{4.3}$$

Because it is based on the percentage of total variance in the data set, in designs containing more than one independent variable the eta squared values associated with each of the effects partitioned in the ANOVA are additive; that is, if one effect accounted for 8 percent of the total variance and another accounted for 12 percent of the total variance, we could say that the two effects together accounted for 20 percent of the total variance.

The eta squared value is the proportion of the total variance as indexed by the sum of squares that is accounted for by the between-groups source of variance. In the present room color example, it can be seen from Table 3.2 that we obtained an eta squared value of .610 ($87.50 \div 143.50 = .610$). If these were the results of an actual study, we would conclude that the color of the room in which students took the mood survey accounted for 61 percent of the variance in the mood scores. It is appropriate to use the Greek letter for eta (η) when reporting it. Thus, the portion of the *F* statement containing eta squared values is written as follows:

$$\eta^2 = .610.$$

4.5.4 PARTIAL ETA SQUARED

In some of its procedures, SPSS reports a statistic known as a partial eta squared value. This is a variation of the eta squared measure already described, focusing directly on the effect and the error associated with it. Keeping with the present one-way design as an example, both eta squared and partial eta squared values use SS_A as the numerator. Eta squared uses the total sum of squares, SS_{Total}, as the denominator; however, to compute partial eta squared we use $SS_A + SS_{S/A}$ as the denominator.

In a one-way between-subjects ANOVA design, where $SS_{Total} = SS_A + SS_{S/A}$, the values for eta squared and partial eta squared will be equal. However, for all of the other designs that we describe in this book, the two measures will yield different values for one of both or the following reasons. First, there will be additional independent variables whose variance will contribute to the denominator of eta squared (the total sum of squares, SS_{Total}) but not contribute to the denominator of partial eta squared. Second, in within-subjects designs described in Chapters 10–12, we will see that there are several different error terms as a result of the variance partitioning process. The partial eta squared measure is computed with the error term (e.g., $SS_{S/A}$) specific to the effect (e.g., SS_A) represented in the denominator. Because of these issues, and unlike eta squared values, partial eta squared values are not additive across the effects, and the interpretation of this strength of effect measure is therefore somewhat different than that of eta squared values. We will focus on the eta squared value rather than the partial eta squared value as the index of strength of effect that we will use in this book.

4.5.5 COHEN'S *d*

Jacob Cohen (1969, 1977, 1988) suggested looking at the potency of the treatment effect by examining what he called "effect size." It bears a similarity to the approach used in calculating the *t* test (discussed in Section 4.8). Cohen proposed that the mean difference can be judged relative to the standard deviations of the groups. The logic for this is as follows. The standard deviation represents the spread of scores around a mean. Think of this score spread as a way of quantifying the "fuzzy boundary" of a group mean. Now mentally place another group mean with its own fuzzy boundary near the first group. If you place the second mean in such a position that the two means with their fuzzy boundaries are visually quite close, then in effect you cannot clearly distinguish between the two means. If the second mean is placed further away from the first, then despite each having a fuzzy boundary, it is possible to more clearly tell them apart.

Cohen's d is a way to quantify how clearly the means can be differentiated. It takes into account the difference in value between the two means with respect to the combined standard deviation of the groups. In equation form, we have

$$d = \frac{\overline{Y}_1 - \overline{Y}_2}{SD}. \tag{4.4}$$

Not only did Cohen suggest this procedure to quantify effect size, he provided general guidelines for interpreting values of d. He proposed that, all else equal, d values of .2, .5, and .8 can be thought of as small, medium, and large effect sizes, respectively. For example, if the mean difference spans a distance of almost 1 SD, then the means of the two groups can be quite easily distinguished, and so we would judge the effect size to be large.

Cohen extended his formulas to the situation in ANOVA where we have more than two means. He used the symbol f to index the effect size for multiple means and showed how d, f, and η^2 are all related. Nonetheless, many writers when referring to Cohen's effect size index will tend to focus on d.

4.6 REPORTING THE RESULTS

Reporting the results of a statistically significant effect of the independent variable involves presenting certain information to the reader. Here is what you are obliged to provide:

- The levels of the independent variable in plain English.
- The means and standard deviations of the groups.
- An indication, again in plain English, of the direction of the difference.
- The F ratio and the degrees of freedom at which it was evaluated (in this order: between-groups degrees of freedom, within-groups degrees of freedom): $F(1, 12) = 18.75$.
- Whether or not the probability reached your alpha level: $p < .05$.
- The strength of the effect and the statistic used to make the evaluation: $\eta^2 = .610$.

This information requirement may seem as though it would involve a huge amount of writing, but the reality is that it can be done, especially with only two groups in the study, in a concise manner. Here is an example of a narrative of the mood study results that provides the necessary information:

Students completing the mood survey in the room painted blue ($M = 19.00$, $SD = 2.16$) reported significantly greater relaxation than students completing the mood survey in the room painted red ($M = 14.00$, $SD = 2.16$), $F(1, 12) = 18.75$, $p < .05$, $\eta^2 = .610$.

4.7 STATISTICAL POWER

Assume that in a given experiment the null hypothesis is not true and should be rejected. That is, assume that we know there is a viable, nonzero difference between the means of two groups. Being able to detect such a true group difference by using a statistical procedure, such as ANOVA, is thought of as *statistical power* (Cohen, 1962, 1969, 1977, 1988). More power corresponds to a greater ability to detect a true effect. An analogy can be made to the process of magnification. Imagine holding a magnifying glass (using a statistical test) to the fine detail of a painting (to examine the data). As you use more powerful magnifiers – going from 2× to 5× to 10×, for example – you can see successively greater detail, distinguishing differences that were not readily apparent to the naked eye or to what was seen under the lower power magnifiers. Increasing the power of a statistical test allows you to increasingly discern differences that were not apparent under a lower power test.

Power is driven by three factors: alpha level, effect size in the population, and sample size. We briefly treat each in turn.

4.7.1 ALPHA LEVEL

Less stringent alpha levels will give us more opportunities to assert that a mean difference is statistically significant. This is because the benchmark separating the *F* distribution into "relatively common" and "relatively rare" areas is shifted toward the left (see Figure 4.2). For example, if our alpha level is set at .10 rather than .05, the benchmark will be placed to the left of what is shown in Figure 4.2, and more *F* ratio values will fall to its right (be identified as statistically significant). In this way, less-stringent alpha levels provide for greater statistical power than more-stringent alpha levels. That consideration needs to be balanced, of course, by the increased likelihood that such a smaller *F* ratio was obtained because the null hypothesis is true. In general, an alpha level of .05 seems to represent a good compromise under most research conditions (Cohen, 1962; Clark-Carter, 1997).

4.7.2   POPULATION EFFECT SIZE

Effect size, in this context, is the mean difference of the groups in the population evaluated against the within-groups population variance. What represents a large effect size has to do with the phenomenon under study, which is by its nature weak, moderate, or strong. There is nothing we researchers can do to affect this reality; rather, it is our responsibility to measure the phenomenon as accurately and objectively as possible and to let the chips fall where they may. The larger the population effect size, that is, the more potent the independent variable is in distinguishing the population group means against a background of error, the greater power we will have in the study.

On the other hand, some researcher judgment may be involved in selecting the levels of the independent variable to use in a study or even in selecting the best measure of behavior to use for the study at hand. For example, if we wanted to determine if preparation time enhanced performance on a final exam of a particular course, we might find a group of students who did not study at all to serve in the control group. If we chose to recruit in an experimental group those who actually studied for half an hour (as a very exaggerated example to make the point), the likelihood of finding an effect of the independent variable (believing very strongly that study time does truly matter) is practically nil, despite the fact that the population effect size is large. Rather, we would surely want to increase the power of our study by selecting an amount of study time for our experimental group that should show a difference if in fact our intuition was correct.

4.7.2   SAMPLE SIZE

Finally, sample size is an important factor in statistical power. The larger the sample size, the greater will be our statistical power. This is true at a basic level in that the closer we come to sampling the entire population, the more precise an estimate we will be able to make of the population mean for each group. At its extreme, if we sampled the entire population, and if we found a difference between the means of the two groups, we would not need a statistical test to determine if they were significantly different – by definition, the means would be statistically different.

In some fields of research, it is common to work with large databases, such as those used in archival research and in educational research. With tens or sometimes hundreds of thousands of cases, virtually any mean difference will turn out to be statistically significant. In such research, criteria other than statistical significance, such as strength of effect, must be used to evaluate the findings lest the researchers be overwhelmed by the amount of statistical power they wield. With very large sample sizes (e.g., $N \geq 1,000$), it is common practice to focus very little on statistical significance and to focus much more on the magnitudes of the effects under study. In fact, there has been a movement in many disciplines to deemphasize or even eliminate tests of significance altogether (see Kline, 2004). We are sympathetic to the arguments that are made on behalf of this movement because by blindly relying on statistical significance testing without considering the magnitudes of the effects we are studying, we turn into automatons rather than researchers. In its defense, we also believe that there is still a useful purpose to be served by statistical significance tests with the relatively small sample sizes that we ordinarily see in research using experimental designs.

4.8 THE LIMITING CASE OF ANOVA: THE *t* Test

A one-way between-subjects ANOVA is designed to compare two or more levels or groups of the independent variable on a single dependent variable. Thus, the ANOVA can handle situations in which the independent variable has two, three, or more levels. Creation of the *t* test predates Fisher's variance ratio and was specifically designed to compare the means of exactly two conditions.

4.8.1 A BIT OF HISTORY

The *t* test was developed by William Sealy Gosset, who was interested in publishing his innovative work. As Salsburg (2001) tells the story, in 1899, Gosset, who had a combined degree in chemistry and mathematics, was hired by the Guinness Brewing Company primarily for his chemistry expertise. But, as it turned out, it was his math skills that made a bigger difference in brewing beer. One of his first ventures into applying mathematics to the brewing of beer was estimating yeast amounts in samples:

When the mash was prepared for fermentation, a carefully measured amount of yeast was used. Yeast are living organisms, and cultures of yeast were kept alive and multiplying in jars of fluid before being put into the mash. The workers had to measure how much yeast was in a given jar in order to determine how much fluid to use. They drew a sample of the fluid and examined it under a microscope, counting the number of yeast cells they saw. (Salsburg, 2001, p. 26)

Obviously, there were measurement errors resulting from sampling the yeast (we would expect different samples to have somewhat different numbers of yeast cells even if they were relatively evenly distributed in the mash) as well as from humans counting the yeast cells. Gosset devised a mathematical model to estimate the error, a critical piece of information needed by the beer brewers, and he had a desire to publish it. The problem was that Guinness had a policy forbidding its employees to publish (fearful that they would reveal corporate secrets). In order to disseminate his work while at the same time avoiding being fired by the young Lord Guinness, in 1904 with the help of Karl Pearson he devised the pseudonym "Student" and published his work in Pearson's *Biometrika* without giving away his true identity. Gosset continued to publish his mathematical work under his pseudonym, and in 1908 published a paper in *Biometrika* that introduced the *t* test.

The sampling distribution of *t*, as is true for *F*, depends on degrees of freedom. The *t* distribution is *leptokurtic* – compared to the normal curve, its tails are higher and the middle portion is a bit more compressed. Gosset actually used *z* for his leptokurtic distribution, but, because that letter came to be associated with the normal distribution, textbook writers began referring to the leptokurtic distribution as "Student's *t*" (taking the last letter from "Student"). It is by this name assigned by the textbook writers that it has become known to generations of students.

4.8.2 THE RELATIONSHIP OF *t* AND *F*

Gosset's *t* test and Fisher's ANOVA are algebraically equivalent, and thus lead researchers to the same conclusions regarding a two-group comparison when the groups have met the assumption of homogeneity of variance (i.e., that the groups have comparable variances). The relationship between the two techniques becomes more obvious when we note that the square of the *t* statistic is equal to the *F* value and, conversely, the square root of *F* is equal to the *t* statistic. Thus,

$$t^2 = F$$

and

$$t = \sqrt{F}. \tag{4.5}$$

Because of the interchangeability of *t* and *F*, Keppel, Saufley, and Tokunaga (1992, p. 129) have suggested that it is useful for students to learn the *t* test primarily "... as an aid in understanding references to the *t* test ... [in] ... the research literature and as a touchstone with the past...."

ANOVA Assumptions

5.1 OVERVIEW

Before conducting an analysis of variance (ANOVA), researchers need to consider three fundamental statistical assumptions that underlie the analysis: (a) the error components associated with the scores of the dependent variable are independent of one another, (b) these errors are normally distributed, and (c) the variances across the levels or groups of the independent variable are equal. Although we discuss these assumptions separately, in practice they are interconnected; a violation of one assumption may and often does affect the others.

A fairly comprehensive literature on the assumptions underlying ANOVA has developed over the past six decades or so. Good general summaries of this work can be found in Glass, Peckham, and Sanders (1972) and Wilcox (1987).

5.2 INDEPENDENCE OF ERRORS

5.2.1 THE CONCEPT OF RESIDUAL OR ERROR VARIANCE

Consider a hypothetical medical research study in which an experimental group in a study received a certain dose of a drug. If each patient in that group showed precisely the same level of improvement, then there would be a zero value remaining when we subtracted the mean of the group from each patient's improvement score. Thus, knowing that patients were under the drug treatment would result in perfect (errorless) prediction of their scores on the dependent variable. Such a situation for the typical patient is given by the following expression:

$$Y_i - \overline{Y_j} = 0, \tag{5.1}$$

where Y_i is a score of a patient in the group and $\overline{Y_j}$ is the mean of that group.

Now consider the situation where the residuals are not zero. Under this condition we appear to have individual differences in reacting to the treatment. In our example, although the drug might have resulted in general improvement in the patients, the fact that they showed differences among themselves suggests that factors in addition to the effects of the drug were affecting their performance. We can think of this situation as representing measurement error for at least two converging reasons:

- Because we cannot identify the factors that are responsible for these individual differences (our data set contains only two pieces of information: the patient's group and his or her score on the dependent variable), this variation in their scores despite the fact that they were under the same drug condition is treated as measurement error.
- Knowing that these patients were under the drug condition would not allow perfect prediction of their scores. Even if the drug had a clear effect, and we knew the average improvement of the patients as a group, there would still be some margin of error involved in predicting their actual scores in the study.

5.2.2 THE MEANING OF INDEPENDENCE OF ERRORS

The first statistical assumption underlying ANOVA is that the residual or error component of the Y_i scores (the difference between these scores and the group mean) is random and independent across individual observations. Hence, no systematic pattern of errors is assumed to be present among the Y_i scores both within and between treatment groups. This means that the measurements representing one case in the study are independent of the data collected from all of the other cases in the study (Hays, 1981).

Dependence occurs when one Y_i contains information about another score. An example of *dependence* – signifying a systematic violation of the independence assumption – would be observed if the dependent variable showed larger residual error for the scores in one group and smaller errors for the scores in another group. This would be indicative of a statistical relationship between the errors and the scores. To the extent that we could predict the values of the residuals better than chance from a knowledge of the group to which a case belonged would indicate that the assumption of independence of errors has been violated. Examples of dependence in research situations are discussed next.

5.2.3 SITUATIONS PRODUCING VIOLATIONS OF ERROR INDEPENDENCE

Several research design situations foster dependence or correlated observations. One important source of potential bias occurs when participants within one treatment condition are tested in small or intact groups or participants enter a treatment condition with previous affiliations (outside of the study) that affect how they may perform as measured on the dependent variable. This "grouping" effect can produce differential error components among the basic scores or observations. Examples of such situations might be evaluating improvement in coping behaviors in patients comprising a single group in group therapy and measuring the effects of a program of instruction in a class of school children.

A second source of dependence can occur when participants in an experimental situation are allowed to communicate with each other about the task demands of the experiment. Such "contamination" may cause participants to influence each others behavior. An example of this is a group problem-solving study.

A third type of dependence situation occurs when the error component of each observation falls into a cyclical pattern typically due to participants' data being collected near each other in time and entered sequentially into the data set. This propinquity of the cases to each other in time, sometimes referred to as *autocorrelation*, may produce residual error components that are not independent of each other.

5.2.4 DETERMINING WHEN YOU HAVE VIOLATED ERROR INDEPENDENCE

Unlike the other two assumptions of normality and equality of variances, determining when your data have violated the assumption of error independence involves more complex diagnostic statistical analysis more typically performed in the domain of multiple regression analysis. For example, Cohen, Cohen, West, and Aiken (2003) discuss various diagnostics for detecting nonindependence of residuals, including plotting the residuals against an ordered numeric variable such as case number and computing the *Durbin–Watson* statistic (see Durbin & Watson, 1950, 1951, 1971).

5.2.5 CONSEQUENCE OF VIOLATING ERROR INDEPENDENCE

Hays (1981) and Stevens (2002) both consider independence to be a "very serious" statistical assumption for ANOVA, one that is often either ignored or dismissed as relatively inconsequential by investigators in the social and behavioral sciences. The consequence of violating the independence assumption is inflation of the targeted alpha (α) level. As Stevens (2002) notes, "Just a small amount of dependence among the observations causes the actual α to be several times greater than the level of significance" (p. 259).

5.2.6 SOLUTIONS TO VIOLATION OF INDEPENDENCE OF ERRORS

There is no easy-to-apply remedy for violating the independence of errors assumption; rather, the goal is to prevent the situation from occurring in the first place. The general maxim that is offered by writers (e.g., Cohen et al., 2003; Keppel & Wickens, 2004) is to randomly sample cases from a population, randomly assign the cases to treatment conditions, and ensure that the treatment conditions are independent of each other. In this regard Winer et al. (1991) have this to say:

Violating the assumption of random sampling of elements from a population and random assignment of the elements to the treatments may totally invalidate any study, since randomness provides the assurance that errors are independently distributed within and between treatment conditions and is also the mechanism by which bias is removed from treatment conditions. (p. 101)

The important point to remember is that careful adherence to proper experimental procedures, random sampling methods, and extra vigilance when using dyads, small groups, or intact groups may eliminate most issues of dependency. But even random sampling or random assignment of cases to groups is no guarantee that the measurement error will be independent (Stevens, 2002).

Should you encounter a research situation in which you believe that the errors are not independent, the alpha level under which you intend to test your statistics (e.g., .05) may in truth be much greater (e.g., .10 or .20). Thus, the most straightforward way of compensating for dependencies in your data is to adopt a more stringent alpha level, for example, shifting from the .05 level to the .01 or even .001 level (Stevens, 2002).

5.3 NORMALITY OF ERRORS

5.3.1 THE NATURE OF THE ASSUMPTION OF NORMALITY OF ERRORS

The second ANOVA assumption is that the error components associated with the Y_i scores are normally distributed. If the residual errors are normally distributed, then the distribution of Y_i scores will follow suit by being distributed in a normal manner. Likewise, if we randomly sample participants from a known normal population, the residual error component of those scores will also be distributed normally (Winer et al., 1991).

5.3.1.1 Normal Distributions

A *normal* distribution is characterized by a bell-shaped curve. It is a symmetrical distribution of scores (both halves of the curve look similar) that produces the same value for the mean, median, and mode. Such a distribution can be seen in Figure 5.1.

The normal curve as an idealized model suggests a symmetrical bell-shaped configuration for scores in a distribution. In practice, most continuous variables (e.g., age, height, test scores, monthly income, attitudes) only approximate this ideal.

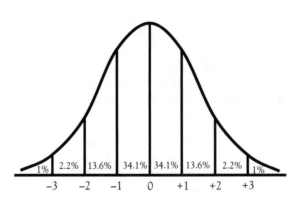

Figure 5.1 Example of a normal distribution.

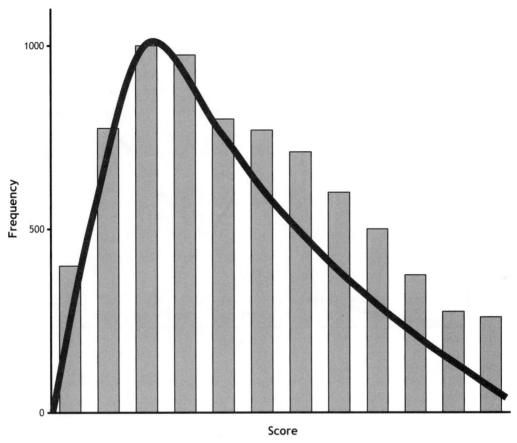

Figure 5.2   A positively skewed distribution.

5.3.1.2   Distributions Exhibiting Skewness

Skewed distributions are asymmetrical, with the most frequent scores clustering around one of the ends of the distribution. Figure 5.2 depicts a *positively skewed* distribution, where the scores are bunched around the lower end; one way to remember a positive skew is that the "longer tail" of the distribution is "pointed" toward the positive (right) side of the *x* axis. Conversely, Figure 5.3 illustrates a *negatively skewed* distribution, where the scores are concentrated on the high end. One way to remember a negative skew is that the "longer tail" of the distribution is "pointed" toward the negative (left) side of the *x* axis. As an organizing principle, remember this mnemonic device: "the tail tells."

5.3.1.3   Distributions Exhibiting Kurtosis

Kurtosis refers to the relative "peakedness" or "flatness" of the scores in a distribution. Figure 5.4 depicts a *leptokurtic* distribution, which tends to have tails that are higher than the normal curve is and bunched in the

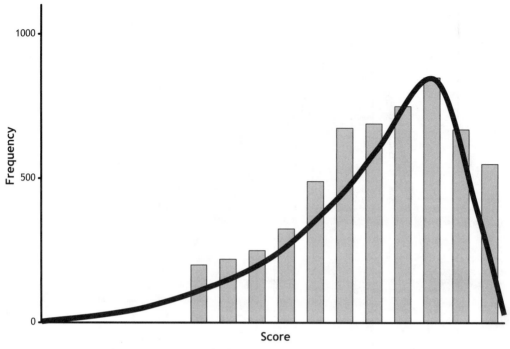

Figure 5.3   A negatively skewed distribution.

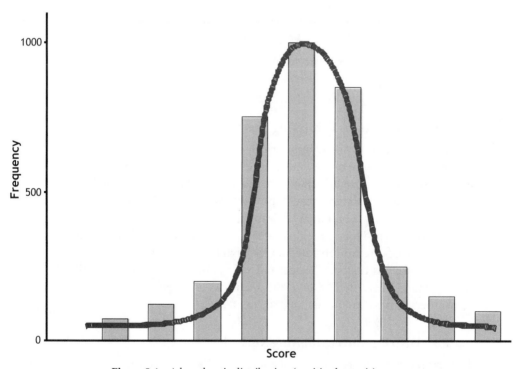

Figure 5.4   A leprokurtic distribution (positive kurtosis).

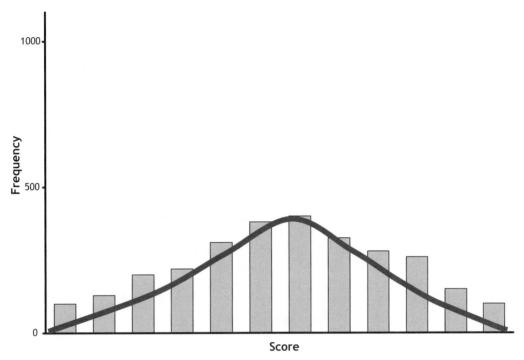

Figure 5.5   A platykurtic distribution (negative kurtosis).

middle, producing a peaked look. Conversely, Figure 5.5 displays a *platykurtic* distribution, which has a flatter, more evenly distributed look than the normal curve.

5.3.2   SITUATIONS PRODUCING VIOLATIONS OF THE ASSUMPTION OF NORMALITY OF ERRORS

ANOVA assumes that the residual error associated with the Y_i scores is normally distributed. However, in practice, we often encounter dependent variables that are not perfectly normal in shape. In fact, much of the time our variables rarely take on the idealized shape depicted in Figure 5.1, but instead reflect the irregularities (lack of symmetry) caused by sampling variability. This variability is particularly evident when sample sizes are small, that is, less than 8–12 (Keppel & Wickens, 2004; Tabachnick & Fidell, 2007), or when outliers are present in the distribution.

Outliers are cases with extreme or unusual values on a particular variable, possibly indicating an exciting serendipitous discovery, but more likely indicative of experimental error (e.g., coding error, participant failure to follow instructions, uncooperative children or rats, fatigue). Outliers should be eliminated unless the researcher deems them to be truly a small part of the population under study.

Detection of outliers can be readily accomplished with computer programs such as SPSS or SAS by converting the values of a variable to standard (z) scores with a $M = 0$ and a $SD = 1$. Standardized values exceeding ± 2.5 can be designated as outliers and are possible candidates for elimination (Meyers et al., 2006).

5.3.3 CONSEQUENCES OF VIOLATING THE ASSUMPTION OF NORMALITY OF ERRORS

ANOVA is considered to be quite resilient or robust to departures from normality. As Winer et al. (1991) suggest, "A reasonable statement is that the analysis of variance F statistic is robust with regard to moderate departures from normality when sample sizes are reasonably large and are equal..." (p. 101).

Confidence in the "robustness" of the normality assumption comes in part from an important statistical principle known as the *central limit theorem*, which indicates that as we increase the sample size (n), the resulting sample mean $\overline{Y}$ will increasingly approximate a normal distribution. Hence, Keppel and Wickens (2004) argue persuasively that "once the samples become as large as a dozen or so, we need not worry much about the assumption of normality" (p. 145).

Nevertheless, assessment of normality violations is certainly a useful endeavor if for no other reason than to intimately acquaint researchers with their data. A number of statistical and graphical approaches are currently available, as noted in Section 5.3.4. For additional details on these approaches to normality violations and also the issue of *multivariate normality*, see Meyers et al. (2006) for an introduction.

5.3.4 ASSESSING NORMALITY OF ERRORS IN DATA
5.3.4.1 Graphical Approaches to Assess Normality

A variety of graphical approaches are available in SPSS and SAS to assess for distributional normality violations. These approaches range from inspection of *histograms* or *stem-and-leaf plots* of the original scores or residuals for each group or level of the independent variable to the construction of *normal probability plots* where scores are rank ordered and plotted against expected normal distribution values. Normality is assumed if the plotted data follow a straight diagonal line. Some of these graphical techniques will be demonstrated in Sections 5.5.4 and 5.6.4.

5.3.4.2 Statistical Approaches to Assess Normality

Most statistical approaches to normality assessment begin with an examination of a variable's skewness and kurtosis. Skewness and kurtosis values that are zero or close to zero indicate a normally distributed variable, and values greater than or less than $+1.0$ and -1.0 are considered indicative of a nonnormally distributed variable.

Somewhat more definitive, but decidedly more sensitive, are the *Kolmogorov–Smirnov* and the *Shapiro–Wilk* tests. Both statistical tests can detect departures from normality, but because of their extreme sensitivity to sample size variations, we recommend using a stringent alpha level of $p < .001$ before concluding a normality violation exists. We will demonstrate these tests with SPSS and SAS in Sections 5.5.4 and 5.6.4.

5.3.5 SOLUTIONS TO VIOLATIONS OF NORMALITY OF ERRORS

One way to reduce nonnormality within a variable is to eliminate outliers that are clearly not representative of the population under study. These outliers tend to skew distributions and distort sample means. A second approach to reducing normality violations is to *trim* the values of the most extreme scores, for example, removing the top and bottom 5 percent of a distribution (Anderson, 2001). Such a procedure reduces variability as well as skewness and kurtosis violations. However, some investigators are opposed to data trimming of extreme scores, arguing that such exclusion will artificially reduce the error term in the ANOVA ($MS_{S/A}$) and afford the subsequent F test a positive bias (Keppel & Wickens, 2004). A third and more common practice is to modify the original Y_i scores in a distribution by means of a mathematical transformation. Typical transformations that can be employed with SPSS and SAS are log, square root, inverse, and arcsine, all of which can be quite effective at reducing situation-specific, distributional skewness but, at the same time, may increase the difficulty of data interpretation.

5.4 HOMOGENEITY OF VARIANCE

5.4.1 NATURE OF HOMOGENEITY OF VARIANCE

The third assumption underlying the analysis of variance is the *homogeneity of variance* or *homoscedasticity* assumption. The homogeneity assumption requires that the distribution of residual errors for each group have equal variances. In practice, this means that the Y_i scores at each level of the independent variable vary about their respective means $\overline{Y_j}$ in a similar (though not identical) fashion.

5.4.2 SITUATIONS PRODUCING VIOLATIONS OF HOMOGENEITY OF VARIANCE

Violation of this assumption, called heterogeneity of variance or heteroscedasticity, has at least three systematic causes (Keppel & Wickens, 2004). First, classification independent variables such as gender or ethnicity may have unique variances associated with the scores on the dependent variable. For example, because White American mental health consumers report fewer incidents of perceived racism in their daily lives than do African American consumers, we might expect greater variability for African Americans on a measure of stress-related racism.

Second, an experimental manipulation of an independent variable can encourage participants to behave more similarly or differently than a control condition, thus producing differences in group variances. That is, motivation to participate in the study, and *floor* or *ceiling effects* (i.e., where participants cannot score any lower or higher on the dependent measure because of intrinsic or extrinsic constraints) can produce extreme variability between and within independent variable groups.

Third, variability on some dependent variables may be related to group size. Heterogeneity can become a serious issue with unequal sample sizes. For example, large group variances associated with small sample sizes tend to produce a liberal F statistic whose nominal alpha level such as .05 is actually operating at a less stringent level such as .10. Conversely, when the situation arises where large sample sizes and large group variances are associated, we produce a conservative F statistic, where the nominal alpha level such as .05 is actually more stringent, for example, .01 (Stevens, 2002).

5.4.3   ASSESSING FOR VIOLATIONS OF HOMOGENEITY OF VARIANCE

There are a variety of ways to detect the presence of heterogeneity of variance. We will review a few of the most popular approaches. Hartley (1950) proposed a simple and elegant test statistic called "F_{MAX}" to gauge the level of homoscedasticity. F_{MAX} is simply a ratio of the largest group variance to the smallest group variance. Thus the formula for F_{MAX} is

$$F_{MAX} = \frac{s^2_{largest}}{s^2_{smallest}}. \tag{5.2}$$

Large values of F_{MAX} are indicative of heterogeneity. Keppel et al. (1992) lobbied for a simple but very conservative assessment criterion (for the introductory statistics curriculum) that identified any F_{MAX} value greater than 3.0 as indicative of heterogeneity; if this was obtained, a more stringent alpha level of .025 was to be applied to the subsequent F value. As Keppel et al. (1992) note,

> research has shown that when F_{MAX} is greater than 3.0, the sampling distribution of F begins to become sufficiently distorted to seriously affect the decision rule based on the undistorted theoretical sampling distribution of F. We can correct for this distortion by selecting a new critical value of F, which we will call F_{ADJ} at a slightly more stringent significance level, namely, $\alpha = .025$. (pp. 119–120)

A less conservative assessment of the F_{MAX} statistic is to use the critical values of F_{MAX} developed by Pearson and Hartley (1970, 1972) and is routinely employed by computer programs, such as SPSS and SAS. Recent discussions by Keppel and Wickens (2004) and Stevens (2002) offer caution concerning the use of the F_{MAX} statistic because of its extreme sensitivity to distributional nonnormality.

More robust detectors of heterogeneity can be found with the test statistics developed by Levene (1960) and Brown and Forsythe (1974). In these measures, the original Y_i scores are converted into absolute deviation scores and then subjected to an analysis of variance. Obtaining a statistically nonsignificant outcome (i.e., failing to reject the null hypothesis) is indicative of equal variances or homogeneity. Note that the subscript i denotes the ordinal position of each case or participant within each group or level of the independent variable (participant number 1, 2, 3, etc.). The j subscript denotes the level or group designation for the independent variable (level 1, 2, 3, etc.).

The Levene and Brown and Forsythe statistics differ in how the absolute difference scores are computed. The Levene test uses an absolute deviation score from the group mean:

$$Z_{ij} = Y_{ij} - \overline{Y}_j. \tag{5.3}$$

The Brown–Forsythe procedure creates a difference score based on each original score's deviation from the group median:

$$Z_{ij} = Y_{ij} - Mdn_j. \tag{5.4}$$

Both of these procedures are commonly used and available in SPSS and SAS; they will be demonstrated in Sections 5.5.5 and 5.6.5. Although both procedures are widely used in the research literature, we agree with the recommendation of Keppel and Wickens (2004), who note that "the Brown–Forsythe procedure is slightly more robust" (p. 151). Specifically, the Brown–Forsythe test should be used when a distribution is skewed.

5.4.4 SOLUTIONS FOR VIOLATIONS OF HOMOGENEITY OF VARIANCE

There are two common approaches to addressing heterogeneity of variance. The first solution is to use a more stringent alpha level, typically the .025 level. Keppel and Wickens (2004), in summarizing the results of their own Monte Carlo simulations, note that

the Type I error will be kept below the 5 percent level by setting $\alpha = .025$. If your concern is to keep the Type I error under control, then a conservative halving of α is the fastest and simplest way to eliminate concerns about heterogeneity. (p. 152)

The second type of remedy for heterogeneity of variance is to transform the original scores. Keppel and Wickens (2004) recommend this approach when the dependent variable consists of reaction time, frequency counts, or proportional measurements. Square root, log, and arcsine transformations can pull or compress a distribution to reduce heterogeneity and also normalize the distribution. SPSS and SAS can easily apply transformations to data. We recommend judicious use of these facilities, as transformed data are sometimes difficult to explain to a reader who is familiar with the original distribution under study.

Table 5.1. Hypothetical data set: Global assessment of functioning by type of therapy

Treatment conditions	Factor A (type of therapy)		
	Brief a_1	Psychodynamic a_2	Cognitive-behavioral a_3
Raw scores	50	35	72
	55	50	69
	68	39	70
	72	55	70
	60	45	71

5.5 SPSS APPLICATIONS: ASSUMPTION VIOLATION DETECTION AND SOLUTION

Consider the hypothetical data in Table 5.1. Assume we are interested in the effects of type of therapy (the independent variable) on global assessment of functioning (GAF), our dependent measure. The three levels of our independent variable (type of therapy) are brief, psychodynamic, and cognitive-behavioral therapy; we coded these therapies as 1, 2, and 3, respectively. The GAF dependent variable is a single score that can vary from 1, indicating *severe dysfunction*, to 100, indicating *good general functioning*. Thus, the higher the score, the better is the mental health functioning of the client. Assume that a total of fifteen adult community mental health clients were assigned randomly and in equal numbers to the three treatment conditions.

In this and other sections throughout Chapters 5–16, we will show in our figures both dialog windows and output tables displayed on the screen by SPSS and SAS. We encourage you to read carefully Appendixes A and B at the back of this book on the basics of these computer statistical packages (SPSS and SAS, respectively) and their usage before attempting these analyses.

5.5.1 SPSS DATA SCREENING

We will begin with a data-cleaning run with SPSS **Frequencies**. This analysis will enable us to check the appropriateness of each variable's set of numerical values. Enter SPSS for Windows and open your SPSS data file by clicking **File ➜ Open ➜ Data** and select the SPSS data file you wish to analyze.

We begin this analysis by clicking **Analyze ➜ Descriptive Statistics ➜ Frequencies**, which produces the SPSS **Frequencies** dialog window. There are only two variables in this file and we have clicked over both of them (**therapy** and **GAFscore**) by highlighting or activating each variable and clicking the **arrow** in the center of the dialog window, as shown in Figure 5.6. The left panel of the dialog window is now empty and the two variables now reside in the right panel labeled **Variable(s)**. The **Display frequency tables** window that produces frequency tables has been checked (this

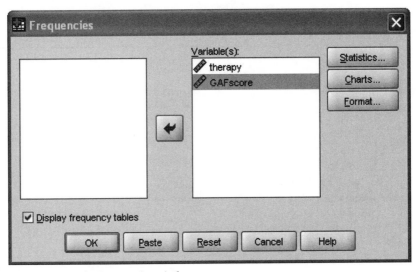

Figure 5.6   Main **Frequencies** window.

is the default). The pushbuttons at the right side of the dialog window (**Statistics, Charts,** and **Format**) provide researchers with a number of output options for each frequency table. We will focus on the **Statistics** pushbutton for the present analysis.

By clicking the **Statistics** pushbutton, the **Frequencies: Statistics** dialog window is produced (see Figure 5.7). This dialog window is composed

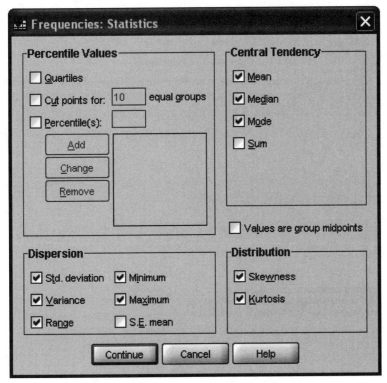

Figure 5.7   Statistics window from a **Frequencies** procedure.

Statistics

		therapy	GAFscore
N	Valid	15	15
	Missing	0	0
Mean		2.00	58.73
Median		2.00	60.00
Mode		1[a]	50[a]
Std. Deviation		.845	12.736
Variance		.714	162.210
Skewness		.000	-.551
Std. Error of Skewness		.580	.580
Kurtosis		-1.615	-1.047
Std. Error of Kurtosis		1.121	1.121
Range		2	37
Minimum		1	35
Maximum		3	72

a. Multiple modes exist. The smallest value is shown

Figure 5.8   Descriptive statistics for our variables.

of four sections: **Percentile Values, Dispersion, Central Tendency,** and **Distribution.** In the present example, we ask for the **Mean, Median, Mode, Standard deviation, Variance, Range, Minimum, Maximum, Skewness,** and **Kurtosis** values for each variable. Some of this information will be used to help make data-cleaning assessments and other information will be used to help assess statistical assumption violations. Clicking the **Continue** pushbutton returns us to the main dialog window. Click on **OK** to obtain the output file with the results.

Figure 5.8 displays the **Statistics** output for both variables (therapy and GAF). Neither variable contained any missing values, nor were there any extreme minimum and maximum values. Figures 5.9 and 5.10 provide a separate **Frequency Table** for therapy and GAFscore, respectively. For the categorical variable therapy, participants were evenly distributed across all three levels. For the continuous variable GAFscore, the scores showed good variability from a low of 35 to a high of 72, indicating reasonable values within published ranges.

Type of Therapy

		Frequency	Percent	Valid Percent	Cumulative Percent
Valid	Brief	5	33.3	33.3	33.3
	Psychodynamic	5	33.3	33.3	66.7
	Cognitive-Behavioral	5	33.3	33.3	100.0
	Total	15	100.0	100.0	

Figure 5.9   Frequency table for type of therapy.

GAFscore

		Frequency	Percent	Valid Percent	Cumulative Percent
Valid	35	1	6.7	6.7	6.7
	39	1	6.7	6.7	13.3
	45	1	6.7	6.7	20.0
	50	2	13.3	13.3	33.3
	55	2	13.3	13.3	46.7
	60	1	6.7	6.7	53.3
	68	1	6.7	6.7	60.0
	69	1	6.7	6.7	66.7
	70	2	13.3	13.3	80.0
	71	1	6.7	6.7	86.7
	72	2	13.3	13.3	100.0
	Total	15	100.0	100.0	

Figure 5.10 Frequency table for GAFscore.

5.5.2 SPSS MISSING VALUES ANALYSIS

Before considering issues of normality, and homogeneity of variance, researchers should first address the companion issues of missing values and outliers. Left unattended, either of these issues could distort the shape of a distribution.

The present example data set contained no missing values for the variables. Had there been one or more missing values for either of the variables, we could have applied one of two strategies: (a) Use the SPSS or SAS listwise deletion default, where the program computes statistics (e.g., means) by omitting cases that have missing values on a particular variable; (b) try to estimate or *impute* the missing value with a statistical replacement value. SPSS offers the **Missing Values Analysis (MVA)** module as a special add-on to the SPSS system. SPSS also offers a mean substitution imputation procedure through the **Transform** facility. Likewise, SAS offers the **MI (Multiple Imputation)** program for imputing missing values. A full discussion of analyzing missing values is beyond the scope of a text on ANOVA like the present book. As a brief introduction we recommend parts of chapters 3A and 3B of Meyers et al. (2006), and for a more complete discussion see Allison (2002) or McKnight, McKnight, Sidani, and Figueredo (2007).

5.5.3 SPSS OUTLIERS ANALYSIS

Recall that outliers refer to cases with extreme values on a particular variable that does not represent the population under study. To assess for outliers on our continuous dependent variable (GAFscore) across levels of our categorical independent variable (therapy), we begin by clicking **Analyze → Descriptive Statistics → Explore**, which opens the **Explore** dialog window (see Figure 5.11).

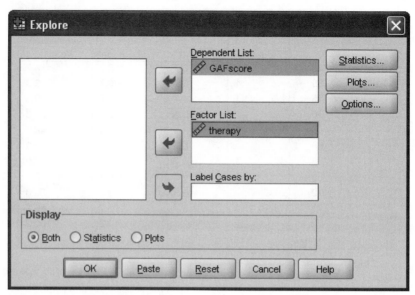

Figure 5.11 Main **Explore** window.

The left panel of the dialog window contains a list of all the variables in the analysis. We click over the continuous dependent variable (GAFscore) to the **Dependent List** window. The **Factor List** window allows you to click over categorical independent variables that will "break" or partition univariate outliers by each group or level of the independent variable. We click over our remaining categorical independent variable (**therapy**) to the **Factor List** window to obtain these profiles. Below the panel of variables (on the left side) is the **Display** panel, which allows you to toggle between a request for **Statistics** (basic descriptive statistics) or **Plots** (window plots or stem-and-leaf plots) for describing each variable. The SPSS default is to display both. On the right side of the dialog window are three pushbuttons: **Statistics**, **Plots**, and **Options**. These three pushbuttons allow you to request additional descriptive statistics, plots, and ways of overriding the default (listwise) missing values option.

Click the **Statistics** pushbutton and the **Explore: Statistics** dialog window is produced (see Figure 5.12). This dialog window is composed of four check windows, two of which concern us here, **Descriptives** (the default) displays basic descriptive statistics, and **Outliers** displays cases with the five largest and smallest **Extreme Values** for each dependent variable. Each of these checkboxes has been clicked. Clicking **Continue** brings you back to the main dialog window.

A click of the **Plots** pushbutton produces the **Explore: Plots** dialog window (see Figure 5.13). The **Boxplots** panel produces boxplots for each group or level of the independent variable. Such plots allow the investigator to see the spread or variability of the scores within each level of the independent variable. The **Plots** panel produces stem-and-leaf and histogram plots of continuous dependent variables partitioned by

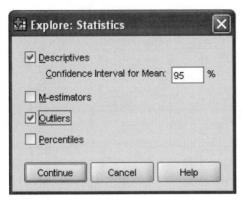

Figure 5.12 The **Statistics** window in **Explore**.

levels or groups of the independent variable. These plots provide both the frequency of occurrence and the actual value of each dependent variable score per level of the independent variable. The **Normality plots with tests** check box produces normal and detrended probability plots that we will not cover in this text, along with the Kolmogorov–Smirnov statistic and Shapiro–Wilk statistic for testing the normality assumption. Because of our focus on ANOVA principles and procedures, we will not elaborate on the use of boxplots, stem-and-leaf plots, or normal probability plots. The interested reader should consult Meyers et al. (2006) for an introduction to these topics.

Clicking **Continue** in the **Explore: Plots** window and **OK** in the Explore window will instruct SPSS to perform the analysis. The SPSS **Explore**

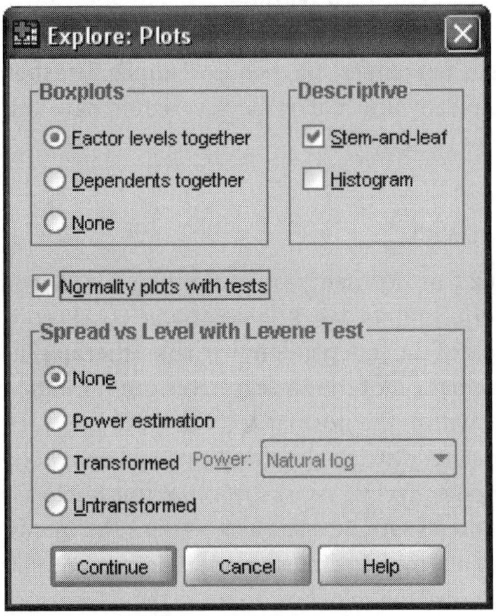

Figure 5.13 The **Plots** window in **Explore**.

Extreme Values[a]

therapy				Case Number	Value
GAFscore	1	Highest	1	4	72
			2	3	68
		Lowest	1	1	50
			2	2	55
	2	Highest	1	9	55
			2	7	50
		Lowest	1	6	35
			2	8	39
	3	Highest	1	11	72
			2	15	71
		Lowest	1	12	69
			2	14	70[b]

a. The requested number of extreme values exceeds the number of data points. A smaller number of extremes is displayed.

b. Only a partial list of cases with the value 70 are shown in the table of lower extremes.

Figure 5.14   Extreme values output from **Explore**.

output as it pertains to cases that have **Extreme Values** on the dependent variable partitioned by levels of the independent variable is shown in Figure 5.14, which lists the two largest and the two smallest cases with **Extreme Values** on our continuous dependent variable (**GAFscore**) broken by levels of the independent variable (**therapy**). Note that the SPSS default is normally the highest and lowest *five* cases, which was truncated as a result of the small sample size. The output in Figure 5.14 provides both the **Extreme Values** and the **Case Number** to facilitate evaluation and identification of potential outliers. In the present example, we deem these values to be within published ranges, and hence we can ignore them. Conversely, had we discovered any unusual outliers, we could have considered them possible candidates for deletion.

5.5.4   SPSS NORMALITY ANALYSIS

We begin our assessment of normality by examining the skewness and kurtosis values of our continuous dependent variable (**GAFscore**) partitioned by the three levels of the independent variable (**therapy**) as shown in Figure 5.15. An examination of the **Descriptives** output indicates that the skewness values are within the normal $+1.0$ to -1.0 range. However, negative kurtosis is associated with the *brief* and *psychodynamic* conditions (-1.869 and -1.557, respectively). We suspect that this kurtosis value is more a function of the small sample size ($n = 5$) than the manipulation itself, and so we do not attempt to transform the dependent variable in the present example. For discussion of how to make these transformations with SPSS, see Meyers et al. (2006).

Descriptives

therapy				Statistic	Std. Error
GAFscore	1	Mean		61.00	4.050
		95% Confidence Interval for Mean	Lower Bound	49.76	
			Upper Bound	72.24	
		5% Trimmed Mean		61.00	
		Median		60.00	
		Variance		82.000	
		Std. Deviation		9.055	
		Minimum		50	
		Maximum		72	
		Range		22	
		Interquartile Range		18	
		Skewness		.071	.913
		Kurtosis		-1.869	2.000
	2	Mean		44.80	3.611
		95% Confidence Interval for Mean	Lower Bound	34.77	
			Upper Bound	54.83	
		5% Trimmed Mean		44.78	
		Median		45.00	
		Variance		65.200	
		Std. Deviation		8.075	
		Minimum		35	
		Maximum		55	
		Range		20	
		Interquartile Range		16	
		Skewness		.052	.913
		Kurtosis		-1.557	2.000
	3	Mean		70.40	.510
		95% Confidence Interval for Mean	Lower Bound	68.98	
			Upper Bound	71.82	
		5% Trimmed Mean		70.39	
		Median		70.00	
		Variance		1.300	
		Std. Deviation		1.140	
		Minimum		69	
		Maximum		72	
		Range		3	
		Interquartile Range		2	
		Skewness		.405	.913
		Kurtosis		-.178	2.000

Figure 5.15   Descriptive statistics from **Explore**.

Our previous suspicion is confirmed by the SPSS **Tests of Normality** output in Figure 5.16. Both **Kolmogorov–Smirnov** and **Shapiro–Wilk** tests were not statistically significant ($p > .01$) for each level of the independent variable. This result indicates that the distribution of error associated with the GAF scores across levels of therapy did not depart significantly from normality.

The normality assumptions are
not violated; the probabilities
(Sig.) are > .05.

Tests of Normality

	therapy	Kolmogorov-Smirnov [a]			Shapiro-Wilk		
		Statistic	df	Sig.	Statistic	df	Sig.
GAFscore	1	.180	5	.200*	.962	5	.819
	2	.164	5	.200*	.975	5	.907
	3	.237	5	.200*	.961	5	.814

*. This is a lower bound of the true significance.

a. Lilliefors Significance Correction

Figure 5.16   Normality tests produced by **Explore**.

5.5.5   SPSS HOMOGENEITY OF VARIANCE ANALYSIS

There are a variety of ways to assess homogeneity of variance (homoscedasticity) within SPSS. We will demonstrate how to assess homogeneity of variance (equal variance) of the dependent variable across levels of the independent variable by using the SPSS **One-Way ANOVA** procedure to obtain Levene's, Welch's, and the Brown–Forsythe tests. Click **Analyze →** **Compare Means → One-Way ANOVA**, which produces the **One-Way ANOVA** dialog window (see Figure 5.17). As can be seen from Figure 5.17, we have clicked over **GAFscore** to the **Dependent List:** and **therapy** to the **Factor:** window.

By clicking the **Options** pushbutton, we produce the **One-Way ANOVA: Options** dialog window (see Figure 5.18) where we checked the **Homogeneity of variance test** (i.e., the Levene test), **Brown–Forsythe** for the Brown–Forsythe test, and **Welch** for the Welch test. Clicking **Continue** brings you back to the main dialog window and clicking **OK** allows the

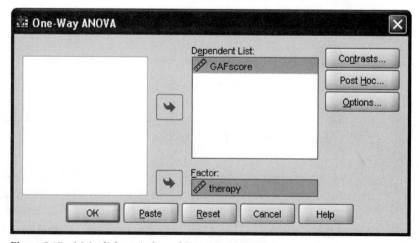

Figure 5.17   Main dialog window of **One-Way ANOVA**.

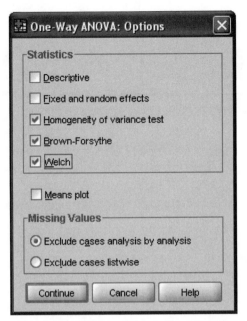

Figure 5.18   Options window in **One-Way ANOVA**.

Test of Homogeneity of Variances

GAFscore

Levene Statistic	df1	df2	Sig.
5.108	2	12	.025

> The equal variances assumption is violated; the probability (Sig.) is < .05.

Figure 5.19   The Levene's test in **One-way ANOVA**.

analysis to be performed. The SPSS Explore procedure can also be used to conduct the Levene and Brown–Forsythe tests, and will be demonstrated in Chapter 6.

Figures 5.19 shows the results of the Levene test. The statistically significant outcome indicates that the assumption of homogeneity has not been met. Instead of evaluating the mean differences with the usual Fisher F test, we must instead use a test such as the Brown–Forsythe test, designed to accommodate group differences in variance. As can be seen in Figure 5.20,

Robust Tests of Equality of Means

GAFscore

	Statistic[a]	df1	df2	Sig.
Welch	24.104	2	5.519	.002
Brown-Forsythe	16.939	2	8.036	.001

a. Asymptotically F distributed.

> The means are significantly different after correcting for the unequal variances across the levels of the independent variable.

Figure 5.20   The Welch and the Brown–Forsythe tests.

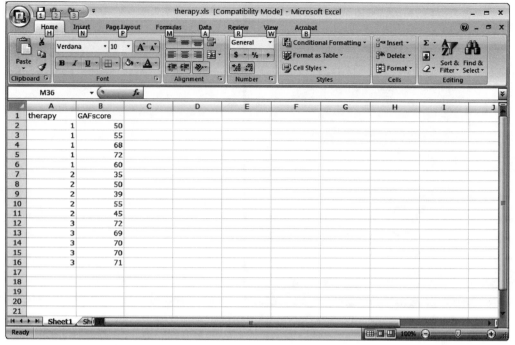

Figure 5.21   The Excel spreadsheet with the data.

the Brown–Forsythe test as well as the more conservative Welch test both indicated that the mean differences are statistically significant despite violating the homogeneity of variance assumption.

5.6 SAS APPLICATIONS: ASSUMPTION VIOLATION DETECTION AND SOLUTION

5.6.1 SAS DATA SCREENING

As recommended in Appendix B, we have entered our data into an Excel spreadsheet and will import the data set into an *SAS Enterprise Guide* project. The Excel spreadsheet is shown in Figure 5.21. Import data from Excel as described in Appendix B. The result of this operation is shown in Figure 5.22.

We will first obtain the frequencies for the occurrence of each value. From the main menu select **Describe → One-Way Frequencies** to reach the screen shown in Figure 5.23. The screen opens on the **Task Roles** tab, which is highlighted at the top of the navigation panel in the far left portion of the screen. Highlight **GAFscore** in the **Variables** panel and using one of the following two techniques place it under **Analysis variables** in the **Roles** panel: (a) drag it to the icon for **Analysis variables**; (b) click on the right facing arrow between the two main panels and select **Analysis**

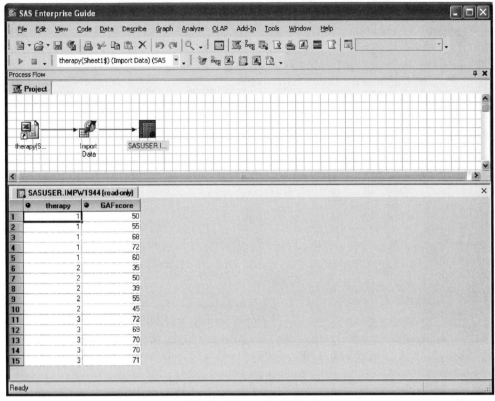

Figure 5.22   The data have been imported to our project from Excel.

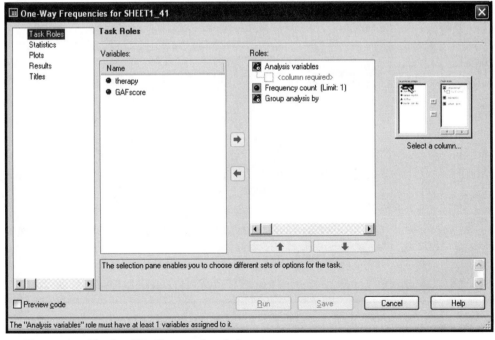

Figure 5.23   The **One-Way Frequencies** window.

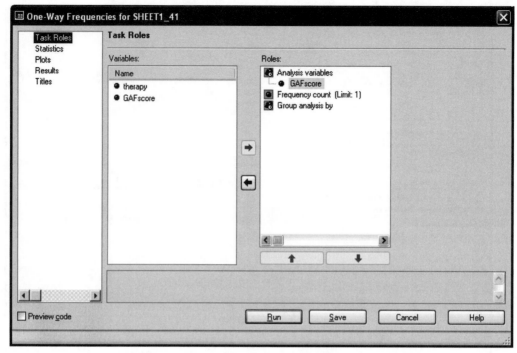

Figure 5.24   GAF scores have been placed in the **Analysis variables** area.

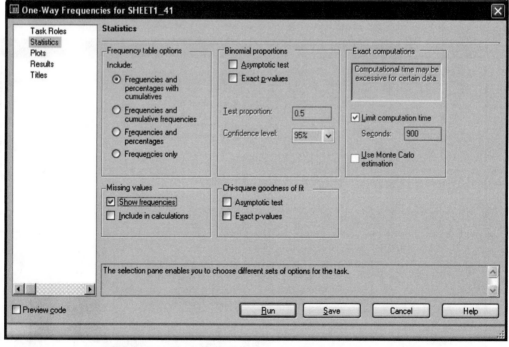

Figure 5.25   The **Statistics** window.

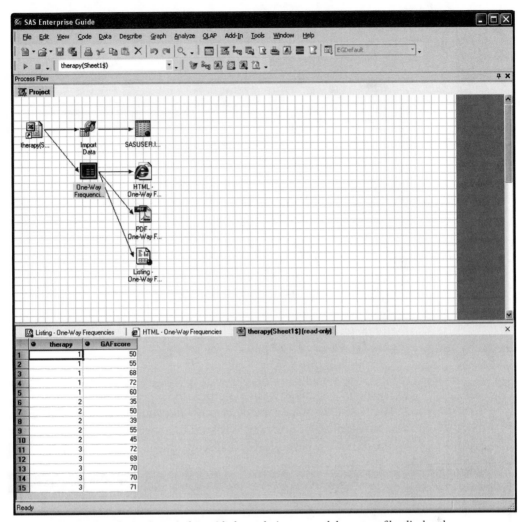

Figure 5.26 The **Project** window with the analysis setup and the output files displayed.

variables from the short menu. When done, your screen will resemble the one shown in Figure 5.24.

Selecting the **Statistics** tab in the navigation panel of the **One-Way Frequencies** procedure brings you to the screen shown in Figure 5.25. Select **Frequencies and percentages with cumulatives** and check the box for **Show frequencies** under **Missing values** (we do not have missing values in this small data file, but in larger data files this is not uncommon and this information will be tabulated in the **Frequencies** output). Click **Run** to perform the analysis.

The **One-Way Frequencies** analysis structure and icons for the output files are now shown in the **Project** window (see Figure 5.26). We had set up the **Options** under the general **Tools** menu to display the results in **HTML**, **PDF**, and **Text** format (you may wish to indicate fewer displays, but we suggest that one of them should be **PDF**); thus, we have three

One-Way Frequencies
Results

The FREQ Procedure

GAFscore				
GAFscore	Frequency	Percent	Cumulative Frequency	Cumulative Percent
35	1	6.67	1	6.67
39	1	6.67	2	13.33
45	1	6.67	3	20.00
50	2	13.33	5	33.33
55	2	13.33	7	46.67
60	1	6.67	8	53.33
68	1	6.67	9	60.00
69	1	6.67	10	66.67
70	2	13.33	12	80.00
71	1	6.67	13	86.67
72	2	13.33	15	100.00

Figure 5.27   Frequencies output.

output files, each providing the same information. The output is shown in Figure 5.27 and is structured in the same way as described for SPSS.

We will now obtain the descriptive statistics. From the main menu select **Describe ➜ Summary Statistics** to arrive at the screen shown in Figure 5.28. We have placed **GAFscore** under **Analysis variables** and

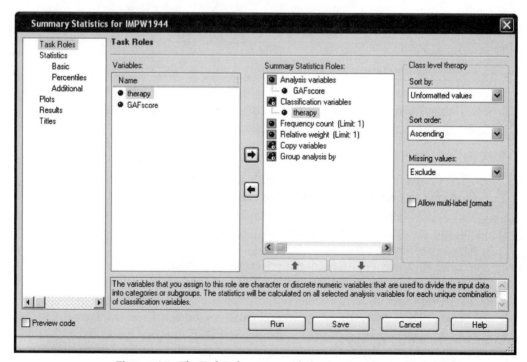

Figure 5.28   The **Task Roles** screen with **GAFscore** as the **Analysis** variable.

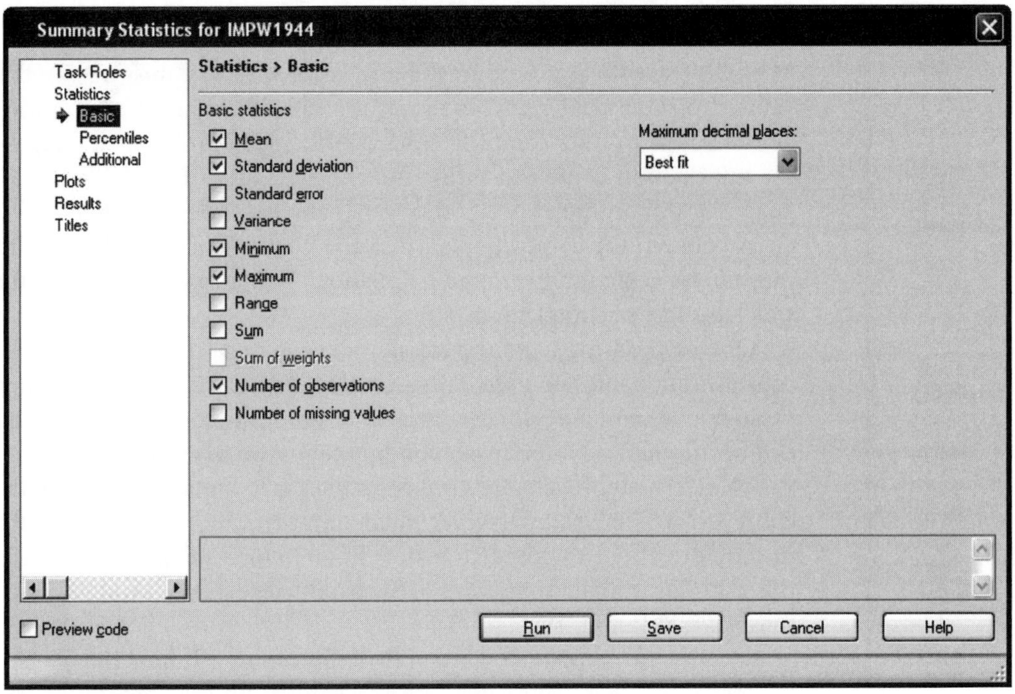

Figure 5.29 The **Basic Statistics** tab.

therapy under **Classification variables** in the **Summary Statistics Roles** panel. By using **therapy** as a classification variable, we will obtain descriptive statistics (on the dependent variable GAF score) for each of the therapy groups separately.

Select the **Basic Statistics** tab. We have opted to display the mean, standard deviation, minimum and maximum scores, and the number of observations as shown in Figure 5.29. Click **Run** to perform the analysis.

The descriptive statistics for each therapy group are displayed in Figure 5.30. The therapy group coded as 1 (brief psychotherapy), for example, had a mean GAF score of 61.00 with a standard deviation of 9.06 (rounded); the lowest score in the group was 50 and the highest was 72.

Summary Statistics
Results

The MEANS Procedure

		Analysis Variable : GAFscore GAFscore				
therapy	N Obs	Mean	Std Dev	Minimum	Maximum	N
1	5	61.0000000	9.0553851	50.0000000	72.0000000	5
2	5	44.8000000	8.0746517	35.0000000	55.0000000	5
3	5	70.4000000	1.1401754	69.0000000	72.0000000	5

Figure 5.30 Output from the **Basic Statistics** procedure.

5.6.2 SAS MISSING VALUES ANALYSIS

Most SAS statistical procedures ignore cases with any missing variable values from the analysis. These cases are called "incomplete." Limiting your analysis to only complete cases is potentially problematic because the information contained in the incomplete cases is lost. Additionally, analyzing only complete cases disregards any possible systematic differences between the complete cases and the incomplete cases. The results may not be generalizable to the population of all cases, especially with a small number of complete cases.

One method for handling missing data is *single imputation*, which substitutes a value for each missing value. For example, each missing value can be imputed with the variable mean of the complete cases, or it can be imputed with the mean conditional on observed values of other variables. This approach treats missing values as if they were known in the complete data analysis. However, single imputation does not reflect the uncertainty about the predictions of the unknown missing values, and the resulting estimated variances of the parameter estimates will be biased toward zero (Rubin, 1987, p. 13).

SAS uses a *multiple imputation* (MI) procedure (Rubin, 1987, 1996) that replaces each missing value with a set of plausible values that represent the uncertainty about the value to impute. MI inference involves three distinct phases:

- The missing data are filled in m times to generate m complete data sets.
- The m complete data sets are analyzed by using standard statistical analyses.
- The results from the m complete data sets are combined to produce inferential results.

To implement the MI procedure, you will need to use SAS syntax "PROC MI." Students interested in using this option need to consult the SAS manual.

5.6.3 SAS OUTLIERS ANALYSIS

To examine the data for outliers, we can obtain box and whiskers plots and can have SAS list the extreme values in each group. We will deal with these one at a time. From the main menu select **Describe → Summary Statistics** to arrive at the **Task Roles** screen. As described earlier, place **GAFscore** under **Analysis variables** and **therapy** under **Classification variables** in the **Summary Statistics Roles** panel. Then click on **Plots** in the navigation panel to arrive at the window shown in Figure 5.31. We have selected **Box and whisker** as the display we wish to view. Because we specified therapy as a classification variable, we will obtain separate plots for each of the therapy groups. Then click **Run** to perform the analysis.

The box and whisker plot is shown in Figure 5.32. The box, actually a vertically displayed rectangle, spans the range from the first quartile

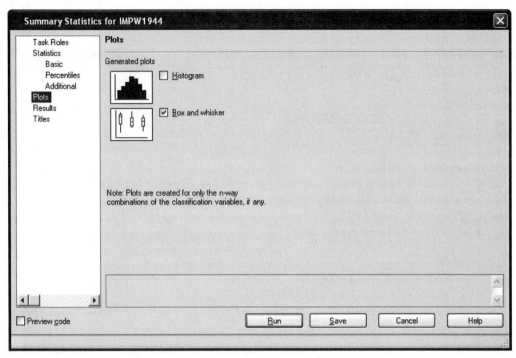

Figure 5.31 The **Plots** screen with **Box and whisker** specified.

Summary Statistics
Box and Whisker Plots

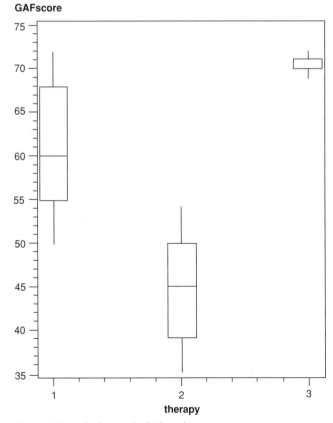

Figure 5.32 The box and whisker plot.

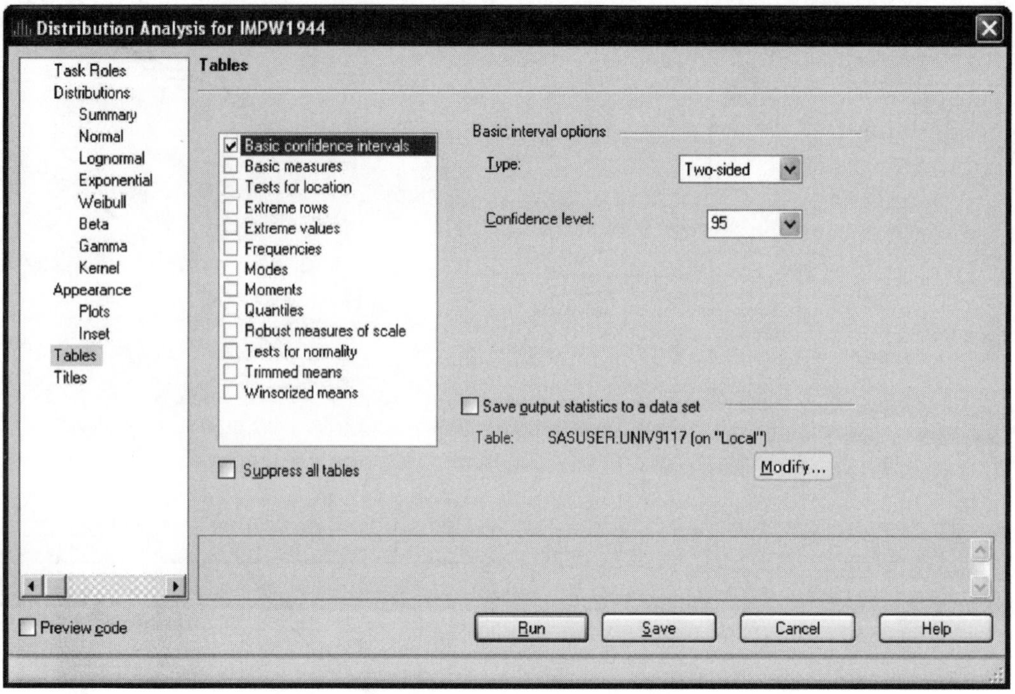

Figure 5.33   Configuring basic confidence intervals.

(the twenty-fifth percentile) to the third quartile (the seventy-fifth percentile) with the median indicated by the horizontal bar inside the rectangle. The whiskers are the vertical bars extending from the top and bottom of the box and ordinarily extend to a distance of 1.5 interquartile range. Outliers would be seen beyond the whiskers; in our very small data set, we have no outliers.

To list the extreme values in the distribution of scores for each group, from the main menu select **Analyze → Distribution Analysis**. In the **Task Roles** window, place **GAFscore** under **Analysis variables** and **therapy** under **Classification variables** in the **Summary Statistics Roles** panel. Then click on **Tables** in the navigation panel to bring you to the screen shown in Figure 5.33. Check **Basic confidence intervals**. On the **Type** drop-down menu, select **Two-sided**; on the **Confidence level** drop-down menu, select **95** (see Figure 5.33). Having configured this window will enable us to make further specifications in other choices.

Then click the checkbox labeled **Extreme values** as shown in Figure 5.34. This activates the **Specify n** panel in which we can indicate how many extreme values we would like to see in the output. We have specified **2** (after all, we have only 5 cases in each group) but with larger data sets you might select a value corresponding to about 5 percent of the sample. The output display will consist of the two lowest and two highest values in each group. Click **Run** to perform the analysis.

A portion of the output for the analysis we just performed is presented in Figure 5.35, where we see the results for the therapy group coded as 1;

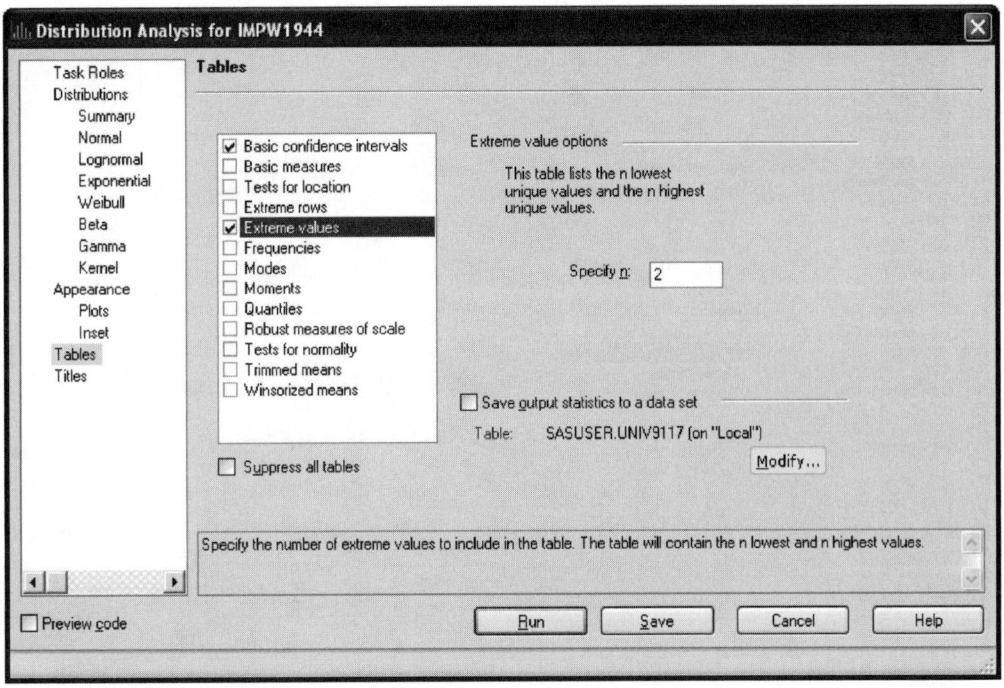

Figure 5.34 Configuring extreme values.

The UNIVARIATE Procedure
Variable: GAFscore (GAFscore)
therapy = 1

Basic Confidence Limits Assuming Normality			
Parameter	Estimate	95% Confidence Limits	
Mean	61.00000	49.75625	72.24375
Std Deviation	9.05539	5.42538	26.02115
Variance	82.00000	29.43476	677.10040

Extreme Observations			
Lowest		Highest	
Value	Obs	Value	Obs
50	1	50	1
55	2	55	2
60	5	60	5
68	3	68	3
72	4	72	4

Figure 5.35 Extreme values output.

Extreme Values			
Lowest		Highest	
Order	Value	Order	Value
1	50	4	68
2	55	5	72

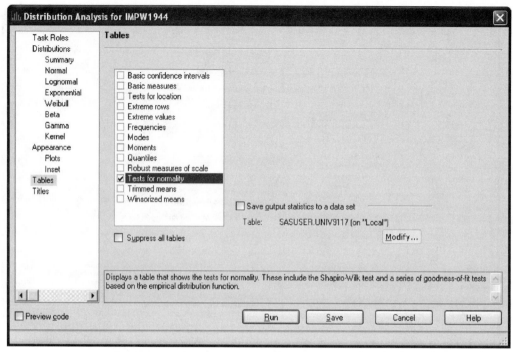

Figure 5.36   Selecting tests for assessing the normality of the distribution.

the other groups have analogous sets of tables in the output. The top table presents the mean, standard deviation, variance, and 95 percent confidence intervals. The next two tables present the extreme values in the distribution. In the middle table, the five highest and five lowest values in therapy group 1 are listed together with the row number in the data file; this is default output in *SAS Enterprise Guide*. Because we only have five cases in the group, the two lists are identical.

The bottom table is the one we requested by selected **Extreme values** in the **Tables** window and specifying an *n* of 2. SAS has displayed in the left portion of the table the two lowest values (under **Value**) and their rank order (under **Order**) and has displayed in the right portion of the table the two highest values (under **Value**) and their rank order (under **Order**). The two highest will always be ranked 1 and 2. The ranks of the highest values will depend on group size; with only five values in the group, the two highest are ranked 4 and 5.

5.6.4   SAS NORMALITY ANALYSIS

To examine the distribution with respect to normality, once again select **Analyze → Distribution Analysis**. In the **Task Roles** window, place **GAF-score** under **Analysis variables** and **therapy** under **Classification variables** in the **Summary Statistics Roles** panel. Then click on **Tables** in the navigation panel to bring you to the screen shown in Figure 5.36. Check **Tests for normality** and click **Run**.

Distribution analysis of: GAFscore

The UNIVARIATE Procedure
Variable: GAFscore (GAFscore)
therapy =   1

Tests for Normality				
Test	**Statistic**		**p Value**	
Shapiro-Wilk	**W**	0.961611	**Pr < W**	0.8192
Kolmogorov-Smirnov	**D**	0.180245	**Pr > D**	>0.1500
Cramer-von Mises	**W-Sq**	0.027473	**Pr > W-Sq**	>0.2500
Anderson-Darling	**A-Sq**	0.186637	**Pr > A-Sq**	>0.2500

> The normality assumptions are not violated; the probabilities (Sig.) are > .05.

Figure 5.37   Output for the normality tests.

The results of our analysis are presented in Figure 5.37. The Shapiro–Wilk and the Kolmogorov–Smirnov tests show results identical to those of SPSS; the other two are not computed by SPSS. As can be seen in the **p Value** column, all four tests have indicated that the distribution of the scores in therapy group 1 does not depart from what would be expected of scores that were distributed normally. That said, we would not encourage you to claim that the five scores are distributed in a normal or any other manner, as five scores are far too few to use as a base for any such conclusions.

5.6.5   SAS HOMOGENEITY OF VARIANCE ANALYSIS

To assess the Assumption of homogeneity of variance, from the main menu select **Analyze → ANOVA → One-Way ANOVA**. In the **Task Roles** window, drag **GAFscore** to the icon for **Dependent variables** and drag **therapy** to the icon for **Independent variable**. This is shown in Figure 5.38.

Select **Tests** in the navigation panel to be presented with the screen shown in Figure 5.39. Select **Welch's variance-weighted ANOVA**, the test we would prefer using to evaluate group differences (instead of the Fisher *F* ratio) if the assumption of equal variances is rejected. *SAS Enterprise Guide* provides three tests of homogeneity of variance (**Bartlett's test**, the **Brown Forsyth** test, and **Levene's** test); choose all three and click **Run**.

Figure 5.40 displays the results of the tests of homogeneity of variance. The Levene test was statistically significant in the SPSS analysis ($p = .025$) but shows a nonsignificant result here ($p = .0627$). This is because SPSS has computed the variability around the mean by using absolute differences between the scores and the mean, a traditional way to perform the calculation. In the **One-Way ANOVA** procedure, SAS uses the squared deviations around the mean, and this produces somewhat different results. The other tests of homogeneity all return statistically significant results consistent with SPSS, and we conclude on that basis that the data do not meet the assumption of homogeneity of variance. Based on that conclusion, we would use the Welch variance-weighted ANOVA in place of the Fisher *F* ratio.

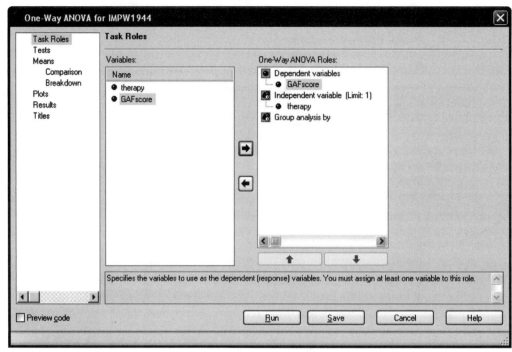

Figure 5.38   The **One-Way ANOVA Task Roles** screen.

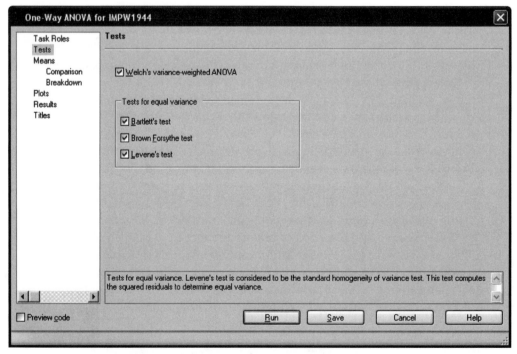

Figure 5.39   The **One-Way ANOVA Tests** screen.

One-Way Analysis of Variance
Results

The ANOVA Procedure

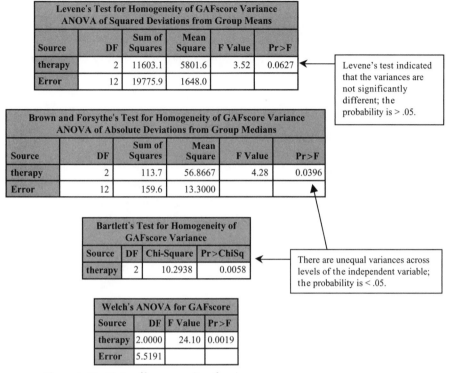

Levene's Test for Homogeneity of GAFscore Variance ANOVA of Squared Deviations from Group Means					
Source	DF	Sum of Squares	Mean Square	F Value	Pr>F
therapy	2	11603.1	5801.6	3.52	0.0627
Error	12	19775.9	1648.0		

Levene's test indicated that the variances are not significantly different; the probability is > .05.

Brown and Forsythe's Test for Homogeneity of GAFscore Variance ANOVA of Absolute Deviations from Group Medians					
Source	DF	Sum of Squares	Mean Square	F Value	Pr>F
therapy	2	113.7	56.8667	4.28	0.0396
Error	12	159.6	13.3000		

Bartlett's Test for Homogeneity of GAFscore Variance			
Source	DF	Chi-Square	Pr>ChiSq
therapy	2	10.2938	0.0058

There are unequal variances across levels of the independent variable; the probability is < .05.

Welch's ANOVA for GAFscore			
Source	DF	F Value	Pr>F
therapy	2.0000	24.10	0.0019
Error	5.5191		

Figure 5.40   Tests of homogeneity of variance.

The Welch results are shown in the bottom portion of Figure 5.40. As can be seen in the column labeled **F value**, the obtained value is 24.10, the same value yielded by SPSS (see Figure 5.20). The probability of this value occurring by chance is .0019. This result is statistically significant (our alpha level is .05), and, as we will see in Chapter 6, indicates that we have reliable group mean differences.

5.7 COMMUNICATING THE RESULTS

Before proceeding with the data analysis, we examined all variables for possible code and statistical assumption violations, as well as for missing values, and outliers, with SPSS Frequencies, Explore, and One-Way ANOVA procedures. The fifteen mental health consumers had no missing values on the one continuous dependent measure, GAF.

Several univariate outliers were detected, none of which were considered extreme or unusual enough to require deletion. Because there was random assignment of participants to treatment conditions and random selection of participants from the population as a whole, we deemed independence between cases to be appropriate. Kolmogorov–Smirnov and Shapiro–Wilk tests of normality of the dependent variable (GAF) across the levels of type of therapy were not statistically significant ($p > .05$), indicating no normality assumption violations. The Bartlett and Brown–Forsythe tests of the equality of error variances assumption were statistically significant ($p < .05$), indicating heterogeneity (unequal variances) across independent variable groups. This heterogeneity was attributed to the small sample size in the present example.

CHAPTER 5 EXERCISES

5.1. Three groups of tae kwon do black belt martial artists were taught a new form that consisted of twenty complicated kicks, punches, and blocks, and under conditions of no feedback, positive feedback and encouragement, and criticism and yelling. Participants were randomly assigned to treatment conditions. After two weeks of "training," students performed their forms and were rated by a Master instructor on a $1 = poor$ to $10 = excellent$ scale.

No feedback	Positive feedback	Criticism
3	7	10
6	8	9
2	8	8
3	9	4
4	10	9

Check for normality and homogeneity of variance violations by using either SPSS or SAS.

5.2. A random sample of five adult community mental health clients were randomly assigned to the following four treatment conditions. Their GAF scores after six weeks of treatment served as the dependent measure. Assume that clients were equated on initial GAF scores at the beginning of the treatment.

Group	Medication only	Brief	Cognitive-behavioral
58	65	50	70
55	60	50	65
60	68	45	60
48	58	40	60
50	60	50	65

Check for normality and homogeneity of variance violations using either SPSS or SAS.

SECTION 3

Between-Subjects Designs

One-Way Between-Subjects Design

6.1 OVERVIEW

In a between-subjects design the levels of the independent variable are represented by different participants. A one-way between-subjects design has just one independent variable. In the room color study described in earlier chapters, for example, the independent variable was the color of the room. The independent variable is said to have levels, which are usually referred to as treatment levels or treatment conditions. A level is the value of the independent variable for a given group. Our room color variable had two levels, red and blue. In the data file, these conditions would be coded arbitrarily as 1 and 2.

We are not limited to having only two levels for an independent variable. Theoretically, we could have a very large number. Practically, however, half a dozen or so levels are about as many as you will ordinarily see. In the example that we will use in this chapter, we will have five levels. Because this is a between-subjects design, each participant is studied under just one of these conditions and contributes just a single score to the data analysis.

6.2 A NUMERICAL EXAMPLE

The SAT is used by a wide range of colleges and universities as part of the application process for college admission. Assume that we are interested in the effect of preparation time on SAT performance. Because it would be unethical to have assigned students randomly to the groups in this particular study, assume that we opted for a quasiexperimental design (the classic descriptions of quasiexperimental and nonexperimental have been presented by Donald T. Campbell and his associates: Campbell, 1957; Campbell & Stanley, 1966; Cook & Campbell, 1979; Shadish, Cook, & Campbell, 2002), where we have found samples of students who have studied for the durations specified by our group structure. In this manner, we have generated five groups of students with each group being composed of seven cases. The groups contain students who have studied for either zero, two, four, six, or eight months prior to taking the SAT.

Our dependent variable is the average of the Verbal and Quantitative SAT scores. The standardized mean on the SAT is 500 with a standard

Table 6.1. Preliminary calculations for a (five-group) one-way ANOVA data set

Treatment conditions or levels	Factor A: Preparation time (in months)					
	0 a_1	2 a_2	4 a_3	6 a_4	8 a_5	Total
Dependent variable or individual scores	370	410	530	560	550	
	380	430	530	580	610	
	420	440	540	590	620	
	420	480	540	620	630	
	430	510	550	640	640	
	430	520	570	650	650	
	440	530	610	660	660	
Sum of scores $(\Sigma Y_{ij}) = (A_j)$	$A_1 = 2{,}890$	$A_2 = 3{,}320$	$A_3 = 3{,}870$	$A_4 = 4{,}300$	$A_5 = 4{,}360$	$\Sigma Y_{ij} = \Sigma A_j = T = 18{,}740$
Sum of squared scores (ΣY^2_{aj})	$Y^2_{a1} = 1{,}197{,}500$	$Y^2_{a2} = 1{,}588{,}400$	$Y^2_{a3} = 2{,}144{,}500$	$Y^2_{a4} = 2{,}650{,}200$	$Y^2_{a5} = 2{,}723{,}600$	$\Sigma Y^2 = 10{,}304{,}200$
Number of cases	$n_1 = 7$	$n_2 = 7$	$n_3 = 7$	$n_4 = 7$	$n_5 = 7$	$(a)(n) = N = 35$
Mean $(\overline{Y}_j)$	$\overline{Y}_1 = 412.86$	$\overline{Y}_2 = 474.29$	$\overline{Y}_3 = 552.86$	$\overline{Y}_4 = 614.29$	$\overline{Y}_5 = 623.86$	$\overline{Y}_T = \dfrac{T}{(a)(n)} = 535.43$
Variance (s^2_j)	723.81	2,295.24	823.81	1,461.91	1,323.81	7,949.08
Standard deviation (s_j)	26.90	47.91	28.70	38.23	36.38	89.16

Table 6.2. Summary table for one-way between-subjects design

| | Between-subjects effects | | | | |
Source	SS	df	MS	F	η^2
Study time (A)	230,496.57	4	57,624.14	43.47*	.853
Error (S/A)	39,772.00	30	1325.73		
Total (T)	270,268.57	34			

*$p < .05$.

deviation of 100. To give you a sense for how to interpret these values, a score of 400 is 1 SD below the mean and is not really competitive in the acceptance model of most college programs, whereas a score of 650 is 1.5 SD above the mean and is considered to be quite competitive in most programs.

The scores of each participant in our example, organized by group, are shown in the top portion of Table 6.1. This table also contains some computational results that we will discuss in Section 6.5. The trend of the data, stated informally, appears to be that greater periods of study generally seem to have resulted in better performance on the SAT.

6.3   PARTITIONING THE TOTAL VARIANCE INTO ITS SOURCES

We have already computed the ANOVA for our numerical example, and the results are shown in the summary table in Table 6.2. In Sections 6.5 and later, we will outline how the ANOVA is performed; for now, our focus is on the results. Because it is a one-way design, the summary table is structured exactly as we have already talked about in Chapter 3. The numbers were calculated by hand and so will differ slightly (from rounding effects) from the output of SPSS and SAS. Briefly, here are the highlights.

The total variance (SS_T) is associated with 34 df (df_T). This is because there are thirty-five observations (elements of data) in the data set (five groups with seven observations in each) and the df_T is calculated as one minus the total number of observations.

Study time is the source of variance representing the independent variable. It is composed of five levels (groups) and so its $df_A = 5 - 1$ or 4. Its mean square (MS_A) as shown in the summary table in Table 6.1 was computed by hand to be 57624.14. This was obtained by dividing the study time sum of squares by the study time degrees of freedom ($230496.57 \div 4 = 57624.286$).

Error or S/A is short for the within-groups source of variance. It represents variability within each group despite the fact that all the students in a particular group studied for the same length of time, and has 30 $df_{S/A}$. This is because each group has seven cases and therefore 6 df. Since there are five groups, there are $6 \times 5 = 30$ df for $df_{S/A}$.

6.4   OMNIBUS AND SIMPLIFYING ANALYSES

6.4.1   THE OMNIBUS ANALYSIS

The summary table for our numerical example shows the results of what is called the *omnibus analysis*. An omnibus analysis is the overall ANOVA that we have been talking about thus far in the book. It evaluates the effect of one or more independent variables.

6.4.2   SIMPLIFYING ANALYSES

In the rare instance in which the experimental design has only one independent variable with exactly two levels, a statistically significant F ratio informs us that the means of the two conditions are reliably different. The reason for this is because there are only two groups in the analysis; thus, there is only one pair of means that are being compared and the statistically significant F ratio pertains to that single paired comparison.

With three or more groups in the analysis, the situation is ordinarily ambiguous. Yes, with a statistically significant F ratio we can assume that there is a pair of means that differ; the question is which pair since there are several alternatives. In a three-group study, for example, the first and second groups may differ, the first and third groups may differ, and/or the second and third groups may differ. Under these circumstances, it is necessary to engage in further analyses to determine between which means the statistically significant mean differences lie. Such a process examines pairs of means to determine if they are significantly different. Think of the omnibus analysis as reflecting a "global," "macroscopic," "general," or "overall" analysis; then the post-ANOVA-looking-at-pairs-of-means procedure can be conceived as a "finer-grained, "microscopic," or "simplifying" level of analysis. It is simplifying in the sense that we analyze separate portions of the data set rather than analyzing the data set as a whole.

In the present example, we have five groups. We have obtained a statistically significant F ratio, but we do not automatically know which means are significantly different from each other. It is therefore necessary to perform an additional simplifying procedure on the data to examine pairs of means. This simplifying procedure in the case of an effect of a single independent variable, such as we have here, is known, in general, as a *multiple comparisons procedure*. However, there are enough different approaches to performing such multiple comparisons that we will devote the entirety of Chapter 7 to this topic, using our numerical example of SAT study time.

6.4.3   EFFECT OF INTEREST IN THIS DESIGN

With only one independent variable in the design, there is only one F ratio in the omnibus analysis that interests us – the F ratio associated with the independent variable. We can see from the summary table in Table 6.2 that the F ratio is statistically significant based on an alpha level of .05.

Eta squared is a whopping .853, indicating that the treatment variable accounted for about 85 percent of the total variance in the data. This rather spectacular finding resulted from contriving the small number of scores in our hypothetical study to get the most out of our numerical example. In the normal course of empirical research that you might be doing, values of $\eta^2 \leq .10$ are not that uncommon.

6.5   COMPUTING THE OMNIBUS ANALYSIS BY HAND

With this overview of the omnibus analysis in place, we can now examine the specific computations involved in the calculation of the various sums of squares, degrees of freedom, mean squares, and F value. Many of you will be using a statistical software application, such as SPSS or SAS, to routinely analyze data you have collected in your research, and we will show you how to run these programs at the end of the chapter. Invoking such an application is an efficient and useful way to analyze your data, but going through the actual arithmetic operations can make it clear what is involved in computing an ANOVA.

The computational process for an ANOVA results in a few key values. We will place a box around these values when we have obtained them with the computations.

6.5.1   TERMINOLOGY AND NOTATION

Consider the previous study on the effects of months of study-preparation time on subsequent SAT performance on a sample of high school seniors. The scores (an individual's average of the Verbal and Quantitative portions of the SAT) comprise the dependent variable and can be arranged as shown in Table 6.1.

We begin by noting that a single independent variable, such as preparation time, can be referred to in generic terms as Factor A. If additional independent variables are involved in the research design (as we will see in Chapters 8–15), we will refer to them as Factors B, C, and so forth (although we will not discuss designs with more than three factors in this book).

Independent variables must have two or more levels or classification groups. In the present example, the five levels of preparation time (zero, two, four, six, and eight months) are designated with a lower case a and a subscripted number to indicate which of the five treatment conditions are being considered. Thus, we have a_1, a_2, a_3, and so forth and these are read as "little 'a' one," "little 'a' two," and "little 'a' three," and so on, respectively.

The dependent variable – the scores on the dependent measure – are designated by the letter Y. As we noted previously in Chapter 3, we add the subscripted letters i and j to denote the ordinal position of each case or participant and treatment group or level, respectively. These scores in each treatment condition can be summed as (ΣY_{i1}, ΣY_{i2}, ΣY_{i3}, ΣY_{i4}, ΣY_{i5}), or more simply (ΣY_1, ΣY_2, ΣY_3, ΣY_4, ΣY_5). We can also represent these sums more simply with a capital letter A and a numerical subscript, which yields A_1, A_2, A_3, and so forth (read as "big 'A' one," "big 'A' two," "big 'A' three"). These sums represent the aggregated total score across all participants within a single treatment condition. By adding these five sums ($A_1 + A_2 + A_3 + A_4 + A_5$) or by summing all participants' scores (ΣY_{ij}) we obtain the grand total, represented by T.

In addition to identifying the individual (Y_{ij} or simply Y) scores, summing them, and generating a grand total, we also need to create a sum of squared scores for each treatment group; this is designated ΣY_{aj}^2 or more simply ΣY_{a1}^2, ΣY_{a2}^2, ΣY_{a3}^2, ΣY_{a4}^2, and ΣY_{a5}^2. These new sums are calculated by squaring the individual Y scores for each treatment and then summing them. The sums of the squared scores are then added together to produce a total sum of squared scores, designated ΣY_{aj}^2 or simply, ΣY^2.

A lower case n with a subscripted number (read "little 'n' one" or "little 'n' two") is used to refer to the number of observations within a particular treatment group. The total number of observations or cases is designated with a capital N (read "big N"). For a one-way between-subjects design with equal group sizes, N can be obtained by multiplying the number of groups by the number of cases contained in each group; thus $N = (a)(n)$. In all the ANOVA examples in this text, sample size (n) will always be equal across independent variable groups, in order to keep the computational procedures and formulas as clear as possible.

As we noted previously, treatment group means are calculated by dividing the treatment sum A_1 by n_1 and A_2 by n_2, and so forth, and is designated $\overline{Y}$ (read "Y bar") with a subscripted number. A grand mean $\overline{Y}_T$ (read as "Y bar T") can also be calculated by dividing the total number of scores by the total number of observations in the study.

6.5.2 SOURCES OF VARIANCE

The summary table for this design is shown in Table 6.2. In a between-subjects design, our focus is on between-subjects variance. We have emphasized this point by showing a general heading for the summary table as "between-subjects effects." SPSS and SAS will use a similar heading for the summary tables that they produce.

The first column of a summary table contains the sources of variance. Total variance is always placed in the last row of the table. Above the total variance are the only two sources we have in a one-way between-subjects design. One source is a known source and represents the independent variable. This is labeled as *between groups* in the summary table and refers, generically, to Factor A. In the example study, this represents the five levels of the independent variable: study preparation time in months (zero, two, four, six, and eight). The other source is the *within groups* or *error*. This represents variation within the groups that is unable to be linked to the independent variable.

6.5.3 SUM OF SQUARES

Throughout this text, we will be providing you with what are called "computational formulas" for computing the various sums of squares by hand. Computational formulas provide a faster and more efficient means of conducting the sometimes intricate calculations involved in producing sums of squares. However, this mathematical efficiency or elegance comes at a cost; computational formulas are not intuitive, nor foster conceptual understanding. That is, it is sometimes difficult to see how the variance is actually partitioned when using these computational procedures. A more intuitive, but more mathematically cumbersome, approach to calculating sums of squares is to use *deviational formulas*. Such formulas depict very concretely how the total variability within the study is broken into between-groups and within-groups variability, all on the basis of how the original raw scores on the dependent variable deviate from the grand mean (total variability), how these same scores deviate from their respective group mean (within-groups variability), and how each group mean deviates from the grand mean (between-groups variability). We will illustrate these deviational formulas with the current numerical example in Appendix E, and keep our discussion focused on computational procedures within each chapter. Remember, either the deviational or computational formula will produce the exact same result.

The sum of squares values (shown as **SS** in Table 6.2) are shown in the second column of Table 6.2. The preliminary calculations and summary statistics shown in Table 6.1 become the ingredients for computing the various sums of squares. In a one-way between-subjects ANOVA, there are three sums of squares that must be calculated:

- Sum of squares between groups or treatment, SS_A, in which the subscript A represents the single independent variable or Factor A.
- Sum of squares within groups or error, $SS_{S/A}$, in which the subscript S/A indicates subjects within each level of the A treatment.
- Sum of squares total, SS_T.

6.5.3.1 Sum of Squares for the Treatment Effect (Between-Groups Variance): SS_A

The formulas for these three sums of squares and the summary of their calculations based on the data from Table 6.1 are as follows:

$$SS_A = \frac{\sum A^2}{n} - \frac{T^2}{(a)(n)}$$

$$= \frac{(2,890)^2 + (3,320)^2 + (3,870)^2 + (4,300)^2 + (4,360)^2}{7}$$

$$- \frac{(18,740)^2}{35} = \boxed{230,496.57}. \tag{6.1}$$

The formulas take the sum of the squared treatment sums and divides this value by n. Then the grand total score is squared and divided by the total number of cases $(a)(n)$. This latter value is subtracted from the previous one to obtain SS_A. The slight discrepancy between the SS_A produced by SPSS and reported in Table 6.1 and Figure 6.8 (230,497.143) and the present value calculated by hand (230,496.57) is due to rounding error, commonly encountered when one uses a calculator.

6.5.3.2 Sum of Squares for the Error Term (Within-Groups Variance): $SS_{S/A}$

$$SS_{S/A} = \sum Y^2 - \frac{\sum A^2}{n}$$

$$= 10{,}304{,}200 - \frac{(2{,}890)^2 + (3{,}320)^2 + (3{,}870)^2 + (4{,}300)^2 + (4{,}360)^2}{7}$$

$$= \boxed{39{,}772}. \tag{6.2}$$

Here each raw score or basic observation is squared and then summed. From this value we then subtract the sum of the squared treatment sums divided by n to produce $SS_{S/A}$.

6.5.3.3 Sum of Squares for the Total Variance: SS_T

$$SS_T = \sum Y^2 - \frac{T^2}{(a)(n)}$$

$$= 10{,}304{,}200 - \frac{(18{,}740)^2}{35} = \boxed{270{,}268.571}. \tag{6.3}$$

Here each raw score or basic observation is squared and then summed. The total sum of squares (SS_T) is thus a function of subtracting the squared grand total score divided by the total number of cases from the sum of the squared scores.

6.5.4 DEGREES OF FREEDOM
6.5.4.1 Degrees of Freedom for Total Variance

The degrees of freedom for each source of variance are contained in the third column of the summary table. To compute the sum of squares for the total variance, we subtract the grand mean from each score or observation in the study. Thus, the degrees of freedom associated with the total variance are the number of observations minus one. Expressed as a formula, our computation is as follows:

$$df_T = (a)(n) - 1 = (5)(7) - 1 = 34, \tag{6.4}$$

where a is the number of groups and n is the number of cases contained in each group.

6.5.4.2   Degrees of Freedom for Between-Groups Variance

The sum of squares for the between-groups variance (SS_A) is based on subtracting the grand mean from each group mean. In our data set, we have five levels of the independent variable and therefore have five group means. The degrees of freedom for the between-groups effect are equal to the number of levels of the independent variable minus one. Expressed as a formula, our computation is as follows:

$$df_A = a - 1 = 5 - 1 = 4, \tag{6.5}$$

where a is the number of groups.

6.5.4.3   Degrees of Freedom for the Within-Groups (Error) Variance

The sum of squares for the unexplained portion of the variance $(SS_{S/A})$ – the variability of the scores within each group – is based on the difference between each score in a given group and its respective mean. In computing the degrees of freedom for the unexplained variance we treat each group in turn. For the zero months group, we have seven scores from which the mean will be subtracted. That gives us 5 df. For the two months group, we also have seven scores from which that mean will be subtracted. That gives us another 5 df. Likewise for the four, six, and eight months groups. All told, the unaccounted for variance has 30 df associated with it. This can be written as the following formula:

$$df_{S/A} = (a)(n - 1) = (5)(7 - 1) = 30. \tag{6.6}$$

Note that we could have figured out the value of the degrees of freedom for the unexplained or error variance based on what we already had computed. Given that there are 34 df in total and that four of them are associated with the between-groups effect, the rest had to be associated with the error variance. We could have thus computed this value as the residual, and this is a common way for researchers to obtain certain degrees of freedom values in the summary table when they are doing the calculations by hand. Of course, by performing the analyses in SPSS or SAS, all of these values are provided to you.

6.5.5   MEAN SQUARES

Mean square values are computed by dividing each sum of squares by its respective degrees of freedom. We only compute mean squares for the between-groups (Factor A) effect and the within-groups or error variance.

For between groups, 230,496.57 divided by 4 is 57,624.14; for error, 39,772 divided by 30 is 1,325.73:

$$MS_A = \frac{SS_A}{df_A} = \frac{230,496.57}{4} = \boxed{57,624.14},$$

$$MS_{S/A} = \frac{SS_{S/A}}{df_{S/A}} = \frac{39,772}{30} = \boxed{1,325.73}.$$

Again, the slight discrepancies between these hand-calculated values and the computer-based computations are due to rounding error.

6.5.6 F RATIO

The F ratio is the mechanism by which we test the null hypothesis that the means of the groups are representative of the same population, that is, that the means of the groups do not differ significantly. In the language of variance, the null hypothesis is that the amount of variance accounted for by the independent variable is not significantly different than zero.

An F ratio is associated with a known source of variance, in this instance the effect of the independent variable. It is computed by dividing the between-groups mean square (MS_A) by the mean square error ($MS_{S/A}$). This ratio captures in one value the effects of the treatment manipulation (i.e., the independent variable and error variability) evaluated against the mean square within groups (also called mean square error or the error term) representing all of the error variability within the study. The formula for computing F for a one-way between-subjects design is as follows:

$$F = \frac{\text{mean square between groups}}{\text{mean square within groups}} = \frac{MS_A}{MS_{S/A}}. \qquad (6.7)$$

In our example study,

$$F = \frac{57,624.14}{1,325.73} = \boxed{43.466}.$$

This F ratio is associated with the degrees of freedom of the sources whose division resulted in the value (the independent variable and the unexplained variance). When we report the F ratio, the degrees of freedom are placed in parentheses immediately following the letter: the degrees of freedom of the mean square in the numerator (between groups) are given first; the degrees of freedom of the mean square in the denominator (within groups) are given second. The two degrees of freedom are separated by a comma and a space. Here is how you would report that piece of the results in the present example:

$$F(4, 30) = 43.47.$$

6.5.7 STATISTICAL SIGNIFICANCE OF F

Unless there is a good reason to do otherwise, researchers set their alpha level at .05. If the probability of the F ratio occurring is .05 or less when the null hypothesis is true, the researchers will reject the null hypothesis and conclude that the means of the groups are significantly different. That is, they will conclude that the independent variable accounts for significantly more than 0 percent of the variance of the dependent variable. The Critical Values table for the F distribution is in Appendix C. In the present case, we have 4 df between groups and 30 df within groups. The table reveals a critical value of F at the .05 level to be 2.69. Since our observed (or computed) F of 43.47 exceeds the value for F corresponding to our preset alpha level of .05, we can conclude that our obtained F value is statistically significant at the .05 alpha level, indicating a statistically significant difference between the five groups.

SPSS and SAS have access to these critical values and make these assessments automatically. For example, when we present the SPSS output for this analysis, we will see a column labeled **Sig** with a value .000. This value is one that SPSS has truncated down to three decimal places. We interpret this outcome to mean that this F value equals or exceeds the tabled critical value of F at the .001 alpha level and would therefore ordinarily occur less than once in every thousand occasions if the null hypothesis is true. Please note that a p value can never actually be .000! A value of zero would indicate that such an F value would never occur and that there is a zero chance of the researcher committing a Type I error in such an instance! Please remember that there is always some chance that the researcher's results are incorrect.

When reporting the F ratio, we need to add a probability element to the F value and its degrees of freedom. In the example analysis, using an alpha level of .05, we would write

$$F(4, 30) = 43.47, \ p < .05.$$

6.5.7 ETA SQUARED

We know from the F ratio being statistically significant that the independent variable can explain some of the variance of the dependent variable. To determine what this percentage is, that is, to determine the strength of the effect of the independent variable, we must calculate the value of eta squared as discussed in Section 4.5.3 of Chapter 4. As you may recall, the formula is as follows:

$$\text{Eta squared} = \frac{\text{sum of squares between groups}}{\text{sum of squares total}}. \tag{6.8}$$

For the example data set:

$$\frac{SS_A}{SS_T} = \frac{230,496.57}{270,268.571} = .853.$$

This eta squared value indicates that roughly 85 percent of the total variance of the dependent variable (SAT Verbal and Quantitative composite score) can be attributed to the independent variable (preparation time)

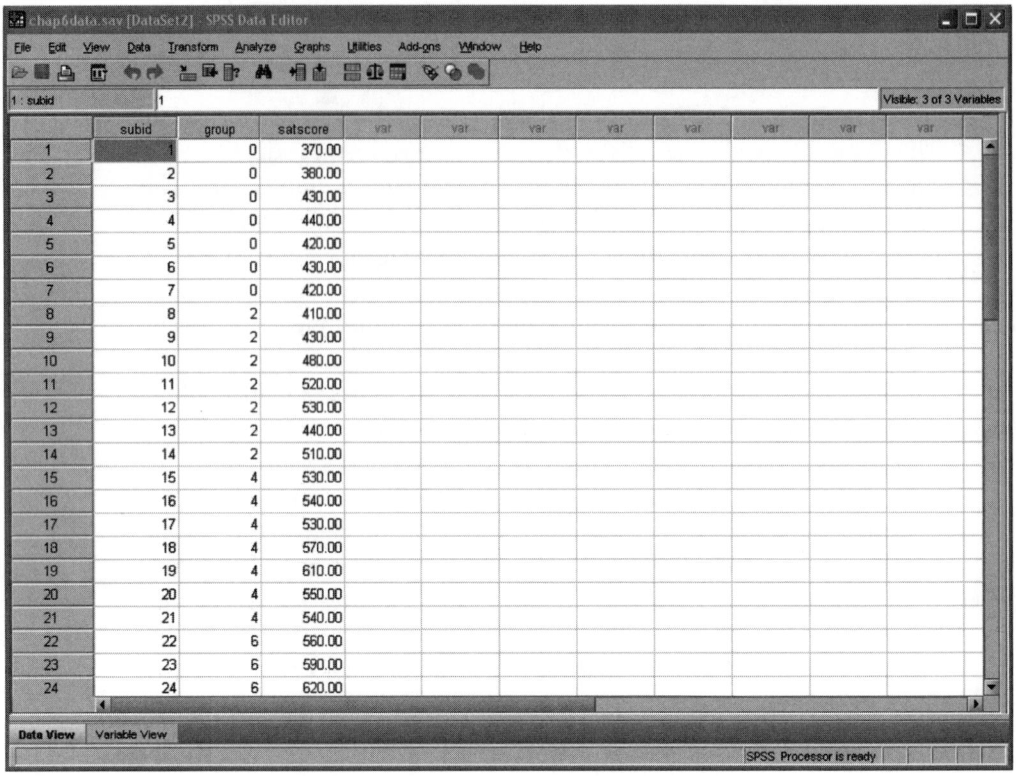

	subid	group	satscore
1	1	0	370.00
2	2	0	380.00
3	3	0	430.00
4	4	0	440.00
5	5	0	420.00
6	6	0	430.00
7	7	0	420.00
8	8	2	410.00
9	9	2	430.00
10	10	2	480.00
11	11	2	520.00
12	12	2	530.00
13	13	2	440.00
14	14	2	510.00
15	15	4	530.00
16	16	4	540.00
17	17	4	530.00
18	18	4	570.00
19	19	4	610.00
20	20	4	550.00
21	21	4	540.00
22	22	6	560.00
23	23	6	590.00
24	24	6	620.00

Figure 6.1 The **SPSS data** file.

and would be considered a large effect size. The full statistical result of the above analysis that should be presented in any written report is as follows:

$$F(4, 30) = 43.47,\ p < .05,\ \eta^2 = .85.$$

6.6 PERFORMING THE OMNIBUS ONE-WAY BETWEEN-SUBJECTS ANOVA IN SPSS

6.6.1 THE DATA FILE

A portion of the data file for this numerical example is shown in Figure 6.1. The first column (not named) contains the line numbers supplied by SPSS. The next column contains a variable named **subid**, which is simply an arbitrary code number assigned to each participant so that we can verify the original data if necessary; **subid** is short for "subject identification," and we use this variable in all of our data files. The variable **group** indicates to which group the cases belong. We have used *value labels* (codes for the **group** variable) of 0, 2, 4, 6, and 8 to represent our groups because the codes inform us directly of the number of months of study the students put in. Finally, the variable **satscore** represents the dependent variable and is, as the name implies, the SAT score (actually the average of Verbal and Quantitative) for each student.

6.6.2   THE SPSS PROCEDURE WE WILL USE

There are several procedures in SPSS through which we can perform various kinds of ANOVAs. Our choice for a one-way between-subjects design is the **One-Way ANOVA** procedure because it is specialized to perform both the omnibus one-way analysis as well as most of the multiple comparison procedures that we will use in Chapter 7. This is a terrific procedure because, by being specialized to analyze one-way between-subjects designs, it can provide to us more variations of the analysis than any other SPSS procedure. Among its limitations are these: (a) it is unable to be used for designs with more than one independent variable, and (b) it cannot accommodate the use of covariates.

6.6.3   OPEN SPSS

If you intend to perform your own analysis as we describe the step-by-step process, then you need to launch SPSS and open the data file corresponding to this example (or the data file that you intend to analyze). The data file must be open before you can specify any SPSS analysis.

6.6.4   OPENING THE MAIN ONE-WAY ANOVA DIALOG WINDOW

From the menu bar at the top of the screen, follow this navigation route: **Analyze → Compare Means → One-Way ANOVA**. That will bring you to the main dialog window for the procedure as shown in Figure 6.2.

Highlight **group** in the left panel and click it over to the **Factor** panel. Then highlight **satscore** and click it over to the panel for **Dependent List**. When you have done this, your window should look like that shown in Figure 6.3.

Note the pushbuttons that we have circled at the right of the SPSS window in Figure 6.3. When selected (clicked), these open subsidiary dialog windows within the main procedure to deal with aspects of the analysis that you can configure to your own specifications. We will work

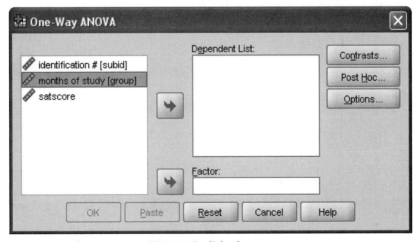

Figure 6.2   The **One-Way ANOVA Main** dialog box.

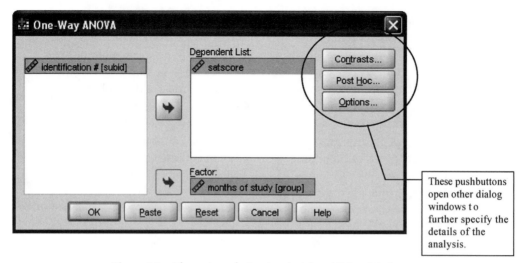

These pushbuttons open other dialog windows to further specify the details of the analysis.

Figure 6.3   The various choices to select for additional analyses.

with the **Options** window now and work with the **Contrasts** and **Post Hoc** windows when we perform our multiple comparisons in the next chapter.

6.6.5   THE OPTIONS WINDOW

Click the **Options** pushbutton. This brings you to the **Options** dialog screen shown in Figure 6.4. In the **Statistics** area we have already checked **Descriptive** (which gives us descriptive statistics) and **Homogeneity of variance test** (which prints out the results of the Levene test). We will presume that our data meet the homogeneity of variance assumption; if the Levene test shows that we have violated that assumption, then we will

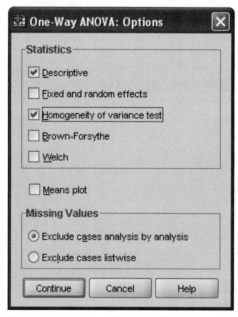

Figure 6.4   The **One-Way ANOVA Options** dialog window.

Descriptives

satscore

	N	Mean	Std. Deviation	Std. Error	95% Confidence Interval for Mean		Minimum	Maximum
					Lower Bound	Upper Bound		
zero months	7	412.8571	26.90371	10.16865	387.9754	437.7389	370.00	440.00
two months	7	474.2857	47.90864	18.10777	429.9776	518.5938	410.00	530.00
four months	7	552.8571	28.70208	10.84837	526.3121	579.4021	530.00	610.00
six months	7	614.2857	38.23486	14.45142	578.9244	649.6471	560.00	660.00
eight months	7	622.8571	36.38419	13.75193	589.2074	656.5069	550.00	660.00
Total	35	535.4286	89.15759	15.07038	504.8019	566.0553	370.00	660.00

Figure 6.5   The first portion of the output from the One-Way ANOVA.

need to rerun the analysis asking for either the **Brown–Forsythe** or the **Welch** significance tests.

In the **Missing Values** area, SPSS uses **Exclude cases analysis by analysis** as a default. The other choice, **Exclude cases listwise**, is applicable when you want to perform many analyses in a single run based on the very same set of cases. Invoking this choice would remove from all of the analyses *any* case with a missing value on *any* dependent measure. Since we have only one analysis, the choice is irrelevant and we can leave the default in place. Click **Continue** to return to the main dialog window. Then click **OK** to run the analysis.

6.7   THE OUTPUT OF THE OMNIBUS ONE-WAY BETWEEN-SUBJECTS ANOVA IN SPSS

6.7.1   DESCRIPTIVE STATISTICS

The descriptive statistics generated by **One-Way ANOVA** are shown in Figure 6.5. The **One-Way ANOVA** procedure provides quite a bit of information: the mean, standard deviation, standard error with its 95 percent confidence interval, and the minimum and maximum scores in each group.

6.7.2   HOMOGENEITY TEST

Figure 6.6 displays the result for Levene's test of homogeneity of variance. The statistic (symbolized as *W* if you are reporting it) has a value of 1.459 and, with 4 and 30 *df*, has a probability of occurrence of .239 if the null hypothesis is true. Against an alpha level of .05 this probability is not statistically significant. We therefore do not reject the null hypothesis that the variances of the five conditions are comparable; in short, it

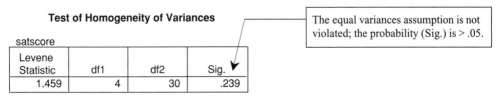

Test of Homogeneity of Variances

> The equal variances assumption is not violated; the probability (Sig.) is > .05.

satscore

Levene Statistic	df1	df2	Sig.
1.459	4	30	.239

Figure 6.6   The results of Levene's homogeneity of variance test.

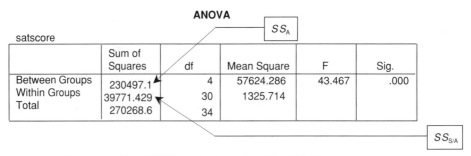

Figure 6.7   The ANOVA results from **One-Way ANOVA**.

appears that the assumption of homogeneity of variance has not been violated.

6.7.3   SUMMARY TABLE

The summary table for the omnibus analysis is presented in Figure 6.7 and is identical to what we showed earlier. The **One-Way ANOVA** procedure does not generate a value for eta square (η^2), but it is easy enough to do on a hand calculator. All we need to do is divide the *between-groups sum of squares* by the *total sum of squares;* this gives us a value of .853 ($230497.143 \div 270268.571 = .853$).

6.8   PERFORMING THE OMNIBUS ONE-WAY BETWEEN-SUBJECTS ANOVA IN SAS

6.8.1   THE DATA FILE

The data file that we are using for the SAS analysis has the same structure as the file we used in the previous SPSS analysis. Figure 6.8 displays the Excel spreadsheet containing our data. In the first column we find **subid** representing the identification code for the participants in the study. In the following columns, we see the group variable with value labels of 0, 2, 4, 6, and 8 representing our **groups** (the codes inform us directly of the number of months of study the students put in), and the dependent variable **satscore**.

6.8.2   IMPORTING THE DATA FILE

To perform the data analysis, you must launch *SAS Enterprise Guide* and import the data file from Excel. Thus, from the main menu select **File → Import Data** and follow the directions provided in Appendix B. Once the data have been imported into the current project, the steps taken will be displayed in icons on the project grid background as shown in Figure 6.9.

6.8.3   PERFORMING THE OMNIBUS ANALYSIS

As was true for SPSS, there are several procedures in *SAS Enterprise Guide* through which we can perform various kinds of ANOVAs. Our choice for a one-way between-subjects design is the **Linear Models** because it is the

subid	group	satscore
1	0	370
2	0	380
3	0	430
4	0	440
5	0	420
6	0	430
7	0	420
8	2	410
9	2	430
10	2	480
11	2	520
12	2	530
13	2	440
14	2	510
15	4	530
16	4	540
17	4	530
18	4	570
19	4	610
20	4	550
21	4	540
22	6	560
23	6	590
24	6	620
25	6	640
26	6	660
27	6	580
28	6	650
29	8	550
30	8	610
31	8	630

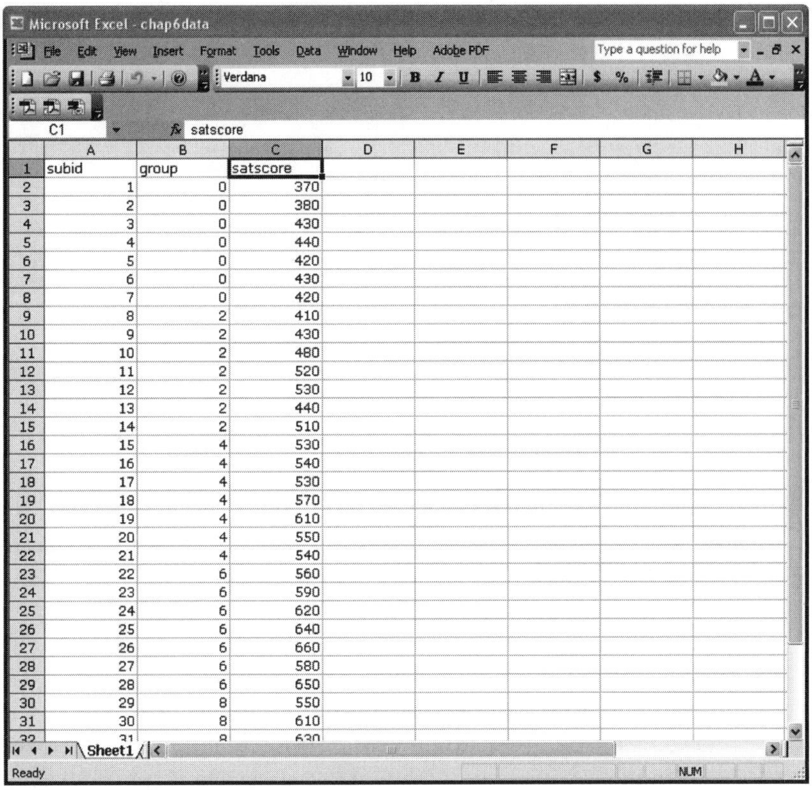

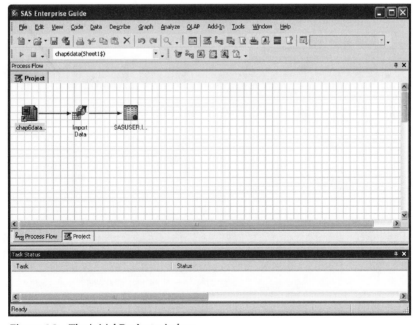

Figure 6.8 The data file in an Excel spreadsheet.

Figure 6.9 The initial **Project** window.

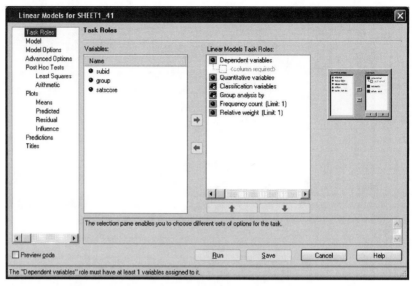

Figure 6.10 The **Task Roles** tab in the **Least Squares** procedure.

procedure we will use for the two-way and three-way between-subjects designs covered in Chapters 8 and 9, respectively.

Select **Analyze ➜ ANOVA ➜ Linear Models**. The window for this procedure opens on the **Task Roles** tab; this is highlighted in the navigation panel in the left portion of the window shown in Figure 6.10. At the top of the panel labeled **Linear Models Task Roles**, there is already a place for a dependent variable being specified. Highlight **satscore** and drag it to the icon for **Dependent variables**. Then drag **group** to the icon for **Classification variables**. At that point your screen should look similar to that shown in Figure 6.11.

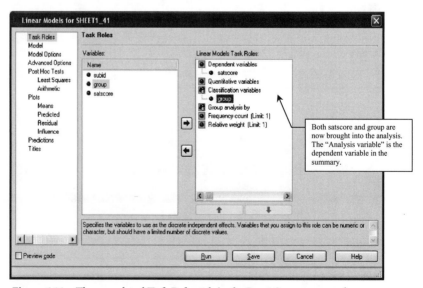

Figure 6.11 The completed **Task Roles** tab in the **Least Squares** procedure.

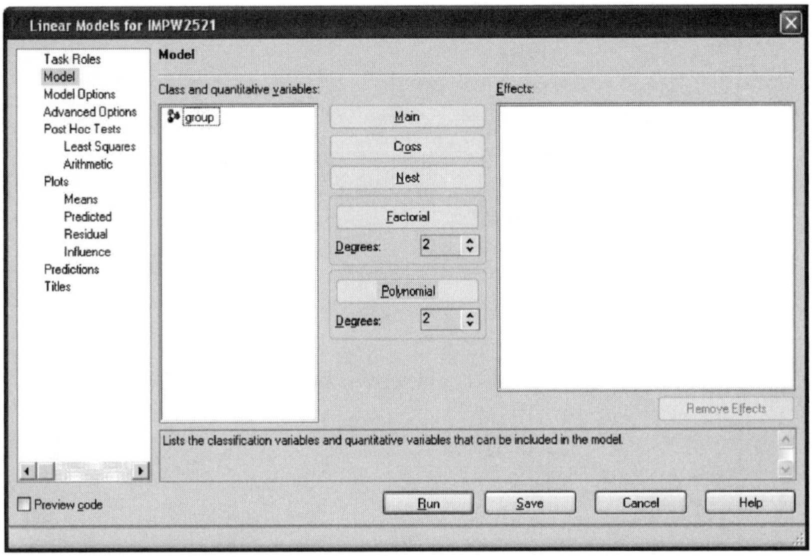

Figure 6.12 The initial **Model** screen.

Click on the **Model** tab to display the screen shown in Figure 6.12. The variable **group**, our independent variable, appears in the **Class and quantitative variables** panel. Highlighting that variable activates the **Main** bar in the middle of the screen; click **Main** to place **group** in the **Effects** panel as shown in Figure 6.13.

Click on the **Model Options** tab. The only specification we need for our analysis is to request output associated with the **Type III** sums of squares (see Figure 6.14).

Descriptive summary statistics can be requested on the **Arithmetic** portion of the **Post Hoc Tests** tab. When you select it, you see the blank

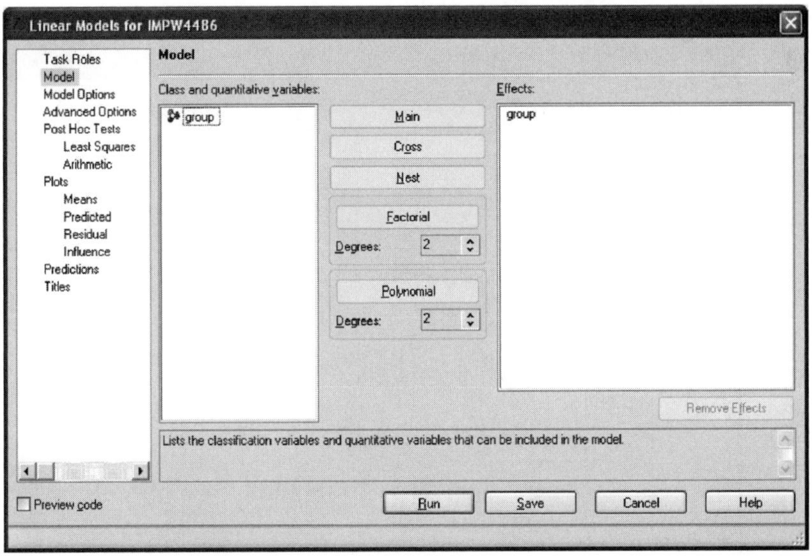

Figure 6.13 The configured **Model** screen.

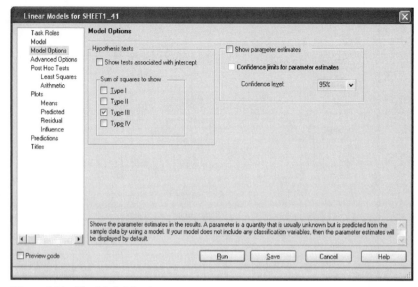

Figure 6.14   The **Model Options** screen.

screen shown in Figure 6.15. Click the **Add** pushbutton and several drop-down menus appear in the **Options for means tests** panel as may be seen in Figure 6.16. To access these menus, click in the right portion of the specification. For example, under **Class effects to use** we see **group** and the specification **False** by it. By clicking **False**, you can display the drop-down menu and select an alternative specification. Change only the following specifications:

- **Class effects to use:** select **True**.
- **Homogeneity of variance:** select **Levene (absolute residuals)**.

Then click **Run** to perform the analysis.

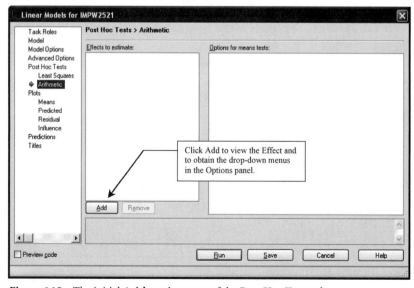

Figure 6.15   The initial **Arithmetic** screen of the **Post Hoc Tests** tab.

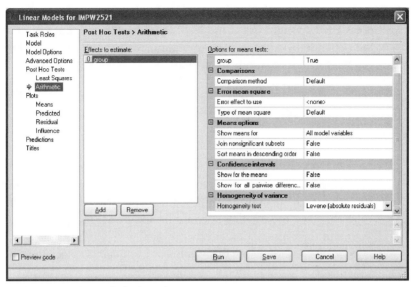

Figure 6.16 The configured **Arithmetic** screen.

6.9 THE OUTPUT OF THE OMNIBUS ONE-WAY BETWEEN-SUBJECTS ANOVA IN SAS

6.9.1 BRIEF OVERVIEW OF THE GENERAL LINEAR MODEL

The output from the **One-Way ANOVA** procedure in SPSS presented in Section 6.7 was structured in a way that matched our hand calculations. But the summary tables produced by SAS and SPSS in certain of their ANOVA procedures contain some entries that we have not yet covered. As you will see in a moment, the summary table from the SAS **Linear Models** procedure that we have just used to generate our analysis has additional wording that we need to introduce.

In its **Linear Models** procedure, SAS computes what is called the *reduced, restricted,* or *corrected model* as opposed to the *full* or *uncorrected model;* SPSS will do likewise in some of its procedures, but there are times when it will present the results of the full model. We will briefly explain at a very surface level the differences between the full and the corrected model here. A more comprehensive description of these models is beyond the scope of this book, but more complete descriptions can be found in Kirk (1995), Maxwell and Delaney (2000), and Myers and Well (1991). However, we need to describe these models to a certain extent so that you can more fully interpret the SPSS and SAS output.

ANOVA is one of many statistical procedures – multiple linear regression, discriminate analysis, and canonical correlation are some other members of this set of procedures – that on the surface may appear quite different but are all manifestations of the *general linear model.* The general linear model statistical procedure is designed to formulate a model to predict for each case the values of the dependent variable from a knowledge of the level or levels of the independent variable each case experienced. The model is linear in that it represents a straight line function. Each predictor is assigned a weight in the model to maximize the accuracy of predicting the dependent variable in combination with the other predictors. The general form of the model is as follows:

$$\text{dependent variable} = a + b_1 X_1 + b_2 X_2 + \cdots + b_n X_n + \text{error}, \tag{6.9}$$

where

- X_1, X_2, and so on are the predictors. In a one-way design, there is only one such predictor and it is the independent variable in the study. In more complex designs, as we will see in subsequent chapters, the predictors could be additional independent variables and their interaction effects.
- b_1, b_2, and so on are the weights or coefficients associated with their respective predictors in the model.
- a is the constant in the model representing the Y intercept, that is, the value at which the straight line function represented by the model crosses the y axis.
- error is measurement error, that is, unaccounted for variance. For example, in a one-way between-subjects design, participants in the same group (those who have been treated in the same way with respect to the "treatment") may still provide different scores on the dependent variable.

The full general linear model shown above has a Y intercept value as well as values for the weights. When SPSS or SAS displays the results for the full model, our output will sometimes display sums of squares associated with the Y intercept and the full (uncorrected) model. The total sum of squares for the full model includes not only the values for the between-groups and within-groups sources of variance but also for the intercept (which is an additional source of variance that is partitioned by the ANOVA procedure).

The corrected or restricted (reduced) model can be conceived of as a partial model in that the variance attributable to the Y intercept is not included. The procedures we use for hand calculating the ANOVA in this book conform to the reduced model. For the purposes of many behavioral and social science students who are learning the details of the ANOVA designs we present here, the results contained in the corrected model are probably sufficient to meet their needs.

6.9.2 DESCRIPTIVE STATISTICS

The descriptive statistics generated by **Linear Models** are shown in Figure 6.17. We requested only the mean, standard deviation, and the number of observations in each group, and that is what is shown in the figure.

Linear Models

The GLM Procedure

Figure 6.17 Descriptive statistics generated by the **Linear Models** procedure.

Level of group	N	satscore	
		Mean	Std Dev
0	7	412.857143	26.9037084
2	7	474.285714	47.9086432
4	7	552.857143	28.7020822
6	7	614.285714	38.2348632
8	7	622.857143	36.3841933

6.9.3   HOMOGENEITY TEST

Figure 6.18 displays the result for Levene's test of homogeneity of variance. The statistic (symbolized as W if you are reporting it) has a value of 1.459 and, with 4 and 30 *df*, has a probability of occurrence, $p = .239$, if the null hypothesis is true. This mirrors the numerical outcome we obtained from SPSS. Against an alpha level of .05 this probability is not statistically significant. We therefore do not reject the null hypothesis that the variances of the five conditions are comparable; in short, it appears that the assumption of homogeneity of variance has not been violated.

6.9.4   SUMMARY TABLE

The summary table for the omnibus analysis is presented in the top portion of Figure 6.19. Note that SAS provides the **Corrected Total** sum of squares, informing us that we are viewing the results of the reduced model. The sum of squares associated with **Model** (see the top table in Figure 6.19) is identical to the sum of squares associated with our **group** effect (see the bottom table in Figure 6.19) because the only variable in the prediction model is the group effect, that is, the **group** effect is the model. Whichever you examine, we see a statistically significant effect numerically matching the output from SPSS. In later chapters where we have between-subjects designs with more than a single independent variable, the **Model** effect will be the cumulation of all effects in the analysis, and our focus will be on the individual effects comprising the model.

Linear Models

The GLM Procedure

Figure 6.18 The result of the Levene test for homogeneity of variance.

Levene's Test for Homogeneity of satscore Variance ANOVA of Absolute Deviations from Group Means					
Source	DF	Sum of Squares	Mean Square	F Value	Pr > F
group	4	1929.8	482.4	1.46	0.2393
Error	30	9919.5	330.7		

Dependent Variable: satscore satscore

Source	DF	Sum of Squares	Mean Square	F Value	Pr > F
Model	4	230497.1429	57624.2857	43.47	<.0001
Error	30	39771.4286	1325.7143		
Corrected Total	34	270268.5714			

$SS_{S/A}$

R-Square	Coeff Var	Root MSE	satscore Mean
0.852845	6.800227	36.41036	535.4286

SS_A

Source	DF	Type III SS	Mean Square	F Value	Pr > F
group	4	230497.1429	57624.2857	43.47	<.0001

Figure 6.19   The output of the ANOVA.

The middle portion of Figure 6.19 presents what SAS calls **R-Square**; this is what we have been labeling as *eta squared* (η^2) and is equivalent to our hand-calculated value. It is keyed to the **Model** sum of squares and is the proportion of the total sum of squares accounted for by the model. In designs with more than a single independent variable, we will wish to compute the separate eta squared values associated with each of the statistically significant effects.

In addition to the **R Square**, SAS provides three other pieces of information in that middle table. The **Coeff Var** is the *coefficient of variation;* it is the ratio of the standard deviation of the sample as a whole to the mean of the sample as a whole and can be useful in comparing the variability of different distributions. The **Root MSE**, the *root mean squared error*, is the square root of the mean square associated with the error variance (i.e., the square root of 1325.7143 is 36.41036). **Satscore Mean** is the overall or grand mean of the sample as a whole.

6.10   COMMUNICATING THE RESULTS

Based on the omnibus ANOVA, we know that the independent variable is associated with a statistically significant and substantial effect size. This portion of the results can be communicated in a simple sentence:

The amount of preparation for the SAT in which students engaged appeared to significantly affect their performance on the test, $F(4, 30) = 43.47$, $p < .05$, $\eta^2 = .853$.

Note that we cannot address the issue of which means differ from which other means with the information obtained from the omnibus analysis. In Chapter 7, we will consider a variety of ways to assess mean differences.

CHAPTER 6 EXERCISES

6.1. Conduct a one-way between-subjects ANOVA by hand and with SPSS or SAS, on the data in Chapter 5, Exercise 5.1. Include a written summary of your findings.

6.2. Conduct a one-way between-subjects ANOVA by hand and with SPSS or SAS, on the data in Chapter 5, Exercise 5.2. Include a written summary of your findings.

6.3. Third grade children were randomly assigned to one of three reinforcement contingency conditions. They were told to study a list of twenty different spelling words for twenty minutes, after which they would be tested. After the test, the children would receive either ice cream and cake, cookies and punch, or celery and carrot juice. The number of correctly spelled words served as the dependent variable. Conduct an ANOVA on these data by hand and with SPSS or SAS, and summarize your results.

Ice cream and cake	Cookies and punch	Celery and carrot juice
17	20	7
18	19	8
19	18	9
20	16	10
20	15	11

Multiple Comparison Procedures

7.1 OVERVIEW

Once we have determined that an independent variable has yielded a significant effect, we must next turn our attention to the differences between the means of the conditions in the study. If there are only two means, then we automatically know that they are significantly different. With three or more means a significant F ratio reveals only that there is a difference between at least one pair of means in the design; in this case we must perform an additional, post-ANOVA *multiple comparison procedure* to determine which of the three or more means differ from which others.

There are a variety of multiple comparison procedures that are available to researchers. Before presenting them, we will first describe some dimensions along which they differ; this will help us, as we go through this chapter, to discuss the differences among them.

7.2 PLANNED VERSUS UNPLANNED COMPARISONS

In an idealized world of research, hypotheses regarding differences between certain of the groups in a design are formulated in advance of the data collection on the basis of the theoretical context out of which the research was generated. Once the study is completed and a statistically significant F ratio is obtained in the omnibus ANOVA (although in this idealized world a statistically significant F ratio may not even be necessary if there are a very few hypothesized mean differences), the researchers then carry out the mean comparisons that they have already specified. Comparisons made under this approach are labeled as *planned* or *a priori* comparisons (Kirk, 1995).

In the practical world of research, most of the time multiple comparisons are performed on an *unplanned* or *a posteriori* basis (Everitt, 2001) without the researchers having formulated specific hypotheses at the outset and/or without restricting themselves to examining only the hypothesized mean differences. Under this strategy, if and only if a statistically significant F ratio is obtained in the omnibus ANOVA, researchers examine a variety of aspects of the results – including all possible pairwise combinations of the group means – to determine which of them reveal statistically significant mean differences.

In defense of the a posteriori strategy, it is very often the case that in behavioral and social science research we may not have sufficiently developed theories to test them as specifically as is called for by using planned comparisons. Related to this defense, much of behavioral and social science research is really exploratory. Therefore, examining many or all of the mean differences to find out what happened in the study may help researchers to develop models of the phenomenon under study, which can then be more rigorously tested in subsequent research studies.

On the other hand, it is not justifiable to use a posteriori multiple comparisons for reasons having nothing to do with the nature of the research; in fact, doing so runs a risk of compromising the integrity of the study. We believe that there are two inappropriate and not necessarily mutually exclusive reasons driving some researchers to use unplanned comparisons: (a) the procedures are so incredibly convenient to use in SPSS and SAS that researchers automatically invoke them under all design conditions whether they are appropriate or not, and (b) researchers sometimes do not know or consider that alternative procedures, perhaps more appropriate to their experimental design and certainly almost as convenient to use as unplanned comparisons, are available to them. We do not endorse using a posteriori comparisons under these two sets of conditions.

7.3 PAIRWISE VERSUS COMPOSITE COMPARISONS

We have been talking about mean differences in Chapter 6 and early in this chapter as though we always compare the mean of a single condition to the mean of some other single condition. That sort of comparison is known as a *pairwise or simple comparison* and, to be sure, is frequently performed. However, other possibilities exist.

It is also possible to compare two means where one or both of them are *weighted linear composites* made up of the means of two or more conditions. These sorts of comparisons are known as *nonpairwise, composite,* or *complex comparisons.* We will be more explicit about the idea of a weighted linear composite later in the chapter when we discuss user-defined contrasts; to keep it simple for now, just think of the composite as an average of two or more means with the means being equally weighted.

We can use our SAT study time example from the last chapter to illustrate a couple of different composite comparisons. Recall that separate groups prepared either zero, two, four, six, or eight months for the test. In the type of nonpairwise comparison known as *one-to-composite comparison,* we could ask if *any* study time made a difference over not preparing at all. The comparison here would take the following form:

```
Group 0 mean versus (combined mean of Groups 2, 4, 6, 8).
```

As you can see, we would address the question of whether any study time made a difference by comparing the group of students who did no preparation at all to the rest of the groups combined. For this set of combined groups, we would probably just compute a simple overall mean for Groups 2, 4, 6, and 8 (each group would be weighted equally) and compare this linear composite to the mean of the students who did not study at all.

In the type of nonpairwise comparison known as *composite-to-composite comparison,* we could ask if a good deal of preparation resulted in higher test scores than some modest amount of preparation. Such a comparison could take this form:

> (combined mean of Groups 2 and 4) versus (combined mean of Groups 6 and 8).

Here, we could combine the groups having studied two and four months, probably weighting them equally, and compare them to a combination of the groups having studied six or eight months, probably weighting them equally as well.

There are at least two major reasons for considering composite comparisons as viable alternatives when you are thinking about your multiple comparisons. First, these may help you better understand the phenomenon under study. That is, even though you took the trouble to distinguish between certain conditions in the original design of the study, a more relevant hypothesis may concern combining certain conditions in the post-ANOVA phase of data analysis. Second, this may be at least in part what the significant F ratio was detecting. We keep saying that a statistically significant F ratio tells you that there is a significant mean difference someplace. Well, the most interesting mean differences are sometimes based on a nonpairwise comparison.

7.4   ORTHOGONAL VERSUS NONORTHOGONAL COMPARISONS

Multiple comparisons may be made on either an orthogonal or nonorthogonal basis. Comparisons that are *orthogonal* are mutually independent of each other. This requires that a single mean or a composite can be used in a stand-alone comparison only once. For example, when we compared Group 0 to all of the other groups in combination and if we were restricting ourselves to orthogonal comparisons, then that would be the only time that Group 0 could be used alone; it could, however, be part of a composite of other groups in a different comparison.

There is a limit to the number of orthogonal comparisons that are possible to perform in any given design. In a one-way design that limit is equal to the degrees of freedom associated with the between-groups source of variance, which is $a - 1$ where a is the number of groups (levels of the independent variable) in the design. In our SAT study time example with five groups, we can make no more than four orthogonal comparisons.

Comparisons that are not orthogonal to each other are not independent of each other. These *nonorthogonal* comparisons subsume the full range of the remaining pairwise and nonpairwise comparisons. Because we can infinitely adjust the weights in the weighted linear combinations of composite comparisons, it is possible to conceive of an infinite number of nonorthogonal comparisons (Kirk, 1995).

When you make comparisons that are not orthogonal to each other, you are doing redundant or overlapping work. In fact, all of these nonorthogonal comparisons " . . . can be expressed as a linear combination of [the orthogonal] contrasts" (Kirk, 1995, p. 115). However, in the end, research interest and not orthogonality is the more important consideration in determining which comparisons to evaluate (Myers & Well, 1991). In this spirit, Toothaker (1993) has argued that the issue of orthogonality may be one of the less important ones in the context of multiple comparison procedures. We therefore bring the issue of orthogonality of comparisons to your attention so that you understand it, but we recommend following the lead of Myers and Well (1991) in performing those comparisons that are most relevant to your research whether or not they are orthogonal.

7.5 STATISTICAL POWER

We discussed the idea of statistical power in Chapter 4. Power reflects the ability to find an effect when one exists. Researchers gain more power when their sample size is larger, when the population effect size is larger, and when their alpha level is less stringent. Power comes into play in the context of multiple comparison procedures. It is very common in the research literature and in statistics textbooks for the authors to speak of and to select multiple comparison procedures on the basis of power.

7.5.1 THE FOCUS OF THE POWER

When we speak of power, the focus is on the mean differences we observe. In our SAT preparation example, we have five means and have obtained a significant F ratio in our omnibus analysis. When we examine our mean differences, they will all be of somewhat different magnitudes. It is these differences that are the focus when we speak of the power of a particular test.

7.5.2 APPLYING THE CONCEPT OF POWER

In an absolute sense, we do not know – nor can we ever know – which means are truly different from which others. Any multiple comparison procedure we use to deal with the question of group mean differences gives us answers in terms of probabilities: this mean difference is large enough to fall in the 5 percent region of the particular statistic's sampling distribution and so we judge that difference to be statistically significant; this other mean difference is too small, falling in the 95 percent region of the particular statistic's sampling distribution, and so we judge that difference to not be statistically significant, and so on.

Here is the conundrum: statistical power is the ability to find an effect if it exists, but we cannot know truly whether the effect actually exists. We therefore apply the concept of power by estimating the relative likelihood of this multiple comparison procedure versus that multiple comparison procedure yielding a statistically significant outcome for a given mean difference with all else equal. And, just in case you are now asking the question, the answer is yes, these tests will from time to time provide different outcomes. And, of course, there is no way to know which ones are "right" when they do yield different results. So we do not worry about it. Instead, we say that Multiple Comparison Procedure X has more power than Multiple Comparison Procedure Y, meaning that Procedure X is more likely than Procedure Y to return a significant result given a certain mean difference with all else equal. An alternative way to view this is that Procedure X is the more powerful of the two if it will find a given mean difference to be statistically significant when Procedure Y will not result in obtaining a significant difference.

7.5.3 THE CONTINUUM OF POWER

At the higher end of the power continuum is, to use colloquial language, *lots of power*. A procedure near the higher power end of the continuum is more likely than most to yield a significant difference for a given magnitude of mean difference (all else equal). At the other end is *low power*. A procedure near the lower power end of the continuum is less likely than most to yield a significant difference for a given magnitude of mean difference.

At least in the world of research and statistics, we generally have a positive regard for power. Who would not admire a procedure that could find an effect that was actually there? But here, as in the general case of power, there are risks as well as benefits. The very powerful procedures – those more likely to find smaller differences to be statistically significant – run an increased risk of leading us to make a false positive (Type I) error (we falsely conclude that a mean difference is statistically significant).

Given the relative ease of finding statistical significance with more powerful tests, another term that can be used to indicate power is the term *liberal*. A test is said to be more liberal (powerful) when it can more easily yield statistically significant mean differences; use of such a test represents a relatively greater amount of risk willing to be taken by researchers in asserting that two means are significantly different. The corresponding label that is used for other end of the continuum is *conservative*. A test is said to be more conservative (less powerful) when it is less likely to evaluate a difference as statistically different. A conservative test needs, all else being equal, greater mean differences before it leads you to conclude that the means are different; it represents a relatively lower level of risk willing to be taken by researchers in asserting that two means are significantly different.

Before you commit to your personal choice, recognize that there is no agreed upon level of power that researchers should wield. Different authors and different instructors have their own personal viewpoints. It is also not unreasonable to select a different level of power in different research situations. As we go along, we will give you more information on which to base your own choices as well as provide you with some recommendations.

7.6 ALPHA INFLATION

7.6.1 NUMBER OF COMPARISONS POSSIBLE

As we have suggested in our discussion thus far, it is possible to examine the results of several comparisons in the post-ANOVA stage of the analysis. The number of pairwise comparisons that can be made, counting both the orthogonal and nonorthogonal ones, is obviously a function of the number of groups (means) that we have in the study; it can be computed by using the following formula (Cohen, 1996, p. 491):

$$\text{number of possible comparisons} = \frac{a(a-1)}{2}. \tag{7.1}$$

In the above formula, a is the number of groups. To illustrate this, if we had three groups in the study, we could make a total of three pairwise comparisons: $(3 \times 2) \div 2 = 6 \div 2 = 3$; if we had five groups in the study, we could make ten pairwise comparisons: $(5 \times 4) \div 2 = 20 \div 2 = 10$; if we had seven groups, we could make twenty-one comparisons: $(7 \times 6) \div 2 = 42 \div 2 = 21$; and so forth.

The main statistical drawback to performing all these comparisons is that if we make enough comparisons at an alpha level of .05, in the long run some of them are going to be found significant by chance alone if the null hypothesis is true. In effect, the comparisons are no longer being made under an alpha level of .05. This is what is meant by *alpha inflation*. If you think of the means from a single study as representing a "family" of means, then we can talk about the *familywise error* increasing beyond our stated alpha level. This issue is also known as *cumulative Type I error*. Alpha inflation – an increase in familywise error beyond our alpha level – can, in turn, lead us to commit one or more Type I (false positive) errors by asserting that a mean difference is "true" (it reflects a "true" difference between the conditions) when such is not the case.

Although less pronounced, we still risk some alpha inflation even when we limit ourselves to making only comparisons that are orthogonal to each other. With $a - 1$ comparisons available to us, we begin on the path toward alpha inflation when we have more than two groups. In our SAT study time example with five groups, for example, we can make four orthogonal comparisons and thus run the risk of at least slightly inflating our alpha level.

7.6.2   EXPECTED CHANGES IN PROBABILITIES

Cohen (1996, pp. 491–492) gives us a way to compute the probability of making at least one Type I error if we perform a certain number (j) of independent comparisons evaluated at the .05 alpha level:

$$\text{probability of at least one Type I error } = 1 - (1 - .05)^j . \quad (7.2)$$

Howell (1997) indicates that this is an expected *familywise error rate*. For example, if we had five groups and made all four orthogonal comparisons ($j = 4$), the probability of making at least one Type I error is about .18 (this is about three times greater than our .05 alpha level); if we had seven groups and made all of the twenty-one pairwise comparisons that were possible ($j = 21$), the probability of making a Type I error is at least .66. Translating this latter case into odds, the odds of incorrectly concluding that at least one mean pair of the twenty-one pairs was significantly different are no worse than 2 to 1. Such a probability of committing a Type I error is unacceptably high.

7.6.3   DEALING WITH ALPHA INFLATION

There are three general situations where we should be very attentive to the possibility of alpha inflation: (a) when we perform a posteriori comparisons, (b) when we make all of the possible orthogonal comparisons, and (c) when we invoke one of the procedures that calls for a relatively large number of the possible comparisons to be made (e.g., performing all pairwise comparisons). Under these three types of circumstances, it is very common for researchers to compensate in one manner or another for alpha inflation; as we will see, most of the commonly used so-called post hoc procedures have some compensation mechanism built in to the calculations.

7.7   GENERAL CATEGORIES OF MULTIPLE COMPARISON PROCEDURES

There are four general categories of multiple comparison procedures that we talk about in this chapter. These categories represent more in the way of arbitrary dividing lines than hard-and-fast different conceptions. Nevertheless, our schema has been partially structured around the way in which researchers generally work with these procedures.

7.7.1   POST HOC COMPARISONS

One approach to multiple comparisons is to use what are often referred to as *post hoc tests*. Post hoc tests are comparisons of pairs of means that were not planned or specified prior to the omnibus ANOVA but are computed only after a statistically significant F ratio has been obtained. Most of the post hoc tests involve performing all possible pairwise comparisons. This approach is atheoretical, exploratory, and most certainly a posteriori. SPSS has a **Post Hoc** dialog window available in almost all of its ANOVA procedures, and SAS has drop-down menus in its ANOVA procedures that contain its post hoc tests. SPSS has eighteen post hoc tests available and SAS has ten. Alpha inflation becomes of great concern here, especially if many comparisons are made. We discuss post hoc tests in Section 7.8.

7.7.2   PRESET CONTRASTS

A second approach to multiple comparisons is the strategy of using what we are calling *preset contrasts*. SPSS has a **Contrasts** dialog window available in its **General Linear Model** procedures; about half a dozen different already structured contrasts are available through that window; SAS has just a couple on its drop-down menu. The approach of using preset contrasts is more restrained than the post hoc test approach in that fewer than the maximum number of comparisons is performed. Some of the comparisons that are available are pairwise and others are composite; furthermore, some are orthogonal and others are nonorthogonal. These contrasts may be used in an atheoretical or exploratory mode, but they may also be used to make comparisons that were planned on the basis of some model prior to the data collection. We discuss preset contrasts in Section 7.15.

7.7.3   POLYNOMIAL CONTRASTS

One type of preset contrast has a sufficiently different emphasis to warrant separate treatment here. It is known as *polynomial contrasts*, and the type of analysis that one performs using these contrasts is commonly called a *trend analysis*. A trend analysis is used to explicate the shape of the function depicting the means of the groups. In plotting this function, groups are represented on the x axis and the dependent variable is represented on the y axis. The means of the groups are then placed in this grid and are then connected to reveal the shape of the function. We discuss polynomial contrasts in Section 7.21.

7.7.4   USER-DEFINED CONTRASTS

A fourth approach to multiple comparisons is to engage in *user-defined contrasts*. This is the most theoretical and often the most rational approach to the issue of post-ANOVA comparisons of means. The researchers specify which groups or combinations of groups will be compared to which others in advance of data collection, and then proceed to carry out that plan usually but not necessarily after the ANOVA. Under this strategy, relatively few comparisons are typically made, and they may be either orthogonal or nonorthogonal. We discuss user-defined contrasts in Section 7.26.

7.8   POST HOC TESTS

7.8.1   A BRIEF HISTORY OF POST HOC TESTS

Kirk (1995) has estimated that there are around thirty post hoc tests that have been presented in the professional literature. Each of these post hoc procedures has acquired the endorsement of some very respected statisticians, and instructors of research and statistics courses almost always have distinct preferences for one or two over most of the others. Generally speaking there appears to be more difference of opinion on the issue of which of the post hoc tests should be used than on most of the other issues covered in this book. Excellent and somewhat detailed treatments of this topic are provided by Jaccard, Becker, and Wood (1984), Kirk (1995), Klockars and Sax (1986), and Toothaker (1993).

A couple of these post hoc tests can be dated back to the 1930s when ANOVA was being popularized. Another wave of development of these tests occurred in the 1950s. Oehlert (2000) provides a short summary of this history that we quote below; do not be concerned if the names of the tests are unfamiliar to you; we will cover the tests that he mentions (and more) in the current section of this book.

The first multiple comparisons technique appears to be the LSD suggested by Fisher (1935)... the next proposal was the SNK (though not so labeled) by Newman (1939). Multiple comparisons then lay dormant till around 1950, when there was an explosion of ideas: Duncan's multiple range procedure (Duncan, 1955), Tukey's HSD (Tukey, 1952), Scheffé's all contrasts method (Scheffé, 1953), Dunnett's method (Dunnett, 1955), and another proposal for SNK (Keuls, 1952). The pace of introduction then slowed again. The REGW procedures appeared in 1960 and evolved through the 1970s (Ryan 1960; Einot and Gabriel 1975; Welsch 1977)... (Oehlert, 2000, p. 107).

Our goals in treating the post hoc tests are as follows:

- To supply you with a listing and description of the tests.
- To give you a very general sense of the distributions underlying these post hoc tests.
- To provide you with some potentially useful guidelines and recommendations in making your choices.
- To illustrate a hand calculation on a selected post hoc test.
- To show you how to perform and interpret a post hoc analysis in SPSS and SAS.

7.8.2   THE SAMPLING DISTRIBUTIONS UNDERLYING POST HOC TESTS

One way to categorize the differences in these post hoc procedures is in terms of the type of the test statistics and sampling distributions that they use (Kirk, 1995). These procedures generally fall into one of the four following categories: Student's t distribution, Studentized range distribution, Studentized maximum modulus distribution, and the F distribution.

7.8.2.1 Student's t Distribution

It is possible to use a t test or a variation of it to compare two means. In the a posteriori post hoc context, this test would be applied to all pairs of means. Thus, with five group means in the design we would conduct ten separate t tests. *Fisher's least significant difference* (LSD) procedure, introduced by Fisher (1935), performs these tests with no correction for alpha inflation. Many writers (e.g., Keppel, 1991) suggest that because of this lack of Type I error protection, the LSD procedure might be best reserved for just a few planned comparisons; otherwise, alpha inflation could cause some false positive decision errors to be made. However, there are still other investigators who support its use (Carmer & Swanson, 1973).

At the other end of the continuum, the *Bonferroni test* (sometimes referred to as the *Dunn* test) is considered to be a good but relatively conservative post hoc test. Pairwise comparisons are made with a t test but statistical significance is evaluated by using an alpha level of .05 divided by the number of multiple comparisons being made. With three comparisons, for example, the effective alpha level is .0167 (.05 ÷ 3 = .0167). A variation on this test was developed by Sidák (1967) to make the test somewhat more powerful (liberal), but even Sidák's variation is still somewhat conservative, yielding corrected alpha levels very close in value to those obtained by using the Bonferroni procedure. In Sidák's procedure, the corrected alpha level is computed by raising the expression $[1 - (1 - .05)]$ to the power of $1/j$ where j is the number of comparisons being made.

7.8.2.2 Studentized Range Distribution

The Studentized range distribution has its origins in Gosset's work, but other statisticians, some of whom have their names associated with a post hoc test in this genre, have worked with this distribution as well. In simplified form, it is possible to determine how much of a mean difference would be needed to achieve statistical significance at a given alpha level. The key to this determination is to obtain an intermediate statistic known as q, the *Studentized range statistic,* which can then be used in a simple formula to work out the mean difference needed (see Appendix F).

The value of q that you need to use has been recorded in table form, and depends on two parameters. One parameter is r, the range of means when they are ordered from lowest to highest. With five groups, $r = 5$. The second parameter is the degrees of freedom associated with the within-groups source of variance from the omnibus ANOVA. Find the coordinate for these two parameters in the Studentized range statistic table, identify your alpha level, and select your q value.

The value of q can also be directly calculated from the ANOVA summary table and a knowledge of the values of the means. Hays (1981) gives the formula as follows:

$$q = \frac{\overline{Y}_{\text{Largest}} - \overline{Y}_{\text{Smallest}}}{\sqrt{\dfrac{MS_{S/A}}{n}}}. \tag{7.3}$$

In the above formula, $\overline{Y}_{\text{Largest}}$ and $\overline{Y}_{\text{Smallest}}$ are the values of the largest and smallest means, respectively, $MS_{S/A}$ is the within-groups mean square from the ANOVA summary table, and n is the size of each group (assuming equal group sizes); with unequal group sizes, Hays (1981) suggests using the harmonic (weighted) mean of the groups (see Section 17.5).

To find your critical value – the mean difference needed in order to assert that the means differ significantly – you engage in three relatively simple calculations (Hays, 1981):

- Divide the within-groups mean square (from the ANOVA summary table) by n, the size of the groups if the group sizes are equal; with unequal group sizes, use the harmonic mean of the group sizes.
- Take the square root of this quantity.
- Multiply the result by q to obtain the so-called *critical difference* (the minimum difference between means that yields statistical significance at your alpha level).

This basic approach has given rise to several variations. The *Tukey Honestly Significant Difference* (HSD) test, for example, applies the computed critical difference to all pairs of means. On the other hand, the *Student–Newman–Keuls* (S–N–K) procedure applies the above critical difference to the pair of means that are most different. If that difference is significant, it *steps down* to look at the difference between the next lowest (or highest) mean and the one most different from it. The value of r, the range parameter, is reduced by 1 (in our study time example, the original value of r was 5; it would now be 4) and the process recomputed. This variation does not do a good job of protecting against Type I error, and some authors (e.g., Keppel et al., 1992) recommend against using the S–N–K procedure for that reason.

7.8.2.3 Studentized Maximum Modulus Distribution

The *Studentized maximum modulus* distribution is related to the Student range statistic and is based on research that Sidák (1967) published in working with variations of the t test. Whereas the range statistic distribution is based on the presumption of equal sample sizes in the groups, the maximum modulus distribution applies to unequal group sizes. Thus,

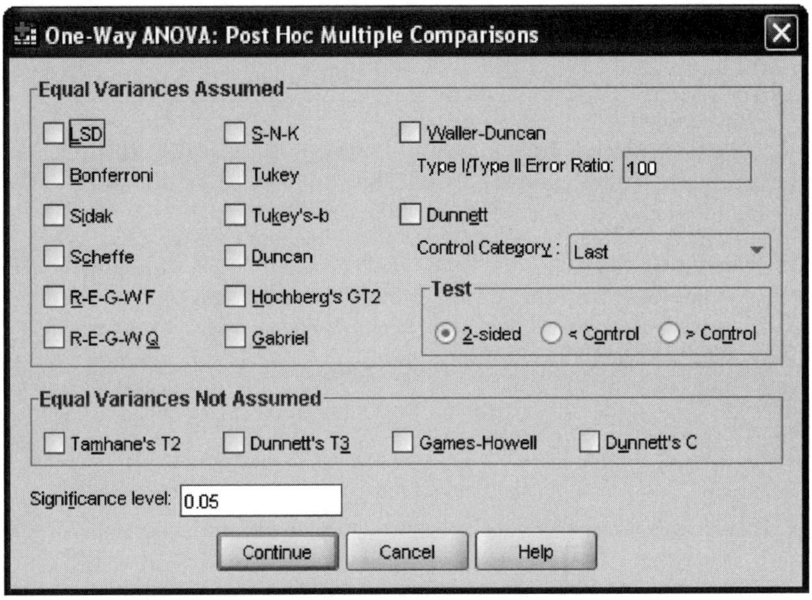

Figure 7.1   The general **Post Hoc** dialog window in SPSS.

the *Hochberg GT2* and *Gabriel* tests, which are based on this latter distribution, can work with unequal sized groups; the *Duncan,* S–N–K, and the two Tukey tests available in SPSS were designed to apply to groups of equal size, at least in hand calculations.

7.8.2.4   *F* Statistic

It is also possible to use an *F* statistic to compare mean differences. Kirk (1995, p. 125) tells us, though, that with the groups having equal sample sizes, *t, q,* and *F* are related to each other as follows.

$$t = \frac{q}{\sqrt{2}} = \sqrt{F}. \tag{7.4}$$

7.8.3   A ROUGH GROUPING OF THE TESTS

The post hoc tests that are available in SPSS (SAS has a subset of these) are shown in Figure 7.1; this is SPSS's **Post Hoc** dialog window. They are conveniently divided into two sets, those assuming equal group or population variances (homogeneity of variance) in the top panel and those not requiring that assumption to be met in the bottom panel. Not that we would recommend the practice but you can select as many as you wish in any one analysis, although with many groups and many tests the output can become quite lengthy.

Table 7.1. Post hoc pairwise comparisons

Abbreviation	Name of test	Distribution	Group size	Strategy
Bonferroni	Bonferroni t	t	Unequal	Single step
Duncan	Duncan Multiple Range	Studentized range	Equal	Step-down
Gabriel		Studentized Maximum Modulus	Unequal	Single step
Hochberg's GT2		Studentized Maximum Modulus	Unequal	Single step
LSD	Fisher's Least Significant Difference	t	Unequal	Single step
R–E–G–W–F	Ryan–Einot–Gabriel–Welch F	F	Equal	Step-down
R–E–G–W–Q	Ryan–Einot–Gabriel–Welch Studentized Range	Studentized range	Equal	Step-down
Scheffé		F	Unequal	Single step
Sidák	Sidák t	t	Unequal	Single step
S–N–K	Student–Newman–Keuls Multiple Range	Studentized range	Equal	Step-down
Tukey	Tukey Honestly Significant Difference	Studentized range	Equal	Single step
Tukey-b		Studentized range	Equal	Step down

7.8.3.1 Homogeneity of Variance Assumed

Table 7.1 presents some of the characteristics that distinguish the majority of tests that assume homogeneity of variance. The tests in Table 7.1 are listed alphabetically. The columns present the following information:

Abbreviation: The abbreviation appearing in the SPSS dialog window (similar abbreviations are used by SAS).

Name: The more complete name of the test.

Distribution: The distribution on which the test is based.

Group Size: Tests that require equal sample sizes to be performed by hand as their creators intended are noted as *Equal*; SPSS and SAS will compensate for unequal group sizes so that the procedures will run under these latter conditions. Tests that were designed to accommodate unequal group sizes are noted as *Unequal*; these tests are capable of working with equal sample sizes as well.

Strategy: A step-down procedure is only applicable to tests that use the Studentized range statistic. When we described the S–N–K procedure in Section 7.4.1.2, we indicated that it repeated its computation each time as it stepped through different values of the range. Thus, means separated by four steps are evaluated differently than those separated by three steps and so on. The procedure starts with the largest range (the means that were most different) and works its way down the line. This is a *step-down* process. Note that the Tukey HSD test also uses the Studentized range statistic but does not use a step-down strategy – all pairs of means are evaluated at the same value of q. The expression "single step" indicates that the test does not take into account by how many steps the means in the pairwise comparison are separated.

Two tests available in SPSS and SAS were not included in Table 7.1 because they did not fit the format we used for the table. The **Waller–Duncan** test in SPSS is based on Bayesian statistical principles, an approach that focuses on various types of probability distributions. If you use it, you must specify a **Type I Error/Type II Error ratio**. As a guide, ratios of 50, 100, and 500 generally match up with alpha levels of .10, .05, and .01. If you do not have equal sample sizes, the result is an approximation. The Dunnett test is conceptually similar to one of the preset contrasts, and we will include it in our discussion of this next group of multiple comparison procedures (see Section 7.15.2.6).

7.8.3.2 Homogeneity of Variance Not Assumed

There are four multiple comparison procedures available in SPSS that can be applied when you cannot assume that the groups have comparable variances. The *Games–Howell, Dunnett T3*, and *Dunnett C* tests are all based on variations of the Studentized range distribution; the *Tamhane T2* is based on the *t* distribution.

7.8.4 SOME GUIDELINES IN SELECTING TESTS

To say that there is a good deal of agreement on which tests to use for a posteriori comparisons would be to adopt a very optimistic attitude in viewing the literature. The major point of agreement is this: If you cannot assume homogeneity of variance, then use one of the tests that do not assume it. After that, the decision process gets complicated.

After they have settled the homogeneity of variance issue, the primary decision basis that many researchers use in deciding which of these tests to apply is power. Their choice would be at least in part determined by the temperament of the researchers – that is, by the level of risk with which the researchers felt comfortable in the particular research context. We will offer our mild suggestions because we do not want to leave you entirely without input but, because most readers are likely to be students, we urge you to talk with your professors about their thoughts on this matter.

We consider first the small group of four tests where homogeneity of variance cannot be assumed. Generally, the Games–Howell test is more powerful than the Dunnett C. With large error degrees of freedom, Dunnett's C is more powerful than his T3, but with small error degrees of freedom that relationship is reversed (Kirk, 1995). The Tamhane T2 test is probably the most conservative of the group. We would give a mild nod to Games–Howell if the situation calls for a bit of power; if your preferences lie toward the other end of the continuum, then we would mildly suggest the more conservative Tamhane T2.

Tests assuming homogeneity of variance outnumber those that do not, and the range of choices is therefore more substantial. For a posteriori tests such as these, the extreme ends of the continuum tend to be generally avoided. *Fisher's LSD* procedure is easily the most liberal (powerful); since it does not control familywise error at all, you run a greater risk of making more false positive judgments than expected of an alpha level of .05. Although using the Fisher LSD procedure based on an *F* ratio (tested at an alpha of .05) has its supporters (e.g., Carmer & Swanson, 1973), we suggest that if you use the LSD procedure at all, you should restrict it to a very small number of planned comparisons. We therefore agree with Keppel (1991) that it should not be used in an a posteriori manner to make several comparisons. On the other side of the spectrum, *Scheffé's* (1953) procedure is easily the most conservative of the available post hoc procedures; many researchers believe that it may be too conservative unless the research context warrants a low, Type I risk level and advise against its general use; we concur.

Keppel and Wickens (2004) recommend against using the S–N–K and Tukey-b tests partly because of their lack of control of familywise error. We would agree and add to that list the Duncan test.

The Bonferroni a very good test but is quite conservative, and it appears that many people pass on it for that reason. Sidák's variation of the Bonferroni test gives it a bit more power but we have not seen a huge endorsement rate for it either.

The two tests that are actively recommended by respected authors (e.g., Howell, 1997; Keppel & Wickens, 2004) are the Tukey HSD test and the *R–E–G–W–Q*. Both hold the expected familywise error rate at no worse than the alpha level. These tests are probably in the midrange of the power spectrum, perhaps shaded somewhat toward the conservative end; the R–E–G–W–Q is more powerful than the Tukey, but the Tukey is the more widely used of the two. If you opt to perform a posteriori pairwise comparisons, then we would encourage you to consider one of them. However, there are alternative strategies covered later in this chapter that you should consider before automatically reaching for any post hoc test.

7.9   COMPUTING A TUKEY HSD TEST BY HAND

Conducting a multiple comparison test is accomplished in a two-step process. The first step involves displaying all possible differences among the means under scrutiny, which we will refer to as creating a *matrix of differences*. A second step involves some simple arithmetic operations using a special statistical formula for the multiple comparison test.

Table 7.2 depicts the five treatment means from our SAT preparation example arranged in ascending order for both the columns and rows of the matrix. Such an arrangement allows us to make (by subtracting one mean from another) all possible pairwise comparisons. For example, we

Table 7.2. Matrix of differences: Treatment means ordered by size

	Factor A				
	a_1	a_2	a_3	a_4	a_5
Means	412.86	474.29	552.86	614.29	623.86
$a_1 = 412.86$	—	61*	140*	201*	211*
$a_2 = 474.29$		—	79*	140*	150*
$a_3 = 552.86$			—	61*	71*
$a_4 = 614.29$				—	10
$a_5 = 623.86$					—

begin by subtracting the zero (a_1) mean from itself (412.86 − 412.86), which of course results in no difference and is indicated with a dash. The next comparison is between the two-month level (a_2) and the zero level (a_1), which results in a difference of 61 (i.e., 474.29 − 412.86). Note that we round up or down to dispense with fractional values. Likewise, the difference between four-month (a_3) and zero (a_1) = 552.86 − 412.86. This process is done for each value across each row. Notice that we do not fill in the matrix values below the diagonal because these values are redundant (mirror images) of the ones above the diagonal. Such a mirror image matrix is called, in mathematical jargon, a *symmetric matrix*. By calculating these difference values, we have generated a pairwise comparison of all of the possible combinations of the five means.

The second step involves calculating the Tukey test formula, which is as follows:

$$D_{\text{Tukey}} = q_t \sqrt{\frac{MS_{S/A}}{n}}, \qquad (7.5)$$

where q_t = a critical value of the Studentized range statistic at alpha = .05 (see Appendix F), $MS_{S/A}$ = mean square error from the overall analysis, and n = sample size of each treatment group ($n = 7$ in the present example).

To determine the value of q_t, we turn to the Studentized range statistic with three necessary components in hand:

- $df_{S/A} = 30$ (in the present example).
- The number of means being compared = 5 (in the present example).
- The alpha level we choose to operate at = .05.

Thus, in the present case, $q_t = 4.10$. We now have all the constituents to complete the formula:

$$D_{\text{Tukey}} = (4.10) \sqrt{\frac{1325.71}{7}} = 56.42 = \boxed{56}.$$

This informs us that any of the five pairwise differences equaling or exceeding 56.42 is a statistically significant difference, and we indicate

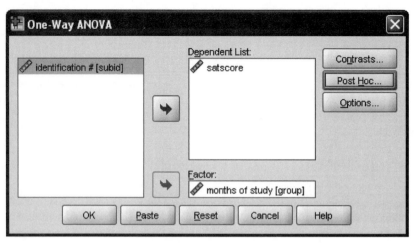

Figure 7.2 The main **One-Way ANOVA** window.

this with an asterisk in the matrix of differences. From this assessment, we conclude that SAT scores were significantly higher as study time in months increased, with the exception of no difference between the six month and eight month study groups.

Another commonly used post hoc test was developed by Fisher (1935) and is known as the *least significant difference* test (LSD). The rationale for its use can be found in Carmer and Swanson (1973). The formula for the LSD is as follows (the matrix of differences should be based on treatment sums and not means):

$$D_{\text{LSD}} = \sqrt{F(1, df_{S/A})}\sqrt{2(n)(MS_{S/A})}. \tag{7.6}$$

7.10   PERFORMING A TUKEY HSD TEST IN SPSS

The omnibus F ratio from our ANOVA was statistically significant, and we are ready to determine where our group differences lie by using the **Tukey HSD** procedure available in the **Post Hoc** window in SPSS. Open your data file. From the main menu, select **Analyze → Compare Means → One-Way ANOVA.** That will bring you to the main dialog window for the procedure as shown in Figure 7.2. Configure it as is shown in Figure 7.2 with **group** as the **Factor** and **satscore** in the **Dependent List** panel.

Click the **Options** pushbutton to reach the dialog window shown in Figure 7.3. Select **Descriptive** but, with the prior analysis already performed, there is no longer a need to run the Levene test. Click **Continue** to return to the main **One-Way ANOVA** window.

Now click the **Post Hoc** pushbutton to reach the list of the post hoc tests available in SPSS. As shown in Figure 7.4, select **Tukey.** Then click **Continue** and in the main window click **OK** to run the analysis.

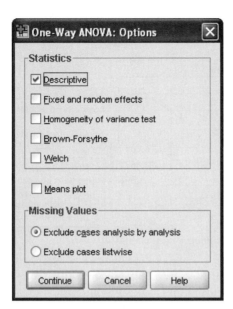

Figure 7.3 The **Options** dialog window.

7.11 THE TUKEY HSD OUTPUT FROM SPSS

You have seen the overall ANOVA output in Chapter 6 and so we will go directly to the results of the post hoc tests. Figure 7.5 shows the output in a form known as *homogeneous subsets*, which is a convenient way to summarize the results of the Tukey pairwise comparisons. The means of the groups are shown across the first five rows with the group names shown at the far left. Each of the last four columns represents a homogeneous subset.

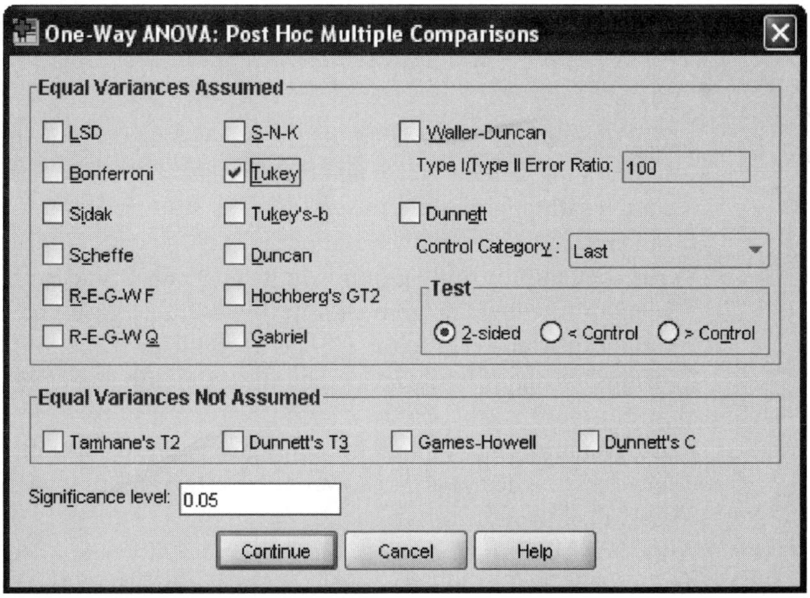

Figure 7.4 The **Post Hoc** dialog window.

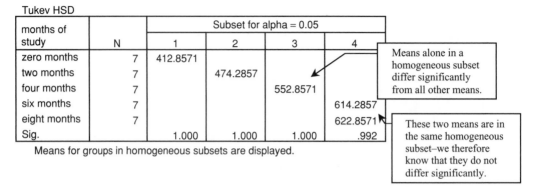

Figure 7.5   The Tukey results in homogeneous subsets.

Here is how to read the display. Means inside a column do not differ from each other. Column 4 shows means of 614.2857 (for the six month study group) and 622.8571 (for the eight month study group). Because these means are in the same column, we know that the Tukey test indicated that they are not significantly different. That is why the term *homogeneous subsets* is used – the subsets are composed of groups whose means are comparable. The other columns each show only a single mean. Thus, there is no mean that is comparable to them – they are significantly different from all of the other means.

Given such a structure, we can read the homogeneous output (using more words than are needed to make everything explicit) from left to right in the following way:

- The zero month study group is in its own subset and has the lowest mean; it has therefore scored significantly lower than all other groups.
- The two month study group is in its own subset and is therefore significantly different from all other groups; that group has scored significantly higher than the zero month study group but significantly lower than the other groups.
- The four month study group is in its own subset and is therefore significantly different from all other groups; that group has scored significantly higher than than the zero month and two month study groups but significantly lower than the six month and eight month study groups.
- The six month and eight month study groups are together in a subset and therefore do not differ; however, those two groups scored higher than all of the other groups.

The homogeneous subset display summarizes the more complete information in the **Multiple Comparisons** output shown in Figure 7.6. The structure of Figure 7.6 is a very common sort of configuration that

Multiple Comparisons

satscore
Tukev HSD

(I) months of study	(J) months of study	Mean Difference (I-J)	Std. Error	Sig.	95% Confidence Interval	
					Lower Bound	Upper Bound
zero months	two months	−61.42857′	19.46216	.028	−117.8807	−4.9765
	four months	−140.00000′	19.46216	.000	−196.4521	−83.5479
	six months	−201.42857′	19.46216	.000	−257.8807	−144.9765
	eight months	−210.00000′	19.46216	.000	−266.4521	−153.5479
two months	zero months	−61.42857′	19.46216	.028	4.9765	117.8807
	four months	−78.57143′	19.46216	.003	−135.0235	−22.1193
	six months	−140.00000′	19.46216	.000	−196.4521	−83.5479
	eight months	−148.57143′	19.46216	.000	−205.0235	92.1193
four months	zero months	−140.00000′	19.46216	.000	83.5473	190.4521
	two months	−78.57143′	19.46216	.003	22.1193	135.0235
	six months	−61.42857′	19.46216	.028	−117.8807	−4.9765
	eight months	−70.00000′	19.46216	.009	−126.4521	−13.5479
six months	zero months	−201.42857′	19.46216	.000	144.9765	257.8807
	two months	−140.00000′	19.46216	.000	83.5479	196.4521
	four months	−61.42857′	19.46216	.028	4.9765	117.8807
	eight months	−8.57143′	19.46216	.992	−65.0235	47.8807
eight months	zero months	−210.00000′	19.46216	.000	153.5479	266.4521
	two months	−148.57143′	19.46216	.000	92.1193	205.0235
	four months	−70.00000′	19.46216	.009	13.5479	126.4521
	eight months	−8.57143′	19.46216	.992	−47.8807	65.0235

*. The mean difference is significant at the 0.05 level.

This value is obtained as follows: zero month mean minus two month mean = 412.8571 − 474.2857 = −61.4286.

The mean difference of −61.4286 is significant based on an alpha of .05.

In this row, the two month study group is compared to each of the other groups.

Figure 7.6 Tukey HSD multiple comparisons.

SPSS uses in a variety of its ANOVA procedures to present the detailed findings of pairwise comparisons.

Each of the major rows of the display focuses on one of the groups. The lines within the major rows present the pairwise comparisons. In the first line of the first major row, for example, the zero month study group is compared to the two months study group. The mean difference is −61.4286. This value was obtained by the following subtraction (the last decimal place may not give the exact results since the printed output gives fewer decimal places than the application uses in its computations):

```
zero month mean minus two month mean = 412.8571 −
474.2857 = −61.4286.
```

Specifically, each of the comparison groups (the ones named in the second column) is subtracted from the focus group (the group in the first column). Note that when the two month study group is the focus group and the zero month study group is the comparison group (the first line of the second major row), the difference is 61.4286 because of the different order of the subtraction. Thus, there is a certain amount of repeated information in the table, which is why it is much larger than the display of the homogeneous subsets. If you continue to read the line where we have the mean difference of −61.4286 between the zero month and the two months study groups, you will reach the column labeled **Sig.** This is the probability of such a

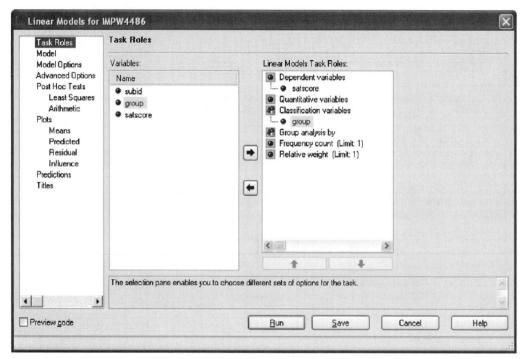

Figure 7.7   The **Linear Models Task Roles** window.

mean difference this large being found if the null hypothesis is true. The value of .028 is less than our alpha value of .05, and so we judge that the two groups differ significantly.

As we scan down the **Sig.** column, we note that all of the pairwise comparisons are statistically significant except for the pairing of the six month and eight month study groups, where we find a value of .992 for the probability of that much of a difference occurring by chance alone, assuming that the null hypothesis is true. This is, of course, what we have already learned from the display of homogeneous subsets. These two displays are just based on different formats and levels of detail to present the same Tukey results, and students can work with whichever they are more comfortable.

7.12   PERFORMING A TUKEY HSD TEST IN SAS

Import your data from Excel into a new project or open *SAS Enterprise Guide* project for Chapter 6. We will assume that you already have the descriptive statistics for the groups and so will deal directly with performing the Tukey post hoc test. From the main menu select **Analyze →
ANOVA → Linear Models.** On the **Task Roles** tab, specify **satscore** as the **Dependent variable** and **group** as the **Classification variable**. This is shown in Figure 7.7.

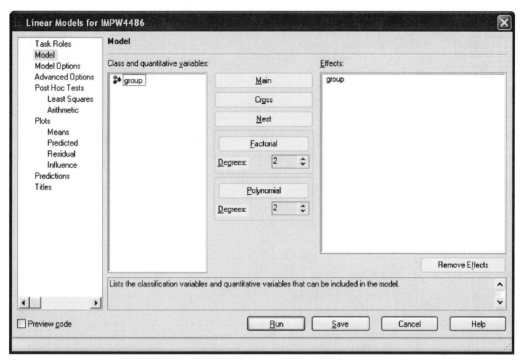

Figure 7.8 The **Linear Models Model** window.

On the **Model** tab, select **group** and click the **Main** bar in the middle of the window. This will place **group** in the **Effects** panel as shown in Figure 7.8.

As we did to perform the omnibus analysis, in the **Model Options** window click only the **Type III sum of squares to show**. Now select **Least Squares** on the **Post Hoc Tests** tab and, when you reach that screen, click the **Add** pushbutton. After Clicking the **Add** pushbutton, four drop-down menus appear in the **Options for means tests** pane as shown in Figure 7.9. When the right-most portion of each horizontal panel is highlighted, a drop-down menu appears, allowing you to choose an alternative to the default that currently shows in the panel. Here is what needs to be done to perform a Tukey post hoc test.

- On the **Class effects to use** drop-down menu now showing **False**, click **False** and the down arrow and select **True**. This will focus on the differences of the means for the **group** variable.
- On the **Comparisons** drop-down menu for **Show p-values for differences** now showing **None**, click in the panel, then click the down arrow, and select **All pairwise differences**.
- On the **Comparisons** drop-down menu for **Adjustment method for comparisons** now showing **Default**, click in the panel, then click the down arrow, and select **Tukey**.
- On the **Show confidence limits** drop-down menu now showing **False**, click **False** and the down arrow and select **True**. This will

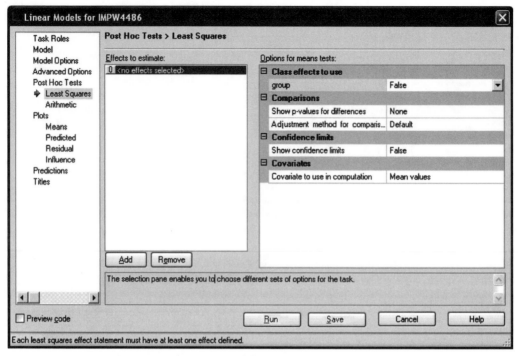

Figure 7.9   The **Linear Models Least Squares** screen.

expose a second drop-down menu for you to specify the **Confidence level** (%). Retain the default of **95**.

- Ignore the **Covariates** drop-down menu because there is no covariate in the analysis.

When you have completed all of these specifications, your screen should be configured as shown in Figure 7.10. Click **Run** to perform the analysis.

7.13   THE TUKEY HSD OUTPUT FROM SAS

The output is shown in Figure 7.11. In the top table, the group designation from the data file occupies the first column and the group means are provided in the second column. Note that SAS has given its own internal-for-the-analysis code numbers to the groups in the third column. It is these code numbers that are used in the bottom table shown in Figure 7.11.

The bottom table in Figure 7.11 displays the results of the pairwise comparisons of the group means. For example, the entry of .0276 for row 1 (the zero month study group) and column 2 (the two month study group) is the p value for this comparison. This is the same value shown by SPSS (it was .028), and indicates that the difference between these two means is statistically significant. As we have already seen from the SPSS analysis, all mean differences except that between the last two groups are significant.

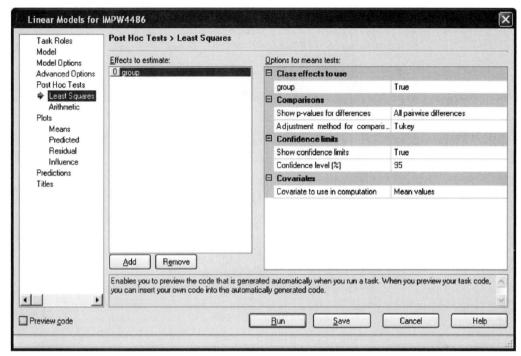

Figure 7.10. The **Least Squares** window is now configured for the Tukey comparisons.

7.14 COMMUNICATING THE TUKEY RESULTS

In reporting the results of the Tukey test, we would probably construct a small table (call it Table 1 for the sake of the write-up) of means and standard deviations rather than reporting them in the text. Given that, one way to report the results is as follows:

The amount of preparation for the SAT in which students engaged appeared to significantly affect their performance on the test, $F(4, 30) = 43.47$, $p < .05$, $\eta^2 = .85$. The means and standard deviations of the groups are presented in Table 1. Results from Tukey's HSD pairwise comparisons indicated that each additional two months of study up to six months was associated with significantly higher SAT scores. However, there was no significant difference in scores between the six month and eight month study groups.

7.15 PRESET CONTRASTS IN SPSS

7.15.1 CHARACTERISTICS OF THE PRESET CONTRASTS

There is a set of contrasts that are already structured in SPSS to make certain comparisons. We have called these preset contrasts but SPSS just calls them **contrasts** in the **Contrasts** dialog window. These preset contrasts are listed in Table 7.3 together with some of their characteristics.

The GLM Procedure
Least Squares Means
Adjustment for Multiple Comparisons: Tukey

group	satscore LSMEAN	LSMEAN Number
0	412.857143	1
2	474.285714	2
4	552.857143	3
6	614.285714	4
8	622.857143	5

Least Squares Means for effect group
Pr > |t| for H0: LSMean(i)=LSMean(j)

Dependent Variable: satscore

i/j	1	2	3	4	5
1		0.0276	<.0001	<.0001	<.0001
2	0.0276		0.0030	<.0001	<.0001
3	<.0001	0.0030		0.0276	0.0093
4	<.0001	<.0001	0.0276		0.9918
5	<.0001	<.0001	0.0093	0.9918	

The difference between group 1 (no study time) and group 2 (two months of study time) is statistically significantly different.

The difference between group 4 (6 months of study time) and group 5 (8 months of study time) is not statistically significantly different.

Figure 7.11   SAS output for the Tukey post hoc test.

We briefly consider these characteristics here; we will then describe the contrasts based on these characteristics in Sections 7.15.1 to 7.16. SAS has nothing analogous to these preset contrasts in point-and-click mode.

7.15.1.1   Dialog Window Containing the Contrast

Five of the six contrasts are available through the **Contrasts** dialog window. This window is not available in the **One-Way ANOVA** procedure we have been using thus far. Rather, we will access **Contrasts** through the **General Linear Model (GLM)** procedure. One of these contrasts, the **Dunnett** procedure, is available in the **Post Hoc** dialog window which can be accessed through either the **One-Way ANOVA** or the **GLM** procedure.

7.15.1.2   Pairwise/Nonpairwise

Pairwise comparisons deal with means of separate conditions. In the non-pairwise comparisons for these contrasts, one of the two means being compared is a linear combination of means from multiple groups.

Table 7.3. Preset contrasts in SPSS dialog windows

Contrast name	Dialog window	Pairwise/ nonpairwise	Orthogonal contrasts	Reference group		Description
				Level	Role	
Deviation	Contrasts	Nonpairwise	No	First/last*	Omitted	Each group except the reference group is compared to the grand mean (which includes the reference group in the calculation).
Difference	Contrasts	Nonpairwise	Yes	Not applicable	Not applicable	Each group except the lowest is compared to the average of the groups coded with lower values (e.g., Level 4 is compared to the average of Levels 1, 2, and 3). This is the mirror image of Helmert contrasts.
Dunnett	Post Hoc	Pairwise	No	First/last*	Focus	Each group is separately compared to the reference group. It can be run two-tailed or directional. This is the same strategy that is used by Simple contrasts.
Helmert	Contrasts	Nonpairwise	Yes	Not applicable	Not applicable	Each group except the highest is compared to the average of the groups coded with higher values (e.g., Level 1 is compared to the average of Levels 2, 3, 4, and 5). This is the mirror image of Difference contrasts.
Repeated	Contrasts	Pairwise	No	Not applicable	Not applicable	Each group is compared to the adjacent group (Level 1 is compared to Level 2, Level 2 is compared to Level 3, and so on).
Simple	Contrasts	Pairwise	No	First/last*	Focus	Each group is separately compared to the reference group. This is the same strategy that is used by the Dunnett test; it has about the same amount of power a directional Dunnett test.

* Using syntax, any group can be used as the reference.

7.15.1.3   Orthogonal Contrasts

We note in Table 7.3 whether or not the contrast is restricted to a set of orthogonal comparisons. As can be seen from the table, only **Difference** and **Helmert** contrasts are orthogonal; the remaining contrasts are not orthogonal.

7.15.1.4   Reference Group

Some of these preset contrasts are structured such that a reference group or category is either defined by the procedure or requires the user to identify the one that will be used as such. In specifying the reference category, SPSS recognizes groups according to the values given to the levels of the independent variable. To illustrate this using our SAT study time example, recall that there are five levels of the study time independent variable. For SPSS, the lowest value code (0 in our example) identifies the first level or group, the next lowest value code (2 in our example) identifies the second level or group, and so on.

For some of these contrasts where the reference group is not defined by the procedure, SPSS provides a way for users to specify the reference group. Using syntax, any group can be specified by users as the reference. In point-and-click mode, users are presented with only two choices for the reference group: the first group or the last group. Clicking the button corresponding to the relevant group specifies the reference group for the analysis.

7.15.2   DESCRIPTION OF THE CONTRASTS

7.15.2.1   Deviation Contrasts

The reference group (which may be either the first or last group) is not considered in isolation in the analysis. Each remaining group is compared to the grand mean of all of the groups (which includes the reference group).

7.15.2.2   Difference Contrasts

Each group except for the first group is compared to the average of the groups coded with lower values (the first group by definition has no groups with value codes below it). In our SAT study time example, the eight-month study group would be compared to the average of the groups having studied for zero, two, four, and six months; the six-month study group would be compared to the average of the groups having studied for zero, two, and four months; and so on. This is the mirror image of **Helmert** contrasts.

7.15.2.3   Dunnett Contrasts

The reference group (first or last) is the focus of the analysis, and is (arbitrarily) labeled in the dialog window as the **Control** group. It is separately compared to each of the remaining groups. You can perform this test in either a two-tailed manner (direction of difference not hypothesized) or in a manner that hypothesizes the control group to have either a higher or lower mean than the other groups. This is the same general strategy that is used by **Simple** contrasts; the directional contrast has about the same amount of power as the **Simple** contrasts procedure.

7.15.2.4   Helmert Contrasts

Each group except for the last group is compared to the average of the groups coded with higher values (the last group by definition has no groups with value codes above it). In our SAT study time example, the zero-month study group would be compared to the groups having studied for two, four, six, and eight months; the two-month study group would be compared to the groups having studied for two, four, and six months; and so on. This is the mirror image of **Difference** contrasts.

7.15.2.5   Repeated Contrasts

Adjacent groups are compared to each other. In our SAT study time example, the zero-month study group would be compared to the two-month study group, the two-month study group would be compared to the four-month study group, and so on.

7.15.2.6   Simple Contrasts

The reference group (first or last group) is the focus of the analysis. It is separately compared to each the remaining groups. This is the same strategy that is used by **Dunnett** contrasts with power that is comparable to a directional Dunnett comparison.

7.16   PERFORMING SIMPLE CONTRASTS IN SPSS

We will use **Simple** contrasts to illustrate how to work with preset contrasts; the setup for the analysis and the output are very similar for the other preset contrasts. For these preset contrasts, we use the **General Linear Model (GLM)** procedure. Once you have opened your data file, follow this navigation route from the menu bar at the top of the screen: **Analyze → General Linear Model → Univariate.** That will bring you to the main dialog window for the procedure as shown in Figure 7.12. You can see in the figure that we have clicked over **satscore** to the **Dependent Variable** panel and have clicked over **group** to the **Fixed Factor(s)** panel. This setup is quite similar to what we have already described in the context of the **One-Way ANOVA** procedure.

Clicking the **Contrasts** pushbutton brings us to the **Contrasts** dialog window shown in Figure 7.13. When it opens, the independent variable is

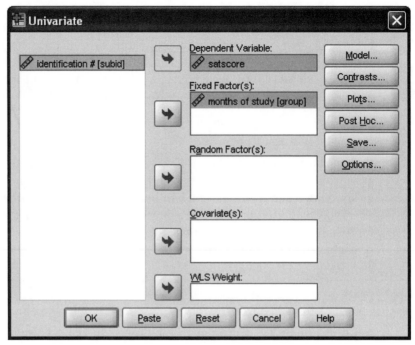

Figure 7.12   The main **GLM Univariate** window.

highlighted with the contrasts set to **None.** Click on the drop-down menu indicated in Figure 7.13; clicking it presents you with the preset contrasts available in SPSS. This is shown in Figure 7.14. Select **Simple** from this drop-down menu; this calls for comparing the reference group to each of the other groups. We will designate the zero month study group as our reference group. Because this is our **first** group (the one with the lowest value code), we must select **First** in the **Reference Group** panel. This is shown in Figure 7.15.

Figure 7.13   Contrasts window.

Figure 7.14 Contrasts window with drop-down menu.

Clicking the **Change** pushbutton to register your choice with SPSS produces what is shown in Figure 7.16. Note that **Simple** is now in parentheses next to the independent variable, with **first** indicated as the reference category. Click **Continue** to return to the main **GLM** window. Then click **OK** to run the analysis.

7.17 THE SIMPLE CONTRASTS OUTPUT FROM SPSS

The output from the contrast analysis is labeled by SPSS as **Custom Hypothesis Tests** and is shown in Figure 7.17. We had specified **First** (zero months of study) as our reference category; this instructed SPSS to compare each of the other groups to this one group. As you can see from the output, the **Contrast Estimate** on the first line of each major row shows the mean difference between our reference group and the group to

Figure 7.15 Simple contrasts with the **First** group as the **Reference** Group.

Figure 7.16 Contrasts window with drop-down menu.

which it is being compared. We saw this same information in the post hoc Tukey analysis.

The fifth line of each major row is labeled as **Sig.** and presents the probability of that mean difference occurring by chance given that the

Contrast Results (K Matrix)

months of study Simple Contrast[a]		Depende... satscore
Level 2 vs. Level 1	Contrast Estimate	61.429
	Hypothesized Value	0
	Difference (Estimate - Hypothesized)	61.429
	Std. Error	19.462
	Sig.	.004
	95% Confidence Interval Lower Bound	21.682
	for Difference Upper Bound	101.176
Level 3 vs. Level 1	Contrast Estimate	140.000
	Hypothesized Value	0
	Difference (Estimate - Hypothesized)	140.000
	Std. Error	19.462
	Sig.	.000
	95% Confidence Interval Lower Bound	100.253
	for Difference Upper Bound	179.747
Level 4 vs. Level 1	Contrast Estimate	201.429
	Hypothesized Value	0
	Difference (Estimate - Hypothesized)	201.429
	Std. Error	19.462
	Sig.	.000
	95% Confidence Interval Lower Bound	161.682
	for Difference Upper Bound	241.176
Level 5 vs. Level 1	Contrast Estimate	210.000
	Hypothesized Value	0
	Difference (Estimate - Hypothesized)	210.000
	Std. Error	19.462
	Sig.	.000
	95% Confidence Interval Lower Bound	170.253
	for Difference Upper Bound	249.747

a. Reference category = 1

This is the mean difference.

This is the probability of the mean difference between Group 0 and Group 2 occurring by chance alone. This probability is less than our alpha level and is therefore statistically significant.

Figure 7.17.   Output of the contrast analysis.

null hypothesis is true. As you can see, all of the groups are significantly different from the reference group.

7.18 PERFORMING SIMPLE CONTRASTS IN SAS

Unlike SPSS, *SAS Enterprise Guide* does not have a large list of preset contrasts, but it does have a **Dunnett** contrast on its menu system. We perform this contrast in the same way that we did to perform a Tukey post hoc test except that we will make somewhat different selections on the **Least Squares** tab. Thus, from the main menu select **Analyze → ANOVA → Linear Models.** On the **Task Roles** tab, specify **satscore** as the **Dependent variable** and **group** as the **Classification variable.** On the **Model** tab, select **group** and click the **Main** bar in the middle of the window to place **group** in the **Effects** panel. In the **Model Options** window click only the **Type III sum of squares to show.**

Now select **Least Squares** on the **Post Hoc Tests** tab and click the **Add** pushbutton to display the **Options for means tests.** Here are the settings for the drop-down menus.

- On the **Class effects to use** drop-down menu now showing **False,** click **False** and the down arrow and select **True.** This will focus on the differences of the means for the **group** variable.
- On the **Comparisons** drop-down menu for **Show p-values for differences** now showing **None,** click in the panel, then click the down arrow, and select **Control using first level.**
- On the **Comparisons** drop-down menu for **Adjustment method for comparisons** now showing **Default,** click in the panel, then click the down arrow, and select **Dunnett.**
- On the **Show confidence limits** drop-down menu now showing **False,** click **False** and the down arrow and select **True.** This will expose a second drop-down menu for you to specify the **Confidence level (%).** Retain the default of **95.**
- Ignore the **Covariates** drop-down menu because there is no covariate in the analysis.

When you have completed all of these specifications, your screen should be configured as shown in Figure 7.18. Click **Run** to perform the analysis.

7.19 THE SIMPLE CONTRASTS OUTPUT FROM SAS

The output for the Dunnett comparisons is shown in Figure 7.19. With the first group defined as the reference group, each of the other groups is compared to that group. Each probability in the last column is based on comparing that particular group to the reference group. Thus, in comparing group 0 to group 2, the probability associated with that statistical test is .0128. As can be seen, all comparisons yielded statistically significant differences.

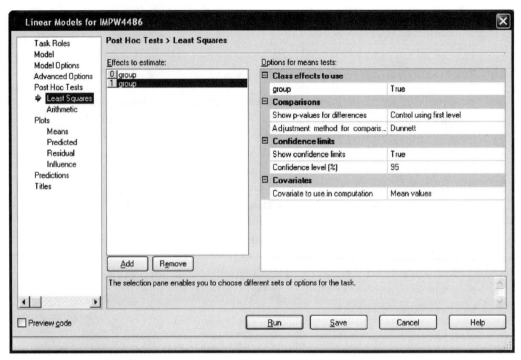

Figure 7.18   The **Least Squares** screen is now configured for the Dunnett comparisons.

7.20   COMMUNICATING THE SIMPLE CONTRAST RESULTS

In reporting the results of the **Simple** contrasts, we would probably construct a small table (call it Table 1 for the sake of the write-up) of means and standard deviations rather than reporting them in the text. Given that, one way to report the results is as follows:

> The amount of preparation for the SAT in which students engaged appeared to significantly affect their performance on the test, $F(4, 30) = 43.47$, $p < .05$, $\eta^2 = .85$. The means and standard deviations of the groups are presented in Table 1. A post-ANOVA set of simple contrasts compared the group with zero months of study to each of the other groups. The results indicated that all of the groups differed significantly from the comparison group; thus, some amount of study resulted in better performance on the SAT than no study at all.

Note that these **Simple** contrasts tell only part of the story revealed by the Tukey tests. This is because the Tukey procedure examined all possible pairwise comparisons and we were able to select which aspect of the comparisons to report. With preset contrasts, you are examining only a particular feature of the results, and the conclusions you are permitted to draw are therefore limited to what you learned about that feature. This may give you some insight as to why many researchers are

The GLM Procedure
Least Squares Means
Adjustment for Multiple Comparisons: Dunnett

group	satscore LSMEAN	H0:LSMean=Control Pr > |t|
0	412.857143	
2	474.285714	0.0128
4	552.857143	<.0001
6	614.285714	<.0001
8	622.857143	<.0001

This is the probability of the mean difference between Group 0 and Group 2 occurring by chance alone. This probability is less than our alpha level and is therefore statistically significant.

Figure 7.19 The output from the Dunnett comparisons.

drawn to the a posteriori pairwise comparisons available as post hoc tests.

7.21 POLYNOMIAL CONTRASTS (TREND ANALYSIS)

7.21.1 POLYNOMIAL FUNCTIONS

As you may remember from your high school algebra class, variables can be raised to a power. For example, in the expression x^2, 2 is the exponent of x; we say that the variable x is raised to the second power (it is squared). In simplified form, a *polynomial* is a sum of such expressions where the exponents are whole positive numbers.

Part of the naming conventions for polynomials is to identify them by the highest appearing exponent appearing in the function. Thus, if the highest exponent is 2, we would label it as *quadratic*, if the highest exponent is 3 we would label it as *cubic*, and if the highest exponent is 4 we would label it as *quartic*.

7.21.2 THE SHAPE OF THE FUNCTIONS

We require for a trend analysis that the levels of the quantitative independent variable approximate interval measurement. If they do meet this requirement, then the shape of the function is interpretable. This is because the spacing of the groups variable on the x axis is not arbitrary and thus the shape of the function is meaningful; if the independent variable did not approach interval measurement, using equal spacing is completely arbitrary (it is appropriate only aesthetically), and the shape of the function is not meaningful since different spacing of the groups along the axis would alter the shape of the function.

Table 7.4 presents some information about examples of polynomial functions on the assumption that the groups on the x axis are spaced on an equal interval basis. Linear functions are straight lines. To draw such a function, all you need to do is to connect two data points (the means of two groups). In the function shown in Table 7.4, x is raised to the first power. Figures 7.20A and B show linear functions.

Table 7.4. Types of polynomial functions

Name	Power of x	Algebraic equation	Characteristics of the function
Linear	1	$y = a + bx$	A straight line function. It requires at least two data points.
Quadratic	2	$y = a + cx^2$	The function has one change in direction (one "peak"). It requires at least three data points.
Cubic	3	$y = a + dx^3$	The function has two changes in direction (two "peaks"). It requires at least four data points.
Quartic	4	$y = a + ex^4$	The function has three changes in direction (three "peaks"). It requires at least five data points.

Quadratic functions have a single bend in the shape of the function. Three data points (the means of three groups) at minimum are required to draw this function. In the function, x is raised to the second power. Figures 7.20C and D show quadratic functions.

Sometimes the shape can be described by more than one function, that is, some curves have components of two or more polynomial functions. Figures 7.20E and F show functions composed of both a linear as well as a quadratic trend. We will discuss this further shortly.

Cubic shapes contain two bends in the function. Four data points (the means of four groups) at minimum are required to draw this function. In the function, x is raised to the third power. Figures 7.20G and H show cubic functions.

The general rule to determine the maximum number of contrasts in an analysis is this: The number of possible polynomial contrasts is equal to $a - 1$ where a is the number of groups. SPSS will not perform polynomial contrasts in excess of the fifth order; that is almost never going to be a problem for you because most of the time you will not wish to deal with more than linear and quadratic contrasts anyway. This is because few theories in the social and behavioral sciences are precise enough to necessitate higher-order polynomials to describe them.

7.21.3   TYING TREND ANALYSIS TO ANOVA

The data points that are schematically plotted in Figure 7.20 are the group means. These are the very same means that an omnibus ANOVA as well as the post hoc or preset contrasts are analyzing. If we have a statistically significant F ratio, then the between-groups variance – this represents the effect of the independent variable – accounts for a significant amount of the total variance of the dependent variable; that is, the statistically significant F ratio indicates that we have mean differences.

A trend analysis begins when a significant effect of the independent variable has been obtained. But now, instead of asking which means differ

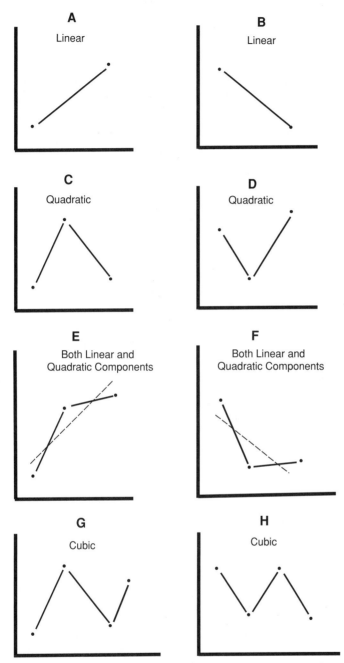

Figure 7.20 Examples of various polynomial functions.

significantly from which others, we ask about the shape of the function. Two points need to be made in this context. First, magnitude differences are reflected in the graph by vertical displacement. We therefore know that the function has a "shape" (it is not a flat or horizontal line) because there is a significant mean difference in the set of means; if the F ratio was not significant, the function would be flat (it would be parallel to the x axis) because the means would not differ significantly in magnitude. Second, to meaningfully address the question of the shape of the function, the groups must be aligned on the x axis in an interval manner (or a reasonable approximation to it). If the independent variable is so measured, as it is in our SAT study time example where the levels are "spaced" in intervals of two months, we can then legitimately ask about the shape of the function.

7.21.4   PARTITIONING THE BETWEEN-GROUPS VARIANCE

In the omnibus ANOVA, we partitioned the total variance into between-groups and within-groups variance. In a trend analysis, we further partition the between-groups variance into its polynomial components by performing polynomial contrasts and testing their significance by using an F ratio. The total of all of the polynomial components is equal to the between-groups variance. We can summarize this as follows, where SS is the sum of squares:

For 2 groups: between-groups $SS =$ linear SS only

$$(SS_A = SS_{\text{linear}})$$

For 3 groups: between-groups $SS =$ linear $SS +$ quadratic SS

$$(SS_A = SS_{\text{linear}} + SS_{\text{quadratic}})$$

For 4 groups: between-groups $SS =$ linear $SS +$ quadratic $SS +$ cubic SS

$$(SS_A = SS_{\text{linear}} + SS_{\text{quadratic}} + SS_{\text{cubic}})$$

For 5 groups: between-groups $SS =$ linear $SS +$ quadratic $SS +$ cubic $SS +$ quartic $SS.$

$$(SS_A = SS_{\text{linear}} + SS_{\text{quadratic}} + SS_{\text{cubic}} + SS_{\text{quartic}}).$$

Let's revisit Figure 7.17 with this information in mind. With only two groups, Figures 7.20A and B show us pure linear trends. All of the between-groups variance is accounted for by the linear component of the polynomial contrast (there is no other polynomial component possible).

Figures 7.20C and D are examples of almost pure quadratic trends. If we tried to fit a straight line through either function, it would be almost parallel to the x axis; such a "flat" slope indicates no mean differences from that linear perspective, and the linear component should not come close to reaching statistical significance.

Figures 7.20E and F are substantially less pure. There is a sufficient bend in each function to most likely generate a statistically significant F ratio for the quadratic trend in the data. But there is a general increase in the values in Figure 7.20E and a general decrease in the values in Figure 7.20F that can be approximated by straight line functions (we have drawn dashed

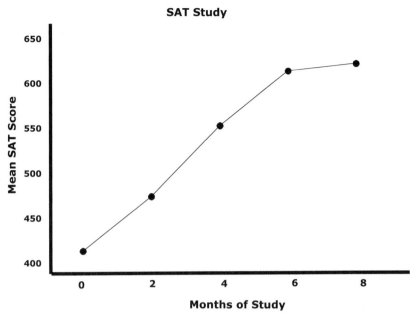

Figure 7.21   Plot of SAT study time means.

lines to make this clear), and this linear component of the functions should also be statistically significant.

Figures 7.20G and H depict relatively pure cubic functions. It is not likely that either the linear or quadratic components would be statistically significant.

7.21.5   OUR EXAMPLE STUDY

Figures 7.21 presents a plot of the results of our hypothetical SAT study time example. As you can see, there is a substantial linear component to the function, which almost certainly would translate to a statistically significant F ratio associated with this linear contrast. At the same time, the function does appear to flatten out at the end, and this could produce a statistically significant quadratic component as well.

We can go even further in our description of the outcome of a trend analysis. Because we are dealing with sums of squares that are additive, we can quantify the strength of each polynomial contrast that is significant with respect to either the between-groups variance or with respect to the total variance. We do this by calculating eta squared values for the particular reference point as follows:

- To calculate the percentage of between-groups variance, a particular polynomial component (e.g., linear, quadratic) accounts for

$$\text{eta squared} = \frac{\text{sum of squares for the polynomial component}}{\text{sum of squares for between groups}}.$$

(7.7)

- To calculate the percentage of the total variance, a particular polynomial component (e.g., linear, quadratic) accounts for

$$\text{eta squared} = \frac{\text{sum of squares for the polynomial component}}{\text{sum of squares for total variance}}.$$

(7.8)

If we were to perform these computations on the results of our trend analysis (which we will do in a moment), we can then speculate that the linear trend would show up as being quite strong whereas the quadratic trend is likely to be quite weak (assuming that it was statistically significant in the first place).

7.22   PERFORMING A TREND ANALYSIS BY HAND

The statistically significant omnibus F test, which we computed previously, on the SAT data suggested that study preparation time affected subsequent SAT scores. As with any omnibus ANOVA employing an independent variable with more than two treatment conditions, we cannot discern the true nature of the relationship between study time and SAT scores, only that there is a statistically significant difference among the means. The plot of the SAT study time means suggests that a straight line fits these data fairly well, indicating the presence of a *linear trend.* The flattening out of this function for the six month and eight month treatment groups may be indicative of a quadratic component to this trend.

Our analysis of linear trend will assess the following statistical hypotheses:

H_0: linear trend is absent (slope of line = zero).

H_1: linear trend is present (slope of line $\neq$ zero).

7.22.1   LINEAR SUM OF SQUARES CALCULATION

To assess for the presence of linear trend we need to calculate a special F ratio. One important constituent of this F ratio is the linear sum of squares ($SS_{A_{\text{linear}}}$). We will now demonstrate the calculation of $SS_{A_{\text{linear}}}$ and its subsequent F ratio.

The formula for $SS_{A_{\text{linear}}}$ is as follows:

$$SS_{A_{\text{linear}}} = \frac{n(\hat{\psi}_{\text{linear}})^2}{\Sigma c^2},$$

(7.9)

where
$n =$ sample size,
$\hat{\psi}_{\text{linear}} =$ (read "psi hat linear") the sum of the treatment means weighted (multiplied) by a set of linear coefficients,
$c =$ linear coefficients.

The linear coefficients are found in a Table of Coefficients of Orthogonal Polynomials found in Appendix E. Coefficients are provided for various components of trend (i.e., linear, quadratic) and are organized by the number of treatment groups or means under scrutiny. In the present example, we need the coefficients that correspond to $a = 5$ treatment levels in the "linear" row:

$$-2 \quad -1 \quad 0 \quad 1 \quad 2$$

Applying the formula for $\hat{\psi}$, we find

$$\begin{aligned}
\hat{\psi}_{\text{linear}} &= (c_1)(\overline{Y}_1) + (c_2)(\overline{Y}_2) + (c_3)(\overline{Y}_3) + (c_4)(\overline{Y}_4) + (c_5)(\overline{Y}_5) \\
&= (-2)(412.86) + (-1)(474.29) + (0)(552.86) \\
&\quad + (1)(614.29) + (2)(623.86) \\
&= -825.72 - 474.29 + 0 + 614.29 + 1247.72 = \boxed{562.00}.
\end{aligned}$$

Additionally, we need to compute the sum of the squared coefficients:

$$\sum c^2 = (-2)^2 + (-1)^2 + (0)^2 + (1)^2 + (2)^2 = \boxed{10}.$$

Thus,

$$SS_{A_{\text{linear}}} = \frac{(7)(562.00)^2}{10} = \boxed{221{,}090.80}.$$

The next steps are to calculate our degrees of freedom, mean squares, and F ratio. Each orthogonal (independent) trend component is associated with 1 df; hence, $df_{A_{\text{linear}}} = 1$.

Thus,

$$MS_{A_{\text{linear}}} = 221{,}090.80/1 = \boxed{221{,}090.80}.$$

Our final step is to calculate the F ratio:

$$F_{A_{\text{linear}}} = \frac{MS_{A_{\text{linear}}}}{MS_{S/A}} = \frac{221{,}090.80}{1{,}325.73} = \boxed{166.77}.$$

This F value is evaluated at 1 and 30 df, and is found to be statistically significant at alpha $= .05$. We can then reject the null hypothesis and conclude that a statistically significant linear relationship exists between the amount of study time and SAT performance.

Again, we note that the hand calculated F value of 166.77 differs slightly from the SPSS calculated value of 165.586 in Figure 7.20 as a result of rounding error.

7.22.2 CALCULATING A NONLINEAR OR QUADRATIC TREND

The most common nonlinear trend component that social and behavioral scientists test for is a *quadratic trend* that indicates a change of direction commonly found in U-shaped or inverted U-shaped functions. To assess for this quadratic trend component, we will again calculate a new sum of squares and subsequent F ratio. We begin this process by consulting our table of Coefficients of Orthogonal Polynomials in Appendix E. Our new set of quadratic coefficients is as follows:

$$2 \quad -1 \quad -2 \quad -1 \quad 2.$$

Hence,

$$
\begin{aligned}
\hat{\psi}_{\text{quadratic}} &= (2)(412.86) + (-1)(474.29) \\
&\quad + (-2)(552.86)(-1)(614.29)(2)(623.86) \\
&= 825.72 - 474.29 - 1,105.72 - 614.29 + 1,247.72 \\
&= \boxed{-120.876}.
\end{aligned}
$$

Thus,

$$
SS_{A_{\text{quadratic}}} = \frac{n(\hat{\psi}_{\text{quadratic}})^2}{\Sigma c^2} = \frac{(7)(-120.86)^2}{14} = \boxed{7,303.57}.
$$

$$(7.11)$$

Next we calculate degrees of freedom, mean squares, and the F ratio. The $SS_{A_{\text{quadratic}}}$ has 1 df associated with it ($df_{A_{\text{quadratic}}} = 1$); thus, $MS_{A_{\text{quadratic}}} = SS_{A_{\text{quadratic}}}/1$ or $MS_{A_{\text{quadratic}}} = SS_{A_{\text{quadratic}}}$. Thus, our F ratio becomes

$$
F_{A_{\text{quadratic}}} = \frac{MS_{A_{\text{quadratic}}}}{MS_{S/A}} = \frac{7,303.57}{1,325.73} = \boxed{5.51}.
$$

$$(7.12)$$

This F value is evaluated with the same degrees of freedom as the $F_{A_{\text{linear}}}(1, 30)$ and is statistically significant at alpha $= .05$. This result indicates a statistically significant quadratic trend is present in the data. SAT scores improve progressively as the number of months of study time increases; however, these scores appear to asymptote by six months of study.

7.23   PERFORMING POLYNOMIAL CONTRASTS (TREND ANALYSIS) IN SPSS

SPSS permits us to perform polynomial contrasts in point-and-click mode, whereas SAS requires that this procedure be accomplished by specifying user-defined contrasts using SAS code (syntax). We will therefore present the SAS procedure to perform a trend analysis in Section 7.34 (after discussing user-defined contrasts).

Open your data file. From the main menu, select **Analyze → Compare Means → One-Way ANOVA**. That will bring you to the main dialog window for the procedure as shown earlier in Figure 7.2. Configure it with **group** as the **Factor** and **satscore** in the **Dependent List** panel.

Click the **Contrasts** pushbutton to reach the **Contrasts** dialog window shown in Figure 7.22. This dialog window is somewhat different from the one we just worked with in the **GLM** procedure, but it allows us to do both polynomial as well as user-defined contrasts very conveniently. Check **Polynomial**. This action makes the **Degree** drop-down menu available.

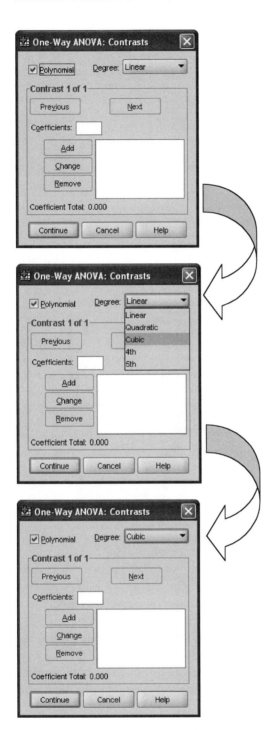

Figure 7.22 Requesting cubic contrasts in the **Contrasts** dialog box.

Select **Cubic** from the drop-down menu (that is complicated enough a polynomial for our purposes); whatever polynomial you choose, you will obtain that and everything below it (linear and quadratic in our case). Click **Continue** to return to the main **One-Way ANOVA** dialog window and click **OK** to run the analysis.

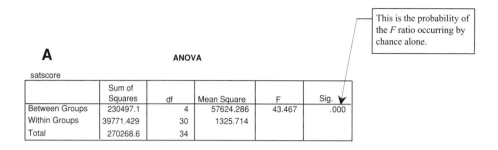

A   ANOVA

satscore

	Sum of Squares	df	Mean Square	F	Sig.
Between Groups	230497.1	4	57624.286	43.467	.000
Within Groups	39771.429	30	1325.714		
Total	270268.6	34			

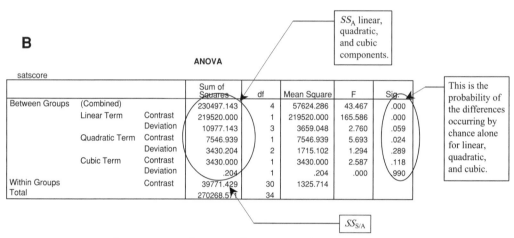

B

ANOVA

satscore

			Sum of Squares	df	Mean Square	F	Sig.
Between Groups	(Combined)		230497.143	4	57624.286	43.467	.000
	Linear Term	Contrast	219520.000	1	219520.000	165.586	.000
		Deviation	10977.143	3	3659.048	2.760	.059
	Quadratic Term	Contrast	7546.939	1	7546.939	5.693	.024
		Deviation	3430.204	2	1715.102	1.294	.289
	Cubic Term	Contrast	3430.000	1	3430.000	2.587	.118
		Deviation	.204	1	.204	.000	.990
Within Groups		Contrast	39771.429	30	1325.714		
Total			270268.571	34			

Figure 7.23 A shows omnibus ANOVA discussed in Chapter 6 and shown in Figure 6.8. B shows output for the trend analysis we just performed.

7.24 OUTPUT FOR POLYNOMIAL CONTRASTS (TREND ANALYSIS) IN SPSS

Recall that we already performed the omnibus ANOVA in the context of Chapter 6. We reproduce the summary table for that analysis in Figure 7.23A so that you can see what the polynomial contrast analysis has accomplished. Figure 7.23B displays the results of the trend analysis that we just performed.

Note first the sum of squares for **Between Groups** from the earlier analysis in Figure 7.23A. It has a value of 230497.143. This value represents the effect of the independent variable of study time. In the trend analysis that we just performed, we see that sum of squares for the **Between Groups (Combined)** also has a value of 230497.143. The reason for why these two values is the same is that they represent the same entity – the overall between-groups variance is the composite of all the polynomial trends; that is, the sum of all the polynomial sums of squares is equal to the between-groups sum of squares.

Our trend analysis has partitioned the between-groups sum of squares (230497.143) into its polynomial portions or contrasts. The summary table shows that 219,520.000 of the 230,497.143 is associated with the linear component or trend. That leaves 10,977.143 remaining for all of the other polynomial contrasts; this leftover is called **Deviation** in the **Linear Term** portion of the summary table.

Of this remaining 10,977.143, we see that 7,546.939 is associated with the quadratic trend. That leaves 3,430.204 for everything else. The cubic trend handles 3,430.000 of that and only the quartic trend is left; it is associated with a sum of squares of .204.

As for the degrees of freedom, we know that with five groups we have 4 *df* associated with the between-groups source of variance. One degree of freedom is associated with each of the polynomial contrasts; with four possible functions (linear, quadratic, cubic, and quartic), we have 4 *df* in total.

Each polynomial trend is associated with an *F* ratio, which in turn is associated with a probability that we evaluate against our alpha level to determine statistical significance. As can be seen from the summary table, both the linear and quadratic components meet a .05 alpha level.

We can now compute eta squared values for our trends, and we will do this with respect to the between-groups variance. The eta squared value for the linear component is .95 ($219,520.000 \div 230,497.143 = .95$) and the eta squared value for the quadratic component is .03 ($7,546.939 \div 230,497.143 = .03$). It does appear that the linear trend is dominant with just a hint of some quadratic component to round out the picture.

7.25 COMMUNICATING THE RESULTS OF THE TREND ANALYSIS

In reporting the results of the trend analysis, we would construct a graph exactly as shown in Figure 7.18 to show the function. Given that, one way to report the results is as follows:

> The amount of preparation for the SAT in which students engaged appeared to significantly affect their performance on the test, $F(4, 30) = 43.47$, $p < .05$, $\eta^2 = .85$. The function relating study time to performance is presented in Figure 7.21. A trend analysis revealed that both the linear, $F(1, 30) = 165.59$, $p < .05$, and quadratic, $F(1, 30) = 5.69$, $p < .05$, components were statistically significant. The linear trend was the dominant one, accounting for 95 percent of the between-groups variance; it largely reflects the increases in SAT scores from zero months of study to six months of study. The quadratic component accounted for 3 percent of the between-groups variance and reflects the relative lack of improvement in SAT scores between six months and eight months of study.

7.26 USER-DEFINED CONTRASTS

User-defined contrasts are planned comparisons, and they ordinarily represent a small subset of all possible comparisons. They can be either

Table 7.5. An orthogonal contrast coefficient table

Group	Coefficients for first contrast	Coefficients for second contrast	Coefficients for third contrast	Product of coefficients for each contrast combination		
				1 vs. 2	1 vs. 3	2 vs. 3
Zero month	4	0	0	0	0	0
Two months	−1	1	1	−1	−1	1
Four months	−1	−1	1	1	−1	−1
Six months	−1	0	−1	0	1	0
Eight months	−1	0	−1	0	1	0
Sum	0	0	0	0	0	0

orthogonal or nonorthogonal, but we will focus here on orthogonal contrasts; nonorthogonal contrasts simply involve following fewer rules.

7.26.1   COEFFICIENTS/WEIGHTS IN GENERAL

Contrasts in general call for weights or coefficients to be assigned to each group mean. When we mentioned weighted linear composites of means earlier, these coefficients are those weights. With preset contrasts, there was no need to explicitly talk about these weights since the procedure automatically assigned them; with user-defined contrasts, you as the user assign them.

One very convenient structure for assigning weights is to lay out a simple structure such as that shown in Table 7.5. We have built the table to demonstrate three contrasts that are all orthogonal to each other. Each group occupies a row in the table and each contrast is represented as a column. You fill in the coefficients (weights) for a particular contrast down that respective column.

Coefficients can take on any values – large numbers, small numbers, fractions – as long as they conform to the few rules that we will outline below. We recommend, and we will use, what we consider to be the easiest values to work with: ones and multiples of one.

7.26.2   RULES FOR ASSIGNING COEFFICIENTS/WEIGHTS

There are five reasonably straightforward rules governing the generation of orthogonal contrasts (see Field, 2005; Keppel & Zedeck, 1989; Pedhazur & Schmelkin, 1991 for a more complete treatment of these rules). They are as follows:

Rule 1: **Groups coded with positive weights (coefficients) are compared with groups given negative weights (coefficients).** It does not matter which groups are coded as positive and which are coded as negative.

Rule 2: **If a group is not involved in a comparison, its weight (coefficient) should be zero.**

Rule 3: **The sum of weights (coefficients) for a given comparison should be zero**. Thus, if one group is compared to a combination of groups and if those groups in the combination are to be weighted equally, then each group in the combination can be assigned a weight of $+1$ or -1 and the group in isolation is assigned a weight of opposite valence equal to the total of the individual weights. For example, if one group is compared to a combined set of 3 groups, each group in the set could be assigned a coefficient of $+1$; the single comparison group is then assigned a coefficient of -3. But, the groups in the composite can be assigned different weights so long as their sum balances to the total of the other group or groups to which the composite is being compared.

Rule 4: **Once a group or set of groups is involved in a contrast, that group or set of groups cannot be used in the same way again.**

Rule 5: **The sum of the products of the coefficients for each combination of contrasts must sum to zero.**

Rule 1 holds that any one contrast in a set of orthogonal contrasts must be "balanced." That is, the groups being compared can be thought of as being on either side of a seesaw. The total weight of one set must cancel the total weight of the other. Because of this, one set of weights must be positive and the other must be negative such that their sum is zero.

Rule 2 indicates that not all of the groups must be involved in any one contrast. Any group not involved in a particular contrast is assigned a zero weight for that contrast; this is not counted as a "use" (see Rule 4).

Rule 3 synthesizes the first two rules by giving you a quick check to make sure that the contrast is balanced by adding to zero. Note that the groups in a combination are not required to be weighted equally, although that is certainly the most convenient and easiest way to combine them. But it is possible that in some research contexts there are good theoretical reasons to weight the groups in an unequal manner. You are free to use any weighting scheme to form a linear combination that makes sense – just make sure that the group or combination of groups to which you are comparing this unequally weighted combination gives you a total of the coefficients that exactly balances the total of the other set of coefficients.

Rule 4 states that you can use a group or a set of groups in an isolated manner in a contrast only one time. Zeros do not count. You must therefore select your contrasts with care.

Rule 5 is probably the key rule in defining orthogonal contrasts (Pedhazur & Schmelkin, 1991), and it is the reason we included the last set of columns in Table 7.5. Multiply the two coefficients on each row of the contrasts on which you are checking for orthogonality to get your products. With five groups, we have five products for each combination of contrasts. Then add these products and obtain zero to verify that the contrasts specified in the table are orthogonal.

7.26.3   THREE EXAMPLE ORTHOGONAL CONTRASTS

Table 7.5 illustrates a set of three orthogonal contrasts. The first contrast compares the zero month study group to all of the others. We have assigned a weight of 4 to the zero month group because we are comparing it to a combination of four groups with each assigned a value of -1 (we are using equal weights here). We could just have easily assigned the zero month group a -4 and the groups in the linear combination each a 1. In truth, we could have used any combination of values that summed to zero, but this set of weights is clear and simple. It is also the case that we assigned each group in the set of four an equal weight; this need not be the case as it is also possible for theoretical reasons to differentially weight the groups. For example, we could have weighted the two, four, six, and eight month groups as -1, -1, -2, and -2. We would have then needed to weight the zero-month group as 6 to "balance" the weights such that they added to zero.

The second contrast compares the two month group to the four month group. One arbitrarily is assigned a weight of 1 and the other a weight of -1, and all of the others are weighted as zero to remove them from the comparison.

The third contrast compares groups that have studied relatively less as a set (the two month and four month study groups combined) to those as a set that have studied relatively more (the six month and eight month study groups). One set has each of its groups equally weighted as 1 and the other set has each of its groups equally weighted as -1. This is because each set contains the same number of groups; had the sets been of unequal size, we would have chosen other numbers that would add to zero. For example, with three groups in one set and two groups in another set, we could have chosen weights such as -2, -2, and -2 versus 3 and 3, respectively.

To verify that these specified contrasts represent orthogonal contrasts, we make sure that each column sums to zero. Then we multiply the weights of pairs of contrasts and add these products to achieve a value of zero; this ensures that each pair of contrasts whose weights we multiplied represents an orthogonal contrast. As you can see from Table 7.5, all of the contrasts are orthogonal to each other.

7.27   PERFORMING USER-DEFINED (PLANNED) COMPARISONS BY HAND

Let's continue with our hypothetical study dealing with student study-time preparation (in months) on SAT performance. Suppose we had the following three hypotheses (or comparisons) that we wanted to assess.

- Students who do not study versus those that do study.
- Students studying for two months versus those studying for four months.

- Students studying for two to four months versus those studying six to eight months.

There are some common steps involved in testing these three hypotheses. The first step involves creating a difference score between the two means being compared. We will refer to this value as $\hat{\psi}$ (psi hat), which statisticians use to refer to a difference between two means. The caret or "hat" above the symbol indicates that this value is an estimate of the difference between the population treatment means.

The second step is to compute a new comparison sum of squares, which we will call $SS_{A_{comp}}$, and a new mean square ($MS_{A_{comp}}$), and form a new F ratio that will produce a new $F_{A_{comp}}$ that can be evaluated to determine if the difference between the two means (captured by $\hat{\psi}$) is statistically significant (see Keppel et al., 1992). The formulas for these three values with their numerical examples for each hypothesis follow.

Sometimes researchers are interested in *simple* comparisons between two treatment means. On other occasions, researchers examine more *complex* or sophisticated hypotheses that require the averaging of two or more groups in order to compare them to another single group or an average of two or more other groups. We will demonstrate the creation of these simple and complex comparisons by weighting the treatment means with special numbers or weights called "coefficients" (see Keppel et al., 1992; Keppel & Wickens, 2004).

7.27.1 HYPOTHESIS 1

The coefficients used to assess the first hypothesis are as follows:

$$4 \quad -1 \quad -1 \quad -1 \quad -1.$$

These coefficients indicate that the first treatment group (zero months of study) is being compared to four other (combined) study groups. This is accomplished by weighting or multiplying each treatment mean by its respective coefficient to give that mean more or less input during the computation of the comparison sum of squares. The formula that generates this weighting ($\hat{\psi}$) is as follows:

$$\hat{\psi}_1 = (c_1)(\overline{Y}_1) + (c_2)(\overline{Y}_2) + (c_3)(\overline{Y}_3) + (c_4)(\overline{Y}_4) + (c_5)(\overline{Y}_5)$$
$$= (4)(412.86) + (-1)(474.29) + (-1)(552.86)$$
$$+ (-1)(614.29) + (-1)(623.86)$$
$$= \boxed{-613.86}. \tag{7.13}$$

With $\hat{\psi}$ calculated, we are now ready to proceed with the formula for the $SS_{A_{comp}}$.

$$SS_{A_{comp}} = \frac{n\left(\hat{\psi}\right)^2}{\Sigma c^2}. \tag{7.14}$$

Note that the denominator refers to the sum of squared coefficients, which we will designate with a lower case c. Substituting our values into the previous equation, we have the following:

$$\begin{aligned} SS_{A_{comp_1}} &= \frac{7(-613.86)^2}{(4)^2 + (-1)^2 + (-1)^2 + (-1)^2 + (-1)^2} \\ &= \frac{2,637,768.60}{20} \\ &= \boxed{131,888.43}. \end{aligned} \tag{7.15}$$

The remaining procedures for computing degrees of freedom, mean squares, and F are straightforward. There is only 1 df in this comparison (even though there are five treatment groups) because we are comparing the zero months of study group with the combined two to eight months of study groups. Hence,

$$df_{A_{comp_1}} = \boxed{1},$$

$$MS_{A_{comp_1}} = \frac{SS_{A_{comp}}}{1} = \frac{131,888.43}{1} = \boxed{131,888.43}.$$

Thus,

$$F_{A_{comp_1}} = \frac{MS_{A_{comp}}}{MS_{S/A}} = \frac{131,888.43}{1,325.73} = \boxed{99.48}. \tag{7.16}$$

This F value is evaluated with 1 and $df_{S/A}$ (which is 30 in the present example). Since our observed F of 99.48 is greater than the critical value of F, at the .05 level (critical value = 4.17), we conclude that this statistically significant F indicates that students in the combined study conditions scored higher on the SAT then did the no-study students.

7.27.2 HYPOTHESIS 2

The second hypothesis contrasts students who studied for two months with those who prepared for four months. The coefficients for this comparison are as follows:

$$0 \quad 1 \quad -1 \quad 0 \quad 0.$$

Thus,

$$\begin{aligned} \hat{\psi}_2 &= (c_1)(\overline{Y}_1) + (c_2)(\overline{Y}_2) + (c_3)(\overline{Y}_3) + (c_4)(\overline{Y}_4) + (c_5)(\overline{Y}_5) \\ &= (0)(412.86) + (1)(474.29) + (-1)(552.86) + (0)(614.29) \\ &\quad + (0)(623.86) = \boxed{-78.57} \end{aligned} \tag{7.17}$$

$$SS_{A_{comp2}} = \frac{n\left(\hat{\psi}\right)^2}{\Sigma c^2} = \frac{7\left(-78.57\right)^2}{2} = \boxed{21{,}606.36} \qquad (7.18)$$

$$MS_{A_{comp2}} = \frac{SS_{A_{comp2}}}{1} = \boxed{21{,}606.36}$$

$$F_{A_{comp2}} = \frac{MS_{A_{comp2}}}{MS_{S/A}} = \frac{21{,}606.36}{1{,}325.73} = \boxed{16.30}. \qquad (7.19)$$

Evaluate at 1 and $df_{S/A}(1, 20)$, $p < .05$. We conclude that students who prepare for four months perform significantly higher on the SAT than do students who study for two months.

7.27.3 HYPOTHESIS 3

The third hypothesis contrasts students who study from two to four months with those who study from six to eight months. The coefficients for this comparison are as follows:

$$0 \quad 1 \quad 1 \quad -1 \quad -1.$$

Thus,

$$\begin{aligned}
\hat{\psi}_3 &= (c_1)(\overline{Y}_1) + (c_2)(\overline{Y}_2) + (c_3)(\overline{Y}_3) + (c_4)(\overline{Y}_4) + (c_5)(\overline{Y}_5) \\
&= (0)(412.86) + (1)(474.29) + (1)(552.86) + (-1)(614.29) \\
&\quad + (-1)(623.86) = \boxed{-211.00}
\end{aligned} \qquad (7.20)$$

$$SS_{A_{comp3}} = \frac{n(\hat{\psi})^2}{\Sigma c^2} = \frac{7(-211.00)^2}{4} = \boxed{77{,}911{,}75} \qquad (7.21)$$

$$MS_{A_{comp3}} = \frac{SS_{A_{comp3}}}{1} = \boxed{77{,}911.75}$$

$$F_{A_{comp3}} = \frac{MS_{A_{comp3}}}{MS_{S/A}} = \frac{77{,}911.75}{1{,}325.73} = \boxed{58.77}. \qquad (7.22)$$

Again, we evaluate at 1 and $df_{S/A}$ (1, 30), $p < .05$. We conclude that students who prepare for six to eight months score higher on the SAT than do students who study only two to four months.

These computed F values correspond directly to the t values that SPSS produces when it conducts **Contrast Tests** as shown later in Figure 7.26. Recall from our discussion in Chapter 6 that the t and F statistics are related: $t^2 = F$ and $\sqrt{F} = t$. Thus, when we square the three SPSS-produced t values, in the **Assume equal variances** section of Figure 7.26 ($-9.958, -4.037, -7.630$), we produce F values (99.16, 16.30, 58.22) that are equivalent to the F values we obtained, less some minor rounding error.

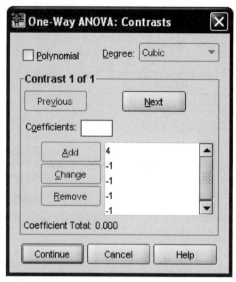

Figure 7.24   The **Contrasts** window.

7.28   PERFORMING USER-DEFINED CONTRASTS IN SPSS

Open your data file. From the main menu, select **Analyze ➜ Compare Means ➜ One-Way ANOVA.** That will bring you to the main dialog window for the procedure as shown earlier in Figure 7.2. Configure it with **group** as the **Factor** and **satscore** in the **Dependent List** panel.

Click the **Contrasts** pushbutton to reach the **Contrasts** dialog window shown in Figure 7.24. Make sure that **Polynomial** is *not* checked. We will be using the large middle area where you see **Contrast 1 of 1** and a **Coefficients** panel with the cursor in it. Here is how this window works:

- **Contrast 1 of 1** counts our contrasts. We have three of them to enter. Each contrast will involve us entering the five weights that we constructed in Table 7.5. We have three contrasts to enter, and we enter them separately contrast by contrast.
- **Coefficients** has a little panel next to it. This is where you enter the coefficients or weights that we have constructed in Table 7.5. Focus on the first contrast. The first weight is 4 and that is what you type into the little panel. Then click **Add**. This will transfer the weight into the large panel in the middle of the window. Type the next weight (-1) and click **Add** to transfer this weight into the large panel. As you do this you will see the weights line up in a column in the large panel as shown in Figure 7.21. Continue in this manner until you have finished with the first contrast.
- SPSS will keep track of the total of the coefficients as you enter them. When you have finished with the first contrast, you should verify that the **Coefficient Total** is 0. This means you have correctly input

Contrast Coefficients

	months of study				
Contrast	zero months	two months	four months	six months	eight months
1	4	−1	−1	−1	−1
2	0	1	−1	0	0
3	0	1	1	−1	−1

Figure 7.25 The **Contrast Coefficients** table.

your weights. In Figure 7.24, we can see that the total is equal to zero (note that SPSS will not multiply your coefficients for you as you enter subsequent contrasts).

At the completion of the first contrast, click the **Next** pushbutton. This will cause the window to now display **Contrast 2 of 2** and (probably to your surprise and perhaps to your distress) to blank out the large middle panel. Do not panic – SPSS will retain your first set of coefficients. In an unfortunate formatting decision, SPSS shows only one set of contrast weights at a time. Repeat what you did for the first contrast, then once again click the **Next** pushbutton, which will now blank out the second set of weights and display **Contrast 3 of 3**, and type in the third contrast. When you finish, click **Continue** to return to the main **One-Way ANOVA** dialog window and click **OK** to run the analysis.

7.29 OUTPUT FROM USER-DEFINED CONTRASTS ANALYSIS IN SPSS

SPSS prints out the coefficients you used for each contrast in a table called **Contrast Coefficients,** which we show in Figure 7.25. The weights for each contrast are shown horizontally in the display but nonetheless are invaluable reminders about what you requested. This is important because SPSS simply labels the contrasts as **1, 2,** and **3** in the analysis proper, which is shown in Figure 7.26.

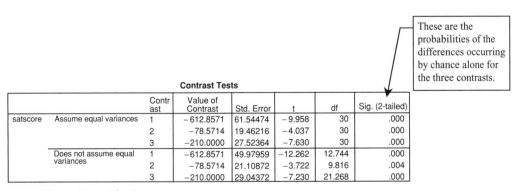

These are the probabilities of the differences occurring by chance alone for the three contrasts.

Contrast Tests

		Contrast	Value of Contrast	Std. Error	t	df	Sig. (2-tailed)
satscore	Assume equal variances	1	−612.8571	61.54474	−9.958	30	.000
		2	−78.5714	19.46216	−4.037	30	.000
		3	−210.0000	27.52364	−7.630	30	.000
	Does not assume equal variances	1	−612.8571	49.97959	−12.262	12.744	.000
		2	−78.5714	21.10872	−3.722	9.816	.004
		3	−210.0000	29.04372	−7.230	21.268	.000

Figure 7.26. The **Contrast** tests.

The results from the **Contrast Tests** are shown in Figure 7.26. Our contrasts are evaluated by t tests. One very nice feature of this analysis is that you can select the tests that apply when equal variances can and cannot be assumed. As you can see from the display, all three contrasts resulted in statistically significant outcomes.

7.30 COMMUNICATING THE RESULTS OF THE CONTRASTS ANALYSES

In reporting the results of these user-defined contrasts, we would probably construct a small table (call it Table 1 for the sake of the write-up) of means and standard deviations rather than reporting them in the text. Given that, one way to report the results is as follows:

The amount of preparation for the SAT in which students engaged appeared to significantly affect their performance on the test, $F(4, 30) = 43.47$, $p < .05$, $\eta^2 = .85$. The means and standard deviations of the groups are presented in Table 1. Three sets of a priori contrasts, all orthogonal to each other, were performed. The results of these yielded the following outcomes. First, the students not studying at all performed more poorly on the SAT than all of those who did study, $t(30) = -9.96$, $p < .05$. Second, those studying for two months performed worse than those studying for four months, $t(30) = -4.04$, $p < .05$. Third, students studying for a relatively shorter period of time (two months and four months combined) performed more poorly than those studying relatively more (six months and eight months combined), $t(30) = -7.63$, $p < .05$.

7.31 PERFORMING USER-DEFINED CONTRASTS (PLANNED COMPARISONS) IN SAS

Planned comparisons are performed in *SAS Enterprise Guide* by adding a couple of lines of command code to the **Linear Models** procedure. When we are using the point-and-click interface to specify the details of a statistical analysis, both SPSS and SAS are translating that into a set of instructions. SPSS calls those instructions command "syntax" and SAS calls those instructions command "code." Either way, they resemble strings of words or abbreviations that these applications refer to when performing an analysis. We will see in Chapter 8 how to add syntax to the SPSS commands so that we may perform our simple effects analysis. Here, we need to add code to our SAS commands in order to perform our planned comparisons.

As we have done before to perform the Tukey post hoc test and the Dunnett comparisons, we select from the main menu **Analyze ➜ ANOVA ➜ Linear Models**. On the **Task Roles** tab, specify **satscore** as the **Dependent variable** and **group** as the **Classification variable**. On the **Model**

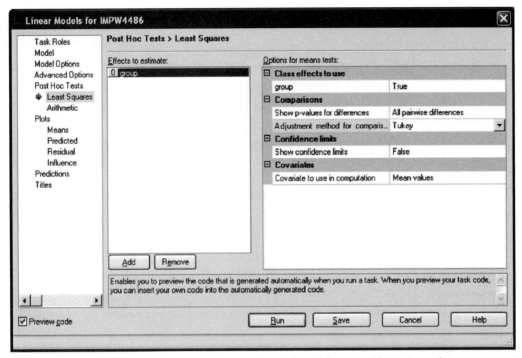

Figure 7.27   The **Least Squares** window configured for a Tukey test with Preview code selected.

tab, select **group** and click the **Main** bar in the middle of the window to place **group** in the **Effects** panel. In the **Model Options** window click only the **Type III sum of squares to show**. Set up the **Least Squares** window as though you are performing a Tukey post hoc test as described in Section 7.12. This is shown in Figure 7.27.

Now click **Preview code** at the bottom left of the window as shown in Figure 7.27. This will display the window containing the code as shown in Figure 7.28.

SAS command code is generally organized into procedures (called "PROCs"). These already have been configured by users through the point-and-click interaction with the application, and inserting code into the middle of a procedure would disrupt it. For this reason, *SAS Enterprise Guide* provides designated areas in the sequence of code for new code to be legally entered. To insert new code, click the **Insert Code** pushbutton located at the top left portion of the window shown in Figure 7.28, which transforms the look of the window. We show this transformation in Figure 7.29, where we have already scrolled down to the bottom of the code.

Select the bar title <**double-click to insert code**> directly above the **LSMEANS group /PDIFF∗ALL ADJUST = TUKEY** line. Double click in that line. This will present you with an **Enter User Code** window, where you can type the command code for the planned comparisons. We have already typed this code in as shown in Figure 7.30.

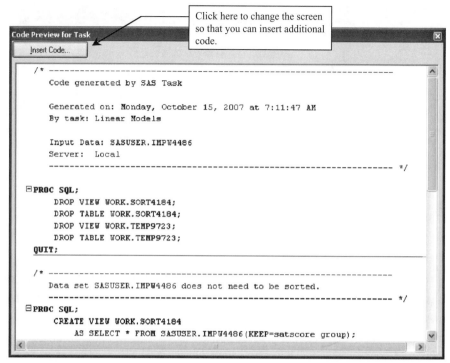

Figure 7.28   The window containing the SAS code for the analysis.

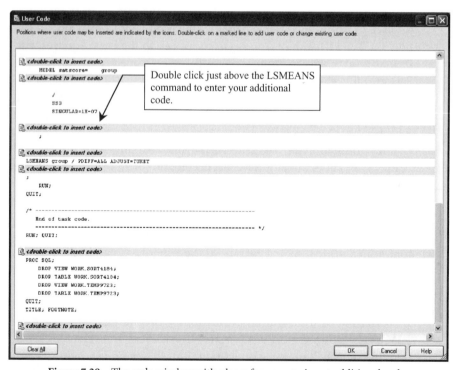

Figure 7.29   The code window with places for users to insert additional code.

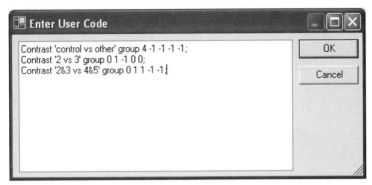

Figure 7.30   The **Enter User Code** window with the planned comparisons specified.

We have already discussed the rationale for the coefficients in these three planned contrasts. The structure of the code is as follows:

- **Contrast** is a keyword in SAS to indicate that we are comparing certain means.
- The characters in single parentheses are "literal strings"; they will be reproduced in the output exactly as we typed them, and will serve as labels for us to help us read the output.
- The word **group** is required to identify the effect that we are using as the focus of the contrasts.
- The coefficients define the particular contrast we are making.
- Each line ends in a semicolon.

After entering the code, click **OK**. This returns you to the **User Code** window where your new code can now be seen once you scroll down to the end of the file (see Figure 7.31). Click **OK** to return to the **Least Squares** window, and click **Run** to perform the analysis.

7.32   OUTPUT FOR PLANNED COMPARISONS IN SAS

The output from the three contrast analyses is presented the final three rows of the output in Figure 7.32. As you can see, each is identified by the label we typed in quotation marks and each is evaluated by its own F ratio. All three yielded statistically significant effects. Note that the F ratio displayed by SAS is translatable to the t value displayed by SPSS because $t^2 = F$. For example, for the first contrast, squaring the t value of 9.958 displayed by SPSS produces the F ratio of 99.16 shown in the SAS output. Thus, we may conclude the following:

- The combined four groups having studied for the SAT performed better than the group that did not prepare.
- The students preparing four months (Group 3) performed better on the SAT than those students preparing for two months (Group 2).
- The students preparing for relatively longer periods of time (six and eight months combined) performed better on the SAT than students

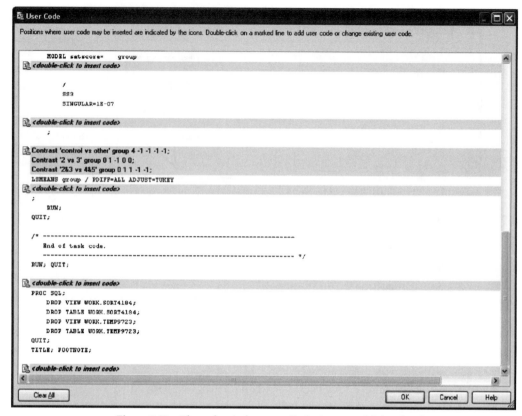

Figure 7.31   The code window now displaying our additional code.

preparing for relatively shorter periods of time (two months and four months combined).

7.33 COMMUNICATING THE RESULTS OF THE PLANNED COMPARISONS

The results of the contrast analyses were provided in Section 7.26; readers are encouraged to consult that section if they need help recalling how to structure the report.

> These are the probabilities of the differences occurring by chance alone for the three contrasts.

Contrast	DF	Contrast SS	Mean Square	F Value	Pr > F
control vs other	1	131457.8571	131457.8571	99.16	<.0001
2 vs 3	1	21607.1429	21607.1429	16.30	0.0003
2 & 3 vs 4 & 5	1	77175.0000	77175.0000	58.21	<.0001

Figure 7.32   The planned comparison output.

7.34 PERFORMING POLYNOMIAL CONTRASTS (TREND ANALYSIS) IN SAS

Polynomial contrasts are performed in *SAS Enterprise Guide* by using the same procedure as we just described for user-defined contrasts. That is, the linear, quadratic, and other polynomial orders (polynomial partitions of the variance) are specified by a set of contrast coefficients.

The various polynomial trends are defined in terms of the means of certain groups being statistically different from the means of other groups. Consider the quadratic function in Figure 7.20C. The second (middle) data point must be significantly different from the other two (combined) for there to be a quadratic trend. The coefficients corresponding to the test of statistical significance would be 1, −2, and 1 for the groups ordered first, second, and third.

It is possible for researchers to generate the appropriate coefficients for various polynomial components (linear, quadratic, cubic, and so on). The values chosen for these coefficients would be affected by the particular polynomial component of interest as well as the number of groups in the study. Rather than having to craft these anew with each study for each research team, tables presenting sets of coefficients applicable to a wide range of group set sizes and polynomial components (for interval level spacing of the groups) were worked out more than half a century ago and are currently widely disseminated. Technical explanations of how these coefficients are derived and how to compute coefficients for unequal intervals can found in several texts, including Keppel (1991) and Kirk (1995).

Tables containing coefficients for testing orthogonal polynomial components of the total variance can be found in Keppel and Wickens (2004, Table A.3) and Kirk (1995, Table E.10) as well as in the public domain (e.g., university Web sites). As we discussed in the context of our hand computations for polynomial contrasts, we present these coefficients from a first-order (a linear function) through a fourth-order polynomial (a quartic function) for up to eight groups in Appendix D. The coefficients are to be read across for the specific case. For example, the contrast coefficients for evaluating a quadratic polynomial function for a set of 5 groups ordered as Groups 1, 2, 3, 4, and 5 are 2, −1, −2, −1, and 2, respectively.

Readers can refer to the figures we presented for planned comparisons if they wish, as we briefly describe the setup here. Select from the main menu **Analyze → ANOVA → Linear Models**. On the **Task Roles** tab, specify **satscore** as the **Dependent variable** and **group** as the **Classification variable**. On the **Model** tab, select **group** and click the **Main** bar in the middle of the window to place **group** in the **Effects** panel. In the **Model Options** window click only the **Type III sum of squares to show**. In the **Least Squares** window, specify **group** as the **Class effect to use** and select **All pairwise differences** for the **Show p-values for differences panel**.

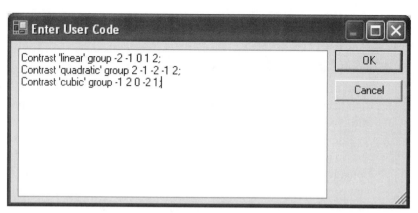

Figure 7.33   The necessary SAS code to perform our polynomial contrasts.

Now click **Preview code** at the bottom left of the **Least Squares** window to display the code. To insert new code, click the **Insert Code** pushbutton located at the top left portion of the window. Scroll down to the bar title <**double-click to insert code**> directly above the **LSMEANS group/PDIFF∗ALL** line. Double click in that line. This will present you with an **Enter User Code** window, where you can type the command code for the planned comparisons. We have already typed this code in, as shown in Figure 7.33.

After entering the code, click **OK**. This returns you to the **User Code** window where your new code can be seen if you wish once you scroll down to the end of the file. Click **OK** to return to the **Least Squares** window, and click **Run** to perform the analysis.

7.35   OUTPUT FOR POLYNOMIAL CONTRASTS (TREND ANALYSIS) IN SAS

The output from the three polynomial contrast analyses is presented in the final three rows of the output in Figure 7.34. As you can see, each is identified by the label we typed in quotation marks in the **Enter User Code** window and each is evaluated by its own F ratio. Both the linear and quadratic components yielded statistically significant effects.

These are the probabilities of the differences occurring by chance alone for linear, quadratic, and cubic.

Contrast	DF	Contrast SS	Mean Square	F Value	Pr > F
linear	1	219520.0000	219520.0000	165.59	<.0001
quadratic	1	7546.9388	7546.9388	5.69	0.0235
cubic	1	3430.0000	3430.0000	2.59	0.1182

Figure 7.34   Output from the polynomial contrasts analysis.

7.36 COMMUNICATING THE RESULTS OF THE POLYNOMIAL CONTRASTS

The results of the contrast analyses were provided in Section 7.20; readers are encouraged to consult that section if they need help recalling how to structure the report.

CHAPTER 7 EXERCISES

7.1. Using the data set from Chapter 5, Exercise 5.1:
   a. Perform an unplanned multiple comparison procedure on the means by using SPSS or SAS.
   b. Conduct a Tukey or LSD test by hand on these data.
   c. Write up your results.

7.2. Using the data set from Chapter 5, Exercise 5.2:
   a. Perform an unplanned multiple comparison procedure on the means using SPSS or SAS.
   b. Conduct a Tukey or LSD test by hand on these data.
   c. Write up your results.

7.3. Using the data set from Chapter 5, Exercise 5.3:
   a. Conduct a planned comparison (of your choice) among the means by using SPSS or SAS.
   b. Conduct this same comparison, using coefficients, by hand.
   c. Write up your results.

Two-Way Between-Subjects Design

8.1 COMBINING TWO INDEPENDENT VARIABLES FACTORIALLY

8.1.1 NAMING THE DESIGN

In Chapter 6, we discussed a between-subjects design that contained a single independent variable (preparation time for the SAT). However, we are not limited to studying the effects of just one independent variable in a research design. In this chapter, we will deal with the inclusion of a second independent variable (note that we still have only one dependent variable). A design containing more than one independent variable is known as a *factorial design* when the variables are combined in a manner described in Section 8.1.2. When those independent variables are between-subjects variables, the design is called a *between-subjects design* or a *between-subjects factorial design*. Designs containing two between-subjects independent variables that are simultaneously varied are *two-way between-subjects (factorial) designs*. These designs are also sometimes referred to as *two-way completely randomized designs* because subjects are assumed to be randomly assigned to the various treatments.

8.1.2 COMBINING INDEPENDENT VARIABLES IN A SINGLE DESIGN

Intertwining two independent variables within the same design is done by combining them in a *factorial* fashion in which each level of one independent variable is combined with each level of the other independent variable. If our independent variables were, for example, gender (female and male) and size of city in which participants resided (large and small), then one combination of the levels of the independent variables might be females living in large cities. This would then be one condition or group in the factorial design. In all, for this example there are four possible combinations and therefore there would be four groups in the study.

Given that in the above example we have four groups in the study, it is theoretically possible to conceive of it and to analyze it as a four-group, one-way design as described in the Chapter 6. So why bother to treat it as a two-way design and make things more complicated? And if it's so important to deal with each independent variable as a separate entity, why not just treat it as two separate one-way designs?

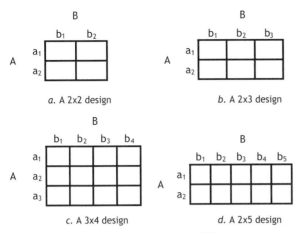

Figure 8.1 Illustration of several factorial designs.

The answer to both of these questions is that we can obtain much more information from conceptualizing it as a two-way design without losing any information available in the one-way designs that compose it. That is, we can still look at gender mean differences as we would in a one-way design with gender as the independent variable and we can still look at city of residence differences as we would with residence as the independent variable. But what we gain in treating the study as a two-way design is the ability to assess the *interaction* of the two independent variables. This interaction effect reflects the possible differential effects of the unique combinations of the levels of the independent variables, and we can evaluate this effect only when we have identified the independent variables within the context of a factorial design. We will discuss this interaction effect in greater detail later in this chapter.

8.1.3 CONFIGURING THE TWO-WAY DESIGN

Using generic labels (*A* and *B*) for the names of the independent variables, we show several different two-way designs in Figure 8.1. It is common for two-way designs to be drawn in a row × column matrix form as we have done in the figure. The rows contain the levels of variable *A* and the columns contain the levels of variable *B*.

Figure 8.1A is a 2 × 2 design (where the "×" is read as the word *by*) in that we have two levels of *A* (a_1 and a_2) and two levels of *B* (b_1 and b_2). Generally, we designate the design in terms of (the number of levels of *A*) × (the number of levels of *B*). Thus, Figure 8.1B depicts a 2 × 3 design, Figure 8.1C depicts a 3 × 4 design, and Figure 8.1D depicts a 2 × 5 design.

8.2 A NUMERICAL EXAMPLE

We will illustrate the general principles of a two-way between-subjects design with a simplified hypothetical example data set. Assume that we

	Large		Small		Rural		
	15		38		42		
	20		43		48		
Female	10	M = 15	33	M = 38	36	M = 42	M = 31.67
	21		46		49		
	9		30		35		
	16		28		10		
	22		34		16		
Male	10	M = 16	22	M = 28	4	M = 10	M = 18
	21		35		15		
	11		21		5		
	M = 15.5		M = 33		M = 26		

Figure 8.2 Data for example.

are interested in feelings of loneliness experienced by adults who are living alone. We sample individuals from three different types of population centers: large metropolitan cities, small municipalities or towns, and rural communities (coded as 1, 2, and 3 in the data file). Recognizing that gender might be an important factor here, we separately code for female and male participants (coded as 1 and 2, respectively).

Further assume that participants are randomly selected from the population of singles representing the respective demographic category. They are each asked to complete a loneliness inventory, the score from which will serve as the dependent variable. A score of zero means that a person has experienced no loneliness whatsoever, whereas the maximum score of 60 reflects repeated and intense feelings of loneliness.

The data for thirty respondents, shown in Figure 8.2, are organized into a 2 × 3 between-subjects design. Each cell contains five individuals of one gender living in one type of community. We show the cell means as well as the means for the rows (levels of *A*) and columns (levels of *B*) in the figure. For example, the average loneliness score for females living in small towns is 38.00, the overall loneliness score for all of the females in the sample is 31.67, and the average loneliness score for all of those in the sample living in rural communities is 26.00.

8.3   PARTITIONING THE VARIANCE INTO ITS SOURCES

8.3.1   THE SUMMARY TABLE

As was true of the one-way between-subjects design, the total variance of the dependent variable in a two-way design is partitioned into its separate sources. These sources are the ones associated with the independent variables and their interaction, which are usually referred to as the effects, and the variance which is not accounted for, which is usually referred to as within-groups variance or error variance. We have computed the ANOVA for the data shown in our numerical example. The summary table of the results may be seen in Table 8.1.

As can be seen from Table 8.1, the total variance is partitioned into four parts. The partitions representing single independent variables are known

Table 8.1. Summary table for two-way between-subjects design

Source	SS	df	MS	F	η^2
Gender (A)	1,400.833	1	1,400.833	38.032*	.267
Residence (B)	1,551.667	2	775.833	21.063*	.296
Gender × Residence ($A \times B$)	1,411.667	2	705.833	19.163*	.269
Error S/AB	884.000	24	36.833		
Total	5,248.167	29			

* $p < .05$

as *main effects*; we thus evaluate the main effect for gender and the main effect for residence. There is also an effect representing the combinations of the levels of these variables. Such an effect is known as an *interaction*; here, we are dealing with the Gender × Residence interaction. We will discuss these three effects in a moment. Finally, we have the unaccounted source of variance known as error variance or within-groups variance.

8.3.2   DEGREES OF FREEDOM

The procedure for computing the sums of squares with a hand calculator will be covered later in this chapter. Degrees of freedom for the total variance is always computed in the same manner:

$$df_{\text{Total}} = \text{total number of observations} - 1.$$

Degrees of freedom for the main effects are determined in the same way as we described in Chapter 6:

$$df_{\text{Main Effect}} = \text{number of levels of the independent variable} - 1.$$

Thus, gender (with two levels) is associated with 1 $df(a - 1 = 2 - 1 = 1)$ and residence (with three levels) is associated with 2 $df(b - 2 = 3 - 1 = 2)$.

Degrees of freedom for the interaction are computed by multiplying the degrees of freedom for each of the main effects:

$$df_{A \times B} = (df_A)(df_B).$$

In our example, gender has 1 *df* and residence has 2 *df*. Multiplying these two values yields 2 *df* for the interaction. Subtracting the degrees of freedom of the two main effects and the interaction from the total degrees of freedom gives us the degrees of freedom associated with the error term.

8.3.3   MEAN SQUARE, F RATIO, AND ETA SQUARED

These values are obtained in the same manner that we described in Chapter 6. The mean square values are computed for each source of variance by dividing the sum of squares for that source by its degrees of freedom. F ratios are computed for the two main effects and the interaction by dividing the respective mean square values by the mean square value for the error term. Eta squared values are determined by dividing the sum of squares for a statistically significant effect by the sum of squares for the total variance.

8.4   EFFECTS OF INTEREST IN THIS DESIGN

In this two-way design, there are three effects of interest to us: the main effect of gender, the main effect of community residence size, and the unique combinations of the levels of gender and residence known as the Gender $\times$ Residence interaction. We look at each in turn.

8.4.1   MAIN EFFECT OF GENDER

One of the effects in which we are interested is that of gender. We want to know if there was a statistically significant difference in feelings of loneliness between the female and male respondents. We thus compare the overall mean of loneliness for females (i.e., across all of the residence levels) with that of males. As can be seen from Figure 8.2, we are thus comparing the mean of 31.67 with the mean of 18.00. The ANOVA output indicates that the F ratio for gender was statistically significant, $F(1, 24) = 38.03$, $p < .05$, $\eta^2 = .267$. With only two levels of gender, we know that the female respondents experienced significantly more loneliness than the male respondents.

8.4.2   MAIN EFFECT OF RESIDENCE

A second effect in which we are interested is the main effect of residence. We thus compare the overall means of participants living in large cities, small towns, and rural communities. It can be seen see from Figure 8.2 that these means are 15.50, 33.00, and 26.00, respectively. The ANOVA output shows a statistically significant F ratio for this main effect, $F(2, 24) = 21.06$, $p < .05$, $\eta^2 = .296$. We therefore presume that at least one pair of means differs significantly but do not know precisely which means differ from which others based on the F ratio information. In this situation, we would need to conduct a post-ANOVA test (post hoc tests, planned comparisons) as described in Chapter 7 to determine which group means differ significantly from which others.

8.4.3   INTERACTION OF THE TWO INDEPENDENT VARIABLES

A third effect that we evaluate in the design is that of the Gender $\times$ Residence interaction. Our focus for this effect is on the cell means, that is, the means representing the unique combinations of the independent variables. The ANOVA output indicates that we have a statistically significant F ratio for the interaction effect, $F(2, 24) = 19.16$, $p < .05$, $\eta^2 = .269$. To explicate this effect we must examine the relationships of the cell means. This involves a post-ANOVA procedure known as *simple effects* analysis. The interest here is to "simplify" the interaction so that it can be properly interpreted.

8.5 THE INTERACTION EFFECT

8.5.1 THE NATURE OF AN INTERACTION

In a general sense, an interaction is said to exist when the pattern of means across one level of an independent variable differs from the pattern across another level of that independent variable. One can also think of this as indicating that the scores on the dependent variable are affected by the particular combination of the levels of our independent variables. It is partially for this reason that it is common to present a line graph of the interaction (even though the independent variables are often categorical variables). Such a presentation usually makes it easier for readers to consolidate the information contained in the interaction. Our graph of the interaction for our numerical example is shown in Figure 8.3.

Our plot is structured in a manner analogous to the way we presented the design in Figure 8.2. The *y* axis always represents the dependent variable – loneliness in this study. We usually try to place the independent variable that is more continuous on the *x* axis but in many studies that turns out to be an arbitrary decision. When the choice as to which independent variable to place on the *x* axis is arbitrary, we suggest placing the independent variable with the most treatments or levels on the *x* axis, thus requiring fewer lines to be drawn in the figure. Here, we have placed the three levels of residence (the columns in Figure 8.2) on the *x* axis, separating them with equal intervals for the sake of aesthetics. The rows in Figure 8.2 represent the two levels of gender, and we have drawn a line in the graph for each of them.

Figure 8.3 Interaction of gender and residence.

To grasp the idea of an interaction, we need to separately examine the pattern of the means for each level of one of the independent variables. It is easiest to describe the strategy we would use for gender because in Figure 8.3 we have drawn separate lines for females and males. We observe the pattern for females: loneliness steadily increases as we look at those who reside in large cities, small towns, and rural communities. That pattern is noticeably different for the male respondents. Loneliness scores are higher for those residing in small towns compared to large cities and rural communities.

Such a difference in pattern – such unique standing of the cell means in the design – as we see in Figure 8.3 is indicative of an interaction. Had the lines been parallel, there would be no difference between the two patterns and there would be no interaction. That is, each cell mean for males would have stood in the same relationship to the other cell means for males as the matching cell mean stood with respect to the other female cell means. Note that the lines need not cross for there to be an interaction – it is quite sufficient for them to be nonparallel for a statistically significant interaction to be obtained.

8.5.2 SIMPLE EFFECTS ANALYSIS

An interaction is an overall assessment of the relationship of the cell means. Just as main effects with three or more levels of the independent variable are omnibus effects that need to be further explicated by post hoc tests or planned comparisons, so too must the omnibus interaction effect be simplified. The need for this simplification process can be illustrated by examining the plot in Figure 8.3.

Consider the female participants. The curve steadily rises from large city residence to rural communities, but it would be of interest to determine if, for example, those in small towns are significantly lonelier than those in large cities or if those living in rural communities are significantly lonelier than those living in small towns.

For males, it is likely (although we do not know for sure at this stage) that those living in rural communities are significantly less lonely than those living in small towns. We also do not currently know if there is a statistically significant difference between those living in large cities and those living in small towns, nor can we tell if there is a statistically significant difference between those living in large cities and those living in rural communities.

The same ambiguity is encountered if we look at the gender comparisons in Figure 8.3. Is the female–male difference statistically significant for those residing in large cities? The difference is minimal and is not likely to be significant. Is there a gender difference for those residing in rural communities? That difference is relatively large and is probably significant. How about female–male differences in small towns? The answer to that cannot be determined from the graph.

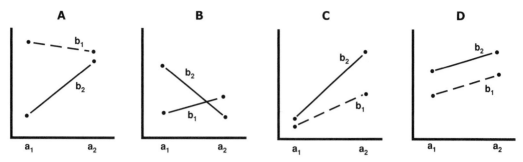

Figure 8.4 Examples of interactions: A, B, and C show interactions that likely would be statistically significant (the functions are not parallel) but D shows an instance where the interaction would not be significant (the functions are parallel).

An analysis of the simple effects for the interaction would answer all of these questions. The general strategy that we use in generating the simple effects is to isolate the levels of a variable and compare the means of those groups. For example, we would isolate the levels of gender. Doing that, we would focus on female respondents and compare the groups of different residents (large city, small town, rural community). We would then do the same for males. Additionally, we could isolate the different levels of residence and compare females and males in each type of community. Knowing which group means differ from which others would permit us to fully explicate the omnibus interaction effect. We will show you how to perform these simple effects analyses later in the chapter.

In general, interactions can usually be diagnosed by examining the plot of the cell means. We have provided schematized plots in Figure 8.4 to illustrate this. The functions in Figure 8.4A, Figure 8.4B, and Figure 8.4C depart from a parallel relationship and would be indicative of an interaction.

- In Figure 8.4A the function for b_1 slants slightly down and the function for b_2 slants sharply up. It is likely that the two functions would not differ significantly at a_2 but would differ significantly at a_1.
- In Figure 8.4B, the functions cross. At a_1 we are likely to find b_2 of significantly greater magnitude than b_1; at a_2, it is possible that b_1 and b_2 may not differ significantly.
- In Figure 8.4C both functions rise but b_2 is rising more sharply than b_1. It is likely that the two functions do not differ significantly at a_1 but would be significantly different at a_2.

Figure 8.4D pictures the two functions as parallel. This would tell us in advance of examining the ANOVA summary table that the interaction effect would not achieve statistical significance.

8.6   PRECEDENCE OF EFFECTS: INTERACTIONS SUPERCEDE MAIN EFFECTS

There is a general rule governing the precedence in which we interpret and report the results of factorial analyses: Interactions supersede main effects. This is true because interactions contain a more complete level of detail than main effects. In our worked example, we obtained a significant main effect of gender – the overall "gender" means revealed that females experienced more loneliness than males. But because the interaction was statistically significant as well, we can see that the main effect is in this instance an overgeneralization. Yes, females by and large reported more loneliness than males, but not universally. For those residing in large cities, male respondents actually reported slightly more loneliness than females. Although this latter difference may not be statistically significant, the point is that the generalization based on the main effect does not hold for the large city environment.

It is further true that while females reported more loneliness than males in both small towns and rural communities, the magnitude of this difference is very different. Once again, the generalization based on the main effect is too broad to capture the nuances of the results.

Because an interaction indicates that the lines are not parallel and that studying the unique combinations of the cells is critical, it will always be true that the details of the relationships existent at the cell mean level will always be more informative regarding the subtleties of the group differences than the main effects. Thus, with a significant interaction, we ordinarily spend most of our energy explicating the cell means based on our simple effects analysis and tend to provide more cursory treatment of the main effects.

8.7   COMPUTING AN OMNIBUS TWO-FACTOR BETWEEN-SUBJECTS ANOVA BY HAND

8.7.1   NOTATION

The computational process for a two-factor between-subjects ANOVA is a direct extension of the computational procedures we covered in Chapter 6 for the single-factor between-subjects ANOVA design. We will continue with the hypothetical example introduced in this chapter. The independent variables are size of residence community (Factor A) and gender (Factor B). The residence independent variable has three levels ($a_1 =$ large, $a_2 =$ small, $a_3 =$ rural) and gender has two levels ($b_1 =$ female, $b_2 =$ male). The dependent variable is a score (from 0 to 60) on a loneliness inventory. This combination of independent variables produces a 3×2 factorial and is depicted in the first matrix of Table 8.2, labeled **Design**.

Respondents' scores on the loneliness inventory (dependent variable) can be arranged into six columns or treatment combinations of the various levels of the two independent variables. Such an arrangement can be seen

Table 8.2. Notation for a 3×2 factorial design

Design				Factor A	
			a_1	a_2	a_3
	Factor B	b_1			
		b_2			

Data Set (Y_{ijk})

$a_1 b_1$	$a_2 b_1$	$a_3 b_1$	$a_1 b_2$	$a_2 b_2$	$a_3 b_2$
Y_{111}	Y_{121}	Y_{131}	Y_{112}	Y_{122}	Y_{132}
Y_{211}	Y_{221}	Y_{231}	Y_{212}	Y_{222}	Y_{232}
Y_{311}	Y_{321}	Y_{331}	Y_{312}	Y_{322}	Y_{332}
Y_{411}	Y_{421}	Y_{431}	Y_{412}	Y_{422}	Y_{432}
Y_{511}	Y_{521}	Y_{531}	Y_{512}	Y_{522}	Y_{532}
AB_{11}	AB_{21}	AB_{31}	AB_{12}	AB_{22}	AB_{32}

AB Matrix of Sums

			Factor A			
			a_1	a_2	a_3	Sum
	Factor B	b_1	AB_{11}	AB_{21}	AB_{31}	B_1
		b_2	AB_{12}	AB_{22}	AB_{32}	B_2
		Sum	A_1	A_2	A_3	T

in the second matrix of Table 8.2, labeled **Data Set (Y_{ijk})**. We refer to the series or observations in this matrix collectively as Y_{ijk} to denote three pieces of information. The subscript i designates the ordinal position of each participant or case within a treatment combination. The subscript j denotes the level of the A treatment and the subscript k represents the level of the B treatment. This notational device reminds us of the exact treatment combination each participant receives.

The six columns of raw data in the Y_{ijk} matrix can each be summed to produce treatment combination totals referred to collectively as AB_{jk} (again, the j subscript refers to the level of the A treatments and the k subscript signifies the level of the B treatments). These six AB_{jk} treatment combination sums can be configured into their own matrix, which we depict in Table 8.2 as the **AB Matrix of Sums**. These AB treatment combination sums can be summed vertically (to obtain the A treatment sums) or horizontally (to obtain the B treatment sums) and will provide the ingredients for our future main effects analyses. The AB matrix of sums can be converted into an AB matrix of means by dividing each AB_{jk} treatment combination sum by n. We have done so at the bottom of Table 8.3.

Before proceeding with the computation of sums of squares, degrees of freedom, mean squares, and F ratios, we need to compute some preliminary calculations (e.g., treatment combination sums, squared sums of the treatment combinations, and treatment combination means, and standard deviations), which are depicted at the bottom of the Y_{ijk} matrix

Table 8.3. Preliminary calculations for a 3×2 ANOVA data set (Y_{ijk} matrix)

		Large female $a_1 b_1$	Small female $a_2 b_1$	Rural female $a_3 b_1$	Large male $a_1 b_2$	Small male $a_2 b_2$	Rural male $a_3 b_2$
		15	38	42	16	28	10
		20	43	48	22	34	16
		10	33	36	10	22	4
		21	46	49	21	35	15
		9	30	35	11	21	5
ΣAB	ΣAB_{jk}	75	190	210	80	140	50
ΣY^2	ΣY_{ijk}^2	1,247	7,398	8,990	1,402	4,090	622
	$\overline{Y}_{jk}$	15.00	38.00	42.00	16.00	28.00	10.00
	s_{jk}	5.52	6.67	6.52	5.52	6.52	5.52

AB matrix (sums):
Size of residence community

Gender	Large (a_1)	Small (a_2)	Rural (a_3)	Marginal sum
Female (b_1)	75	190	210	475
Male (b_2)	80	140	50	270
Marginal sum	155	330	260	745

AB matrix (means):
Size of residence community

Gender	Large (a_1)	Small (a_2)	Rural (a_3)	Marginal sum
Female (b_1)	15.00	38.00	42.00	31.67
Male (b_2)	16.00	28.00	10.00	18.00
Marginal sum	15.50	33.00	26.00	

of Table 8.3. Also provided in Table 8.3 is an *AB* matrix of sums that we will also be using in our computation of the sums of squares. The *AB* matrix of means (at the bottom of Table 8.3) is useful when plotting treatment means or examining simple effects analyses.

We are now ready to proceed with the analysis. You can examine Table 8.3 to see where the actual values we are using in the computational formulas come from. We owe a special debt of thanks for the early work by Keppel (1991), Keppel et al. (1992), and Keppel and Wickens (2004) on the development of the computational formulas we are about to use.

8.7.2 CALCULATING SUM OF SQUARES

There are five sums of squares to be calculated in a two-way between-subjects ANOVA:

- Sum of squares main effect of Factor *A* (SS_A).
- Sum of squares main effect of Factor *B* (SS_B).
- Sum of squares interaction effect of Factors *A* and *B* ($SS_{A \times B}$).

- Sum of squares within groups or error ($SS_{S/AB}$) in which the subscript S/AB indicates subjects within each level of the AB treatment combinations.
- Sum of squares total (SS_T).

The formulas for these five sums of squares and the summary of their calculations based on the data in Table 8.3 are as follows:

$$SS_A = \frac{\Sigma A^2}{(b)(n)} - \frac{T^2}{(a)(b)(n)}$$

$$= \frac{(155)^2 + (330)^2 + (260)^2}{(2)(5)} - \frac{(745)^2}{(3)(2)(5)}$$

$$= \frac{24{,}225 + 108{,}900 + 67{,}600}{10} - \frac{555{,}025}{30}$$

$$= 20{,}052 - 18{,}500.83 = \boxed{1{,}551.17}; \qquad (8.1)$$

$$SS_B = \frac{\Sigma B^2}{(a)(n)} - \frac{T^2}{(a)(b)(n)}$$

$$= \frac{(475)^2 + (270)^2}{(3)(5)} - \frac{(745)^2}{(3)(2)(5)}$$

$$= \frac{225{,}625 + 72{,}900}{15} - \frac{555{,}025}{30}$$

$$= 19{,}901.67 - 18{,}500.83 = \boxed{1{,}400.84}; \qquad (8.2)$$

$$SS_{A \times B} = \frac{\Sigma(AB)^2}{n} - \frac{\Sigma A^2}{(b)(n)} - \frac{\Sigma B^2}{(a)(n)} + \frac{T^2}{(a)(b)(n)}$$

$$= \frac{(75)^2 + (190)^2 + (210)^2 + (80)^2 + (140)^2 + (50)^2}{5}$$

$$- 20{,}052 - 19{,}901.67 + 18{,}500.83$$

$$= 22{,}865 - 20{,}052 - 19{,}901.67 + 18{,}500.83 = \boxed{1{,}412.16}; \qquad (8.3)$$

$$SS_{S/AB} = \sum Y^2 - \frac{\Sigma(AB)^2}{n}$$

$$= 1{,}247 + 7{,}398 + 8{,}990 + 1{,}402 + 4{,}090 + 622 - 22{,}865$$

$$= 23{,}749 - 22{,}865 = \boxed{884.00}; \qquad (8.4)$$

$$SS_T = \sum Y^2 - \frac{T^2}{(a)(b)(n)}$$

$$= 23{,}749 - 18{,}500.83 = \boxed{5{,}248.17}. \qquad (8.5)$$

8.7.3   CALCULATING DEGREES OF FREEDOM

Below are the formulas for the degrees of freedom associated with each sum of squares, and the simple computations involved based on our numerical example:

$$df_A = a - 1 = 3 - 1 = \boxed{2}$$
$$df_B = b - 1 = 2 - 1 = \boxed{1}$$
$$df_{A \times B} = (a - 1)(b - 1) = (3 - 1)(2 - 1) = (2)(1) = \boxed{2}$$
$$df_{S/AB} = (a)(b)(n - 1) = (3)(2)(5 - 1) = (3)(2)(4) = \boxed{24}$$
$$df_T = (a)(b)(n) - 1 = (3)(2)(5) - 1 = \boxed{29}.$$

8.7.4   CALCULATING MEAN SQUARES AND F RATIOS

Mean squares are calculated by dividing each sum of squares by its respective degrees of freedom. Note that we do not calculate a mean square total value. Three F ratios are calculated in a two-way between-subjects ANOVA. These three F ratios assess the three sources of between-subjects variability found in the study. Thus, F_A assesses the main effect of Factor A, F_B assesses the main effect of Factor B, and $F_{A \times B}$ assesses the interaction effect. Each F ratio is formed by dividing the respective mean square by the mean square error ($MS_{S/AB}$). The calculation of these mean squares and F ratios follows:

$$MS_A = \frac{SS_A}{df_A} = \frac{1{,}551.17}{2} = \boxed{775.59}$$

$$MS_B = \frac{SS_B}{df_B} = \frac{1{,}400.84}{1} = \boxed{1{,}400.84}$$

$$MS_{A \times B} = \frac{SS_{A \times B}}{df_{A \times B}} = \frac{1{,}412.16}{2} = \boxed{706.08}$$

$$MS_{S/AB} = \frac{SS_{S/AB}}{df_{S/AB}} = \frac{884.00}{24} = \boxed{36.83}$$

$$F_A = \frac{MS_A}{MS_{S/AB}} = \frac{775.59}{36.83} = \boxed{21.06}$$

$$F_B = \frac{MS_B}{MS_{S/AB}} = \frac{1{,}400.84}{36.83} = \boxed{38.04}$$

$$F_{A \times B} = \frac{MS_{A \times B}}{MS_{S/AB}} = \frac{706.08}{36.83} = \boxed{19.17}.$$

8.7.5   EVALUATING THE F RATIOS

We test the null hypothesis for each of our three observed (calculated) F ratios by evaluating (or comparing) each F value with critical values of F

(see Appendix C) at the following degrees of freedom: $(df_{effect}, df_{S/AB})$, at a particular alpha level. Thus, for the main effect of Factor A, our F_A value was 21.06 and is evaluated with (2, 24) degrees of freedom. The critical value of F at these degrees of freedom (at the .05 level) is 3.40. Because our F_A is greater than (or equal to) the critical value, we reject the null and conclude that size of residential community affects loneliness scores.

Likewise, for the main effect of Factor B, our F_B was 38.04 and is evaluated with (1, 24) degrees of freedom. The critical value of F is 4.26 at the .05 level. Because our F_B is greater than the critical value, we reject the null and conclude that gender also affects loneliness appraisals.

Lastly, we evaluate the interaction effect ($F_{A \times B} = 19.17$) with (2, 24) degrees of freedom. We note that our $F_{A \times B}$ is greater than the critical value of F (3.40); thus, we reject the null and conclude that the two independent variables (residence and gender) combine to produce a statistically significant unique joint effect. We will next turn our attention to the computational details of these statistically significant main effects and interactions.

8.8   COMPUTING SIMPLE EFFECTS BY HAND

Calculating the simple effects of a two-factor design is comparable to conducting several single-factor analyses, one at each level of one of the independent variables. The goal of such analyses is to try and isolate or identify the source of the statistically significant interaction effect found in the omnibus analysis. Figure 8.5 depicts the relationship between a

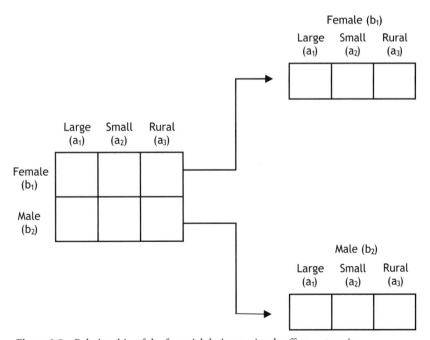

Figure 8.5   Relationship of the factorial design to simple effects extraction.

full factorial design and its constituent components that comprise the simple effects analyses. In Figure 8.5, the full factorial arrangement of the two independent variables (residential community size and gender) is broken down into two single factor analyses of residential community size, collapsed across or conducted at each level of gender (b_1: females and b_2: males). Such an analysis is referred to as the simple effects analysis of Factor A at b_1 and Factor A at b_2. An alternative approach, which we will not describe in detail, is to conduct simple effects analyses by collapsing across levels of Factor A. Such an approach would be referred to as the simple effects analysis of Factor B at a_1, Factor B at a_2, and Factor B at a_3. Either simple effects approach (A at b_k or B at a_j) will provide the researcher with the necessary information to interpret the interaction. Computational details on the latter approach can be found in Keppel et al. (1992) and Keppel and Wickens (2004).

Let us now turn our attention to conducting the simple effects of Factor A at b_1 and A at b_2 based on our previous numerical example examining loneliness as a function of residential community size and gender. Note that we will be modifying slightly our formulas for sum of squares, degrees of freedom, and mean squares because we are collapsing our analyses across each level of Factor B (i.e., b_1 and b_2) to conduct two single-factor ANOVAs.

The following computational steps will be based on the sums found in the AB Matrix of Sums in Table 8.3. We begin with the first row of sums (b_1) to conduct the simple effects of A at b_1, and then proceed to the second row of sums (b_2) to conduct the simple effects of A at b_2.

8.8.1   SIMPLE EFFECTS OF A AT b_1

$$SS_{A\,at\,b1} = \frac{\Sigma(AB_{j1})^2}{n} - \frac{(B_1)^2}{(a)(n)}$$

$$= \frac{(75)^2 + (190)^2 + (210)^2}{(5)} - \frac{(475)^2}{(3)(5)}$$

$$= \frac{85,825}{5} - \frac{225,625}{15}$$

$$= 17,115 - 15,041.67 = \boxed{2,123.33} \tag{8.6}$$

$$df_{A\,at\,b1} = a - 1 = 3 - 1 = \boxed{2} \tag{8.7}$$

$$MS_{A\,at\,b1} = \frac{SS_{A\,at\,b_1}}{df_{A\,at\,b1}} = \frac{2,123.33}{2} = \boxed{1,061.67} \tag{8.8}$$

$$F_{A\,at\,b1} = \frac{MS_{A\,at\,b_1}}{MS_{S/AB}} = \frac{1,061.67}{36.83} = \boxed{28.83}. \tag{8.9}$$

We evaluate our $F_{A\,at\,b1}$ with $df = df_{A\,at\,b1}$, and $df_{S/AB} = (2, 24)$. The critical value of F (see Appendix C) at the 5 percent level is 3.40. Because our observed F value exceeds the critical value of F, we reject the null hypothesis and conclude that there are statistically significant differences among the female residential size loneliness scores.

This statistically significant F should be followed up with planned or unplanned (post hoc) comparisons. We refer you to our previous discussions in Chapter 7 for the computational details of these analyses. Remember to substitute $MS_{S/AB}$ and $df_{S/AB}$ for $MS_{S/A}$ and $df_{S/A}$ in the Tukey and LSD formulas of Equations 7.5 and 7.6.

8.8.2 SIMPLE EFFECTS OF A AT b_2

$$
\begin{aligned}
SS_{A \text{ at } b_2} &= \frac{\Sigma(AB_{j2})^2}{n} - \frac{(B_2)^2}{(a)(n)} \\
&= \frac{(80)^2 + (140)^2 + (50)^2}{(5)} - \frac{(270)^2}{(3)(5)} \\
&= \frac{28{,}500}{5} - \frac{72{,}900}{15} \\
&= 5{,}700 - 4{,}860 = \boxed{840.00}
\end{aligned}
\tag{8.10}
$$

$$
df_{A \text{ at } b_2} = a - 1 = 3 - 1 = \boxed{2}
\tag{8.11}
$$

$$
MS_{A \text{ at } b_2} = \frac{SS_{A \text{ at } b_2}}{df_{A \text{ at } b_2}} = \frac{840.00}{2} = \boxed{420.00}
\tag{8.12}
$$

$$
F_{A \text{ at } b_2} = \frac{MS_{A \text{ at } b_2}}{MS_{S/AB}} = \frac{420.00}{36.83} = \boxed{11.40}.
\tag{8.13}
$$

Our evaluation criteria for this simple effect analysis remains the same as the previous analysis because the degrees of freedom remain unchanged. Thus, we conclude that this observed F is statistically significant, indicating loneliness differences among the three residential community groups for males. Again, planned or post hoc comparisons should be conducted next to determine what is producing the interaction effect observed with our omnibus ANOVA.

One last computational note: If the interaction in your omnibus ANOVA is not statistically significant, but one or both of your main effects is statistically significant, you could perform the following LSD post hoc tests on the marginal sums of A or B.

$$
\begin{aligned}
&D_{\text{LSD}-\text{A}}(\text{significant main effect of A}) \\
&= \sqrt{F(1, df_{S/AB})}\sqrt{(2)(b)(n)(MS_{S/AB})}.
\end{aligned}
\tag{8.14}
$$

$$
\begin{aligned}
&D_{\text{LSD}-\text{B}}(\text{significant main effect of B}) \\
&= \sqrt{F(1, df_{S/AB})}\sqrt{(2)(a)(n)(MS_{S/AB})}.
\end{aligned}
\tag{8.15}
$$

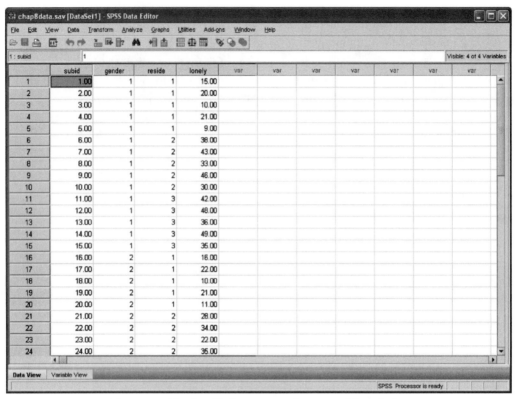

Figure 8.6   The data file for a two-way between-subjects design.

8.9   PERFORMING THE OMNIBUS ANALYSIS IN SPSS

8.9.1   STRUCTURING THE DATA FILE

The SPSS data file containing the variables for the analysis of the numerical example we have been using throughout this chapter is shown in Figure 8.6. That data file is structured as follows. The first variable (column) is a subject identification code that we have named **subid**. Next are the independent variables. We have given them arbitrary codes (values) to represent our categories. For **gender**, females are coded as 1 and males are coded as 2; for **reside**, large city, small town, and rural community are coded as 1, 2, and 3, respectively. The final column holds the dependent variable for the analysis. We have named this variable **lonely**. In this column are the scores on the loneliness inventory for each participant.

8.9.2   STRUCTURING THE ANALYSIS

We will use the SPSS **GLM Univariate** module to perform the analysis. From the main menu, select **Analyze ➔ General Linear Model ➔ Univariate** to reach the main dialog window shown in Figure 8.7. Click **lonely**

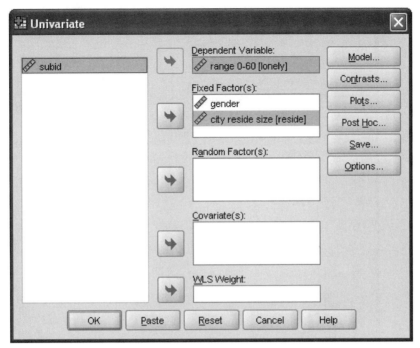

Figure 8.7   The main **GLM Univariate** dialog window.

over to the **Dependent Variable** panel and click **gender** and **reside** over to the **Fixed Factor(s)** panel. We will first run the omnibus analysis to determine which effects are significant. Based on those results we will then determine which post-ANOVA analyses we will perform.

Click the **Options** pushbutton to reach the **Options** dialog window shown in Figure 8.8. The top half of the window displays the **Estimated Marginal Means**. Such means are unweighted in that they represent the average of the group means without taking into account the sample sizes on which those means were based. SPSS presents these in separate tables and uses these estimated marginal means to plot the interaction. We will click over the two main effects and the interaction (**gender, reside, gender∗reside**) from the left panel labeled **Factor(s) and Factor Interactions** to the right panel labeled **Display Means for**.

The bottom half of the **Options** window displays the observed means. Under the **Display** heading, we check **Descriptive statistics** to obtain the observed means and standard deviations of the groups. It is also necessary to determine if the data meet the homogeneity of variance assumption, that is, the assumption that the conditions in the study (the combinations of the levels of the independent variables) have equal variances on the dependent variable), and so we check **Homogeneity tests**.

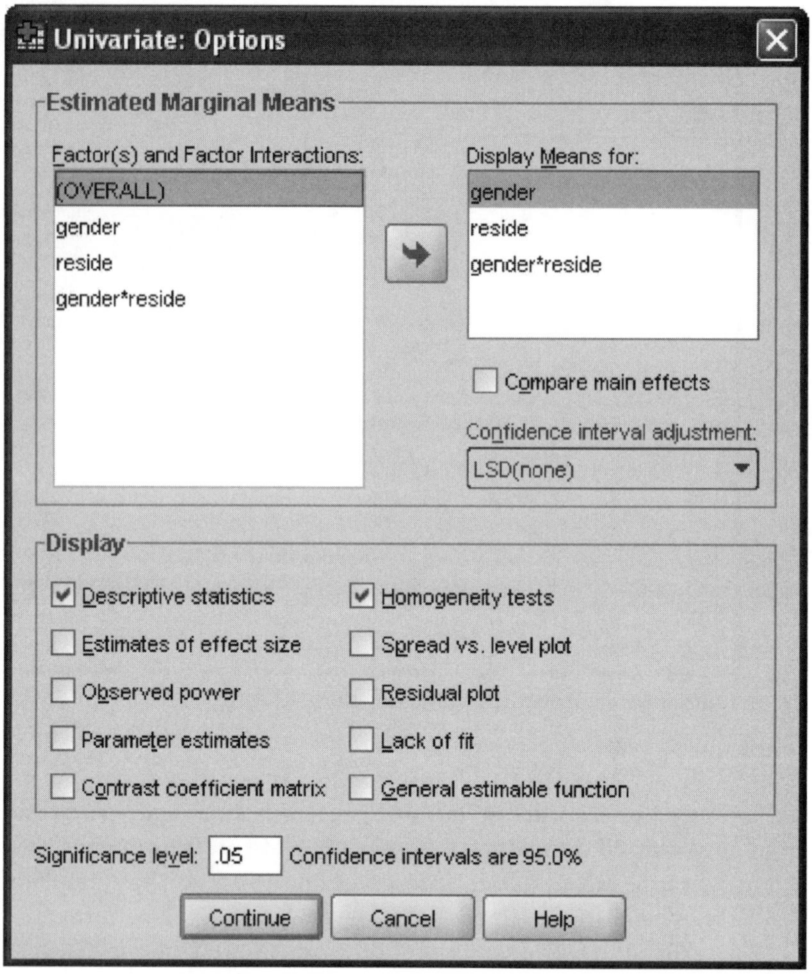

Figure 8.8   The **GLM Univariate Options** dialog window.

Clicking the **Plots** pushbutton presents the dialog window shown in Figure 8.9. From the **Factors** panel, click **reside** over to the **Horizontal Axis** panel and click **gender** over to the **Separate Lines** panel. Then click **Add** to register the plot with SPSS. Doing so will place **gender∗reside** in the **Plots** panel at the bottom of the window. Clicking **Continue** returns us to the main dialog window. Now click **OK** to run the omnibus analysis.

8.10   SPSS OUTPUT FOR THE OMNIBUS ANALYSIS

We present here selected output of the analysis. The descriptive statistics that we requested in the **Options** dialog window is shown in Figure 8.10. Since we originally clicked over the independent variables in the main dialog window in the order **gender** and then **reside** (see Figure 8.7), SPSS kept that ordering in its output. Because of this ordering, the main rows are organized by **gender**. Within each gender level the residence sizes are

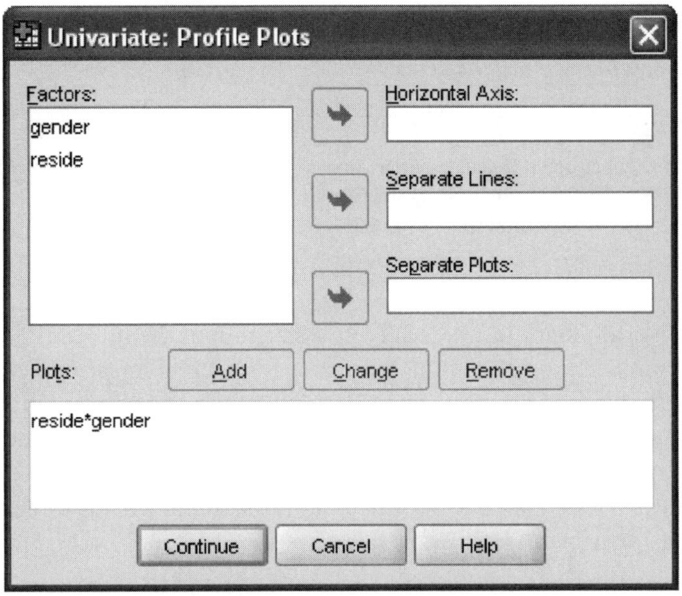

Figure 8.9 **GLM Univariate Plots** dialog window.

listed. Had we clicked them over in the other order, the main rows would have been organized by size of city and male and female means would have been shown for large cities, then small cities, then rural towns.

The **GLM Univariate** procedure uses the Levene test to determine if the assumption of homogeneity of variance has been met. From Figure 8.11 we see that the F ratio for that test is not significant; we may thus conclude that the data meet the homogeneity of variance assumption.

The summary table for the omnibus analysis is shown in Figure 8.12, and duplicates the information we presented in Table 8.1. As we already know, the main effects of **gender** and **reside** as well as the interaction are

Descriptive Statistics

Dependent Variable: range 0-60

gender	city reside size	Mean	Std. Deviation	N
female	large	15.0000	5.52268	5
	small	38.0000	6.67083	5
	rural	42.0000	6.51920	5
	Total	31.6667	13.61022	15
male	large	16.0000	5.52268	5
	small	28.0000	6.51920	5
	rural	10.0000	5.52268	5
	Total	18.0000	9.46422	15
Total	large	15.5000	5.23344	10
	small	33.0000	8.15135	10
	rural	26.0000	17.80137	10
	Total	24.8333	13.45256	30

Figure 8.10 Descriptive statistics for the cells.

Levene's Test of Equality of Error Variances[a]

Dependent Variable: range 0-60

F	df1	df2	Sig.1
.123	5	24	.986

Tests the null hypothesis that the error variance of the dependent variable is equal across groups.

a. Design: Intercept+gender+reside+gender * reside

> The Levene test yielded an F ratio that was not statistically significant. Thus, our data have met the assumption of homogeneity of variance.

Figure 8.11   Test of homogeneity of variance.

all statistically significant. Recall our discussion of the full and corrected models in Section 6.9.1 in the context of the general linear model. What we see in Figure 8.12 are the results for the full model that includes the intercept term. For our purposes, we can ignore the intercept and focus on the corrected model.

The entries in the SPSS summary table relative to what is contained in Table 8.1 may be understood as follows:

Corrected Model combines all of the effects attributable to the independent variables into one sum of squares value. In this design, adding the sums of squares associated with the main effect of **gender**, the main effect of **reside**, and the Gender × Reside interaction yields the corrected model sum of squares. The footnote on the value in the summary table is used to report the eta squared (called **R squared** by SPSS) for the corrected model (4,364.167 divided by 5,248.167 is .832). We ordinarily compute separate eta squared values for each of the effects separately as shown in Table 8.1.

Corrected Total corresponds to the total variance from Table 8.1. It is based on the "corrected" model that excludes the regression solution.

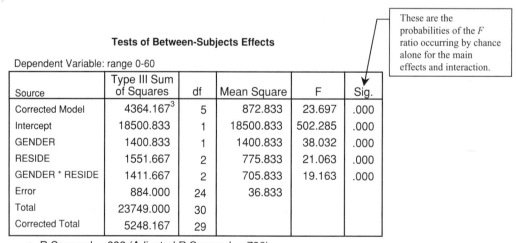

> These are the probabilities of the F ratio occurring by chance alone for the main effects and interaction.

Tests of Between-Subjects Effects

Dependent Variable: range 0-60

Source	Type III Sum of Squares	df	Mean Square	F	Sig.
Corrected Model	4364.167[3]	5	872.833	23.697	.000
Intercept	18500.833	1	18500.833	502.285	.000
GENDER	1400.833	1	1400.833	38.032	.000
RESIDE	1551.667	2	775.833	21.063	.000
GENDER * RESIDE	1411.667	2	705.833	19.163	.000
Error	884.000	24	36.833		
Total	23749.000	30			
Corrected Total	5248.167	29			

a. R Squared = .832 (Adjusted R Squared = .796)

Figure 8.12   Summary table for the omnibus analysis.

Intercept (referring to the Y intercept) represents the height of the regression function from the x axis corresponding to the value of 0 of the independent variable. It is not of particular interest to us since we are not taking a regression approach to ANOVA, but the displacement of the function is sufficient to be associated with a substantial sum of squares value.

Total represents the total sum of squares for the "uncorrected" model including the regression solution. To obtain this value, SPSS adds the sums of squares of the corrected model, the error term, and the intercept.

8.11 PERFORMING THE POST-ANOVA ANALYSES IN SPSS

The omnibus analysis indicated that all of the effects were statistically significant. In practice, we would tend not to bother with the main effects because we have obtained a statistically significant interaction in the omnibus analysis. However, to illustrate how we would approach the main effects, we will discuss them here.

Only one main effect can be unambiguously interpreted as it stands – because there were only two levels of **gender**, we know that the females felt significantly more loneliness than the males. For the other two effects (the main effect of **reside** and the interaction), we need additional information before we can render an interpretation of the results.

8.11.1 POST HOC TEST FOR THE MAIN EFFECT OF RESIDE

The independent variable **reside** has three levels – large city, small town, rural community – and in order to determine which groups differ from which, we will perform a post hoc test. From the main menu select **Analyze ➔ General Linear Model ➔ Univariate** to reach the main dialog window and configure your analysis as you did above (if you are working in the same session and have not run an intervening analysis, SPSS will retain your prior configuration).

Click the **Post Hoc** pushbutton, which will bring you to the window shown in Figure 8.13. We have checked the **Tukey** procedure in the set of tests grouped under the heading **Equal Variances Assumed** because the Levene test indicated that our data have met the assumption of homogeneity of variance. Click **Continue** to return to the main dialog window (we will ask you to click the **Paste** pushbutton in a moment).

8.11.2 SIMPLE EFFECTS ANALYSES FOR THE INTERACTION

Performing the simple effects analyses requires us to add a few words to the syntax underlying the ANOVA. In the main dialog window, click **Paste**. If the **Syntax** window is not immediately activated, find it on your desktop (it may be minimized in the Windows operating environment), and click on it to make it active. It should be similar to what is shown in Figure 8.14.

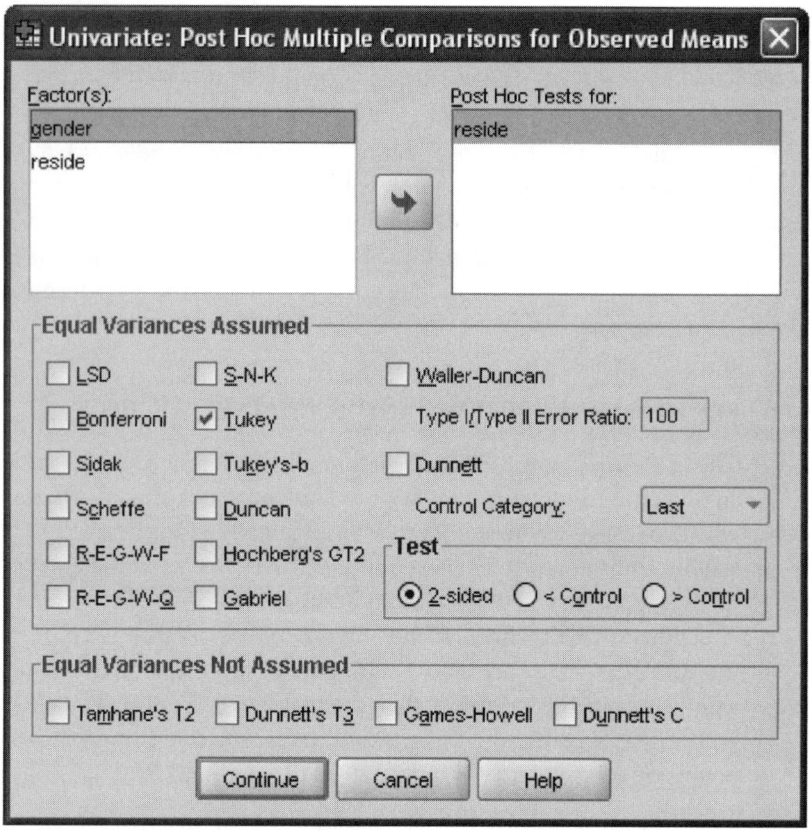

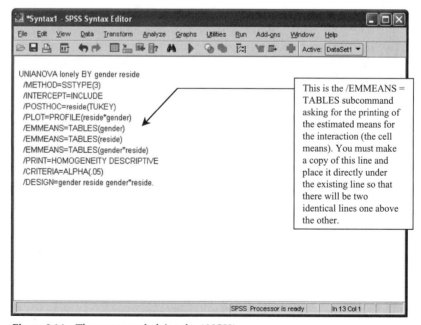

Figure 8.13 GLM Univariate **Post hoc** dialog box.

Figure 8.14 The syntax underlying the ANOVA.

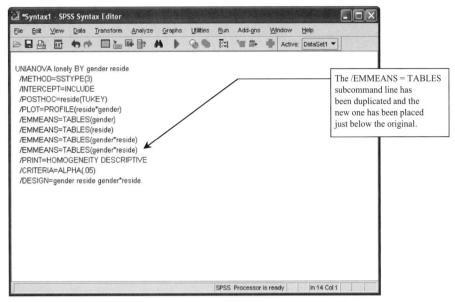

Figure 8.15 The syntax with the /**EMMEANS = TABLES** line for the interaction duplicated and placed immediately below the original.

What you see in Figure 8.14 is the command structure that actually drives the data analysis. Every time you move a variable into a panel or select a choice offered to you by clicking an option on the screen, SPSS translates your point-and-click activity into a form that it understands: at least at one level, what it actually uses to guide the analysis is this syntactic structure. Not that long ago, it was necessary for researchers to write this structure themselves in order to perform their analyses on mainframe computer systems.

Locate the indented line reading /**EMMEANS = TABLES(gender* reside)** and duplicate it. An easy way to accomplish this is to highlight the line, select **Copy** from the **edit** menu, place the cursor at the left margin for the line, and select **Paste** from the **edit** menu. When you have finished this operation the **Syntax** window should look similar to that shown in Figure 8.15.

We are now going to add a few words of syntax to the two interaction lines as shown in Figure 8.16. These two lines can be added in any order. Arbitrarily, we will write the **gender** comparison first and the **reside** comparison second.

Place your cursor at the end of the first interaction line and, after typing a space, type the following all in lower case:

```
compare (gender) adj (bonferroni).
```

This *command structure* tells SPSS to **compare** the two levels of **gender** (female and male) at each level of the other independent variable(s) named

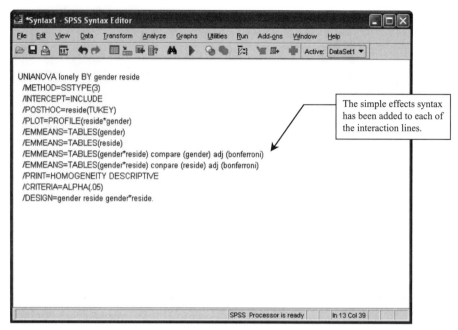

Figure 8.16   The simple effects syntax has been added to the interaction lines.

in the parentheses after **TABLES**. Since the only other variable there is **reside**, SPSS will compare females and males for each level of **reside**. The text **adj (bonferroni)** instructs SPSS to perform these comparisons of the **gender** means using a Bonferroni adjustment such that the alpha level for the comparisons remains at .05 and thus avoids alpha inflation. In the output, a **Pairwise Comparisons** table will be presented for this set of comparisons.

Now place your cursor at the end of the second interaction line and, after typing a space, type the following all in lower case:

```
compare (reside) adj (bonferroni).
```

This command structure tells SPSS to **compare** the various levels of **reside** (large city, small town, rural community) at each level of **gender**. Thus, we will be comparing the three residence sizes for the female respondents and again for the male respondents. Note that each of these comparisons involves three groups but that presents no problem for SPSS – all of the comparisons will be performed. And because we have added the **adj (bonferroni)** syntax, these comparisons will be made at a Bonferroni adjusted alpha level of .05.

To perform these post-ANOVA analyses, from the main menu select **Run → All**. SPSS will highlight the entire set of syntax and run the analysis. It will then deposit the output in an **Output** window.

These are the probabilities of the differences occurring by chance alone for the pairwise comparisons with reside as the first order.

Pairwise Comparisons

Dependent Variable: range 0-60

city reside size	(I) gender	(J) gender	Mean Difference (I-J)	Std. Error	Sig.[a]	95% Confidence Interval for Difference[a]	
						Lower Bound	Upper Bound
large	female	male	−1.000	3.838	.797	-8.922	6.922
	male	female	1.000	3.838	.797	-6.922	8.922
small	female	male	10.000[*]	3.838	.016	2.078	17.922
	male	female	−10.000[*]	3.838	.016	-17.922	-2.078
rural	female	male	32.000[*]	3.838	.000	24.078	39.922
	male	female	−32.000[*]	3.838	.000	-39.922	-24.078

Based on estimated marginal means

*. The mean difference is significant at the .05 level.

a. Adjustment for multiple comparisons: Bonferroni.

Figure 8.17   Simple effects comparisons of **gender** for each level of **reside**.

8.12   SPSS OUTPUT FOR THE POST-ANOVA ANALYSES

8.12.1   SIMPLE EFFECTS ANALYSES: GENDER

Figure 8.17 shows the **Pairwise Comparisons** for the first simple effects analysis in which we asked SPSS to compare males and females at each level of **reside**. There are three main horizontal rows in the **Pairwise Comparisons** table, the first for **large**, the next for **small**, and the last for **rural**. Each is divided into two portions but with the same information provided for each of those portions. Reading across the very top row, for those living in large cities, the female loneliness mean (15.00) minus the male loneliness mean (16.00) is −1.00. That difference is not statistically significant ($p = .797$).

Consider the next major row. For those residing in small towns, the difference between the female mean (38.00) and the male mean (28.00) is 10.00, a difference that is statistically significant ($p = .016$). We can also see that the female–male difference is significant for those residing in rural communities.

8.12.2   SIMPLE EFFECTS ANALYSES: RESIDE

Figure 8.18 shows the **Pairwise Comparisons** for the second simple effects analysis in which we asked SPSS to compare the places of residence at each level of **gender**. There are two main horizontal rows in this output, one for **females** and another for **males**. Each is divided into three portions corresponding to the three levels of **reside**.

Let's examine the first division of the first row of the table. We are working with the loneliness scores of females. The first comparison is between those living in large cities and those living in small towns. The difference between these means (large − small) is −23.00, a difference that

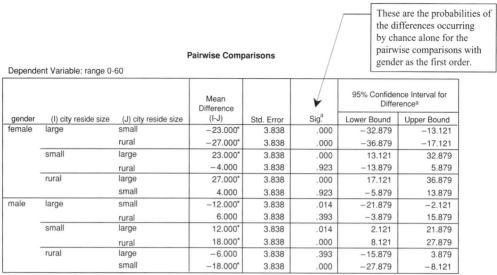

These are the probabilities of the differences occurring by chance alone for the pairwise comparisons with gender as the first order.

Pairwise Comparisons

Dependent Variable: range 0-60

gender	(I) city reside size	(J) city reside size	Mean Difference (I-J)	Std. Error	Sig.ᵃ	95% Confidence Interval for Differenceᵃ	
						Lower Bound	Upper Bound
female	large	small	−23.000*	3.838	.000	−32.879	−13.121
		rural	−27.000*	3.838	.000	−36.879	−17.121
	small	large	23.000*	3.838	.000	13.121	32.879
		rural	−4.000	3.838	.923	−13.879	5.879
	rural	large	27.000*	3.838	.000	17.121	36.879
		small	4.000	3.838	.923	−5.879	13.879
male	large	small	−12.000*	3.838	.014	−21.879	−2.121
		rural	6.000	3.838	.393	−3.879	15.879
	small	large	12.000*	3.838	.014	2.121	21.879
		rural	18.000*	3.838	.000	8.121	27.879
	rural	large	−6.000	3.838	.393	−15.879	3.879
		small	−18.000*	3.838	.000	−27.879	−8.121

Based on estimated marginal means

*. The mean difference is significant at the .05 level.

a. Adjustment for multiple comparisons: Bonferroni.

Figure 8.18 Simple effects comparisons of **reside** for each level of **gender**.

is statistically significant at the .05 level ($p = .001$). Just below that is the comparison of those living in large cities and those living in rural communities. The difference between these means (large − rural) is −27.00, a difference that is also statistically significant.

As we study the table, we can see that for females there is no reliable difference between those living in small towns and rural communities and that for males there is no reliable difference between those living in large cities and rural communities. All other comparisons yielded means that were significantly different.

8.12.3 MAPPING SIMPLE EFFECTS ANALYSES TO THE GRAPH OF THE INTERACTION

Obtaining the simple effects analyses is a necessary step in understanding the results of a study but it is not sufficient. The next step is to conceptually map these statistical tests back to the overall interaction effect. This can most easily be done by viewing the graph of the interaction (drawn in Figure 8.3) and mentally superimposing on it these simple effects outcomes. Such processing of the output is needed to write an informative and accurate results section.

Focusing first on the lines of the graph, we learned from the simple effects analysis that for females there was no significant difference between the small town and rural community respondents. Thus, we can conclude that females are significantly lonelier in small towns and rural communities than in large cities. Males, on the other hand, are significantly lonelier in small towns than in either large cities or rural communities.

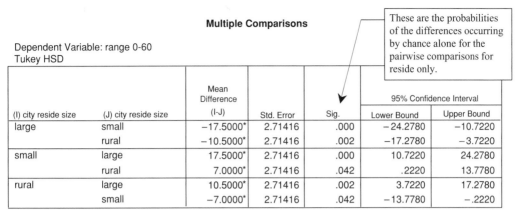

Multiple Comparisons

Dependent Variable: range 0-60
Tukey HSD

(I) city reside size	(J) city reside size	Mean Difference (I-J)	Std. Error	Sig.	95% Confidence Interval	
					Lower Bound	Upper Bound
large	small	−17.5000*	2.71416	.000	−24.2780	−10.7220
	rural	−10.5000*	2.71416	.002	−17.2780	−3.7220
small	large	17.5000*	2.71416	.000	10.7220	24.2780
	rural	7.0000*	2.71416	.042	.2220	13.7780
rural	large	10.5000*	2.71416	.002	3.7220	17.2780
	small	−7.0000*	2.71416	.042	−13.7780	−.2220

> These are the probabilities of the differences occurring by chance alone for the pairwise comparisons for reside only.

Based on observed means.

*. The mean difference is significant at the .05 level.

Figure 8.19 Post hoc analysis of the **reside** main effect.

Focusing next on the residence location of our respondents, we learned from the simple effects analysis that there was no difference between females and males in large cities but that the other two locations were associated with statistically significant differences between the genders. Of particular note is the finding that females are quite lonely in rural communities but that their male counterparts are remarkably not lonely.

8.12.4 MAIN EFFECTS POST HOC ANALYSES: RESIDE

Figures 8.19 and 8.20 present the results of the post hoc Tukey HSD test on the main effect of **reside**. We discussed the structure of these tables in the last chapter. As can be seen, each level differs significantly from each other. But, because this variable is involved in a statistically significant interaction, the story that is told by these main effect results is clearly far

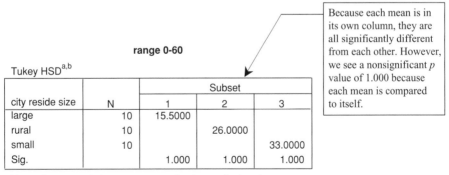

range 0-60

Tukey HSD[a,b]

city reside size	N	Subset		
		1	2	3
large	10	15.5000		
rural	10		26.0000	
small	10			33.0000
Sig.		1.000	1.000	1.000

> Because each mean is in its own column, they are all significantly different from each other. However, we see a nonsignificant *p* value of 1.000 because each mean is compared to itself.

Means for groups in homogeneous subsets are displayed.
Based on Type III Sum of Squares
The error term is Mean Square(Error) = 36.833.

a. Uses Harmonic Mean Sample Size = 10.000.

b. Alpha = .05.

Figure 8.20 Homogeneous subgroups based on the outcome of the Tukey multiple comparisons test.

from complete. In a written presentation of the results, these post hoc comparisons would contribute little if anything to the dialog and should be omitted.

8.13   PERFORMING THE OMNIBUS ANALYSIS IN *SAS ENTERPRISE GUIDE*

8.13.1   IMPORTING THE DATA FILE

The Excel file containing the data for the numerical example we have been using throughout this chapter is shown in Figure 8.21. That data file is structured as follows. The first variable (column) is a subject identification

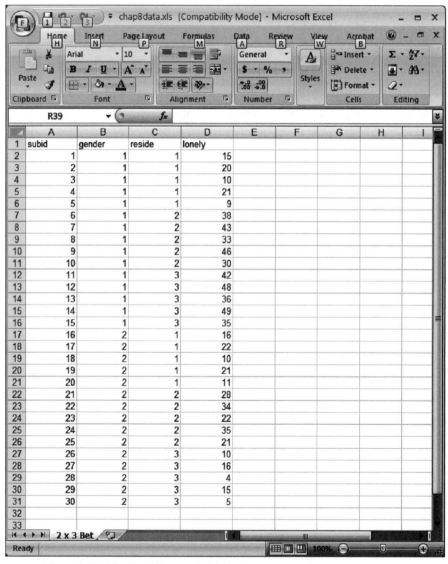

Figure 8.21   The Excel file containing the data.

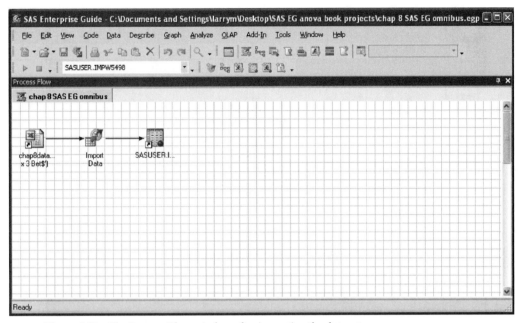

Figure 8.22 The **Process Flow** window after importing the data set.

code that we have named **subid**. Next are the independent variables. We have given them arbitrary codes (values) to represent our categories. For **gender**, females are coded as 1 and males are coded as 2; for **reside**, large city, small town, and rural community are coded as 1, 2, and 3, respectively. The final column holds the dependent variable for the analysis. We have named this variable **lonely**. In this column are the scores on the loneliness inventory for each participant. As described in Appendix B, import this data into *SAS Enterprise Guide*. When you have imported the data file, your window should resemble the one shown in Figure 8.22.

8.13.2 STRUCTURING THE ANALYSIS

Select **Analyze → ANOVA → Linear Models**. The window for this procedure opens on the **Task Roles** tab; this is highlighted in the navigation panel in the left portion of the window shown in Figure 8.23. At the top of the panel labeled **Linear Models Task Roles** there is already a place for a dependent variable being specified. Highlight **lonely** and drag it to the icon for **Dependent variables** (you can also highlight **lonely**, click the right-facing arrow, and select **Dependent variables** as the destination). Then drag **gender** and **reside** to the icon for **Classification variables** (or highlight them, click the arrow, and select your destination). When finished, your screen should look similar to the screen shown in Figure 8.24.

Click on the **Model** tab. The variables **group** and **reside** appear in the **Class and quantitative variables** panel. Highlight **gender** and click the **Main** bar to place **gender** in the **Effects** panel. Then do the same with **reside**. Finally, highlight **gender** and, while depressing the **Control** key, highlight **reside**; both variables should now be highlighted. Clicking the

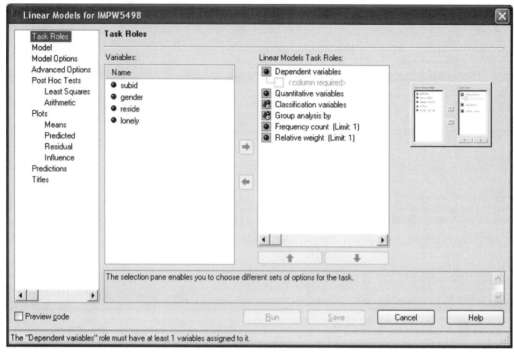

Figure 8.23   The **Task Roles** window.

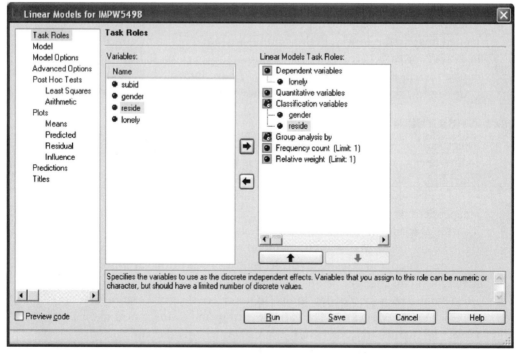

Figure 8.24   The **Task Roles** screen configured.

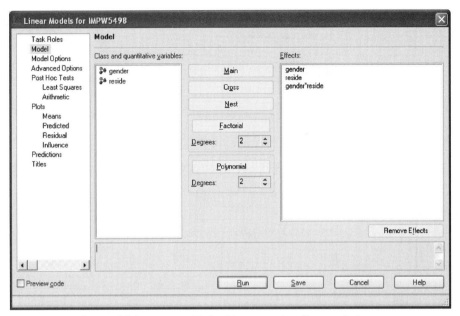

Figure 8.25 The model is now specified with two main effects and the two-way interaction.

Cross bar while the two are highlighted will cause them to be brought over to the **Effects** panel as an interaction effect. The final result of this operation is shown in Figure 8.25.

Click on the **Model Options** tab. The only specification we need for our analysis is to request output associated with the **Type III sums of squares to show**. This is shown in Figure 8.26.

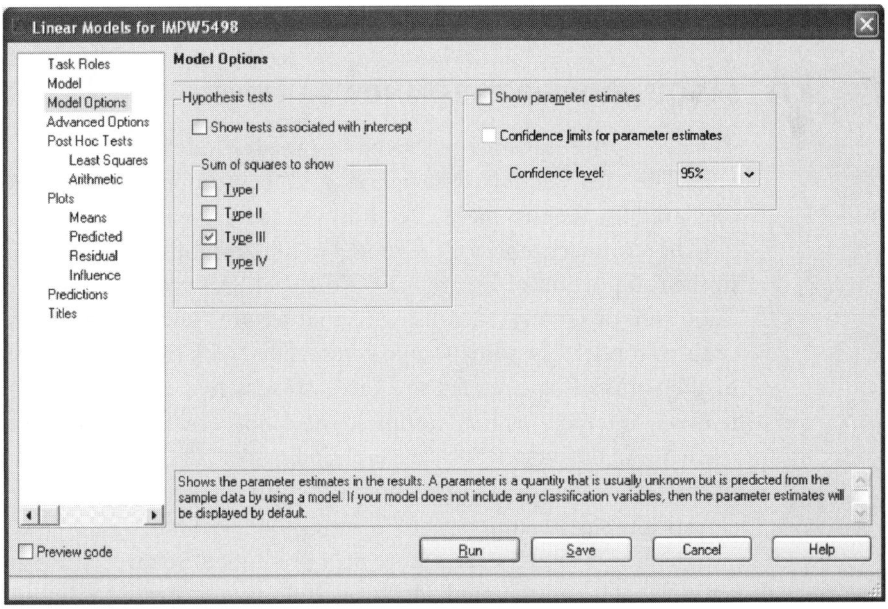

Figure 8.26 We have specified only computing the Type III sum of squares.

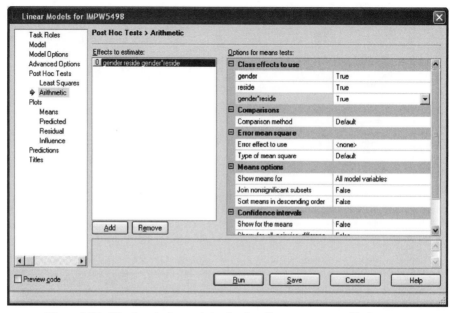

Figure 8.27   The descriptive statistics for the effects are now specified.

Descriptive summary statistics can be requested on the **Arithmetic** portion of the **Post Hoc Tests** tab. This was described in Section 6.8.3, and we will summarize what you should specify here. Clicking **Add** displays a set of drop-down menus, only a few of which need modifying. For **Class effects to use**, select **True** for **gender, reside**, and **gender∗reside**. We will not request a **Homogeneity of variance** test as SAS does not compute this for factorial models. The specifications that we have selected are displayed in Figure 8.27. Then click **Run** to perform the analysis.

8.14   SAS OUTPUT FOR THE OMNIBUS ANALYSIS

The descriptive statistics generated by **Linear Models** are shown in Figure 8.28. The mean, standard deviation, and the number of observations are displayed for the two main effects as well as for the interaction.

The summary table for the model in the omnibus analysis is presented in the top portion of Figure 8.29. Note that SAS provides the **Corrected Total** sum of squares, informing us that we are viewing the results of the reduced model. The sum of squares associated with **Model** is a compilation of the sums of squares for the three effects (two main effects and one two-way interaction) that comprise the model (each effect is treated as a predictor in the model). The last table in Figure 8.29 shows the partitioning of the effects comprising the model. As can be seen, all three effects are statistically significant.

The middle portion of Figure 8.29 presents **R Square**; this is an alternative label for eta squared. However, **R Square** is computed based on the

The GLM Procedure

| Level of gender | N | lonely | |
		Mean	Std Dev
1	15	31.6666667	13.6102202
2	15	18.0000000	9.4642183

| Level of reside | N | lonely | |
		Mean	Std Dev
1	10	15.5000000	5.2334395
2	10	33.0000000	8.1513462
3	10	26.0000000	17.8013732

| Level of gender | Level of reside | N | lonely | |
			Mean	Std Dev
1	1	5	15.5000000	5.55268051
1	2	5	38.0000000	6.67083203
1	3	5	42.0000000	6.51920241
2	1	5	16.0000000	5.52268051
2	2	5	28.0000000	6.51920241
2	3	5	10.0000000	5.52268051

Figure 8.28 Descriptive statistics output.

Linear Models

The GLM Procedure

Dependent Variable: lonely lonely

Source	DF	Sum of Squares	Mean Square	F Value	Pr > F
Model	5	4364.166667	872.833333	23.70	<.0001
Error	24	884.000000	36.833333		
Corrected Total	29	5248.166667			

R-Square	Coeff Var	Root MSE	lonely Mean
0.831560	24.43912	6.069047	24.83333

Source	DF	Type III SS	Mean Square	F Value	Pr > F
gender	1	1400.833333	1400.833333	38.03	<.0001
reside	2	1551.666667	775.833333	21.06	<.0001
gender*reside	2	1411.666667	705.833333	19.16	<.0001

These are the probabilities of the *F* ratio occurring by chance alone for the main effects and interaction.

Figure 8.29 ANOVA results.

model with all three effects combined (added) together. In the context of ANOVA, we ordinarily wish to obtain the eta squared value for each separate effect. To do this, as we had to do with SPSS, we must perform the hand calculation, dividing each sum of squares by the total sum of squares (**Corrected Total**) shown in Figure 8.29. The coefficient of variation (the ratio of the standard deviation of the sample as a whole to the mean of the sample as a whole), the root mean square error, and the grand mean of the dependent variable are also displayed in that middle table.

8.15   PERFORMING THE POST-ANOVA ANALYSES IN SAS

The omnibus analysis indicated that all of the effects were statistically significant. In practice, we would tend not to bother with the main effects because we have obtained a statistically significant interaction in the omnibus analysis. However, to illustrate how we would approach the main effects, we will discuss them here.

Only one main effect can be unambiguously interpreted as it stands – because there were only two levels of **gender**, from our descriptive statistics we know that the females felt significantly more loneliness than the males. For the other two effects (the main effect of **reside** and the interaction), we need additional information before we can render an interpretation of the results.

We wish to perform comparisons of the means for (a) the main effect of **reside** and (b) the means of the groups for the interaction, that is, we wish to perform tests of simple effects. If we wanted to use a Tukey test for the main effect and Bonferroni tests for the simple effects, we would need to perform the analysis twice, once for the Tukey procedure and again for the Bonferroni procedure. For our purposes, we opt here to use a Bonferroni corrected multiple comparisons test for both of these so that we can accomplish the specifications in one step.

Configure the analysis as described above. Then select the **Least Squares** portion of the **Post Hoc Tests** tab. This brings you to the screen shown in Figure 8.30. For **Class effects to use**, set **reside** and **gender*reside** to **True**; this will generate, respectively, the post hoc tests for the main effect of **reside** and the simple effects tests for the interaction.

Least squares means are the same values that SPSS calls estimated marginal means as described in Section 8.9.2. By placing the post hoc tests in the context of least squares, it makes it clear that these post hoc tests are performed in the least square means. When working with the estimated marginal means in SPSS, we performed Bonferroni corrected pairwise comparisons; for this reason, we will also perform Bonferroni comparisons here in *SAS Enterprise Guide*. Thus, for **Comparisons**, set **Show p-values for differences** to **All pairwise differences** and set **Adjustment method for comparison** to **Bonferroni**. Then click **Run** to perform the comparisons.

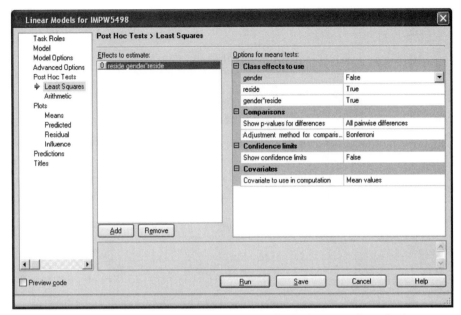

Figure 8.30 We have now specified Bonferroni corrected pairwise comparisons for the main effect of **reside** and for the simple effects of the two-way interaction.

8.16 SAS OUTPUT FOR THE POST ANOVA ANALYSES

8.16.1 POST HOC ON THE MAIN EFFECT

Figure 8.31 shows the multiple comparisons for the main effect of **reside**. The upper display presents the levels of **reside** in the first column, the means in the second column, and in the third column the code number

Linear Models

The GLM Procedure
Least Squares Means
Adjustment for Multiple Comparisons: Bonferroni

reside	lonely LSMEAN	LSMEAN Number
1	15.5000000	1
2	33.0000000	2
3	26.0000000	3

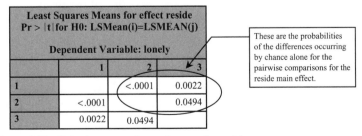

Least Squares Means for effect reside Pr > |t| for H0: LSMean(i)=LSMEAN(j) Dependent Variable: lonely		
	1	2	3
1		<.0001	0.0022
2	<.0001		0.0494
3	0.0022	0.0494	

These are the probabilities of the differences occurring by chance alone for the pairwise comparisons for the reside main effect.

Figure 8.31 Pairwise comparison of means for **reside**.

Linear Models

The GLM Procedure
Least Squares Means
Adjustment for Multiple Comparisons: Bonferroni

gender	reside	lonely LSMEAN	LSMEAN Number
1	1	15.0000000	1
1	2	38.0000000	2
1	3	42.0000000	3
2	1	16.0000000	4
2	2	28.0000000	5
2	3	10.0000000	6

Least Squares Means for effect gender * reside Pr > |t| for H0: LSMean(i)=LSMEAN(j) Dependent Variable: lonely					
i/j	1	2	3	4	5	6
1		<.0001	<.0001	1.0000	0.0365	1.0000
2	<.0001		1.0000	<.0001	0.2328	<.0001
3	<.0001	1.0000		<.0001	0.0192	<.0001
4	1.0000	<.0001	<.0001		0.0688	1.0000
5	0.0365	0.2328	0.0192	0.0688		0.0014
6	1.0000	<.0001	<.0001	1.0000	0.0014	

The difference between group 1 (urban female) and group 6 (rural male) is not statistically significantly different.

The difference between group 5 (small city male) and group 6 (rural male) is statistically significantly different.

Figure 8.32   Tests of simple effects.

assigned by the **Linear Model** procedure to each level of **reside**. The lower display uses the code numbers in presenting a matrix showing the probability values associated with each mean difference. For example, the mean difference of the groups coded 1 and 3 (the large city and rural groups) has a value of $p = .0022$ and that mean difference is therefore statistically significant.

8.16.2   SIMPLE EFFECTS ANALYSES

Figure 8.32 displays the pairwise comparisons of the means of the interaction structured in the same way as we just described for the main effect. The upper table gives code numbers to the groups and the lower table shows the p values associated with the pairwise comparisons. Recall that females were coded as 1 and males as 2, and that large cities, small towns, and rural communities were coded as 1, 2, and 3, respectively. Thus, the group coded as 2 represents gender = 1 and reside = 2, that is, females living in small towns; the group coded as 5 represents gender = 2 and reside = 2, that is, males living in small towns.

Examining the lower table shows that the p value associated with that mean difference is .2328, a value that we would judge to be not statistically significant. Translated into more meaningful prose, males and females living in small towns do not differ in their degree of expressed loneliness. As indicated in Section 8.12.3, results of the simple effects tests need to be examined with respect to the graph of the means so that these comparisons can be taken in context.

8.17 COMMUNICATING THE RESULTS

A written summary of the results is as follows:

A 2×3 between subjects design using gender (male and female) and size of residential community (large city, small town, and rural community) evaluated the degree to which participants experienced feelings of loneliness. While the main effect of gender, $F(1, 24) = 38.03$, $p < .05$, $\eta^2 = .27$, and the main effect of residential community, $F(2, 24) = 21.06$, $p < .05$, $\eta^2 = .30$, were both significant, these were superceded by the significant Gender $\times$ Residential Community interaction, $F(2, 24) = 19.16$, $p < .05$, $\eta^2 = .27$.

The interaction is presented in Figure 1. Simple effects tests were performed by using a Bonferroni adjustment to hold the alpha level at .05. Females were lonelier in both small towns and rural communities than they were in large cities. Males were lonelier in small towns than in either large cities or rural communities. Furthermore, although both groups were relatively lonely in small towns, females were significantly lonelier in that environment than were males. In contrast, females were quite lonely in rural communities, but males in that environment were not very lonely at all.

CHAPTER 8 EXERCISES

8.1. Male and female community mental health clients (Factor a) were randomly assigned to the following three treatment groups (Factor b): Psychoanalytic, cognitive-behavioral, and brief psychotherapy. GAF scores at the end of four weeks of treatment served as the dependent measure. The data were as follows:

$a_1 b_1$	$a_1 b_2$	$a_1 b_3$	$a_2 b_1$	$a_2 b_2$	$a_2 b_3$
55	62	65	55	70	75
45	60	65	55	65	70
40	65	70	60	60	65
35	70	70	50	60	60
52	68	60	55	65	60
50	60	60	53	65	65
48	60	65	50	70	70
50	55	65	38	68	70

a.   Conduct an ANOVA on these data by hand and with SPSS or SAS.

b. If the interaction is statistically significant, conduct simple effects analyses, and any necessary multiple comparisons tests by hand and with SPSS or SAS.

8.2. Three breeds of dogs (Factor a: German Shepherds, Chihuahuas, and Poodles) were exclusively fed three types of dog food (Factor b: Krunchies, Meaties, and Nibblets) during the course of a one month dog obedience training seminar. At the end of the course each dog was rated on a $1 = very$ *disobedient* to $10 = very$ *obedient* Likert-type scale. The data are as follows:

$a_1 b_1$	$a_1 b_2$	$a_1 b_3$	$a_2 b_1$	$a_2 b_2$	$a_2 b_3$	$a_3 b_1$	$a_3 b_2$	$a_3 b_3$
8	10	10	5	10	5	8	8	8
9	10	9	5	9	5	7	5	9
10	10	8	6	8	4	8	4	7
10	9	10	6	10	6	7	7	8
9	7	10	5	10	5	8	5	8
9	9	9	6	9	6	9	5	7

a. Conduct an ANOVA on these data by hand and with SPSS or SAS.
b. If the interaction is statistically significant, conduct simple effects analyses, and any necessary multiple comparisons tests by hand and with SPSS or SAS.

8.3. Market researchers were interested in testing the effectiveness of two new hand lotion products (a_1 = Softy and a_2 = Mushy) on adult women from four geographical regions of the U.S. (b_1 = West, b_2 = Midwest, b_3 = South, and b_4 = East). The dependent measure was lotion perceived effectiveness with $1 = not$ *very effective* and $10 = very$ *effective*. The data are as follows:

$a_1 b_1$	$a_1 b_2$	$a_1 b_3$	$a_1 b_4$	$a_2 b_1$	$a_2 b_2$	$a_2 b_3$	$a_2 b_4$
7	8	5	9	4	5	2	5
8	9	6	9	4	5	1	7
9	10	6	9	4	5	2	5
10	8	6	10	3	6	1	7
7	9	7	7	4	4	2	5
8	10	7	7	2	4	3	6
7	10	7	5	4	4	4	7
7	10	8	7	5	4	2	5

a. Conduct an ANOVA on these data by hand and with SPSS or SAS.
b. If the interaction is statistically significant, conduct simple effects analyses, and any necessary multiple comparisons tests by hand and with SPSS or SAS.

Three-Way Between-Subjects Design

9.1 A NUMERICAL EXAMPLE OF A THREE-WAY DESIGN

In Chapter 8 we added a second independent variable into a between-subjects design to generate a two-way factorial. At this point you probably realize that we are not limited to combining just two between-subjects independent variables in research designs (although we are still limiting ourselves to analyzing a single dependent variable). Theoretically, we could combine many such variables together, despite the fact that the complexities of such designs grow exponentially as the designs get more complex. It is possible to see in the research literature some ANOVA designs using five independent variables and a few using four variables; however, three-way designs are the common limit for most research questions. If you understand the logic of analyzing a three-way design, you can invoke the same strategies to handle those with four or five independent variables.

We will illustrate the general principles of a three-way between-subjects design by using the following simplified hypothetical example data set. Assume that we wish to measure citizen satisfaction with the public school system. We sample individuals representing three different political preferences: liberal, moderate, and conservative (coded 1, 2, and 3, respectively, in the data file). We also code for whether or not these individuals voted in last election (yes coded as 1 and no coded as 2), using voting as an indicator of political involvement. Finally, we code for whether or not there are school-aged children living in the home (yes coded as 1 and no coded as 2).

To assess satisfaction with the public school system, we administer a survey to each of the participants and compute a total score. Satisfaction scores near zero reflect little satisfaction with the system; scores near or at the maximum of 40 reflect considerable satisfaction with the public school system. These scores comprise our dependent variable.

The design, together with the data, is presented in Figure 9.1. There are five participants per cell. Note that with three independent variables, drawing the configuration in such a way that we can show all of the data points is not a trivial manner. In order to schematically present the design, we need to show it in segments by isolating the levels of one of the independent variables. We have chosen to display the data by separating the participants on the basis of whether or not school-aged children are

Kids in Home

	Liberal		Moderate		Conservative	
Vote Yes	36		14		10	
	40		21		18	
	32		7		2	
	39		20		15	
	33	36	8	14	5	10
Vote No	35		32		15	
	40		39		21	
	30		25		9	
	39		38		22	
	31	35	26	32	8	15

Satisfaction with
Public Schools
Range 0 - 40

Kids Not in Home

	Liberal		Moderate		Conservative	
Vote Yes	33		35		20	
	40		40		27	
	26		30		13	
	40		39		26	
	26	33	31	35	14	20
Vote No	28		18		36	
	35		25		40	
	21		11		32	
	36		22		40	
	20	28	14	18	32	36

Figure 9.1   Design and data for our three-way between-subjects example.

present in the home, although we could just as legitimately have broken the data apart by one of the other independent variables. Once we select the variable to split apart, it becomes a bit more difficult to conceptually compare the levels of this variable because each of its levels is in separate tables.

Based on how we schematically presented the design in Figure 9.1, we can say that this design is a $2 \times 3 \times 2$, although the order of specifying the independent variables is arbitrary. The number of separate cells in the design – the number of unique combinations of levels of the independent variables – is known by multiplying the number of levels of each variable. Thus, there are twelve cells in the design ($2 \times 3 \times 2 = 12$), each representing a unique combination of the three independent variables. This tier of analysis – working with the twelve cell means – is the essence of the design and represents the *highest order* effect (i.e., the effect with the most detail).

9.2   PARTITIONING THE VARIANCE INTO ITS SOURCES

9.2.1   THE SUMMARY TABLE

As was true of the all the designs that we have been and will be discussing, the total variance of the dependent variable is partitioned into its separate sources. These sources are the effects that are associated with the independent variables and the error variance.

Table 9.1. Summary table for three-way between-subjects design

Source	SS	df	MS	F	η^2
Kids home (A)	326.667	1	326.667	9.289	0.046
Voter (B)	106.667	1	106.667	3.033	0.015
Politics (C)	1,672.500	2	836.250	23.780	0.237
$A \times B$	326.667	1	326.667	9.289	0.046
$A \times C$	1,060.833	2	530.417	15.083	0.151
$B \times C$	490.833	2	245.417	6.979	0.070
$A \times B \times C$	1,375.833	2	687.917	19.562	0.195
S/ABC	1,688.000	48	35.167		
Total	7,048.000	59			

We have computed the ANOVA for the data shown in our numerical example. The summary table for the corrected model is shown in Table 9.1. As can be seen from Table 9.1, the total variance is partitioned into many more parts than we have thus far encountered in the previous chapters. We look at these below.

9.2.2 THERE ARE THREE MAIN EFFECTS

With three independent variables in the design, each is associated with its own main effect on the dependent variable. Thus, we evaluate the main effects for children in the home, having voted in the last election, and political preference. Take the main effect of having children in the home as an example. We ask: Are participants who have school-aged children at home more or less satisfied than those who do not have children at home? Using Figure 9.1 as a frame of reference, we would compute the mean of all those in the upper panel (children at home) and compare it to the mean of all those in the lower panel (no children at home). Such computations yield a mean satisfaction score for those with children in the home of 23.67 and a mean satisfaction score for those without children in the home of 28.33. This difference is evaluated by the F ratio and, as shown in the summary table, those means are statistically significant, $F(1, 48) = 9.289$, $p < .05$, $\eta^2 = .046$.

9.2.3 THERE ARE THREE TWO-WAY INTERACTIONS

Three-way designs also allow us to evaluate two-way interactions. Uniquely combining three independent variables two at a time can be done in three different ways, which represent the three interactions: Children in Home × Political Preference, Children in Home × Voting, and Voting × Political Preference. Each of these is separately evaluated.

For example, consider the Voting × Political Preference interaction. To evaluate this interaction, we collapse across the third independent variable, children in home. The Voting × Political Preference interaction thus represents a 2 × 3 design (two levels of voting and three levels of political preference) in which each cell contains data pooled together from participants with and without children in the home.

	Political Preference		
	Liberal	Moderate	Conservative
Vote Yes	34.50	24.50	15.00
Vote No	31.50	25.00	25.50

Figure 9.2 Means for the Voting × Political Preference interaction.

We show the means for this interaction effect in Figure 9.2. For each cell in this 2 × 3 configuration, we combined the data for the two children in the home cells. For example, the mean for liberals who voted in the last election is 34.50. This is the average of liberals who voted in the last election who have children in the home (mean = 36.00) and liberals who voted in the last election who do not have children in the home (mean = 33.00).

Visual inspection of the means for this Voting × Political Preference interaction suggests that there might very well be a significant effect. Looking across the top row of means shows a steady drop in satisfaction from liberal to moderate to conservative participants, but looking across the bottom row shows a drop in satisfaction from liberals to moderates that levels off such that conservatives and moderates are approximately equally satisfied. Our visual inspection is confirmed by the statistical analysis. As we can see from the summary table in Table 9.1, this interaction effect is statistically significant, $F(2, 48) = 6.979$, $p < .05$, $\eta^2 = .070$.

9.2.4　THE THREE-WAY INTERACTION

Because we have three independent variables, it is possible for the unique combinations of all three to be differentially associated with participant performance; thus, one of the effects evaluated in the design is the three-way interaction of Children in Home × Voting × Political Preference. This is the highest order (most complex) effect in the design; that is, this effect deals with the cell means (twelve cell means in this case) at the most detailed tier of the design.

The means that are evaluated for this three-way interaction are shown in Figure 9.1. As you can see, this is the most fundamental tier of the design in which each combination of the levels of the independent variables is represented by its own mean.

The summary of the analysis in Table 9.1 indicates that the three-way interaction is statistically significant, $F(2, 48) = 19.562$, $p < .05$, $\eta^2 = .195$. Although it is much more complex than the two-way interaction we discussed in the previous chapter, to understand the three-way interaction we approach it in much the same way that we dealt with the two-way interaction. That is, we ultimately interpret statistically significant three-way interactions by breaking them down into a set of two-way interactions. We do this by splitting one of the independent variables by its levels and examining each level separately, leaving the other two independent variables in factorial (two-way) combinations. When we examine each level, we will find at least one of these two-way interactions to be statistically significant; in interpreting the three-way interaction, we narrate each of the two-way analyses.

Applying our interpretation strategy to the present numerical example, we must first split apart one of the independent variables. This is already done in Figure 9.1 as we have separate matrices of means for those with children in their home and for those who do not have children in their home. It is therefore convenient for us to use this breakdown here. This split has us examining the two-way interaction of Voting × Political Preference for each level of children in home. Thus, we are looking at two two-way interactions. In your own research, the choice of which variable to use as a "breaking" variable will likely be governed by the theory on which the research is based. If there is no theoretical reason for selecting the breaking variable, then the choice should be based on the ease of explication needed to communicate the results.

In the process of visually examining the three-way interaction we next look at the pattern of means (the two-way interaction of Voting × Political Preference) in each portion of the split. The fact that the three-way interaction was statistically significant informs us in advance that the patterns we will see (i.e., the two-way interactions) will be different. This examination can be done either by looking at the values of the means in Figure 9.1 or by graphing the means. We will work with the graphic representation, and we show the plots in Figure 9.3.

Figure 9.3 shows both two-way interactions, one for those who have children in the home (top graph) and another for those with no children in the home (bottom graph). As you can see, the patterns are different, something that is to be expected based on the significant three-way interaction. By the way, such pattern differences would be seen regardless of which independent variable we chose to split; once we examined the two-way interactions, we would always see different patterns in the plots.

As was the case with the two-way interaction discussed in the last chapter, visual inspection of the plots is useful to gain an impression of the pattern of the means, but it needs to be supplemented with the simple effects analyses to determine which means in the plots were and were not significantly different from each other.

9.3 COMPUTING THE PORTIONS OF THE SUMMARY TABLE

The degrees of freedom, mean squares, F ratios, and eta squared values are computed in the same manner that we described in Chapters 6 and 8. Thus, the degrees of freedom for the triple interaction, as is true for all interactions, is computed by multiplying the degrees of freedom of those main effects contained in it. All F ratios involve dividing the mean square of the respective effect by the mean square of the error term. Eta squared values represent the division of the sum of squares for the effect by the sum of squares for the total variance.

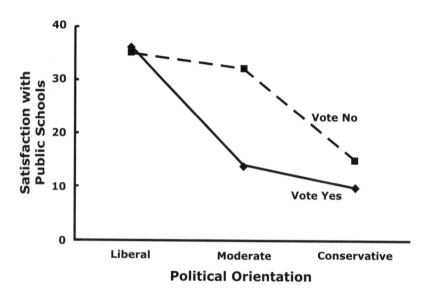

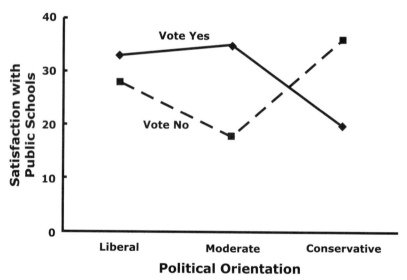

Figure 9.3 Plots of the three-way interaction.

9.4 PRECEDENCE OF EFFECTS: HIGHER-ORDER INTERACTIONS, LOWER-ORDER INTERACTIONS, MAIN EFFECTS

We have already discussed in Chapter 8 the general precedence rule that interactions supersede main effects. The logic underlying that rule can now be extended to three-way (and more complex) designs. Most generally, higher-order interactions supersede lower-order interactions, and lower-order interactions supersede main effects.

The highest order interaction in a design is the one in which the unique combinations of all of the independent variables are accounted for. In a three-way design, it would be the three-way interaction. This three-way interaction contains the most complete level of detail regarding the means of the groups. If that effect is statistically significant, then we recognize that the combinations of all three independent variables needs to be known in order to best predict the value of the dependent variable. Effects representing fewer variables – two-way interactions and main effects – will, by definition, ignore the differences in the levels of one or two of the independent variables, respectively, and will therefore present a less detailed and thus somewhat incomplete picture of the result.

If the three-way interaction in a three-way design is statistically significant, you should focus on that effect. Most researchers would tend to ignore any statistically significant two-way interactions and may or may not give cursory attention to the statistically significant main effects, depending on, among other things, the theoretical framework within which the research was generated and the strength of the effects as indexed by eta squared.

If the three-way interaction is not statistically significant, we next examine the two-way interactions. These interactions are all at the same tier of the design and are independent of one another. Thus, it is appropriate to present in detail any and all two-way interactions that are statistically significant.

Main effects represent the most global tier of a design. Logically, if a variable is involved in a statistically significant interaction then the interpretation of any statistically significant main effect associated with it will result in some degree of overgeneralization, and researchers must be careful about how they report and communicate main effects. However, it is possible that in a three-way design only two independent variables may be involved in a statistically significant two-way interaction (e.g., only one two-way interaction turns out to be statistically significant). Under that condition, we would certainly want to fully explicate a statistically significant main effect on the third independent variable that is not involved in an interaction (e.g., we would perform post-ANOVA tests if there were more than two levels of that independent variable).

9.5   COMPUTING BY HAND THE OMNIBUS THREE-WAY BETWEEN-SUBJECT ANALYSIS

The computational procedures for a three-way between-subjects analysis of variance are a direct extension of the computations we covered previously in Chapter 8 for the two-factor case. As we have discussed within this chapter, a three-factor between-subjects design has the following sources of variance: the main effects of factors A, B, and C and the within-subjects or error component of *S/ABC*. Additional sources of variation include the four interaction effects: $A \times B$, $A \times C$, $B \times C$, and $A \times B \times C$.

We agree with the sentiments offered by Keppel and Wickens (2004) when they note that "you will almost certainly use a computer to perform the brute numerical work when you analyze a three-factor design.... Nevertheless, it is instructive to look at the computational procedure" (p. 476). Because of the tedium involved in hand calculations with three or more factors and the questionable pedagogical payoff, we will forego our usual elaboration of the numerical computations of this and all future types of three-factor designs in this text.

However, we will provide you with the computational formulas for conducting a three-way between-subjects design by hand for those intrepid students and instructors who desire this additional level of detail. These formulas can be found in Table 9.2. Further discussion of this topic can be found in Keppel (1991) and Keppel and Wickens (2004).

9.6 PERFORMING THE OMNIBUS ANALYSIS IN SPSS

9.6.1 STRUCTURING THE DATA FILE

A portion of the SPSS data file containing the variables for the analysis of the numerical example we have been using throughout this chapter is presented in Figure 9.4. That data file is structured as follows. The first

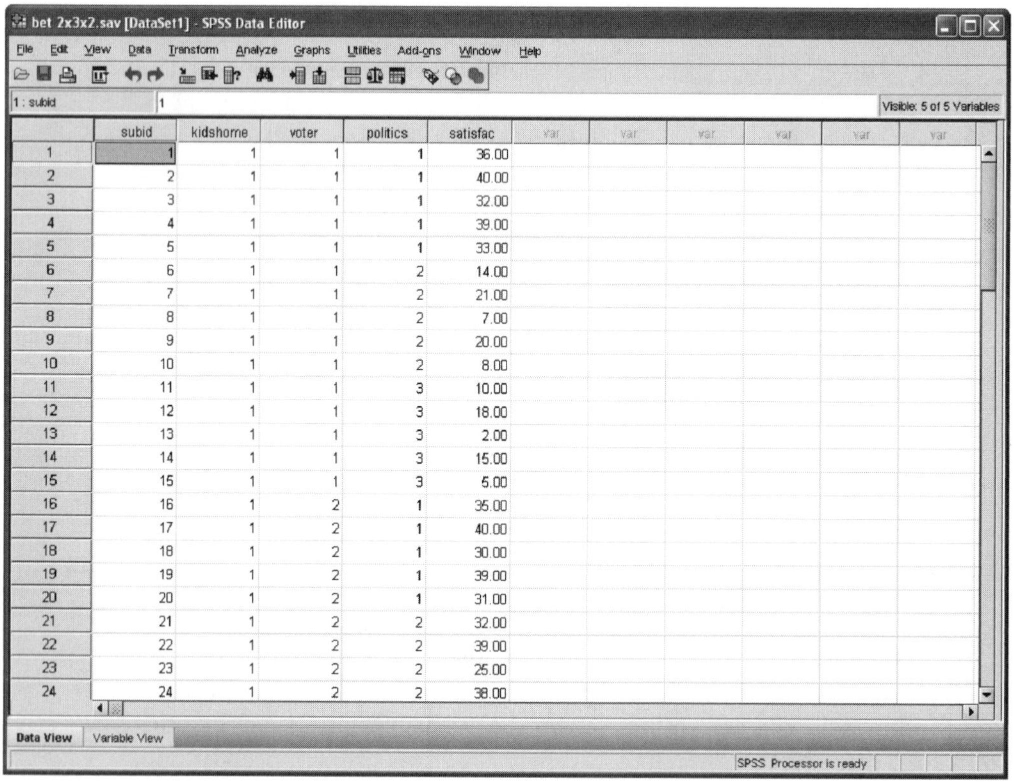

	subid	kidshome	voter	politics	satisfac
1	1	1	1	1	36.00
2	2	1	1	1	40.00
3	3	1	1	1	32.00
4	4	1	1	1	39.00
5	5	1	1	1	33.00
6	6	1	1	2	14.00
7	7	1	1	2	21.00
8	8	1	1	2	7.00
9	9	1	1	2	20.00
10	10	1	1	2	8.00
11	11	1	1	3	10.00
12	12	1	1	3	18.00
13	13	1	1	3	2.00
14	14	1	1	3	15.00
15	15	1	1	3	5.00
16	16	1	2	1	35.00
17	17	1	2	1	40.00
18	18	1	2	1	30.00
19	19	1	2	1	39.00
20	20	1	2	1	31.00
21	21	1	2	2	32.00
22	22	1	2	2	39.00
23	23	1	2	2	25.00
24	24	1	2	2	38.00

Figure 9.4 The data file for the three-way numerical example.

Table 9.2. Computational formulas for the three-way between-subjects design

Source	df	SS	MS	F
A	$a-1$	$\dfrac{\Sigma A^2}{(b)(c)(n)} - \dfrac{T^2}{(a)(b)(c)(n)}$	$\dfrac{SS_A}{df_A}$	$\dfrac{MS_A}{MS_{S/ABC}}$
B	$b-1$	$\dfrac{\Sigma B^2}{(a)(c)(n)} - \dfrac{T^2}{(a)(b)(c)(n)}$	$\dfrac{SS_B}{df_B}$	$\dfrac{MS_B}{MS_{S/ABC}}$
C	$c-1$	$\dfrac{\Sigma C^2}{(a)(b)(n)} - \dfrac{T^2}{(a)(b)(c)(n)}$	$\dfrac{SS_C}{df_C}$	$\dfrac{MS_C}{MS_{S/ABC}}$
$A \times B$	$(a-1)(b-1)$	$\dfrac{\Sigma AB^2}{(c)(n)} - \dfrac{\Sigma A^2}{(b)(c)(n)} - \dfrac{\Sigma B^2}{(a)(c)(n)} + \dfrac{T^2}{(a)(b)(c)(n)}$	$\dfrac{SS_{A \times B}}{df_{A \times B}}$	$\dfrac{MS_{A \times B}}{MS_{S/ABC}}$
$A \times C$	$(a-1)(c-1)$	$\dfrac{\Sigma AC^2}{(b)(n)} - \dfrac{\Sigma A^2}{(b)(c)(n)} - \dfrac{\Sigma C^2}{(a)(b)(n)} + \dfrac{T^2}{(a)(b)(c)(n)}$	$\dfrac{SS_{A \times C}}{df_{A \times C}}$	$\dfrac{MS_{A \times C}}{MS_{S/ABC}}$
$B \times C$	$(b-1)(c-1)$	$\dfrac{\Sigma BC^2}{(a)(n)} - \dfrac{\Sigma B^2}{(a)(c)(n)} - \dfrac{\Sigma C^2}{(a)(b)(n)} + \dfrac{T^2}{(a)(b)(c)(n)}$	$\dfrac{SS_{B \times C}}{df_{B \times C}}$	$\dfrac{MS_{B \times C}}{MS_{S/ABC}}$
$A \times B \times C$	$(a-1)(b-1)(c-1)$	$\dfrac{\Sigma ABC^2}{n} - \dfrac{\Sigma AB^2}{(c)(n)} - \dfrac{\Sigma AC^2}{(b)(n)} - \dfrac{\Sigma BC^2}{(a)(n)} + \dfrac{\Sigma A^2}{(b)(c)(n)} + \dfrac{\Sigma B^2}{(a)(c)(n)} + \dfrac{\Sigma C^2}{(a)(b)(n)} - \dfrac{T^2}{(a)(b)(c)(n)}$	$\dfrac{SS_{A \times B \times C}}{df_{A \times B \times C}}$	$\dfrac{MS_{A \times B \times C}}{MS_{S/ABC}}$
S/ABC	$(a)(b)(c)(n-1)$	$\Sigma Y^2 - \dfrac{\Sigma(ABC)^2}{n}$	$\dfrac{SS_{S/ABC}}{df_{S/ABC}}$	
Total	$(a)(b)(c)(n)-1$	$\Sigma Y^2 - \dfrac{T^2}{(a)(b)(c)(n)}$		

variable is a subject identification code (**subid**). The next three variables are the independent variables in the analysis; each has been assigned arbitrary codes to represent the categories. These variables are **kidshome** (children present in the home coded as 1, children not in the home coded as 2), **voter** (having voted in the last election coded as 1, not voting in the last election coded as 2), and **politics** (liberal, moderate, and conservative coded as 1, 2, and 3, respectively). The last variable in the data file is our dependent variable, named **satisfac**, the score on the survey assessing satisfaction with the public school system.

9.6.2   STRUCTURING THE ANALYSIS

From the main menu select **Analyze → General Linear Model → Univariate** to reach the main dialog window shown in Figure 9.5. Click **satisfac** over to the **Dependent Variable** panel and in this order click **kidshome**, **voter**, and **politics** over to the **Fixed Factor(s)** panel. We will first run the omnibus analysis to determine which effects are significant. Based on those results we will then determine which post-ANOVA analyses we will perform.

The order in which we bring the variables into the **Fixed Factor(s)** panel is important when we have three or more independent variables in the analysis. Recall our discussion in Chapter 8 about the way in which the observed means are displayed in the **Descriptive Statistics** output table. We saw that earlier specified variables change more slowly than later specified variables. Having structured the presentation of the example in

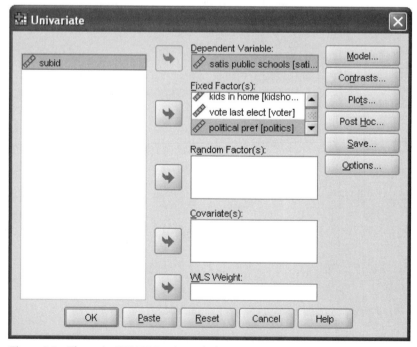

Figure 9.5   The main **GLM Univariate** window.

Figure 9.1 such that our major split was on the **kidshome** variable, we knew to select this independent variable first to have it change most slowly in the output showing the observed means. With **politics** on the *x* axis of our plots, we opt for this variable to change most rapidly and thus move it into the **Fixed Factor(s)** panel last. That leaves **voter** to be moved over second.

What if you did not know how in advance of the data analysis how you would present the design and the results? If you obtained a statistically significant three-way interaction, we advise you to record all of the cell means first. These are all available from the **Descriptive Statistics** table, although the means for the variable that is changing more rapidly will be a bit easier to quickly read. Once you examine all of the cell means you can determine how you want to structure your report of the results. That, in turn, will suggest to you the order that you will probably want to move the independent variables into the **Fixed Factor(s)** panel. As we will see, this decision will also affect which simple effects analyses you will perform (you can also **Paste** the syntax and move these variables to the desired positions on the main command line).

Click the **Options** pushbutton to reach the **Options** dialog window shown in Figure 9.6. Click over all of the effects (the three main effects, the three two-way interactions, and the three-way interaction) from the left

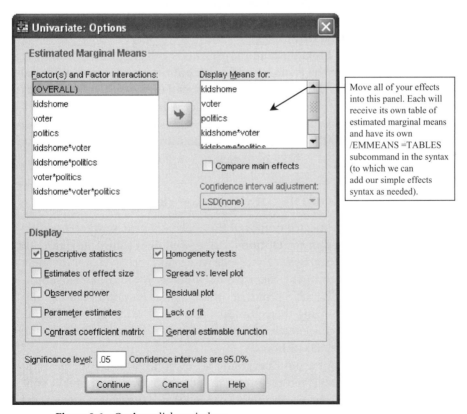

Figure 9.6 **Options** dialog window.

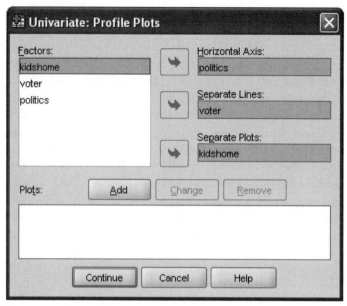

Figure 9.7   **Plots** dialog window.

panel labeled **Factor(s) and Factor Interactions** to the right panel labeled **Display Means for**. The bottom half of the **Options** window concerns the observed means. Under the **Display** heading, check **Homogeneity tests** and **Descriptive statistics**.

Clicking the **Plots** pushbutton presents the dialog window shown in Figure 9.7. From the **Factors** panel, click **politics** over to the **Horizontal Axis** panel, click **voter** over to the **Separate Lines** panel, and then click **kidshome** under separate plots. Then click **Add** to register the plot with SPSS. Doing so will place **politics∗voter∗kidshome** in the **Plots** panel at the bottom of the window. Clicking **Continue** returns us to the main dialog window. Now click **OK** to run the omnibus analysis.

9.7   SPSS OUTPUT FOR THE OMNIBUS ANALYSIS

We present here selected output of the analysis. The descriptive statistics that we requested in the **Options** dialog window is shown in Figure 9.8 as a **Descriptive Statistics** table. Since we originally clicked over the independent variables in the main dialog window in the order **kidshome**, **voter**, and **politics**, SPSS kept that ordering in its output. Because of this ordering, in reporting the cell means and standard deviations, **liberal**, **moderate**, and **conservative** are shown for each level of **voter** (yes and no) for each level of **kidshome**.

The means for the **politics** variable are most easily seen in the **Descriptives Statistics** table, but the means for all of the effects are actually contained in this output. For example, the top third of the table represents the

Descriptive Statistics

Dependent Variable: satis public schools

kids in home	vote last elect	political pref	Mean	Std. Deviation	N
yes	yes	liberal	36.0000	3.53553	5
		moderate	14.0000	6.51920	5
		conservative	10.0000	6.67083	5
		Total	20.0000	12.97800	15
	no	liberal	35.0000	4.52769	5
		moderate	32.0000	6.51920	5
		conservative	15.0000	6.51920	5
		Total	27.3333	10.64134	15
	Total	liberal	35.5000	3.86580	10
		moderate	23.0000	11.30388	10
		conservative	12.5000	6.75360	10
		Total	23.6667	12.24276	30
no	yes	liberal	33.0000	7.00000	5
		moderate	35.0000	4.52769	5
		conservative	20.0000	6.51920	5
		Total	29.3333	8.90960	15
	no	liberal	28.0000	7.51665	5
		moderate	18.0000	5.70088	5
		conservative	36.0000	4.00000	5
		Total	27.3333	9.38591	15
	Total	liberal	30.5000	7.33712	10
		moderate	26.5000	10.18986	10
		conservative	28.0000	9.85450	10
		Total	28.3333	9.04904	30
Total	yes	liberal	34.5000	5.46199	10
		moderate	24.5000	12.26784	10
		conservative	15.0000	8.15135	10
		Total	24.6667	11.92313	30
	no	liberal	31.5000	6.91616	10
		moderate	25.0000	9.36898	10
		conservative	25.5000	12.18606	10
		Total	27.3333	9.85877	30
	Total	liberal	33.0000	6.25763	20
		moderate	24.7500	10.62705	20
		conservative	20.2500	11.43805	20
		Total	26.0000	10.92967	60

This is the mean satisfaction score for those voting in the last election that have children in the home.

This is the mean satisfaction score for everyone who has children in the home.

This is the mean satisfaction score for moderates who did not vote in the last election and who have no children in the home.

This is the mean satisfaction score for those voting in the last election.

Figure 9.8 The descriptive statistics output.

data from those who have children in the home. Thus, on that **Total** row, we see the mean of **liberal** to be 35.50. This represents liberal respondents with children in the home. As another example, the bottom third of the table is given the label **Total** under the **kids in home** column. Thus, the means in this portion of the table combine the data from those with and without children in the home. On the row labeled as **no** under the **vote last elect** column we see the mean for **conservative** to be 25.50. This represents conservative participants with no children in the home. It may take a bit of work, but it is possible to discern all of the means for all of the effects from this table.

The **GLM Univariate** procedure uses the Levene test to determine if the assumption of homogeneity of variance has been met. In Figure 9.9, we see that the F ratio for that test is not significant; we may thus conclude that the data meet the homogeneity of variance assumption.

Levene's Test of Equality of Error Variances[a]

Dependent Variable: satis public schools

F	df1	df2	Sig.
.731	11	48	.704

The Levene test yielded an *F* ratio that was not statistically significant. Thus, our data have met the assumption of homogeneity of variance.

Tests the null hypothesis that the error variance of the dependent variable is equal across groups.

a. Design: Intercept+kidshome+voter+politics+kidshome * voter+kidshome * politics+voter * politics+kidshom voter * politics

Figure 9.9   The results of the Levene test.

The summary table for the omnibus analysis is shown in Figure 9.10, and duplicates the information we presented in Table 9.1. As we already know, all of the effects except for the main effect of **voter** are statistically significant. The **Corrected Model** combines all of our effects (the main effects and interactions); the **R Squared** associated with it (shown in the footnote to the table) is a complete eta squared for these combined effects, and so is not useful for our purposes (we compute separate eta squared values for each of the statistically significant effects).

9.8   PERFORMING THE POST-ANOVA ANALYSES IN SPSS

The omnibus analysis indicated that most of the effects were statistically significant. Our focus will be on the main effect of **politics** (because it has more than two levels) and the three-way interaction.

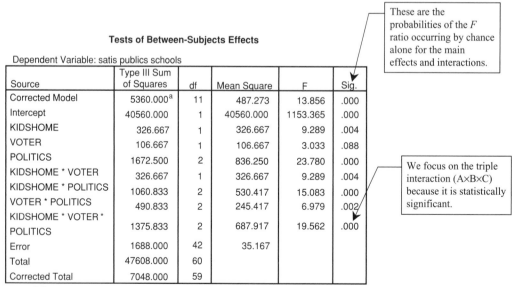

Tests of Between-Subjects Effects

Dependent Variable: satis publics schools

Source	Type III Sum of Squares	df	Mean Square	F	Sig.
Corrected Model	5360.000[a]	11	487.273	13.856	.000
Intercept	40560.000	1	40560.000	1153.365	.000
KIDSHOME	326.667	1	326.667	9.289	.004
VOTER	106.667	1	106.667	3.033	.088
POLITICS	1672.500	2	836.250	23.780	.000
KIDSHOME * VOTER	326.667	1	326.667	9.289	.004
KIDSHOME * POLITICS	1060.833	2	530.417	15.083	.000
VOTER * POLITICS	490.833	2	245.417	6.979	.002
KIDSHOME * VOTER * POLITICS	1375.833	2	687.917	19.562	.000
Error	1688.000	42	35.167		
Total	47608.000	60			
Corrected Total	7048.000	59			

These are the probabilities of the *F* ratio occurring by chance alone for the main effects and interactions.

We focus on the triple interaction (A×B×C) because it is statistically significant.

a. R Squared = .760 (Adjusted R Squared = .705)

Figure 9.10   The summary table produced by SPSS.

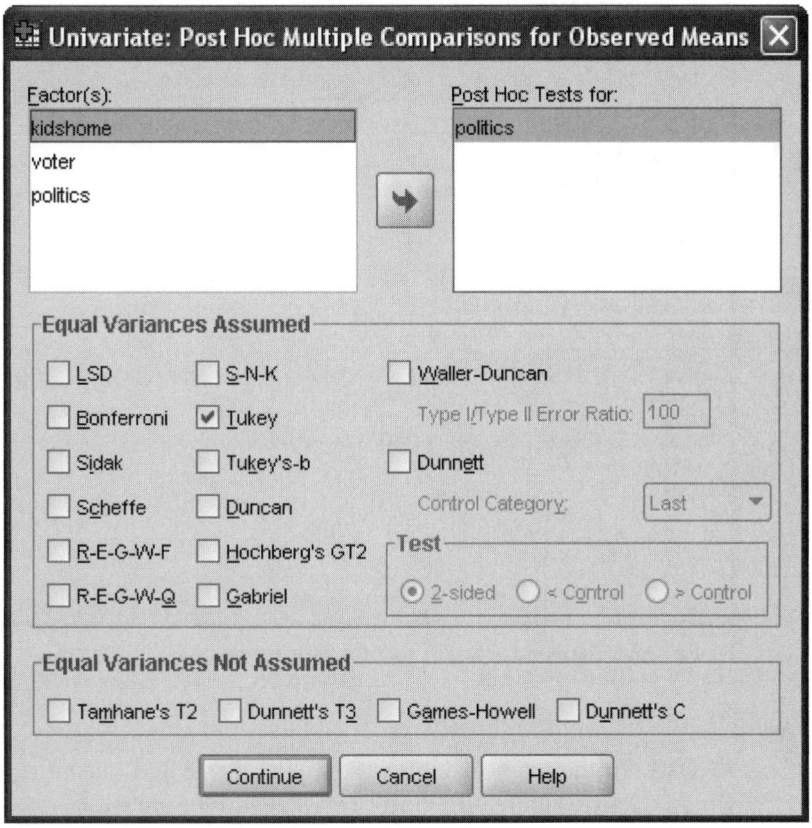

Figure 9.11 The **Post Hoc** dialog box.

9.8.1 POST HOC TEST FOR THE MAIN EFFECT OF POLITICS

The independent variable **politics** has three levels (liberal, moderate, and conservative), and in order to determine which groups differ from which we will perform a post hoc test. From the main menu select **Analyze → General Linear Model → Univariate** to reach the main dialog window and configure your analysis as you did for the omnibus analysis.

The one difference in configuring the analysis between the omnibus analysis and the simple effects analysis is that you will ask SPSS to perform a Tukey post hoc test. Click the **Post Hoc** pushbutton that will bring you to the window shown in Figure 9.11. Click over **politics** from the **Factor(s)** panel to the **Post Hoc Tests for** panel, and check the **Tukey** procedure in the set of tests grouped under the heading **Equal Variances Assumed** because the Levene test indicated that our data have met the assumption of homogeneity of variance. Click **Continue** to return to the main dialog window.

9.8.2 SIMPLE EFFECTS ANALYSES FOR THE THREE-WAY INTERACTION

As you know from the previous chapter, performing the simple effects analyses requires us to add a few words to the syntax underlying the

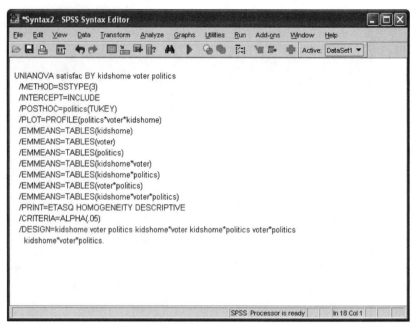

Figure 9.12   The **Syntax** window with the pasted syntax.

ANOVA. In the main dialog window, click **Paste** and make the **Syntax** window active. Figure 9.12 illustrates what you should see.

The syntax is structured in the same way as the two-way ANOVA discussed in Chapter 7 except that there are seven **/EMMEANS = TABLES** subcommands (one for each of the seven effects of interest) rather than the three that we saw in Chapter 8. Each one of these subcommands generates a table of estimated marginal means and each one can accept simple effects syntax.

Our focus will be on the simple effects for the three-way interaction. Note that the order of the variables in parentheses on the **/EMMEANS = TABLES** subcommand is the order in which we specified the variables originally (see the top line of syntax to verify this). In the breakdown of the three-way interaction we know that we will be interested in the two **voter∗politics** interactions, one for participants who have children in the home and the other for those who have no children in the home. This confirms that **kidshome** should be the first mentioned variable of the three in parentheses. It also means that we do not need to run the simple effects analysis for **kidshome** because that variable is not graphed in the interactions we have plotted (it is the breaking or splitting variable).

We will now make one copy of this three-way interaction line and add the simple effects syntax to each as shown in Figure 9.13. To the first line we add **compare (politics) adj (bonferroni)**. This will compare the three

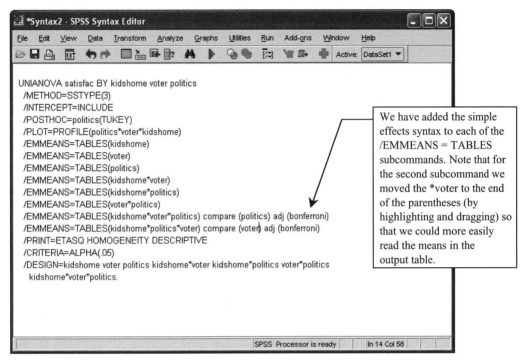

Figure 9.13 The **Syntax** window with the simple effects analysis specified.

levels of **politics** (liberal, moderate, and conservative) for each combination of the other two independent variables specified in parentheses following **/EMMEANS = TABLES**. That is, this comparison will be made for voters with children in the home, nonvoters in the home, voters with no children in the home, and nonvoters with no children in the home. We therefore obtain a great deal of information with just four additional words of syntax.

The second three-way interaction line will be used to perform pairwise comparisons on **voter**. To make the table containing the estimated marginal means for this comparison consonant with the **Pairwise Comparisons** table, we will cut and paste (or highlight and drag) ***voter** from its middle position in the parentheses on the **/EMMEANS = TABLES** subcommand to last place in the parentheses. Then we add our simple effects syntax **compare (voter) adj (bonferroni)**. This will compare the two levels of **voter** (those who voted in the last election and those who did not) for each combination of the other two independent variables specified in parentheses. For example, the comparison will be made for liberals with children in the home, conservatives with no children in the home, and so on.

To run these post-ANOVA analyses, select **Run ➜ All** from the main menu. The analyses will be placed in an **Output** window for you to examine.

Estimates

Dependent Variable: satis public schools

kids in home	vote last elect	political pref	Mean	Std. Error	95% Confidence Interval	
					Lower Bound	Upper Bound
yes	yes	liberal	36.000	2.652	30.668	41.332
		moderate	14.000	2.652	8.668	19.332
		conservative	10.000	2.652	4.668	15.332
	no	liberal	35.000	2.652	29.668	40.332
		moderate	32.000	2.652	26.668	37.332
		conservative	15.000	2.652	9.668	20.332
no	yes	liberal	33.000	2.652	27.668	38.332
		moderate	35.000	2.652	29.668	40.332
		conservative	20.000	2.652	14.668	25.332
	no	liberal	28.000	2.652	22.668	33.332
		moderate	18.000	2.652	12.668	23.332
		conservative	36.000	2.652	30.668	41.332

Pairwise Comparisons

Dependent Variable: satis public schools

kids in home	vote last elect	(I) political pref	(J) political pref	Mean Difference (I-J)	Std. Error	Sig.[a]	Lower Bound	Upper Bound
yes	yes	liberal	moderate	− 22.000*	3.751	.000		
			conservative	26.000*	3.751	.000		
		moderate	liberal	− 22.000*	3.751	.000	− 3	
			conservative	4.000*	3.751	.875	− 5.304	13.304
		conservative	liberal	− 26.000*	3.751	.000	− 35.304	− 16.696
			moderate	− 4.000*	3.751	.875	− 13.304	5.304
	no	liberal	moderate	3.000*	3.751	1.000	− 6.304	12.304
			conservative	20.000*	3.751	.000	10.696	29.304
		moderate	liberal	− 3.000*	3.751	1.000	− 12.304	6.304
			conservative	17.000*	3.751	.000	7.696	26.304
		conservative	liberal	− 20.000*	3.751	.000	− 29.304	− 10.696
			moderate	− 17.000*	3.751	.000	− 26.304	− 7.696
no	yes	liberal	moderate	− 2.000*	3.751	1.000	− 11.304	7.304
			conservative	13.000*	3.751	.003	3.696	22.304
		moderate	liberal	2.000*	3.751	1.000	− 7.304	11.304
			conservative	15.000*	3.751	.001	5.696	24.304
		conservative	liberal	− 13.000*	3.751	.003	− 22.304	− 3.696
			moderate	− 15.000*	3.751	.001	− 24.304	− 5.696
	no	liberal	moderate	10.000*	3.751	.031	.696	19.304
			conservative	− 8.000*	3.751	.114	− 17.304	1.304
		moderate	liberal	− 10.000*	3.751	.031	− 19.304	− .696
			conservative	− 18.000*	3.751	.000	− 27.304	− 8.696
		conservative	liberal	8.000*	3.751	.114	− 1.304	17.304
			moderate	18.000*	3.751	.000	8.696	27.304

> These are the probabilities of the differences occurring by chance alone for the pairwise comparisons with the independent variables in order of kids in home, voting record, and political reference.

Based on estimated marginal means

*. The mean difference is significant at the .05 level.

a. Adjustment for multiple comparisons: Bonferroni.

Figure 9.14 Simple effects analysis comparing the levels of political preference.

9.9 SPSS OUTPUT FOR THE POST-ANOVA ANALYSES IN SPSS

9.9.1 SIMPLE EFFECTS ANALYSES: POLITICAL PREFERENCE

Figure 9.14 shows the estimated marginal means in the table labeled **Estimates** and the **Pairwise Comparisons** for the first simple effects analysis in which we asked SPSS to compare the levels of **politics** at each level of **voter and kidshome**. These represent the comparisons of the points on our line graphs (see Figure 9.3). The first main row (top quarter of the table) isolates those who have children in the home and who voted in the last election. The top portion of that row compares liberals first to moderates (the difference is 22.00, $p = .000$) and then to conservatives (the difference is 26.00, $p = .001$); in both cases, the mean difference is statistically significant. The next portion of that row compares moderates first to liberals (which has already been made) and then to conservatives; the difference is 4.00, $p = .875$, and is not statistically significant. The last portion compares conservatives to the other two groups; both of these comparisons have been made above but are included by SPSS to follow through on the format it is using.

The next quarter of the table isolates those with children in the home who failed to vote in the last election. We can see from the comparisons that liberals and moderates did not differ significantly but both differ from conservatives. In like manner, we can interpret the mean differences for participants who do not have children in the home in the lower half of the table.

9.9.2   SIMPLE EFFECTS ANALYSES: VOTER

Figure 9.15 shows the estimated marginal means and the **Pairwise Comparisons** for the second simple effects analysis in which we asked SPSS to compare those who did and did not vote in the last election at each level of **politics** and **kidshome**. There are two main horizontal rows in this output, one for those with children in the home and another for those without children in the home. Each is divided into three portions corresponding to the three levels of **politics**.

The first comparison for those with children in the home is for the liberal participants. The difference between those who voted and those who did not vote was 1.00 and is not statistically significant ($p = .791$). Moderate participants differ by 18.00, a significant difference ($p = .001$), but conservatives differ by a nonsignificant 5.00 ($p = .189$). For those with no children in the home, whether or not people voted in the last election affected satisfaction scores for moderates and conservatives but not for liberals.

9.9.3   MAPPING SIMPLE EFFECTS ANALYSES TO THE GRAPH OF THE INTERACTION

Mapping the simple effects to the plot of the interaction (see Figure 9.3) requires a little more effort than it did for what may now appear to be the relatively simple two-way interaction of Chapter 8's example. But the task may be accomplished one step at a time by viewing the graphs and consulting the **Paired Comparison** tables.

We focus first on the top graph dealing with those who have children in the home. For those who voted in the last election (the solid line in

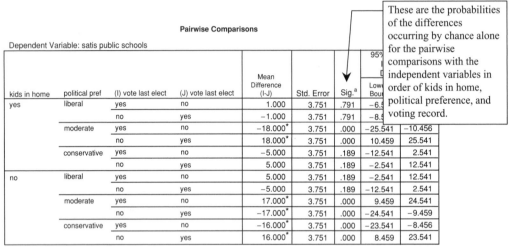

Estimates

Dependent Variable: satis public schools

kids in home	political pref	vote last elect	Mean	Std. Error	95% Confidence Interval Lower Bound	Upper Bound
yes	liberal	yes	36.000	2.652	30.668	41.332
		no	35.000	2.652	29.668	40.332
	moderate	yes	14.000	2.652	8.668	19.332
		no	32.000	2.652	26.668	37.332
	conservative	yes	10.000	2.652	4.668	15.332
		no	15.000	2.652	9.688	20.332
no	liberal	yes	33.000	2.652	27.668	38.332
		no	28.000	2.652	22.668	33.332
	moderate	yes	35.000	2.652	29.668	40.332
		no	18.000	2.652	12.668	23.332
	conservative	yes	20.000	2.652	14.668	25.332
		no	36.000	2.652	30.668	41.332

Pairwise Comparisons

Dependent Variable: satis public schools

These are the probabilities of the differences occurring by chance alone for the pairwise comparisons with the independent variables in order of kids in home, political preference, and voting record.

kids in home	political pref	(I) vote last elect	(J) vote last elect	Mean Difference (I-J)	Std. Error	Sig.[a]	95% Low Bou	Bou
yes	liberal	yes	no	1.000	3.751	.791	−6.5	
		no	yes	−1.000	3.751	.791	−8.5	
	moderate	yes	no	−18.000*	3.751	.000	−25.541	−10.456
		no	yes	18.000*	3.751	.000	10.459	25.541
	conservative	yes	no	−5.000	3.751	.189	−12.541	2.541
		no	yes	5.000	3.751	.189	−2.541	12.541
no	liberal	yes	no	5.000	3.751	.189	−2.541	12.541
		no	yes	−5.000	3.751	.189	−12.541	2.541
	moderate	yes	no	17.000*	3.751	.000	9.459	24.541
		no	yes	−17.000*	3.751	.000	−24.541	−9.459
	conservative	yes	no	−16.000*	3.751	.000	−23.541	−8.456
		no	yes	16.000*	3.751	.000	8.459	23.541

Based on estimated marginal means

*. The mean difference is significant at the .05 level.

a. Adjustment for multiple comparisons: Bonferroni.

Figure 9.15   Simple effects analysis comparing the levels of voter.

the figure), liberals were more satisfied than either of the other two political groups. For those who did not vote (dashed line), conservatives were less satisfied than either of the other two groups. Furthermore, moderates who voted were less satisfied with the public schools than those who did not vote; for the other two political groups, whether people voted or not did not matter with respect to their level of satisfaction.

Now we focus on the bottom graph, which presents the results for those who do not have children in the home. For those who voted in the last election, conservatives were less satisfied with the public schools than the other political groups. For those who did not vote, moderates were less satisfied than the other two groups. Furthermore, moderates who voted in the last election were more satisfied than those who did not vote; for the other two political groups, whether people voted or not did not matter with respect to their level of satisfaction.

A. Political Preference

Multiple Comparisons

Dependent Variable: satis public schools
Tukey HSD

(I) political pref	(J) political pref	Mean Difference (I-J)	Std. Error	Sig.	95% Confidence Interval Lower Bound	Upper Bound
liberal	moderate	8.2500*	1.87528	.000	3.7147	12.7853
	conservative	12.7500*	1.87528	.000	8.2147	17.2853
moderate	liberal	- 8.2500*	1.87528	.000	-12.7853	- 3.7147
	conservative	4.5000	1.87528	.052	- .0353	9.0353
conservative	liberal	-12.750*	1.87528	.000	-17.2853	- 8.2147
	moderate	- 4.5000	1.87528	.052	-9.0353	.0353

Based on observed means.
*. The mean difference is significant at the .05 level.

B. Homogeneous Subsets

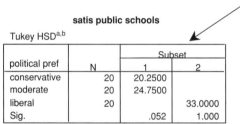

satis public schools

Tukey HSD[a,b]

political pref	N	Subset 1	2
conservative	20	20.2500	
moderate	20	24.7500	
liberal	20		33.0000
Sig.		.052	1.000

Means for groups in homogeneous subsets are displayed
Based on Type III Sum of Squares
The error term is Mean Square(Error) = 35.167.
a. Uses Harmonic Mean Sample Size = 20.000.
b. Alpha = .05.

> The mean for liberal is in its own column indicating that it is significantly different from both the conservative and moderate political preference groups. However, we see a nonsignificant p value of 1.000 because the mean is compared to itself. Conservative and moderate are not significantly different from each other (Sig. = .052).

Figure 9.16   Post hoc results on the political preference main effect.

9.9.4   MAIN EFFECTS POST HOC ANALYSES: RESIDE

Figure 9.16 presents the results of the post hoc Tukey HSD test on the main effect of **politics**. As can be seen, liberals were more satisfied with the public school system than were the other two political groups. Since the eta squared value associated with this main effect was relatively high compared to the other values, it is probably worthwhile to mention this analysis in the results.

9.10   PERFORMING THE OMNIBUS ANALYSIS IN *SAS ENTERPRISE GUIDE*

9.10.1   STRUCTURING THE DATA FILE

A portion of the Excel file containing the data for the numerical example we have been using throughout this chapter is shown in Figure 9.17. That data file is structured as follows. The first variable is a subject identification

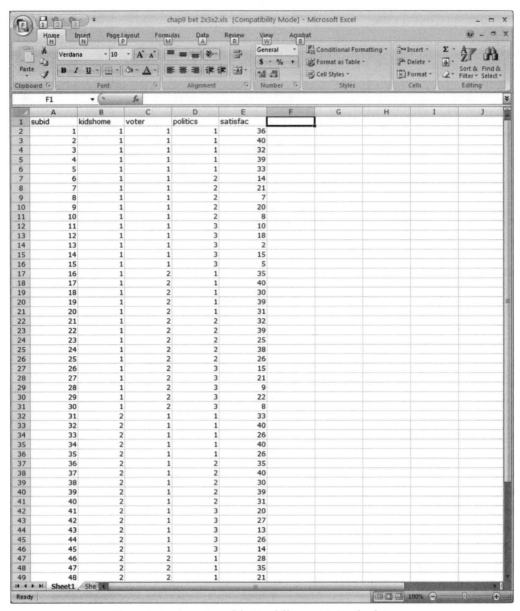

Figure 9.17 A portion of the Excel files containing the data.

code (**subid**). The next three variables are the independent variables in the analysis; each has been assigned arbitrary codes to represent the categories. These variables are **kidshome** (children present in the home coded as 1, children not in the home coded as 2), **voter** (having voted in the last election coded as 1, not voting in the last election coded as 2), and **politics** (liberal, moderate, and conservative coded as 1, 2, and 3, respectively). The last variable in the data file is our dependent variable, named **satisfac**, the score on the survey assessing satisfaction with the public school system.

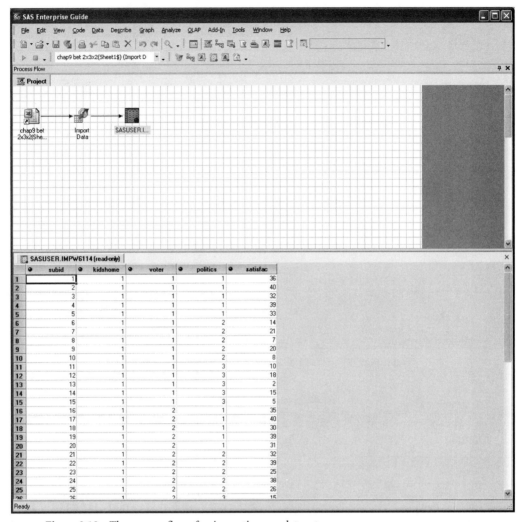

Figure 9.18 The process flow after importing our data set.

When you have imported the data file in *SAS Enterprise Guide*, your window should resemble the one shown in Figure 9.18.

9.10.2 STRUCTURING THE ANALYSIS

Select **Analyze → ANOVA → Linear Models**. The window for this procedure opens on the **Task Roles** tab; this is highlighted in the navigation panel in the left portion of the window shown in Figure 9.19. At the top of the panel labeled **Linear Models Task Roles** there is already a place for a dependent variable being specified. Highlight **satisfac** and drag it to the icon for **Dependent variables** (or highlight **satisfac**, click the right-facing arrow, and select **Dependent variables**). Then, drag **kidshome**, **voter**, and **politics** to the icon for **Classification variables**. At that point, your screen should look similar to the screen shown in Figure 9.20.

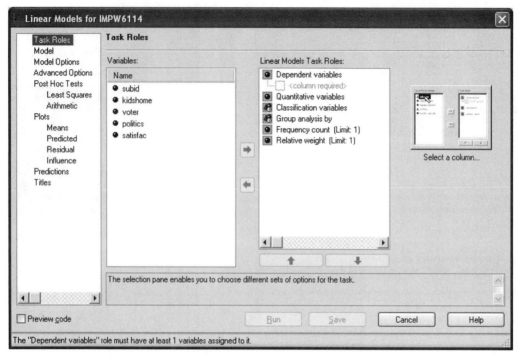

Figure 9.19   The **Task Roles** window.

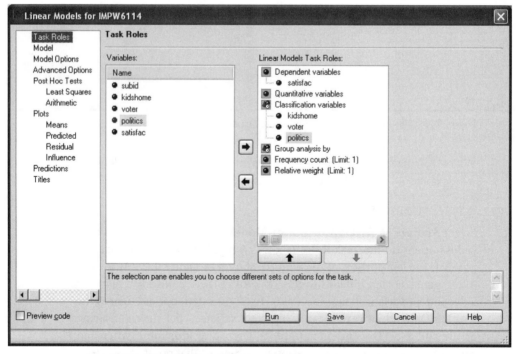

Figure 9.20   The **Task Roles** screen configured.

Click on the **Model** tab. The variables **kidshome, politics,** and **voter** appear in the **Class and quantitative variables** panel. Note that *SAS Enterprise Guide* has a penchant for listing effects in alphabetical order, overriding the order that we used in specifying the variables in the **Task Roles** window; however, the important issue is that the proper effects are in the model, and we will not quibble with this particular idiosyncrasy of the software.

Highlight each of the individual independent variables one at a time, each time placing them in the **Effects** panel by clicking the **Main** bar. Then highlight **kidshome** and **voter** while depressing the Control key; this will allow both to become highlighted. Clicking the **Cross** bar while the two are highlighted will cause them to be brought over to the **Effects** panel as an interaction effect. Repeat this process for the remaining two-way interactions and for the three-way interaction. The final result of this operation is shown in Figure 9.21.

Click on the **Model Options** tab. The only specification we need for our analysis is to request output associated with the **Type III** sums of squares. This is shown in Figure 9.22.

Descriptive summary statistics can be requested on the **Arithmetic** portion of the **Post Hoc Tests** tab. Clicking **Add** displays a set of drop-down menus, only some of which need modifying. For **Class effects to use**, select **True** for all of the effects as shown in Figure 9.23. Then click **Run** to perform the analysis.

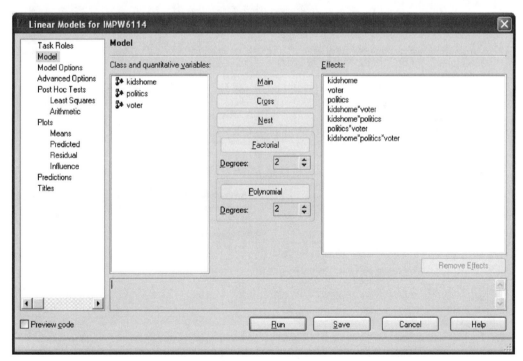

Figure 9.21   The model is now specified with three main effects, three two-way interactions, and the three-way interaction.

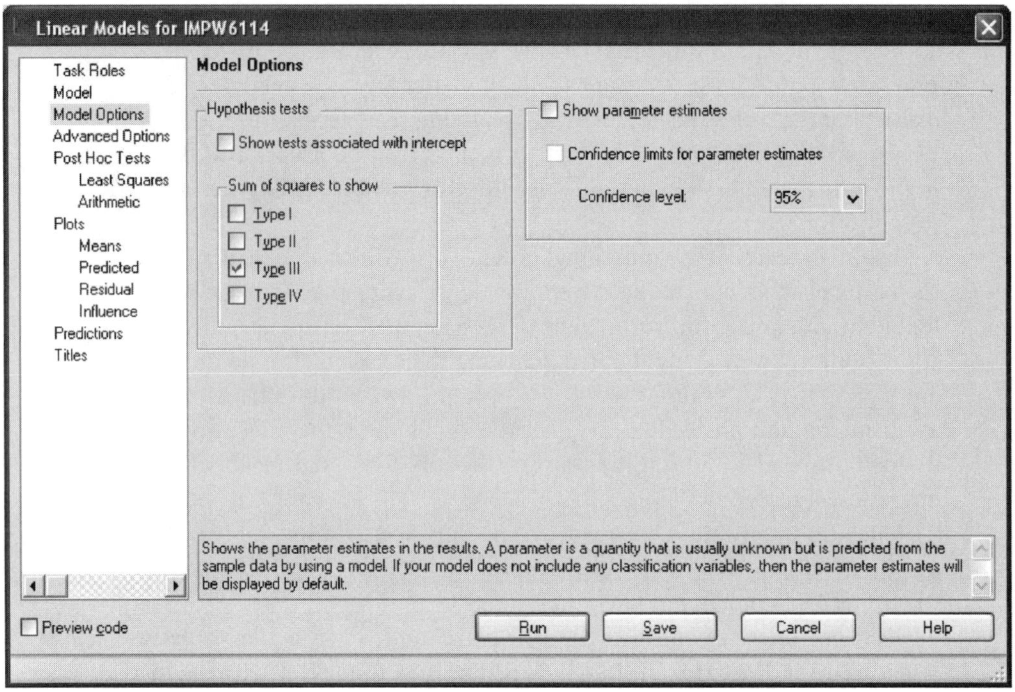

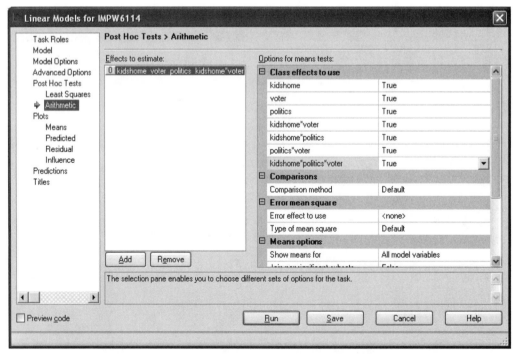

Figure 9.22   We have specified only computing the Type III sum of squares.

Figure 9.23   The descriptive statistics for the effects are now specified.

Linear Models

The GLM Procedure

Level of voter	Level of politics	N	satisfac	
			Mean	Std Dev
1	1	10	34.5000000	5.4619899
1	2	10	24.5000000	12.2678441
1	3	10	15.0000000	8.1513462
2	1	10	31.5000000	6.9161646
2	2	10	25.0000000	9.3689795
2	3	10	25.0000000	12.1860576

This is the mean satisfaction score for conservatives who voted in the last election.

Level of kidshome	Level of voter	Level of politics	N	satisfac	
				Mean	Std Dev
1	1	1	5	36.0000000	3.53553391
1	1	2	5	14.0000000	6.51920241
1	1	3	5	10.0000000	6.67083203
1	2	1	5	35.0000000	4.52769257
1	2	2	5	32.0000000	6.51920241
1	2	3	5	15.0000000	6.51920241
2	1	1	5	33.0000000	7.0000000
2	1	2	5	35.0000000	4.52769257
2	1	3	5	20.0000000	6.51920241
2	2	1	5	28.0000000	7.51664819
2	2	2	5	18.0000000	5.70087713
2	2	3	5	36.0000000	4.00000000

This is the mean satisfaction score for moderates who did not vote in the last election and who have no children in the home.

Figure 9.24 Descriptive statistics output.

9.11 SAS OUTPUT FOR THE OMNIBUS ANALYSIS

A portion of the descriptive statistics generated by **Linear Models** is shown in Figure 9.24. The mean, standard deviation, and the number of observations are displayed for each effect.

The summary table for the omnibus analysis is presented in the top portion of Figure 9.25. Note that SAS provides the **Corrected Total** sum of squares, informing us that we are viewing the results of the reduced model. The sum of squares associated with **Model** is a compilation of the sums of squares for the seven effects (three main effects, three two-way interactions, and one three-way interaction) that comprise the model (each effect is treated as a predictor in the model). These seven effects are evaluated individually in the bottom table. As can be seen, all of the effects except for **voter** are statistically significant.

Linear Models

The GLM Procedure

Dependent Variable: satisfac satisfac

Source	DF	Sum of Squares	Mean Square	F Value	Pr > F
Model	11	5360.000000	487.272727	13.86	<.0001
Error	48	1688.000000	35.166667		
Corrected Total	59	7048.000000			

R-Square	Coeff Var	Root MSE	satisfac Mean
0.760499	22.80827	5.930149	26.00000

These are the probabilities of the *F* ratio occurring by chance alone for the main effects and interactions.

Source	DF	Type III SS	Mean Square	F Value	Pr > F
kidshome	1	326.666667	326.666667	9.29	0.0037
voter	1	106.666667	106.666667	3.03	0.0880
politics	2	1672.500000	836.250000	23.78	<.0001
kidshome*voter	1	326.666667	326.666667	9.29	0.0037
kidshome*politics	2	1060.833333	530.416667	15.08	<.0001
voter*politics	2	490.833333	245.416667	6.98	0.0022
kidsho*voter*politic	2	1375.833333	687.916667	19.56	<.0001

Figure 9.25   ANOVA results.

The middle portion of Figure 9.25 presents **R-Square**; this is an alternative label for eta squared. However, **R-Square** is computed based on the model with all seven effects combined (added) together. In the context of ANOVA, we ordinarily wish to obtain the eta squared value for each separate effect. To do this, we must perform the hand calculation, dividing each sum of squares by the total sum of squares (**Corrected Total**) shown in Figure 9.25. The coefficient of variation (the ratio of the standard deviation of the sample as a whole to the mean of the sample as a whole), the root mean square error (the square root of the mean square associated with the error term), and the grand mean of the dependent variable are also displayed in that middle table.

9.12   PERFORMING THE POST-ANOVA ANALYSES IN SAS

The omnibus analysis indicated that all but one of the effects were statistically significant. The significant three-way interaction supersedes all of the two-way interactions and so we will not follow up on these. Furthermore, we ordinarily would not bother with the main effects either because we have obtained a statistically significant three-way interaction in the omnibus analysis. However, to illustrate how we would approach the statistically significant main effects, we will discuss them here.

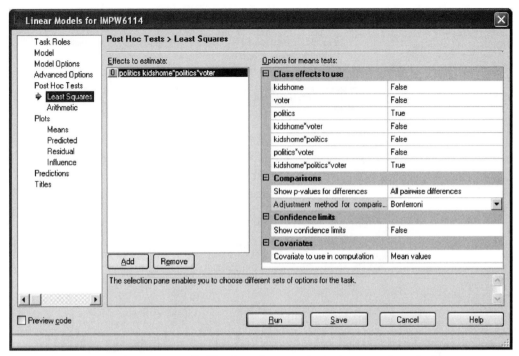

Figure 9.26   We have now specified Bonferroni corrected pairwise comparisons for the main effect of political preference and for the simple effects of the three-way interaction.

Because there were only two levels of **kidshome**, we know from our descriptive statistics that families with children in the home (coded as 1) were less satisfied than families without children in the home (coded as 2). The main effect of **voter** was not significant, indicating that these two overall means can be treated as being comparable. Political preference, however, has three levels, requiring us to perform a multiple comparison procedure to determine which pairs of means were reliably different.

We wish to perform comparisons of the means for (a) the main effect of **politics** and (b) the means of the groups for the three-way interaction; that is, we wish to perform tests of simple effects. As we did for the two-way design discussed in Chapter 8, we will use a Bonferroni corrected multiple comparisons test for both of these so that we can accomplish the specifications in one step.

Configure the analysis as described above for the omnibus analysis. Then select the **Least Squares** portion of the **Post Hoc Tests** tab. This brings you to the screen shown in Figure 9.26. For **Class effects to use**, set **politics** and **kidshome*politics*voter** to **True**; this will generate the post hoc tests for the main effect of **politics** and the simple effects tests for the interaction, respectively. For **Comparisons**, set **Show p-values for differences** to **All pairwise differences** and set **Adjustment method for comparison** to **Bonferroni**. Then click **Run** to perform the comparisons.

Linear Models

The GLM Procedure
Least Squares Means
Adjustement for Multiple Comparisons: Bonferroni

politics	satisfac LSMEAN	LSMEAN Number
1	33.0000000	1
2	24.7500000	2
3	20.2500000	3

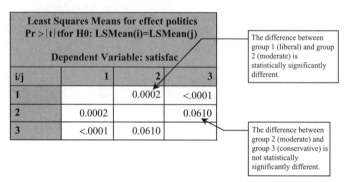

Least Squares Means for effect politics Pr > |t| tfor H0: LSMean(i)=LSMean(j) Dependent Variable: satisfac			
i/j	1	2	3
1		0.0002	<.0001
2	0.0002		0.0610
3	<.0001	0.0610	

The difference between group 1 (liberal) and group 2 (moderate) is statistically significantly different.

The difference between group 2 (moderate) and group 3 (conservative) is not statistically significantly different.

Figure 9.27   Pairwise comparison of means for politics.

9.13 SAS OUTPUT FOR THE POST-ANOVA ANALYSES IN SAS

9.13.1 MAIN EFFECT ANALYSES

Figure 9.27 shows the multiple comparisons for the main effect of **politics**. The upper display shows the levels of **politics** in the first column (liberal was coded as 1, moderate was coded as 2, and conservative was coded as 3), the means in the second column, and the code number assigned to each in the third column. The lower display uses the code numbers in presenting a matrix showing the probability values associated with each mean difference. All differences were statistically significant except for the means of the groups coded 2 (moderates) and 3 (conservatives), which have a value of $p = .0610$.

9.13.2 SIMPLE EFFECTS ANALYSES

Figure 9.28 displays the pairwise comparisons of the means of the inter-action structured in the same way as we just described for the main effect. The upper table gives code numbers to the groups and the lower table shows the p values associated with the pairwise comparisons.

Recall that children in and not in the home were coded as 1 and 2, respectively, that having voted and not having voted in the last election was coded as 1 and 2, respectively, and that liberal, moderate, and conservative were coded as 1, 2, and 3, respectively. To illustrate how to read the pairwise comparison table in Figure 9.28, consider families that had children in the

Linear Models

The GLM Procedure
Least Squares Means
Adjustement for Multiple Comparisons: Bonferroni

kidshome	voter	politics	satisfac LSMEAN	LSMEAN Number
1	1	1	36.0000000	1
1	1	2	14.0000000	2
1	1	3	10.0000000	3
1	2	1	35.0000000	4
1	2	2	32.0000000	5
1	2	3	15.0000000	6
2	1	1	33.0000000	7
2	1	2	35.0000000	8
2	1	3	20.0000000	9
2	2	1	28.0000000	10
2	2	2	18.0000000	11
2	2	3	36.0000000	12

Least Squares Means for effect kidsho*voter*politics
$Pr > |t|$ for H0: LSMean(i)=LSMean(j)

Dependent Variable: satisfac

i/j	1	2	3	4	5	6	7	8	9	10	11	12
1		<.0001	<.0001	1.0000	1.0000	<.0001	1.0000	1.0000	0.0061	1.00		
2	<.0001		1.0000	<.0001	0.0011	1.0000	0.0004	<.0001	1.0000	0.0331	1.0000	<.0001
3	<.0001	1.0000		<.0001	<.0001	1.0000	<.0001	<.0001	0.6878	0.0011	1.0000	<.0001
4	1.0000	<.0001	<.0001		1.0000	0.0002	1.0000	1.0000	0.0144	1.0000	0.0026	1.0000
5	1.0000	0.0011	<.0001	1.0000		0.0026	1.0000	1.0000	0.1611	1.0000	0.0331	1.0000
6	<.0001	1.0000	1.0000	0.0002	0.0026		0.0011	0.0002	1.0000	0.0741	1.0000	<.0001
7	1.0000	0.0004	<.0001	1.0000	1.0000	0.0011		1.0000	0.0741	1.0000	0.0144	1.0000
8	1.0000	<.0001	<.0001	1.0000	1.0000	0.0002	1.0000		0.0144	1.0000	0.0026	1.0000
9	0.0061	1.0000	0.6878	0.0144	0.1611	1.0000	0.0741	0.0144		1.0000	1.0000	0.0061
10	1.0000	0.0331	0.0011	1.0000	1.0000	0.0741	1.0000	1.0000	1.0000		0.6878	1.0000
11	0.0011	1.0000	1.0000	0.0026	0.1611	1.0000	0.0144	0.0026	1.0000	0.6878		0.0011
12	1.0000	<.0001	<.0001	1.0000	1.0000	<.0001	1.0000	1.0000	0.0061	1.0000	0.0011	

> The difference between group 1 (Kids in home, voted, liberals) and group 6 (Kids in home, not voted, conservative) is statistically significantly different.

> The difference between group 7 (Kids not at home, voted, liberal) and group 9 (Kids not at home, voted, conservative) is not statistically significantly different.

Figure 9.28 Tests of simple effects.

home and that consider themselves to be liberal. Those who voted in the; last election (**LSMEAN Number 1**) had a mean satisfaction with the schools of 36; those who did not vote in the last election (**LSMEAN Number 4**) had a mean satisfaction with the schools of 35. These two groups do not differ from each other in the degree of satisfaction they reported ($p = 1.000$).

9.13.3   MAPPING SIMPLE EFFECTS ANALYSES TO THE GRAPH OF THE INTERACTION

Mapping the simple effects to the plot of the interaction (see Figure 9.3) was discussed in the context of SPSS in Section 9.9.3, and the reader is referred to that discussion if needed to determine which data points in the plot significantly differ from which others.

9.14   COMMUNICATING THE RESULTS

A written summary of the results is as follows.

A $2 \times 3 \times 2$ between-subjects design investigated citizen attitudes toward the public school system. The independent variables were whether or not there were children in the home (yes or no), political preference (liberal, moderate, or conservative), and whether or not participants had voted in the last election (yes or no). Overall, there were significant main effects for whether there were children living in the home, $F(1, 48) = 9.29$, $p < .05$, $\eta^2 = .05$, and political preference, $F(2, 48) = 9.29$, $p < .05$, $\eta^2 = .24$. For this latter effect, a Tukey test indicated that liberals ($M = 33.00$, $SD = 6.26$) were more satisfied with the public school system than either moderates ($M = 24.75$, $SD = 10.63$) or conservatives ($M = 20.25$, $SD = 11.44$).

These main effects were obtained in the context of a significant three-way interaction, $F(2, 48) = 19.56$, $p < .05$, $\eta^2 = .20$, the plot for which is presented in Figure 1. Bonferroni adjusted simple effects tests holding the alpha level at .05 revealed the following.

For those who voted in the last election, liberals were more satisfied than either of the other two political groups. For those who did not vote conservatives were less satisfied than either of the other two groups. Furthermore, moderates who voted were less satisfied with the public schools than those who did not vote; for the other two political groups, whether people voted or not did not matter with respect to their level of satisfaction.

For those who voted in the last election, conservatives were less satisfied with the public schools than the other political groups. For those who did not vote, moderates were less satisfied than the other two groups. Furthermore, moderates who voted in the last election were more satisfied than those who did not vote; for the other two political groups, whether people voted or not did not matter with respect to their level of satisfaction.

CHAPTER 9 EXERCISES

9.1. Assume that male (a_1) and female (a_2) community mental health clients are randomly assigned to a special antianxiety medication (b_1) or a placebo (b_2) control condition, in addition to one of three types of psychotherapy conditions ($c_1 =$ brief, $c_2 =$ cognitive-behavioral, and $c_3 =$ psychoanalytic). GAF scores at the end of six weeks of treatment served as the dependent measure. The $2 \times 2 \times 3$ factorial data are as follows:

	c_1				c_2				c_3		
$a_1 b_1$	$a_1 b_2$	$a_2 b_1$	$a_2 b_2$	$a_1 b_1$	$a_1 b_2$	$a_2 b_1$	$a_2 b_2$	$a_1 b_1$	$a_1 b_2$	$a_2 b_1$	$a_2 b_2$
40	35	50	45	60	70	65	50	40	50	48	50
45	40	55	48	65	70	65	55	45	55	45	45
48	38	58	60	68	65	60	60	38	52	45	40
40	35	50	44	65	65	60	55	30	55	60	45
45	35	48	50	65	55	66	50	35	50	55	50
45	35	50	50	70	60	70	55	40	50	50	55

a. Using SPSS or SAS, conduct an ANOVA on these data.
b. If the interaction(s) is(are) statistically significant, conduct the necessary simple effects and multiple comparisons tests with SPSS or SAS.

9.2. Below is a hypothetical $3 \times 2 \times 2$ factorial experiment, where $n = 5$. Participants were randomly assigned to the following treatment conditions.

					Treatments						
a_1	a_1	a_1	a_1	a_2	a_2	a_2	a_2	a_3	a_3	a_3	a_3
b_1	b_1	b_2	b_2	b_1	b_1	b_2	b_2	b_1	b_1	b_2	b_2
c_1	c_2	c_1	c_2	c_1	c_2	c_1	c_2	c_1	c_2	c_1	c_2
4	5	6	3	6	7	8	7	9	7	8	10
3	5	6	3	7	5	5	5	10	9	9	9
8	9	7	2	6	7	8	7	9	10	10	8
7	5	7	4	7	5	8	7	8	10	8	8
5	2	5	4	7	6	6	8	8	9	9	7

a. Conduct an ANOVA on these data with SPSS or SAS.
b. If the interaction(s) is(are) statistically significant, conduct a simple effects analysis and any necessary multiple comparisons tests.

9.3. Consider the following hypothetical $3 \times 2 \times 2$ factorial experiment in which college freshmen were randomly assigned to one of three type of music conditions ($a_1 =$ heavy metal, $a_2 =$ easy listening, and $a_3 =$ rap) and in which they listened to prerecorded music in their spare time with ipod technology. They also participated in either in-class (b_1) or online (b_2) lectures for all courses taken during the semester, and they were also assigned to either live on campus (c_1) or off campus (c_2). Students' semester grade point average served as the dependent measure. The data are as follows:

Treatments

a_1	a_1	a_1	a_1	a_2	a_2	a_2	a_2	a_3	a_3	a_3	a_3
b_1	b_1	b_2	b_2	b_1	b_1	b_2	b_2	b_1	b_1	b_2	b_2
c_1	c_2	c_1	c_2	c_1	c_2	c_1	c_2	c_1	c_2	c_1	c_2
2.66	3.01	2.87	2.16	3.16	3.35	3.16	3.09	3.30	3.40	3.50	3.60
2.98	3.00	2.99	2.24	3.25	3.40	3.27	3.10	3.35	3.45	3.50	3.65
2.96	1.95	2.65	3.11	3.08	3.45	3.53	3.15	3.40	3.50	3.55	3.70
3.01	2.90	3.05	2.94	3.00	3.41	3.13	3.20	3.45	3.55	3.60	3.80
3.15	2.25	3.10	2.55	3.34	3.08	3.27	3.25	3.50	3.60	3.65	3.75

a. Conduct an ANOVA on these data with SPSS or SAS.
b. If the interaction(s) is(are) statistically significant, conduct simple effects analyses and multiple comparisons tests with either SPSS or SAS.
c. Write up your results.

SECTION 4

Within-Subjects Designs

One-Way Within-Subjects Design

10.1 THE CONCEPT OF WITHIN-SUBJECTS VARIANCE

In between-subjects designs, participants contribute a single value of the dependent variable, and different participants are represented under each level of the independent variable. The situation is different for within-subjects designs. Here, we register a value of the dependent variable in each and every condition of the study for every one of the participants or cases in the study. That is, each participant is measured under each level of the independent variable.

Within-subjects designs are still univariate studies in that there is only one dependent variable in the design. As is true for between-subjects designs, the dependent variable is operationalized in a particular manner. What differentiates between-subjects and within-subjects designs is the number of such measurements in the data file that are associated with each participant: In between-subjects designs, researchers record just one data point (measurement) for each participant; in within-subjects designs, researchers record as many measurements for each participant as there are conditions in the study.

Within-subjects designs can theoretically contain any number of independent variables combined in a factorial manner. Practically, because all participants must contribute data under each level of each independent variable, the logistics of the data collection procedures can become quite challenging with several independent variables in the design. Also, there is the possibility that subjects could be affected or changed by exposure to one level of an independent variable, a situation known as *carry-over effects,* such that it may not be appropriate to expose them to more than that single level; the likelihood of eliminating all of the possible carry-over effects diminishes with each additional independent variable included in the design (we will discuss carry-over effects more in Section 10.4). For these reasons, it is relatively uncommon to encounter within-subjects designs incorporating more than three independent variables.

10.2 NOMENCLATURE

Within-subjects designs are named by using the same strategy as is used for between-subjects designs: we talk about *n*-way designs where *n* is a count of the number of independent variables we have. Thus, a one-way within-subjects design contains only a single independent variable, a two-way within-subjects design contains two independent variables, and so on. We will explicate the one-way design in this chapter and the two-way and three-way designs in Chapters 11 and 12, respectively.

Within-subjects designs are also known as *repeated measures designs* because the dependent variable is repeatedly measured, that is, it is measured under each research condition for each participant. The one-way within-subjects design is also sometimes called a Treatment × Subjects design because all of the levels of the independent variable (the treatment) are crossed with (are administered to) all of the subjects (participants) in the study.

10.3 NATURE OF WITHIN-SUBJECTS VARIABLES

Within-subjects variables are those in which participants are measured under each of the conditions (i.e., cases are repeatedly measured). Participants are thus represented in each and every research condition in the study. Generally, there are two structural forms that this repeated measurement can take: (a) it can mark the passage of time, or (b) it can be unrelated to the time of the measurement but simply indicate the conditions under which participants were measured. Although these two forms of within-subjects designs do not affect either the fundamental nature of the design or the data analysis, they do imply different research data collection procedures to create the measurement opportunities.

10.3.1 VARIABLES MARKING TIME

A within-subjects variable marking the passage of time is one in which the first level of the variable is measured at one point in time, the next level of the variable is measured at a later point in time, the third level of the variable is assessed at a still later period of time, and so on. The most commonly cited example of a time-related within-subjects design is the *pretest–posttest* study. While not a true experimental design (Campbell & Stanley, 1963; Cook & Campbell, 1979; Shadish, Cook, & Campbell, 2001), partly because there is no control condition, participants are measured at least twice, once at the time designated as the pretest and once at the time designated as the posttest. In such a study the treatment is ordinarily administered between the two measurements and researchers would ordinarily have hypothesized either an increase or a decrease in the characteristic assessed by the dependent variable from the pretest to the posttest.

Each measurement (pretest and posttest) is performed in exactly the same manner by using the same methodology. Thus, there is just one dependent variable, but individuals are measured at Time 1 and again at Time 2. In this manner, time completely covaries with the measurement and it is not uncommon for researchers to talk about the within-subjects variable in terms of time. For example, the terms *pretest* and *posttest* have been chosen because the temporal relationship of the measurement has been conveniently built into the terminology.

A pretest–posttest design can have more than two levels. For example, it might be important to have two or more pretests to establish a stable performance baseline prior to the administration of some treatment condition. It is also possible that researchers will want to have two or more posttests, perhaps to establish the degree to which the treatment effect will diminish (or exhibit itself) over time or perhaps to determine how long it takes for the treatment to maximize its effect.

10.3.2 VARIABLES UNRELATED TO TIME

A within-subjects variable does not have to be time related to measure participants under all of the research conditions. To illustrate this, consider a simple hypothetical study in which we ask students to classify briefly presented selections of music as quickly as they can. We use two types of music: rock and classical. Students are asked to press one key if they think that the music is rock and another key if they think that the selection is classical. We measure reaction time as well as whether the student is correct on each trial. We administer 100 trials to the students, 50 for each music condition, such that the rock and classical music selections are presented to each student in a different random order. At the end of this data collection procedure in this simplified example, we average the reaction times for rock and classical music for those trials where the students answered correctly; thus, reaction time is our dependent variable. This gives us two reaction time scores for each student, one for rock music and one for classical music. These measurements represent a within-subjects independent variable whose two levels are rock music and classical music.

10.4 THE ISSUE OF CARRY-OVER EFFECTS

A carry-over effect is the change in a later measure resulting from exposure to a previous condition. This is a common occurrence in our everyday lives. Having learned to roller skate, we can more efficiently learn how to ice skate; having taken our first statistics course, we are better prepared for the next statistics course. In a research study, having experienced a particular treatment condition may result in differences in performance in some subsequent task compared to those not having experienced it. These are all carry-over effects and, generally speaking, they are expected and desirable.

Some carry-over effects in the context of research design are not desirable. As an extreme example, if we wished to evaluate the efficacy of comparing cognitive-behavioral therapy to psychoanalysis for a set of clients, we would not first provide seven years of psychoanalysis to them and then give them some cognitive-behavioral therapy (or the reverse). We would obviously not be able to observe the "pure" effects of cognitive-behavioral therapy under these circumstances – what we would see is the effects of cognitive-behavioral therapy after clients have experienced extensive psychoanalysis. Having received one type of therapy would presumably carry over to the next research condition. The adverse consequence of this carry-over would be to confound our measurements of cognitive-behavioral therapy success, rendering this particular design untenable.

This potential problem of carry-over effects is something that researchers considering a within-subjects study must directly confront. If experiencing one level of the independent variable – the repeated measure – will adversely carry over to the experience of another level of the independent variable, then a within-subjects study should not be run. For example, if we were studying the effectiveness of applying a memorization strategy in trying to remember sequences of numbers (digit span), having taught participants to group the numbers into sets of twos would change their performance if we wished to test them when they did not use the grouping strategy. On the other hand, if we tested them first without any instruction of strategy, we might be able to use that as a baseline in testing their performance with the grouping strategy being used. Even then, we would want to have a control group given a second set of trials without a strategy in place because having had some "practice" with the digit span test might affect subsequent performance.

The issue of carry-over is usually less of a concern when we work with time-related within-subjects variables because it is precisely the carry-over effects that we intend to measure when we design such a study. In a simple pretest–posttest design, for example, the treatment effect is carried across the time between the measurements. As long as our actual pretest measurement did not produce the changes in the values of the dependent measure that we see in the posttest, and as long as some other event occurring at the same time as the treatment or any natural changes in the participants over the time period did not cause the changes, and assuming that a control condition cannot be measured, then a pretest–posttest design may be better than not running the study at all.

A within-subjects variable that is unrelated to time is more susceptible to adverse carry-over effects. Because there is no natural temporal sequence in which the levels of the independent variable must be experienced, the sequence is usually under the control of the researchers. Here it is vital that the researchers are able to reasonably eliminate carry-over effects as potential causes of differences between the means of the independent variable. In the music identification study described earlier, for example, it can be argued that if the students received fifty successive trials in identifying rock music before they received their fifty trials for classical music, a number of confounds would present themselves (e.g., they are well trained by the time they are exposed to classical music; they may recognize that the trials are "blocked," assume that the next trial will be the same type of music, and respond on the basis of their guess of music order rather than reacting to the music itself). By randomizing the order of music trials and presuming that having just heard a snippet of classical or rock music will not affect the students on the next trial, a within-subjects design appears to be viable. Actually, it becomes the design of choice for the study in that individual differences can be statistically accounted for in a within-subjects design (because all subjects are measured under all conditions) but cannot be so accounted for in a between-subjects design (because different subjects are measured under each level of the independent variable). Thus, in a within-subjects design, we can identify a type of variance known as within-subjects variance.

10.5   BETWEEN- VERSUS WITHIN-SUBJECTS VARIANCE

10.5.1   BETWEEN-SUBJECTS VARIANCE IN A BETWEEN-SUBJECTS DESIGN

In a one-way design, whether it is between subjects or within subjects, we are interested in comparing the distributions of scores under each level of the independent variable by using the means (and standard deviations) as proxies for the distributions. For between-subjects designs, each level of the independent variable is associated with different participants; we therefore refer to the levels as "groups," and presume, if individuals were randomly assigned to the groups, that at least in the long run differences between participants will have been "randomized out" and will not confound any mean differences we might observe between the treatment conditions. The variance that we partition in such an ANOVA is therefore called between-subjects variance.

Chapters 6 through 9 have presented *n*-way between-subjects designs. The variance of the dependent variable that we have partitioned has all been between-subjects variance. That is, because a subject contributed just one score to the analysis, and because different subjects were measured under each level of the independent variable, talking about differences between the sets of scores was tantamount to talking about differences between the sets of subjects. When we compared the means of different groups, we were also comparing the differences in the performance of different subjects. The situation is more complex in a within-subjects design.

10.5.2   BETWEEN-SUBJECTS VARIANCE IN A WITHIN-SUBJECTS DESIGN

Within-subjects designs contain multiple participants who are measured repeatedly on the dependent variable. With multiple subjects, one set of scores (a set of repeated measures) will have resulted from the measurement of Subject 1, another set of scores will have resulted from the measurement of Subject 2, and so on. When we compare these sets of scores to each other, we are comparing the subjects to each other. In comparing different subjects to each other, we are dealing with between-subjects variance. Thus, even when using a within-subjects design, some

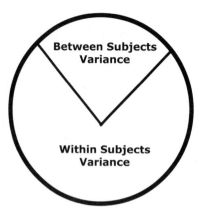

Figure 10.1 Total variance of the dependent variable partitioned into between-subjects and within-subjects variance.

portion of the variance of the dependent variable can be attributed to differences between the subjects.

The between-subjects variance in a within-subjects design captures the extent to which *individual differences* were observed between the subjects in the study. Once the individual differences are accounted for (once the variance due to between-subjects variance is identified) in the ANOVA, then the remaining variance is within-subjects variance. This is illustrated in Figure 10.1. The circle represents the total variance of the dependent variable. We have partitioned the total variance into two general regions: between-subjects variance and within-subjects variance.

10.5.3   WITHIN-SUBJECTS VARIANCE IN A WITHIN-SUBJECTS DESIGN

A within-subjects design requires all of the participants to be measured under each study condition. With the very same participants represented at each level of the independent variable, any mean differences that we observe between the conditions cannot be attributed to differences in the participants experiencing those conditions (i.e., all conditions are experienced by exactly the same people). Thus, in comparing one level of the independent variable to its other levels, there are no between-subjects differences (there is no between-subjects variance) with which to contend. Because the same participants are involved in such comparisons of means, the variance represented by the means of the different conditions represents the within-subjects variance. In that sense, it is often said that *in within-subjects designs, participants serve as their own controls.* In this way, within-subjects designs are very powerful. They statistically remove (account for) a potentially major source of error, namely, individual differences among the participants.

If within-subjects designs are so powerful, why are they not used all or most of the time? The answer is that many of the independent variables we study are not amenable to a within-subjects manipulation because of the adverse carry-over effects that would be involved. That is, for many of the constructs we study, having exposed people to one condition (one level of the independent variable) sufficiently changes them so that they are no longer naïve with respect to any other research condition. Furthermore,

Table 10.1. Raw data for a (five-group) one-way within-subjects ANOVA

	Factor A: Symptom intensity measured before and after drug therapy					
	One month before	One week before	One week after	One month after	One year after	
Subjects	a_1	a_2	a_3	a_4	a_5	Sum (S)
s_1	12	12	7	5	5	41
s_2	9	10	5	6	5	35
s_3	9	8	7	5	6	35
s_4	8	6	6	4	4	28
s_5	8	9	6	4	5	32
s_6	9	10	8	6	7	40
s_7	12	10	7	5	4	38
s_8	6	7	5	7	5	30
	$M = 9.125$	$M = 9.000$	$M = 6.375$	$M = 5.250$	$M = 5.125$	

in other research approaches, such as archival research where the data are already collected, the best we may often be able to do is to compare different cases whose dependent variable scores are recorded under different levels of the independent variable.

10.6 A NUMERICAL EXAMPLE OF A ONE-WAY WITHIN-SUBJECTS DESIGN

The simplified data set that will serve as our example uses a time-based independent variable in a pretest–posttest configuration. In our hypothetical study, it is hoped that a certain drug will alleviate the intensity of symptoms of a certain disease. Symptom intensity was evaluated on a twelve-point scale with values toward twelve indicating more intense symptoms. Two pretreatment baseline measurements are made, the first a month prior to treatment and the second a week prior to treatment. Posttreatment measures of symptom intensity are made after one week, one month, and one year following administration of the drug. In this simplified example, we have eight patients in the study. As can be seen from the means of the conditions in Table 10.1, symptom intensity hovers around a value of 9 in the pretests, drops to a value somewhat in excess of 6 after the first week, and then appears to level off at a value just a little more than 5 after a month.

10.7 EFFECT OF INTEREST IN THIS DESIGN

10.7.1 OMNIBUS EFFECTS

In a within-subjects design, the total variance of the dependent variable is partitioned into between-subjects and within-subjects variance. The between-subjects variance indexes individual subject differences, and is ordinarily not of interest other than having it account for its portion of the variance.

With only one independent variable in the design, there is only one F ratio in the omnibus analysis that interests us – the F ratio associated with the independent variable. This is a within-subjects variable and therefore is a component of the within-subjects variance. In the numerical example we use here, this F ratio will be statistically significant, indicating that the drug has significantly reduced symptom intensity.

10.7.2 MULTIPLE COMPARISONS

With a statistically significant effect, a post-ANOVA test is needed to determine which means differ significantly from which others. Unlike a between-subjects design, we cannot perform post hoc tests in SPSS – this software is programmed such that the **Post Hoc** window is available only for between-subjects independent variables. Instead, we can perform the same type of comparisons (i.e., paired comparisons) that are used in the simple effects analyses that we have described in Chapter 7. SPSS and SAS make this quite easy, as you will see when we describe how to perform the ANOVA and multiple comparisons procedures in SPSS and SAS.

10.8 THE ERROR TERM IN A ONE-WAY WITHIN-SUBJECTS DESIGN

The error term for a one-way within-subjects design must be understood from a different perspective than we understand the error term in a between-subjects design. We can illustrate the conception of within-subjects error in Figure 10.2, where we have graphed the individual data

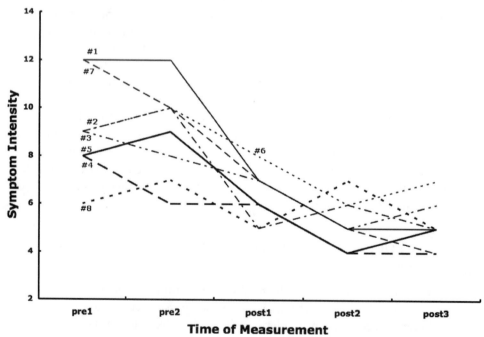

Figure 10.2 Pattern of symptom intensity over the repeated measurements for each participant.

for each participant in our numerical example. For the sake of convenience, we have spaced the conditions equally on the x axis. Each line or function represents the data for a single subject. Differences in the heights of the functions represent different degrees of symptom intensity.

The display is a little cluttered with all eight subjects depicted, but the important feature to note in Figure 10.2 is that the lines are not parallel. This indicates that symptom intensity across the time periods presents a different pattern for the different patients in the study and is the key to understanding the error term in the ANOVA. Consider for a moment the unlikely situation where the lines for each patient were actually parallel but of different elevations in the graph. What would this suggest? It would suggest that although the patients differ in the intensity of their symptoms, the "progression" or change in symptom intensity over the period during which our measurements occurred would be the same for everyone. Thus, knowing the level of their initial intensity (or the intensity at any one point), the symptom intensity of the patients at every measurement occasion would be perfectly predictable. With such perfect predictability, there would be no variability in the way that patients exhibited symptom intensity over the year or so of our measurement. Such consistency would in turn suggest that there is no unaccounted for variability or error variance concerning this progression over time.

The data shown in Figure 10.2 are more typical of what would be observed in an empirical study in that they are far from being error free. The degree of inconsistency of the study participants is visually represented by their different patterns of symptom intensity across time. To the extent that the functions are not parallel, we can say that we have unaccounted for variance, that is, we can say that we have *error variance* observed in the data with respect to the consistency of patients reporting symptom intensity over time. Specifically, despite the fact that all of the individuals were exposed to the same treatment and were measured at the same time, we observe quite different patterns in their reports of symptom intensity. Note that this effect is not the same as different levels of symptom intensity observed between the patients. Although it is true that some patients experienced more intense symptoms than others, our focus here is on the differences in the pattern of symptom intensity for the different patients.

At this point, some of you might be thinking, "It is not terribly surprising that different people are exhibiting different patterns, so why make so much out of it? Different people have different body chemistries, different reactions to having illness, and so on." We agree that this finding is not surprising. The key issue is that all of these factors that produce different patterns are *unknown* in the context of the research study. Without having measured these factors (variables), we have no way to examine the way in which they covary with the dependent measure. And so these different patterns exhibited by the subjects in the study represent, from a measurement perspective, *error variance*.

Statistically, nonparallel functions are indicative of an *interaction*, and that is the case here. Because the interaction is indicative of the differences in patterns of reactions to the treatment, and because differences in patterns of reactions are indicative of error variance, the interaction effect is defined as the error term in the ANOVA. The two variables that interact are (a) time of measurement, which represents our within-subjects independent variable – our treatment effect – and (b) the subjects in the study. Because this interaction effect is the error variance, the mean square associated with the Treatment × Subjects interaction is used in the ANOVA as the denominator (it is used as the error term) in the F ratio assessing the effect of the within-subjects independent variable.

10.9   COMPUTING THE OMNIBUS ANALYSIS BY HAND

The procedures for conducting a one-way within-subjects ANOVA have both some parallels to the one-way between-subjects design and some very real differences. Table 10.2 depicts the basic observations of our hypothetical drug therapy study, which we will refer to as the *AS* data matrix. As can be seen, the eight subjects (or participants) are measured on the dependent variable (symptom intensity) on five separate occasions (or treatments). Each subject is designated with a lower case *s* and subscript (read "little '*s*' one," etc.), and the levels of Factor *A* continue to be designated a_1, a_2, a_3, and so forth. Sums (A_j), sums of squared scores (Y_{aj}^2), and means ($\overline{Y}_j$) are computed for each treatment condition, as in the single-factor between-subjects case we discussed in Chapter 6. A new computational procedure, required of within-subjects designs, is to sum each subject's score on the dependent variable across all treatment conditions. We will designate these subject sums as S (read "big 'S'"), and they can be seen on the right side of Table 10.2. The *A* treatment sums and the *S* subject sums both sum to the grand sum *T*.

10.9.1   SUM OF SQUARES

There are four sums of squares that we need to calculate for a one-way within-subjects ANOVA, all of which are produced by manipulating the three types of sums found in this design: *A*, *S*, and *T*.

- Sum of squares between groups or treatments (SS_A) in which the subscript *A* represents the variability among the treatments of independent variable or Factor *A*.
- Sum of squares subjects (SS_S) in which the subscript *S* reflects the average variability among the subjects.
- Sum of squares interaction ($SS_{A \times S}$) (read "sum of squares *A* by *S*") reflects the treatments by subjects variability or how the individual subjects respond to the various treatments. The larger this sum of squares, the greater is the variability among the subjects.
- Sum of squares total (SS_T).

Table 10.2. *AS* data matrix and preliminary calculations for a (five-group) one-way within-subjects ANOVA

| Subjects | Factor A: Symptom intensity measured before and after drug therapy | | | | | |
	One month before a_1	One week before a_2	One week after a_3	One month after a_4	One year after a_5	Sum (S)
s_1	12	12	7	5	5	41
s_2	9	10	5	6	5	35
s_3	9	8	7	5	6	35
s_4	8	6	6	4	4	28
s_5	8	9	6	4	5	32
s_6	9	10	8	6	7	40
s_7	12	10	7	5	4	38
s_8	6	7	5	7	5	30
Sum of scores $\Sigma Y = \Sigma A_j$:	$A_1 = 73$	$A_2 = 72$	$A_3 = 51$	$A_4 = 42$	$A_5 = 41$	$\Sigma Y = \Sigma A = T = 279$
Sum of squared scores ΣY_{aj}^2:	$Y_{a_1}^2 = 695$	$Y_{a_2}^2 = 674$	$Y_{a_3}^2 = 333$	$Y_{a_4}^2 = 228$	$Y_{a_5}^2 = 217$	$\Sigma Y_{aj}^2 = 2,147$
Number of cases:						$N = 8$
Mean $(\overline{Y}_j)$:	$\overline{Y}_1 = 9.13$	$\overline{Y}_2 = 9.00$	$\overline{Y}_3 = 6.38$	$\overline{Y}_4 = 5.25$	$\overline{Y}_5 = 5.13$	$\overline{Y}_T = \dfrac{T}{(a)(n)} = 6.98$
Variance (s_j^2):	$s_1^2 = 4.13$	$s_2^2 = 3.71$	$s_3^2 = 1.13$	$s_4^2 = 1.07$	$s_5^2 = 0.98$	
Standard deviation (s_j):	$s_1 = 2.03$	$s_2 = 1.93$	$s_3 = 1.06$	$s_4 = 1.04$	$s_5 = 0.99$	

The formulas for these four sums of squares and the summary of their calculations based on data from Table 10.2 are as follows:

Sum of Squares for the Treatment Effect (Within-Groups Variance): SS_A

$$SS_A = \frac{\Sigma A^2}{n} - \frac{T^2}{(a)(n)}$$
$$= \frac{(73)^2 + (72)^2 + (51)^2 + (42)^2 + (41)^2}{8} - \frac{(279)^2}{(5)(8)}$$
$$= 2,069.88 - 1,946.03 = \boxed{123.85}, \qquad (10.1)$$

$$SS_S = \frac{\Sigma S^2}{a} - \frac{T^2}{(a)(n)}$$
$$= \frac{(41)^2 + (35)^2 + (35)^2 + (28)^2 + (32)^2 + (40)^2 + (38)^2 + (30)^2}{5}$$
$$- \frac{(279)^2}{(5)(8)}$$
$$= 1,976.60 - 1,946.03 = \boxed{30.57}, \qquad (10.2)$$

$$SS_{A \times S} = \Sigma Y_{aj}^2 - \frac{\Sigma A^2}{n} - \frac{\Sigma S^2}{a} + \frac{T^2}{(a)(n)}$$
$$= 2{,}147 - 2{,}069.88 - 1{,}976.60 + 1{,}946.03 = \boxed{46.55},$$
$$\tag{10.3}$$

$$SS_T = \Sigma Y_{aj}^2 - \frac{T^2}{(a)(n)}$$
$$= 2{,}147 - 1{,}946.03 = \boxed{200.97}. \tag{10.4}$$

As you will see Sections 10.12.3 and 10.12.4 (see Figures 10.13 and 10.14), the SPSS output for the sum of squares (under **Type III Sum of Squares**) directly corresponds to our final hand computations. Specifically, SS_A and $SS_{A \times S}$ can be seen in Figure 10.13 under the sources labeled **PREPOST** and **Error (PREPOST)**, respectively. Likewise, SS_S can be found in Figure 10.14 under the source labeled **Error**. We will explain these figures in more detail at the end of this chapter.

10.9.2 CALCULATING DEGREES OF FREEDOM

Below are the formulas for the degrees of freedom associated with each sum of squares, and the simple computations involved based on our numerical example:

$$df_A = a - 1 = 5 - 1 = \boxed{4}$$
$$df_S = n - 1 = 8 - 1 = \boxed{7}$$
$$df_{A \times S} = (a - 1)(n - 1) = (5 - 1)(8 - 1) = (4)(7) = \boxed{28}$$
$$df_T = (a)(n) - 1 = (5)(8) - 1 = 40 - 1 = \boxed{39}.$$

10.9.3 CALCULATING MEAN SQUARES AND F RATIO

We calculate three mean squares by dividing each sum of squares by its respective degrees of freedom. The F ratio is formed by dividing the mean square treatments by the error component mean square treatments by subjects. The calculation of these mean squares and the F ratio follows:

$$MS_A = \frac{SS_A}{df_A} = \frac{123.85}{4} = \boxed{30.96} \tag{10.5}$$

$$MS_S = \frac{SS_S}{df_S} = \frac{30.57}{7} = \boxed{4.37} \tag{10.6}$$

$$MS_{A \times S} = \frac{SS_{A \times S}}{df_{A \times S}} = \frac{46.55}{28} = \boxed{1.66} \tag{10.7}$$

$$F = \frac{MS_A}{MS_{A \times S}} = \frac{30.96}{1.66} = \boxed{18.65}. \tag{10.8}$$

10.9.4 EVALUATING THE F RATIO AND TREATMENT EFFECT MAGNITUDE

We test the null hypothesis for our observed (calculated) F ratio by evaluating (or comparing) the F value with critical values of F (see Appendix C) at the following degrees of freedom: $(df_A, df_{A \times S})$, at a particular alpha level. Thus, for our present numerical example we have $F(4, 28)$ df and the critical value is 2.95 at the .05 alpha level. Because our observed F of 18.65 exceeds the critical value of F, we reject the null hypothesis and conclude that a statistically significant treatment effect is present. More concretely, symptom intensity appears to be a function of time of test.

To determine how much variance we are accounting for by our treatment manipulation (time of test), we will compute eta squared and partial eta squared.

The formulas for these two measures of treatment effect magnitude, based on the present numerical example, follow:

$$\eta^2 = \frac{SS_A}{SS_T} = \frac{123.85}{200.97} = \boxed{0.616} \text{ or } 62\% \qquad (10.9)$$

$$\text{partial } \eta^2 = \frac{SS_A}{SS_A + SS_{A \times S}} = \frac{123.85}{170.04} = \boxed{0.728} \text{ or } 73\%. \quad (10.10)$$

Because the partial eta squared statistic is typically generated by computer programs like SPSS and SAS, it gets reported more often than other measures of treatment magnitude (e.g., omega squared, partial omega squared, and eta squared). From this assessment, we can conclude that our statistically significant F value accounts for about 73 percent of the total variance. These results are conveniently depicted in Table 10.3.

10.10 PERFORMING USER-DEFINED (PLANNED) COMPARISONS BY HAND

10.10.1 DESIGNATING THE COMPARISONS OF INTEREST

Our statistically significant F in our previous numerical example informs us that there is a real difference between or among the five treatment

Table 10.3. Summary table for one-way within-subjects design

Source	SS	df	MS	F	η^2
			Within-subjects effects		
Symptom intensity (A)	123.85	4	30.96	18.65*	0.73
Subject (S)	30.57	7	4.37		
$A \times S$	46.55	28	1.66		
Total	200.97	39			

Note: Partial eta squared $= SS_A/SS_T - SS_S = 123.85/170.40 = 0.727$.
*$p < .05$.

Table 10.4. Comparison matrix for a pairwise within-groups comparison

Subjects	One month before a_1	One year after a_2	Sum (S)
s_1	12	5	17
s_2	9	5	14
s_3	9	6	15
s_4	8	4	12
s_5	8	5	13
s_6	9	7	16
s_7	12	4	16
s_8	6	5	11
Sum (A):	73	41	114
Sum of squared scores ($\Sigma Y_{a_j}^2$):	695	217	912
Mean ($\overline{Y}_j$):	9.13	5.13	7.13

group means. Conducting a set of planned comparisons will allow us to discover what is "driving" this treatment effect and also will allow us to directly test specific research hypotheses that we may have formulated at the start of the study. We encourage the interested reader to review our previous general discussion of this topic in Chapter 7, Section 7.14.

In the between-subjects designs that we covered previously, analytical comparisons are evaluated with the omnibus error term (e.g., $MS_{S/A}$, $MS_{S/AB}$). In the within-subjects case the overall within-subjects error term ($MS_{A \times S}$) is generally not appropriate, particularly if the assumptions of the analysis have been violated (Keppel et al., 1992; Keppel & Wickens, 2004). This is because the overall within-subjects error term captures information from all of the conditions; when we perform planned comparisons, we are very often focusing on only a subset of the conditions and need an error term that represents only those conditions. Thus, comparisons should be made with a separate or unique error term that is based only on the treatments currently being compared. There are a variety of approaches to develop these separate error terms, some of which are reviewed in Keppel and Wickens (2004).

We will demonstrate perhaps the simplest approach to making a within-subjects comparison offered by Keppel et al. (1992) that produces a unique error term. Based on our previous numerical example, suppose we are interested in comparing groups a_1 (one month before) and a_5 (one year after) in order to examine the most extreme preeffects and posteffects on symptom intensity. The analysis begins by creating what we will call a *Comparison Matrix* based on the two treatment groups we are interested in, and the necessary A and S sums (see Table 10.4).

As can be seen, Table 10.4 is a recapitulation of Table 10.2 but with only two treatment groups. Notice that former treatment group a_5 is now designated a_2. Thus, we are now poised to recompute our analysis with these two groups only. Hence, our $a = 2$ and $n = 8$. We can now proceed computing our sum of squares, degrees of freedom, mean squares, and F for this pairwise comparison.

10.10.2 COMPUTATIONS FOR A WITHIN-GROUP PAIRWISE COMPARISON

$$SS_{A_{\text{comp}}} = \frac{\Sigma A^2}{n} - \frac{T^2}{(a)(n)} = \frac{(73)^2 + (41)^2}{8} - \frac{(114)^2}{(2)(8)} = \boxed{64.00}$$

$$SS_s = \frac{\Sigma S^2}{a} - \frac{T^2}{(a)(n)}$$

$$= \frac{(17)^2 + (14)^2 + (15)^2 + (12)^2 + (13)^2 + (16)^2 + (16)^2 + (11)^2}{2}$$

$$- \frac{(114)^2}{(2)(8)} = \boxed{15.75}$$

$$SS_{A_{\text{comp}} \times S} = \Sigma Y_{aj}^2 - \frac{\Sigma A^2}{n} - \frac{\Sigma S^2}{a} + \frac{T^2}{(a)(n)}$$

$$= 912 - 876.25 - 828 + 812.25 = \boxed{20.00}$$

$$SS_T = \Sigma Y_{aj}^2 - \frac{T^2}{(a)(n)} = 912 - 812.25 = \boxed{99.75}$$

$$df_a = a - 1 = 2 - 1 = \boxed{1}, \, df_s = n - 1 = 8 - 1 = \boxed{7}$$

$$df_{A_{\text{comp}} \times S} = (a - 1)(n - 1) = (1)(7) = \boxed{7}, \, df_T = (a)(n) - 1$$

$$= 16 - 1 = \boxed{15}$$

$$MS_{A_{\text{comp}}} = \frac{SS_{A_{\text{comp}}}}{df_{A_{\text{comp}}}} = \frac{64.00}{1} = \boxed{64.00}$$

$$MS_S = \frac{SS_S}{df_S} = \frac{15.75}{7} = \boxed{2.25}$$

$$MS_{A_{\text{comp}} \times -S} = \frac{SS_{A_{\text{comp}} \times S}}{df_{A_{\text{comp}} \times S}} = \frac{20.00}{7} = \boxed{2.86}$$

$$F_{A_{\text{comp}}} = \frac{MS_{A_{\text{comp}}}}{MS_{A_{\text{comp}} \times S}} = \frac{64.00}{2.86} = \boxed{22.38}.$$

This observed F is evaluated at $(df_{A_{\text{comp}}}, df_{A_{\text{comp}} \times S}) = F(1, 7)$. The critical value at $\alpha = .05$ is 5.59. Our observed F of 22.38 exceeds the critical value. We therefore reject the null and conclude there is a statistically significant difference between the two means. Specifically, symptom intensity was significantly reduced between the first and last tests.

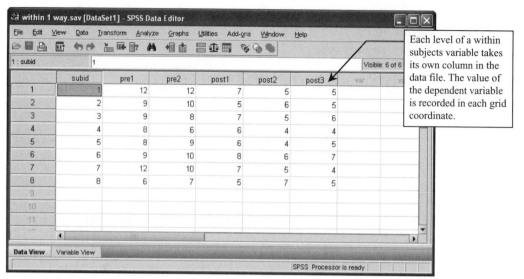

Figure 10.3   Data file for our one-way within-subjects example.

10.11   PERFORMING THE OMNIBUS ANALYSIS IN SPSS

10.11.1   STRUCTURING THE DATA FILE

The data file for a within-subjects design is structured in a similar manner to the between-subjects data files that we have described thus far –
there are just more measurements on the dependent variable to include.
We show the SPSS data file for our five-level pretest–posttest example in
Figure 10.3. The first column, as always, is used for our participant identification number; we have named this variable **subid**. The next five columns
are devoted to the levels of our within-subjects independent variable
listed in the temporal order that the data were collected. Each level of
a within-subjects independent variable receives its own column in the
data file. Recorded under each level are the measurements. For example,
the patient whose **subid** is 1 reported a symptom intensity of 12 at the
first pretest and an intensity of 12 at the second pretest. That person's
posttest intensity scores for one week, one month, and one year following
treatment were 7, 5, and 5, respectively.

Each row in the data file contains data for a different participant. But
because the cases are repeatedly measured, we have several measurements
on the dependent variable recorded in the data file for each case. SPSS
ordinarily treats each column as a separate, stand-alone variable. And
although we used variable names that probably provide strong cues to
the nature of these entries, if we viewed the data file (the spreadsheet or
grid) from the standpoint of a naïve observer it might not be obvious that
the variables **pre1**, **pre2**, **post1**, **post2**, and **post3** were intended by the
researchers to make up a single within-subjects variable with five levels.
And from the standpoint of a computer-based data analysis application
such as SPSS, each column represents a separate variable until it is told

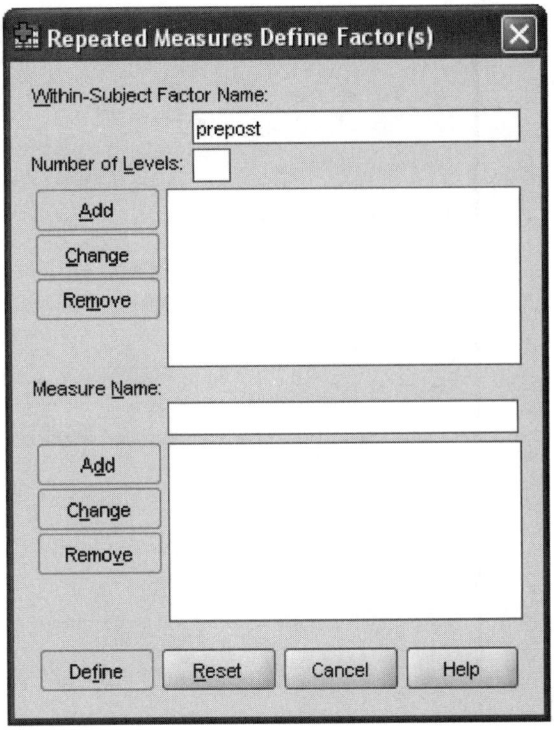

Figure 10.4 Initial dialog window for **GLM Repeated Measures**.

differently. It is for this reason that the initial dialog windows we encounter in the SPSS data analysis process will be different from what we have seen thus far.

10.11.2 DEFINING THE WITHIN-SUBJECTS VARIABLE

Once your data file is constructed, from the main SPSS menu select **Analyze → General Linear Model → Repeated Measures**. This will open the dialog window shown in Figure 10.4.

You will be presented with an initial (preliminary) window to **Define** the within-subjects factor(s) in the study by providing a name for the factor and by specifying the number of levels associated with it. By providing a name for the within-subjects factor, SPSS will be ready to accept some variables to represent its levels. How many? You simply type in the number of levels associated with the within-subjects factor.

Highlight the default name **factor1** and type in a meaningful name for your variable. We have named our variable **prepost**. Next, type in the number of levels for your within-subjects variable. We have specified that it contains five levels. This is shown in Figure 10.5. Then click **Add**. Your variable will appear in the panel to the right of the **Add** button as shown in Figure 10.6.

If we had a second or third within-subjects variable in the study (as we will in Chapters 11 and 12), we would repeat these steps for these additional repeated measures. When finished naming our variables (one

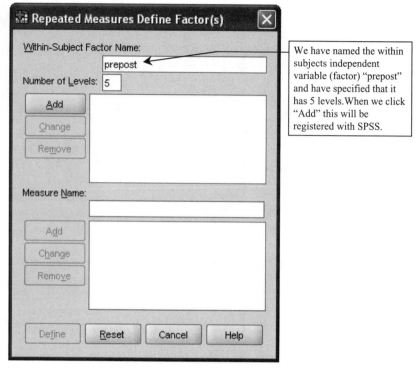

We have named the within subjects independent variable (factor) "prepost" and have specified that it has 5 levels. When we click "Add" this will be registered with SPSS.

Figure 10.5   Initial dialog window for **GLM Repeated Measures** with our within-subjects independent variable named and the number of levels it has specified.

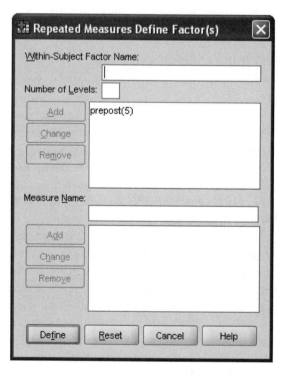

Figure 10.6 Initial dialog window for **GLM Repeated Measures** showing the prepost variable with five levels ready to be defined.

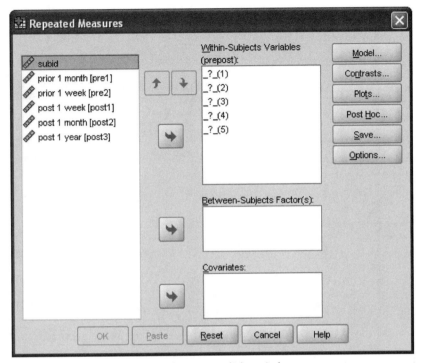

Figure 10.7   Main **GLM Repeated Measures** dialog window.

variable for this present example), click the **Define** pushbutton to reach the main **GLM Repeated Measures** dialog window.

The main dialog window for **GLM Repeated Measures** is shown in Figure 10.7. Note that the name of the within-subjects factor, **prepost**, is already known by SPSS and is placed at the very top of the dialog window. The first job required of us by SPSS is to define the structure of the within-subjects variable, that is, to indicate the variables in the data file that comprise the levels of the within-subjects factor. In the main panel is a list of five variable placeholders with question marks (there are five because we have specified five levels in the initial window, and there are question marks because we have yet to indicate which variables comprise the levels), indicating that we need to identify five variables for those slots.

Click over the variables to the panel labeled **Within-Subjects Variables**. The order in which the variables are selected is important here. The first variable you click over will be placed in the first blank to replace the first question mark, the second variable clicked over will be placed in the second blank, and so on. After clicking the variables over in the proper order, our window is shown in Figure 10.8.

Once the variables have been moved over to the placeholders in the top panel, the metamorphosis of the five stand-alone variables into a single within-subjects variable consisting of five levels has been completed. With this within-subjects variable now in place, we can structure our one-way within-subjects ANOVA.

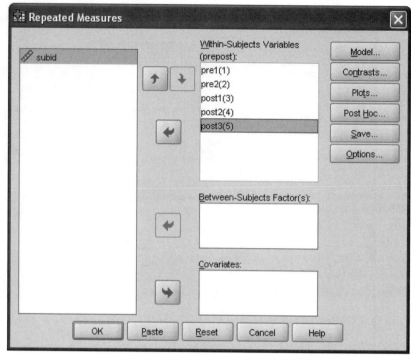

Figure 10.8 Main **GLM Repeated Measures** dialog window with our prepost within-subjects variable now defined.

10.11.3 STRUCTURING THE OMNIBUS ANOVA

Click the **Options** pushbutton to reach the dialog window shown in Figure 10.9. The top portion of the window is devoted to **Estimated Marginal Means**. We will return to this option to perform our multiple comparisons procedure if the effect of the treatment variable is statistically significant. For now, our focus is on the omnibus ANOVA.

In the bottom portion of the window labeled **Display**, check **Descriptive statistics** to output the observed means and standard deviations. Click **Continue** to return to the main dialog window and **OK** to run the omnibus analysis.

10.12 SPSS OUTPUT FOR THE OMNIBUS ANALYSIS

10.12.1 CODING AND DESCRIPTION

Figure 10.10 shows the **Within Subjects Factors** table. It presents the levels of the repeated measure with the value codes that SPSS has assigned to each; it is a good way to verify that you have set up the analysis correctly. These codes will be used by SPSS in presenting the paired comparisons that we will ask for if we obtain a statistically significant effect of our **prepost** variable.

The descriptive statistics are presented in Figure 10.11. They include the observed means, standard deviations, and sample sizes for each condition.

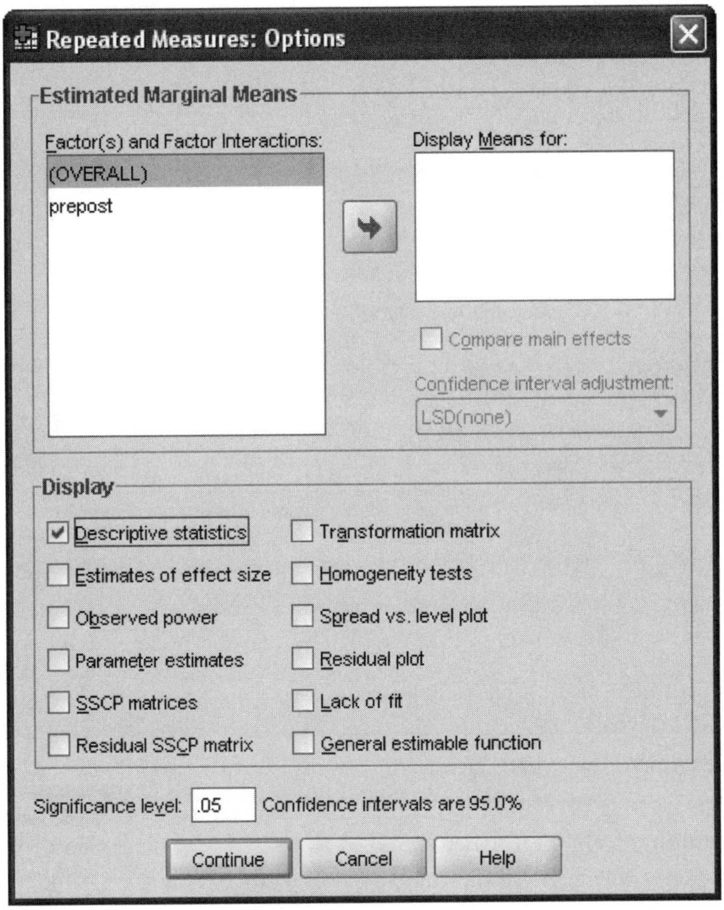

Figure 10.9 The **Options** dialog window.

Within-Subjects Factors

Measure: MEASURE_ 1

prepost	Dependent Variable
1	pre1
2	pre2
3	post1
4	post2
5	post3

Figure 10.10 SPSS' values for the within-subjects variable.

Descriptive Statistics

	Mean	Std. Deviation	N
prior 1 month	9.13	2.031	8
prior 1 week	9.00	1.927	8
post 1 week	6.38	1.061	8
post 1 month	5.25	1.035	8
post 1 year	5.13	.991	8

Figure 10.11 Descriptive statistics.

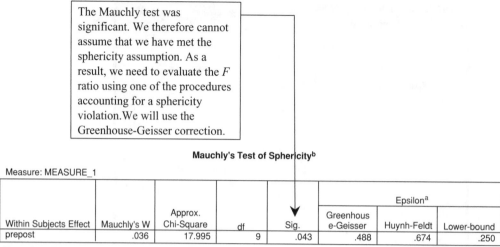

The Mauchly test was significant. We therefore cannot assume that we have met the sphericity assumption. As a result, we need to evaluate the F ratio using one of the procedures accounting for a sphericity violation. We will use the Greenhouse-Geisser correction.

Mauchly's Test of Sphericity[b]

Measure: MEASURE_1

Within Subjects Effect	Mauchly's W	Approx. Chi-Square	df	Sig.	Epsilon[a]		
					Greenhouse-Geisser	Huynh-Feldt	Lower-bound
prepost	.036	17.995	9	.043	.488	.674	.250

Tests the null hypothesis that the error covariance matrix of the orthonormalized transformed dependent variables is proportional to an identity matrix.

a. May be used to adjust the degrees of freedom for the averaged tests of significance. Corrected tests are displayed in the Tests of Within-Subjects Effects table.

b. Design: Intercept
Within Subjects Design: prepost

Figure 10.12 Mauchly's test of sphericity.

10.12.2 EVALUATING SPHERICITY

Figure 10.12 presents the results of **Mauchly's Test of Sphericity** (this is automatically performed by **GLM Repeated Measures**). John William Mauchly was a physicist who, together with John Eckert, built the first general-purpose electronic digital computer, named ENIAC, in 1946. He published the sphericity test in 1940. Mauchly's test simultaneously evaluates two assumptions: (a) that the levels of the within-subjects variable have equal variances (i.e., that there is homogeneity of variance); and (b) that the pairs of levels of the within-subjects variable are correlated to the same extent (the null hypothesis states that the correlations between the levels do not differ from each other).

It is worth noting that because of this latter evaluation, it is necessary to have more than two levels of the within-subjects variable in order for a Mauchly test to be meaningfully interpreted. That is, the test must compare at least one correlation (e.g., the correlation of Level 1 and Level 2) with another correlation (e.g., the correlation of Level 1 with Level 3). If there are only two levels of the within-subjects variable, there would be only one correlation without another to which we could compare it.

With three or more levels of the within-subjects variable, there are more than two correlations, and so Mauchly's sphericity test becomes enabled. For example, if we had three levels, then we could compare the correlations of Levels 1 and 2, Levels 2 and 3, and Levels 1 and 3. The null hypothesis here is that the correlations are of comparable magnitude. If the Mauchly test is significant, indicating that the assumption of sphericity has not been met, then you should use one of the corrections automatically provided by SPSS.

The Mauchly test, evaluated against a chi square distribution, was statistically significant, $W(9) = 0.036, p < .05$. This indicates that we have violated the assumption of sphericity. Mauchly's W is evaluated with 9 df because there are ten pairwise comparisons of the correlations (five levels taken two at a time allows for ten combinations) and the degrees of freedom are computed as the number of comparisons minus one.

Given a violation of sphericity, we would use one of the available corrections shown under the heading of **Epsilon** in Figure 10.12. The two most commonly used corrections are *Greenhouse–Geisser* and the *Huynh–Feldt*, probably because the *lower bound* correction is considered by many to be too conservative. All three corrections work by multiplying the degrees of freedom for the F ratio by the respective epsilon value shown in the table and evaluating the significance of F against these adjusted degrees of freedom.

10.12.3   EVALUATING THE EFFECT OF THE WITHIN-SUBJECTS VARIABLE

Figure 10.13 displays the summary table for the within-subjects variance. With a significant sphericity test, we use the **Greenhouse–Geisser** adjustment. Note that multiplying the Greenhouse–Geisser **Epsilon** value shown in Figure 10.12 of 0.488 by the degrees of freedom for **Sphericity Assumed,** we obtain the degrees of freedom shown in Figure 10.13. Thus, (4) (0.488) = 1.951 and (28)(0.488) = 13.657. This results in a more stringent test of significance because we have fewer degrees of freedom (less power) in evaluating the F ratio.

There are two sources of within-subjects variance that may be seen in Figure 10.13, one for the independent variable of prepost and one for the within-subjects error (which is the Treatment × Subjects interaction). Note that SPSS ties the error variance directly to the effect by specifying **Error(PREPOST)** in its label; technically, the full label for this error term

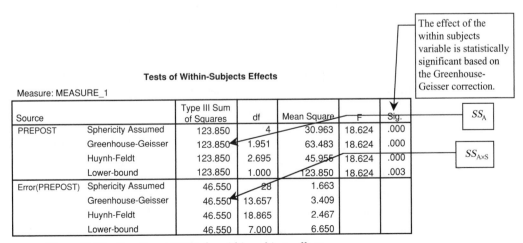

Figure 10.13   Omnibus ANOVA for within-subjects effects.

Tests of Between-Subjects Effects

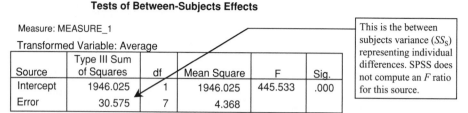

Measure: MEASURE_1

Transformed Variable: Average

Source	Type III Sum of Squares	df	Mean Square	F	Sig.
Intercept	1946.025	1	1946.025	445.533	.000
Error	30.575	7	4.368		

This is the between subjects variance (SS_S) representing individual differences. SPSS does not compute an F ratio for this source.

Figure 10.14   Omnibus ANOVA for between-subjects effects.

is the Prepost × Subjects interaction but SPSS leaves out the "× Subjects" portion of the label for all of its within-subjects error terms. SPSS provides the effect in parentheses because in analyses with more than one within-subjects variable each effect will have its own error term; such labeling will be very useful for helping us read the summary table in those circumstances.

As can be seen from Figure 10.13, the within-subjects variable was statistically significant under the Greenhouse–Geisser correction, $F(1.951, 13.657) = 18.624$, $p < .05$, $\eta^2 = 0.727$. With more than two conditions as we have here, this significant effect is our cue to perform a multiple comparisons for the means.

10.12.4   EVALUATING THE BETWEEN-SUBJECTS VARIANCE

The between-subjects portion of the variance is shown in Figure 10.14. The **Intercept** focuses on the full model which we do not deal with here. What SPSS calls **Error** is the variance representing individual differences; it does not calculate an F ratio for this effect.

10.13   PERFORMING THE POST-ANOVA ANALYSIS IN SPSS

Define your variables just as you did for the omnibus analysis so that you have a main dialog window configured as we showed in Figure 10.8. Click the **Options** pushbutton to arrive at the **Options** dialog window. Clear the bottom (**Display**) part of the window if anything is selected, since that information was produced in the omnibus analysis and is presumably available to you. We will now work with the top portion of that window. The top portion of the **Options** dialog window focuses on **Estimated Marginal Means**. We have used this in prior chapters in structuring the simple effects tests having obtained a significant interaction. We do something comparable here.

Using a point-and-click method to perform the paired comparisons procedure, follow the steps shown in Figure 10.15: Click the **prepost** variable over into the **Display Means for** panel. Place a checkmark in the box for **Compare main effects** and select **Bonferroni** from the drop-down menu that will become active when you check the box. This will produce Bonferroni corrected paired comparisons to be performed just as though we were performing a simple effects analysis. Click **Continue** to return to the main dialog window and click **OK** to run the analysis.

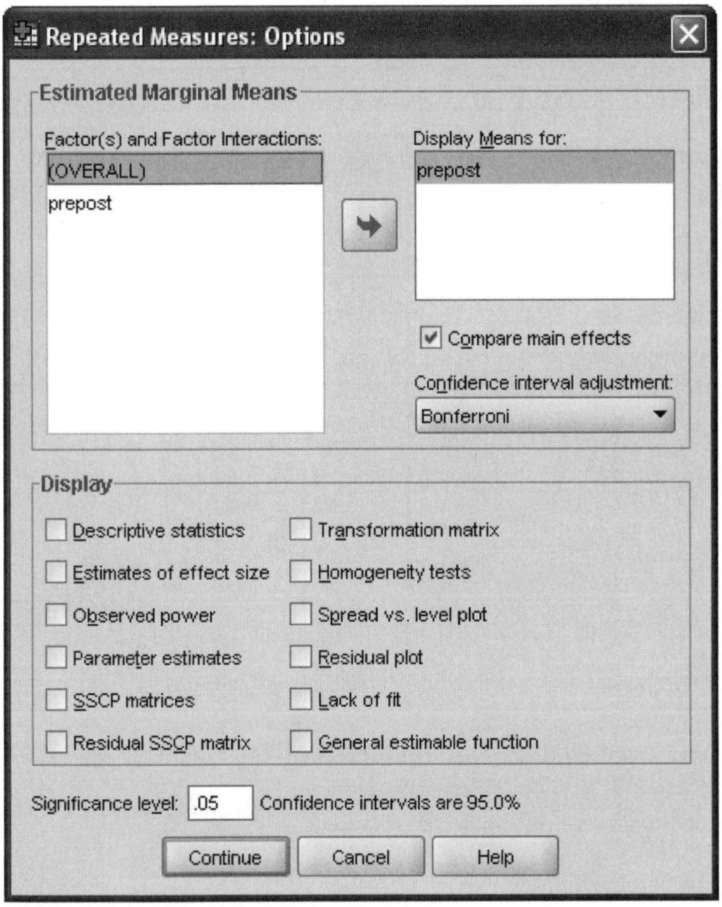

Figure 10.15 The **Options** dialog window set up to have SPSS perform paired comparisons on the levels of the within-subjects variable.

10.14 SPSS OUTPUT FOR THE POST-ANOVA ANALYSIS

The estimated marginal means are shown in Figure 10.16. With equal sample sizes in repeated measures designs, these means are equal to the observed means shown in the **Descriptive statistics** output. The value

Estimates

Measure: MEASURE_1

| prepost | Mean | Std. Error | 95% Confidence Interval | |
			Lower Bound	Upper Bound
1	9.125	.718	7.427	10.823
2	9.000	.681	7.389	10.611
3	6.375	.375	5.488	7.262
4	5.250	.366	4.385	6.115
5	5.125	.350	4.296	5.954

Figure 10.16 Estimated marginal means.

Pairwise Comparisons

Measure: MEASURE_1

(I) prepost	(J)prepost	Mean Difference (I-J)	Std. Error	Sig.[a]	95% Confidence Interval for Difference[a]	
					Lower Bound	Upper Bound
1	2	.125	.479	1.000	-1.807	2.057
	3	2.750*	.590	.023	.372	5.128
	4	3.875*	.895	.034	.268	7.482
	5	4.000*	.845	.021	.595	7.405
2	1	-.125	.479	1.000	-2.057	1.807
	3	2.625*	.625	.040	.107	5.143
	4	3.750*	.750	.016	.728	6.772
	5	3.875*	.693	.008	1.084	6.666
3	1	-2.750*	.590	.023	-5.128	-.372
	2	-2.625*	.625	.040	-5.143	-.107
	4	1.125	.581	.938	-1.214	3.464
	5	1.250	.366	.112	-.225	2.725
4	1	-3.875*	.895	.034	-7.482	-.268
	2	-3.750*	.750	.016	-6.772	-.728
	3	-1.125	.581	.938	-3.464	1.214
	5	.125	.398	1.000	-1.479	1.729
5	1	-4.000*	.845	.021	-7.405	-.595
	2	-3.875*	.693	.008	-6.666	-1.084
	3	-1.250	.366	.112	-2.725	.225
	4	-.125	.398	1.000	-1.729	1.479

Based on estimated marginal means

*.The mean difference is significant at the .05 level.

a. Adjustment for multiple comparisons: Bonferroni.

Figure 10.17   Pairwise comparisons of means.

codes used by SPSS in the pairwise comparisons are presented in the first column of the output.

The **Pairwise Comparisons** table is shown in Figure 10.17. We have seen this sort of table from our simple effects analyses in Chapters 8 and 9. There are five major rows in the table because each of the five levels of the repeated measure is given its own focus. The conditions are designated by their value codes rather than by any descriptive name or label.

We will talk you through the first major row, which focuses on the first level **pre1**. The value code of **1** in the upper left-hand corner under the column (**I**) **prepost** stands for **pre1**, the first level of the prepost within-subjects factor. The value codes under the column (**J**) **prepost** stand for the other levels of **prepost** (e.g., Level 2 is **pre2**, Level 3 is **post1**, and so on). The difference between the means of **pre1** and **pre2** (**pre1** − **pre2**) is 0.125 (9.125 − 9.00 = 0.125) and is not statistically significant ($p = 1.000$ as shown in the **Sig** column). The difference between **pre1** and **post1** (Levels 1 and 3 using value codes) is 2.750 (9.125 − 6.375 = 2.750) and is statistically significant. In like manner ($p = .023$), **pre1** differs significantly from **post2** and **post3**.

Overall, the two pretest means do not differ significantly from each other and the last two posttest means do not differ from each other. The other differences are statistically significant.

10.15   SPSS AND SAS DATA FILE STRUCTURES

10.15.1   THE MULTIVARIATE STRUCTURE USED BY SPSS

One major difference between SPSS and SAS is in the way that they require the data file to be structured in order to handle a repeated measure (a within-subjects variable) in the design. As we have seen, SPSS uses the same data structure for any analysis. This type of data structure, which we have not needed to label before now, is often referred to as *multivariate* or *wide form.*

In multivariate structure, all of the information for each case is contained in one row, using each column to hold one piece of information. In a repeated measures design, participants will have multiple scores on the dependent variable. For example, having five levels of a within-subjects variable (e.g., **pre1, pre2, post1, post2,** and **post3**) means that each level will have its own column in the row; in each column, we will see the result of the measurement (the value of the dependent variable) obtained under each condition. The data structure is called multivariate because the row for each case contains multiple (more than one) instances of the dependent variable.

10.15.2   THE UNIVARIATE (STACKED) STRUCTURE USED FOR REPEATED MEASURES BY *SAS ENTERPRISE GUIDE*

SAS Enterprise Guide uses a structure known variously as *univariate, narrow,* or *stacked* form. In univariate or stacked column format, each row is permitted to contain only one score on the dependent variable, and is the defining feature of univariate format. In a between-subjects design, where we have only one score per case, multivariate and univariate data structures look identical. In designs where each participant provides more than a single score, that is, in designs containing one or more within-subjects variables, the required data structure to meet stacked format differentiates itself from multivariate format.

A portion of the stacked data file entered into an Excel spreadsheet that we will import to *SAS Enterprise Guide* is presented in Figure 10.18. Note that the first five lines (below the variable names) represent the information for the participant identified as **subid** 1. This is because each case has five different scores, one for each level of the within-subjects variable.

Under the univariate requirement that only one score may appear on any given row, we must use five rows to capture the measurements for that person. Each row contains information relevant to the score it contains. Let's look across the row and discuss the columns that are represented.

The first column identifies the particular subject whose data is contained in the row. The identifier variable is named **subid** in the Excel file. The second variable, which we have named **Time**, represents the particular level of the within-subjects variable whose score we are providing. The variable in the third column is our effort to circumvent the proclivity of *SAS Enterprise Guide* to order the levels of the within-subjects variables

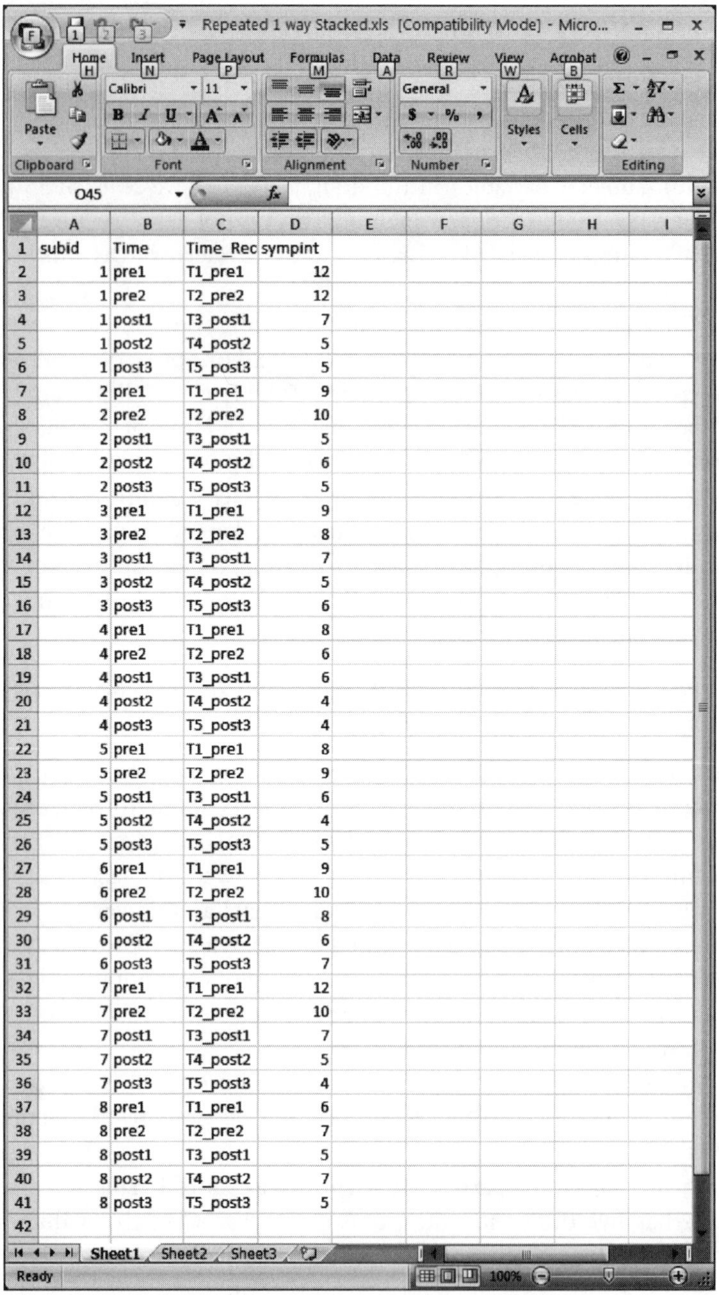

subid	Time	Time_Rec	sympint
1	pre1	T1_pre1	12
1	pre2	T2_pre2	12
1	post1	T3_post1	7
1	post2	T4_post2	5
1	post3	T5_post3	5
2	pre1	T1_pre1	9
2	pre2	T2_pre2	10
2	post1	T3_post1	5
2	post2	T4_post2	6
2	post3	T5_post3	5
3	pre1	T1_pre1	9
3	pre2	T2_pre2	8
3	post1	T3_post1	7
3	post2	T4_post2	5
3	post3	T5_post3	6
4	pre1	T1_pre1	8
4	pre2	T2_pre2	6
4	post1	T3_post1	6
4	post2	T4_post2	4
4	post3	T5_post3	4
5	pre1	T1_pre1	8
5	pre2	T2_pre2	9
5	post1	T3_post1	6
5	post2	T4_post2	4
5	post3	T5_post3	5
6	pre1	T1_pre1	9
6	pre2	T2_pre2	10
6	post1	T3_post1	8
6	post2	T4_post2	6
6	post3	T5_post3	7
7	pre1	T1_pre1	12
7	pre2	T2_pre2	10
7	post1	T3_post1	7
7	post2	T4_post2	5
7	post3	T5_post3	4
8	pre1	T1_pre1	6
8	pre2	T2_pre2	7
8	post1	T3_post1	5
8	post2	T4_post2	7
8	post3	T5_post3	5

Figure 10.18 The Excel file containing the data in stacked format.

in alphabetic order in the output; our use of **T1** through **T5** will make it easier for us to view the results (it does not change the data or the results in any way). We will use this **Time_Recode** variable in the analysis instead of the **Time** variable. The last column, named **sympint**, is the actual symptom intensity score of the participant under the specified condition.

To illustrate how to read the data file (or construct one), consider the participant coded as **subid** 1. That person's information occupies Rows 2 through 6 in the Excel file shown in Figure 10.18. He or she showed symptom intensities of twelve in the first pretest, twelve in the second pretest, seven in the first posttest, five in the second posttest, and five in the last posttest. The next five rows represent the same information in the same order for the second participant. Each participant's information is recorded in the same way throughout the data file. More information on data file structure may be found in Davis (2007) and Slaughter and Delwiche (2006).

10.16   PERFORMING THE OMNIBUS ANALYSIS IN SAS

Import your data file from Excel into a new *SAS Enterprise Guide* project. From the main menu, select **Analyze → ANOVA → Mixed Models**. The window opens on the **Task Roles** tab as shown in Figure 10.19. From the **Variables to assign** pane, select **sympint** and drag it to the icon for **Dependent variable** in the **Task roles** panel. Select **Time_Recode** and drag it over to the icon for **Classification variables**. Select **subid** and drag it over to the area under **Classification variables**.

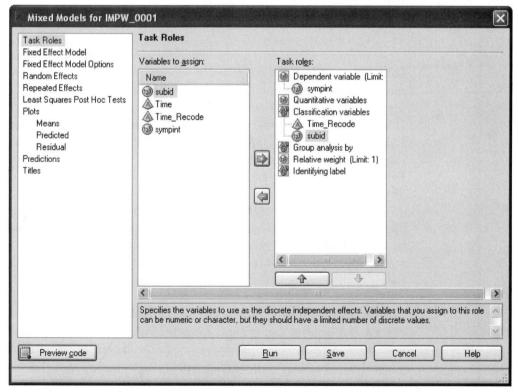

Figure 10.19   The **Task Roles** window.

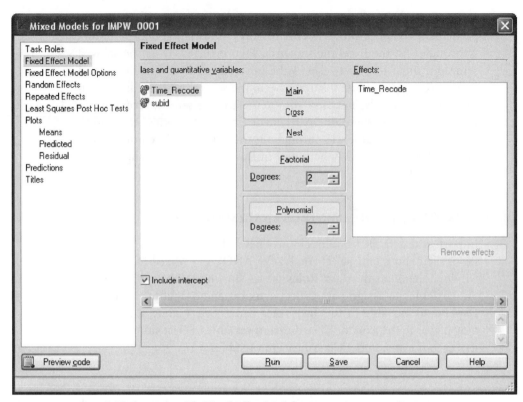

Figure 10.20 The **Fixed Effect Model** screen.

In the navigation panel at the left of the screen, select **Fixed Effect Model.** In the **Class and quantitative** window, select **Time_Recode** and then select the **Main** pushbutton; **Time_Recode** will automatically appear in the **Effects** window as shown in Figure 10.20.

In the navigation panel at the left of the screen, select **Fixed Effect Model Options.** Select **Type 3** under **Hypothesis test type**, **Residual maximum likelihood** under **Estimation method** (this is the default), and **Between and within subject portions** under **Degrees of freedom method**. This is illustrated in Figure 10.21.

Selecting the **Random Effects** tab in the navigation tab brings you to the blank initial screen shown in Figure 10.22. Click the **Add** pushbutton and two displays will be presented. First, the expression **<no effects selected>** will appear in the **Random effects to estimate** panel. Second, several panels in the **Random effects and options** panel will become available. This is shown in Figure 10.23.

Our goal in this window is to specify our **subid** variable as a random effect and to register that this is the way we have identified our subjects in the data file. A *random effect* is associated with an independent variable whose levels are not systematically manipulated by the researchers or which are the results of the sampling plan. We typically wish to generalize beyond the specific levels present in the research to a wider population of

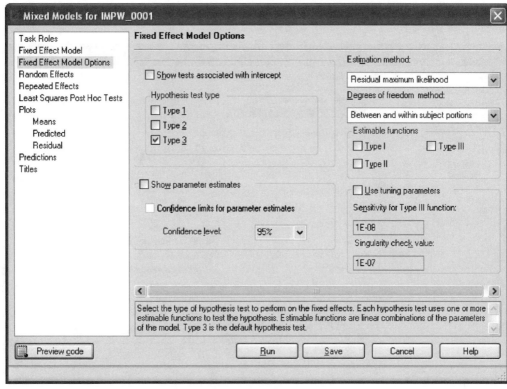

Figure 10.21 The **Fixed Effect Model Options** screen.

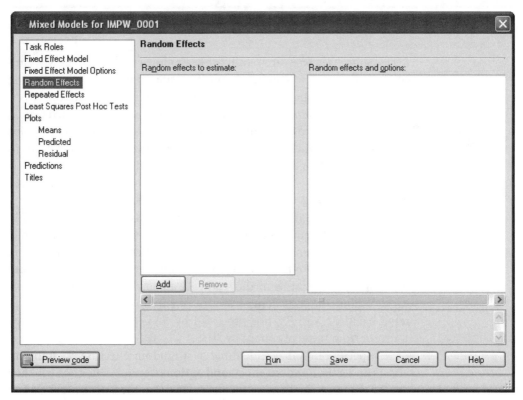

Figure 10.22 The initial **Random Effects** screen.

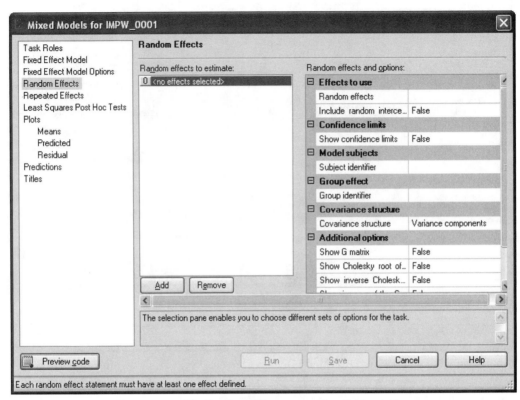

Figure 10.23   The **Random Effects** screen after clicking **Add**.

levels. Subjects in a research study are examples of such variables. We sample them as best as we can, randomly if possible, but we wish to generalize beyond our participants to the population they represent. Random effects are contrasted with *fixed effects*, where we have specifically selected certain levels or conditions of an independent variable so that we can describe their effect with much less concern about generalizing to other conditions that are not included in the study. This topic is discussed in Chapter 17. More extensive discussions of random effects may be found in Maxwell and Delaney (2000) and Keppel and Wickens (2004).

Although SPSS did not broach this issue in its **GLM** procedure, *SAS Enterprise Guide* requires that we treat subjects as a random effect when we are using within-subjects designs. When independent variables are treated as random effects in an ANOVA, different error terms are used to compute the F ratios in the study. As a consequence of this, at least for the within-subjects designs we describe in this chapter and in Chapters 11 and 12, the F ratios produced by *SAS Enterprise Guide* will be different from those produced by SPSS.

To specify our subject identifier as a random effect, select **Random effects** under the **Random effects and options** frame; a dialog box will then appear at the far right end of the menu. Position the cursor over this box and click, and a new window will appear as shown in Figure 10.24. Select

Figure 10.24 Specifying ID as a random effect.

subid and then Click the **Main** Pushbutton and **subid** will automatically appear in the **Random Effects** pane. Select the **OK** pushbutton.

With **subid** now specified as the random effect, click the **Subject identifier** under the **Model subjects** frame and a box will appear at the far right end of that menu (see Figure 10.25). Position the cursor over this box and click, and a new window will appear as shown in Figure 10.26. Select **subid** and then Click the **Main** Pushbutton; **subid** will automatically appear in the **Random Effects** pane. Select the **OK** pushbutton.

Selecting the **Least Squares Post Hoc Tests** tab in the navigation tab brings you to the screen shown in Figure 10.27. Once again we see frames with selection menus. Select **Time_Recode** in the **Effects to use** frame. A drop-down menu will appear; select **True** as shown in Figure 10.27. Click the **Run** pushbutton to perform the omnibus analysis that we have just specified.

10.17 SAS OUTPUT FOR THE OMNIBUS ANALYSIS

The means of the five conditions are shown in the lower display in Figure 10.28, and the F ratio for **Time_Recode** is shown in the upper display. The F ratio produced by *SAS Enterprise Guide* is somewhat numerically different from the value produced by SPSS, although both lead us to

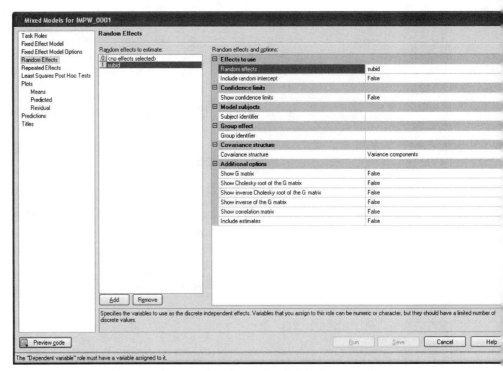

Figure 10.25   Selecting **Subject identifier** under **Model subjects**.

conclude that the intensity of symptoms differed across the time periods measured. Such a numerical difference is due to different algorithms used by SAS and SPSS to analyze within-subjects variables. This difference is due to the use of subjects as a random effect in the general model that SAS builds in while it performs the statistical analysis; SPSS does not identify subjects as a random effect in its model.

10.18   PERFORMING THE POST-ANOVA ANALYSIS IN SAS

With a statistically significant effect obtained for the five-level within-subjects variable, we need to perform a multiple comparisons procedure to determine which means are significantly different from which others. Set up the analysis exactly as described for the omnibus analysis. In the **Least Squares Post Hoc Tests** window, click on **Show p-values for differences** and select **All pairwise differences**. Just below that is the **Adjustment method for comparison** menu. We will perform a **Bonferroni** correction on our comparisons. This is shown in Figure 10.29. Click **Run** to perform this analysis.

10.19   SAS OUTPUT FOR THE POST-ANOVA ANALYSIS

The pairwise mean comparisons are shown in Figure 10.30. The conditions that are being compared are in the second and third columns in the display. We can also see Student's *t* value analyzing that mean difference

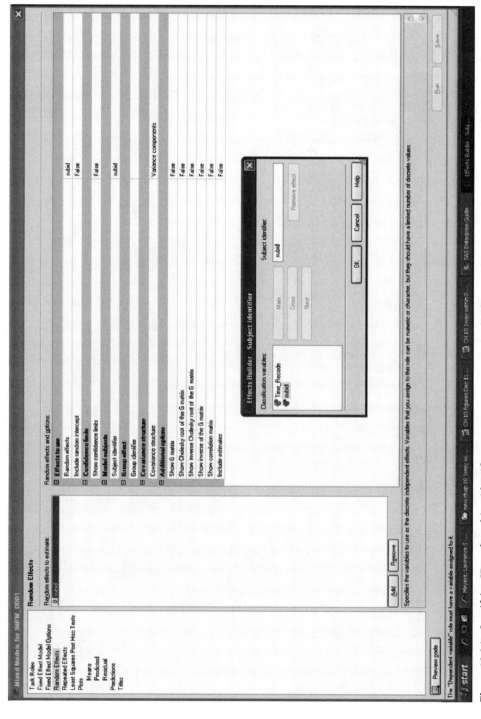

Figure 10.26 Specifying ID as the subject identifier.

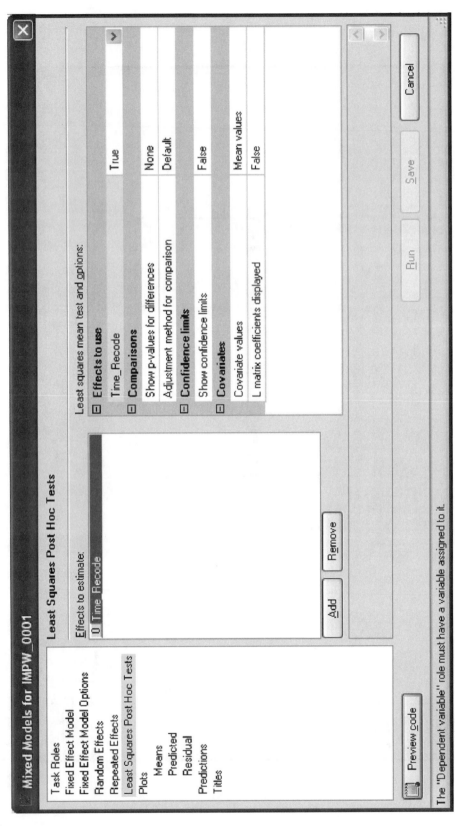

Figure 10.27   The Least Squares Post Hoc Tests window.

Type 3 Tests of Fixed Effects				
Effect	Num DF	Den DF	F Value	Pr > F
Time_Recode	4	28	18.62	<.0001

> The effect of the within subjects variable is statistically significant.

Least Squares Means						
Effect	Time_Recode	Estimate	Standard Error	DF	t Value	Pr > |t|
Time_Recode	T1_pre1	9.1250	0.5248	28	17.39	<.0001
Time_Recode	T2_pre2	9.0000	0.5248	28	17.15	<.0001
Time_Recode	T3_post1	6.3750	0.5248	28	12.15	<.0001
Time_Recode	T4_post2	5.2500	0.5248	28	10.00	<.0001
Time_Recode	T5_post3	5.1250	0.5248	28	9.77	<.0001

Figure 10.28 The means of the conditions and the F ratio for **Time_Recode**.

(labeled **t Value**) in a middle column of the table. We also obtain from *SAS Enterprise Guide* in any analysis involving one or more within-subjects variables the uncorrected probability of that t value occurring by chance, assuming that the null hypothesis is true (labeled **Pr > |t|**).

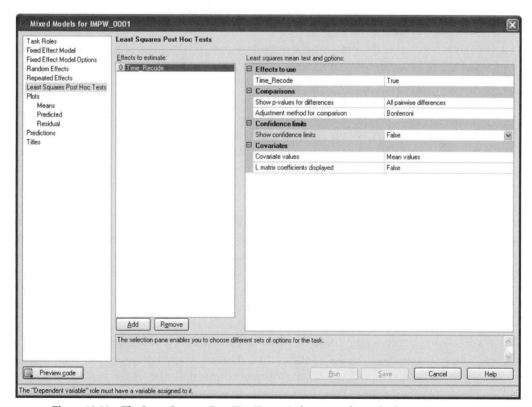

Figure 10.29 The **Least Squares Post Hoc Tests** window set up for multiple comparisons.

Differences of Least Squares Means										This Table presents both the unadjusted *p*-value and the Bonferroni adjustment.
Effect	Time_Recode	Time_Recode	Estimate	Standard Error	DF	t Value	Pr > |t|	Adjustment	Adj P	
Time_Recode	T1_pre1	T2_pre2	0.1250	0.6447	28	0.19	0.8477	Bonferroni	1.0000	
Time_Recode	T1_pre1	T3_post1	2.7500	0.6447	28	4.27	0.0002	Bonferroni	0.0021	
Time_Recode	T1_pre1	T4_post2	3.8750	0.6447	28	6.01	<.0001	Bonferroni	<.0001	
Time_Recode	T1_pre1	T5_post3	4.0000	0.6447	28	6.20	<.0001	Bonferroni	<.0001	
Time_Recode	T2_pre2	T3_post1	2.6250	0.6447	28	4.07	0.0003	Bonferroni	0.0035	
Time_Recode	T2_pre2	T4_post2	3.7500	0.6447	28	5.82	<.0001	Bonferroni	<.0001	
Time_Recode	T2_pre2	T5_post3	3.8750	0.6447	28	6.01	<.0001	Bonferroni	<.0001	
Time_Recode	T3_post1	T4_post2	1.1250	0.6447	28	1.75	0.0919	Bonferroni	0.9195	
Time_Recode	T3_post1	T5_post3	1.2500	0.6447	28	1.94	0.0627	Bonferroni	0.6265	
Time_Recode	T4_post2	T5_post3	0.1250	0.6447	28	0.19	0.8477	Bonferroni	1.0000	

Figure 10.30   Output of the **Multiple Comparisons** procedure.

For designs involving a repeated measure, *SAS Enterprise Guide* displays the Bonferroni corrected probabilities in the final column labeled **Adj P**. These values are somewhat different from those produced by SPSS. To calculate the Bonferroni adjusted probabilities, *SAS Enterprise Guide* multiplies the uncorrected probabilities by the number of paired mean comparisons. This is a very conservative approach to reducing the familywise error rate. When the result of this multiplication yields a value greater than 1, *SAS Enterprise Guide* displays a value of 1.0000 in the **Adj P** column.

In the present study, we have five conditions and a total of ten possible pairwise comparisons. Thus, for example, the uncorrected probability associated with the mean difference between **post1** and **post3** is 0.0627 (9th row in Figure 10.30) and the Bonferroni corrected adjusted probability is 0.6265 (because SAS keeps more decimal places than it typically displays, simply multiplying the tabled values will lead to an inexact result).

SPSS does not have a post-ANOVA procedure in which we can select "all pairwise comparisons." The number of paired comparisons SPSS assumes are being made at any one time are generally fewer than the number presumed by SAS, especially, as we will see in the remaining chapters, when we perform simple effects analyses to explicate interaction effects. For these reasons, SPSS will almost always produce Bonferroni adjusted probabilities that are different (and less conservative) than those obtained from *SAS Enterprise Guide*. One way to offset the extreme conservatism of SAS is use a different multiplier to compute the Bonferroni correction if the number of comparisons you intend to make is fewer than the total number of such comparisons displayed by SAS. In the present study, a case could be made to compare each of the posttests to the final pretest score for a total of three pairwise comparisons. Under this circumstance it would be appropriate to multiply the uncorrected probability by 3 (rather than 10) to obtain the Bonferroni corrected probabilities for those particular comparisons.

For this data set, the differences between the SPSS and SAS results are small and, even with the severe Bonferroni correction employed by SAS, lead us to the same conclusions: (a) the two pretest means do not differ significantly from each other, (b) the last two posttest means do not differ from each other, and (c) the other differences are statistically significant.

10.20   COMMUNICATING THE RESULTS

Eight participants suffering from Disease X received Drug ABC in an effort to reduce the intensity of their symptoms. A baseline one month and one week prior to treatment yielded a comparable baseline at a symptom intensity of 75.00. Mean intensities measured one week, one month, and one year following treatment yielded means of 75.00, 55.00, and 35.00. Mauchly's test indicated that the data violated the assumption of sphericity, $W(9) = 0.04$, $p < .05$. Based on a Greenhouse–Geisser correction for the violation of sphericity, a one-way within-subjects ANOVA revealed a significant difference in the pretest and posttest means, $F(1.951, 13.657) = 18.62$, $p < .05$, within-subjects $\eta^2 = .73$. Pairwise comparisons using a Bonferroni correction to maintain an alpha level of .05 revealed that intensity remained constant from the two pretest baseline measures, significantly decreased after a week following drug therapy, and further significantly decreased after one month. Symptom intensity did not significantly differ from that level at the end of a year.

CHAPTER 10 EXERCISES

10.1 Assume we are interested in the effects of room color on the problem solving capabilities of fourth grade children. Accordingly, children are asked to solve ten anagram word puzzles in red, yellow, and white colored rooms. Assume that the order of the colored rooms was appropriately counterbalanced. Children spent ten minutes in each room solving the anagrams. The dependent variable was the number of correct solutions. The data are as follows:

Room color

Subject	Red a_1	Yellow a_2	White a_3
1	2	4	7
2	2	3	7
3	1	5	5
4	1	5	8
5	3	6	7

a.   Conduct an ANOVA by hand and with SPSS or SAS.
b.   Perform a set of pairwise comparisons by hand and with SPSS or SAS.

10.2 Assume we are interested in the effects of conversational affective intonation on conversation memory. Participants were presented with three brief videotaped conversation conditions, which were randomly counterbalanced: happy, neutral, or angry. The content of each conversation was kept the same for each condition, but the intonation and pausal contours of the dialogue were varied for particular utterances to create the three conversational treatments. The dependent variable was the total number of idea units recalled from the conversation. The data are as follows:

Type of conversation

Subject	Happy a_1	Neutral a_2	Angry a_3
1	19	4	12
2	18	4	12
3	20	4	13
4	19	3	18
5	15	5	10
6	18	2	13

a.   Conduct an ANOVA on these data by hand and with SPSS or SAS.
b.   Perform a set of pairwise comparisons by hand and with SPSS or SAS.

Two-Way Within-Subjects Design

11.1 COMBINING TWO WITHIN-SUBJECTS FACTORS

A two-way within-subjects design is one in which the levels of two within-subjects variables are combined factorially. Recall from Chapter 11 that within-subjects independent variables can be either time related or not. There are three possible combinations of types of within-subjects variables that can be represented in a two-way design: (a) both can be time independent, (b) one can be time independent and the other time based, and (c) both can be time based. This latter combination is not often found in the research literature; the other two are more common, and we provide illustrations of them here.

- *Both variables time independent*: Students can be asked to indicate if short bursts of music are either rock or classical (music variable) under either silent or noisy conditions (noise variable). In such a situation, participants would be exposed randomly to the various combinations of the independent variables. The dependent variable could be speed of recognition of the music.
- *One variable time independent and the other variable time dependent*: Students can be asked to indicate if short bursts of music are either rock or classical (music variable); these judgments could be made after they had been awake for either four, twelve, or twenty-four hours (a time-since-sleep variable). At each level of time since sleep, the students would be exposed to the two kinds of music in a random order. The time-since-sleep variable could be operationalized by either (a) measuring the students over the course of a day first at four hours, then at twelve hours, and finally at twenty-four hours after sleep, or (b) measuring the students on one day at one of these levels (determined randomly among the three), measuring them on a second day at one of the two remaining levels (determined randomly), and measuring them on a third day at the one remaining level of the time-since-sleep variable. The dependent variable could still be speed of recognition of the music.

Subid	Car/No Drinks	Car/One Drink	Car/Three Drinks	SUV/No Drinks	SUV/One Drink	SUV/Three Drinks
1	1	4	6	2	5	4
2	1	5	7	1	6	7
3	2	6	9	1	6	5
4	2	5	10	1	4	6
5	0	3	7	2	4	5
6	0	1	6	1	2	3
Mean	1.00	4.00	7.50	1.33	4.50	5.00

Figure 11.1   Raw data for numerical example.

11.2   A NUMERICAL EXAMPLE OF A TWO-WAY WITHIN-SUBJECTS DESIGN

The simplified data that will serve as our example use two time-independent variables as shown in Figure 11.1. We are studying driving errors that are made in emergency handling situations on a standardized route. Participants are six students who are given extensive experience with the route so that it is familiar to them. The dependent variable is the number of errors made by the students during a loop around the route.

There are two independent variables in the study. For each jaunt the students will be tested after they have consumed either zero, one, or three alcoholic beverages. They will drive a sport sedan (car) on one type of driving occasion and a family-sized, sport utility vehicle (SUV) on another type of occasion. Because this is a within-subjects design, each student will be tested under all of the experimental conditions although the order will be randomized for each student to balance the influence of any carry-over effects on the conditions. That is, each student will drive the car after consuming no alcoholic drinks, also after consuming one alcoholic drink, and also after consuming three alcoholic drinks; he or she will also drive the SUV after consuming no alcoholic drinks, also after consuming one alcoholic drink, and also after consuming three alcoholic drinks.

11.3   PARTITIONING THE VARIANCE INTO ITS SOURCES

The two summaries of the ANOVA, one for the between-subjects variance and another for the within-subjects variance, are shown in Table 11.1. We will discuss the *between-subjects effects* first.

11.3.1   BETWEEN-SUBJECTS EFFECTS

The upper portion of Table 11.1 shows the between-subjects effects; it summarizes the between-subjects source of variance. Some students generally make fewer driving errors than other students. This represents the individual differences among the six participants; this source of variance is known as *error* (it is actually *between-subjects error variance*) rather than between-subjects variance. With six participants there are 5 *df*. No *F* ratio is ordinarily computed for individual differences.

Table 11.1. Summary table for within-subjects and between-subjects effects for the two-way numerical example

Between-subjects effects

Source	SS	df	MS
Subject differences (S)	30.222	5	6.044

Within-subjects effects

Source	SS	df	MS	F	Sig.	η^2 Partial	Complete*
Vehicle (A)	2.778	1	2.778	1.761	0.242		
Error ($A \times S$)	7.889	5	1.578				
Drinks (B)	157.389	2	78.694	52.658	0.000	.913	.774
Error ($B \times S$)	14.944	10	1.494				
$A \times B$	17.056	2	8.528	26.017	0.000	.839	.084
Error ($A \times B \times S$)	3.278	10	0.328				
Total within subjects	203.334						

* Complete based on the total within-subjects sum of squares.

11.3.2 WITHIN-SUBJECTS EFFECTS

The lower portion of Table 11.1 shows the *within-subjects effects* and summarizes the within-subjects partitioning of the variance. Several aspects of the table are worthy of note; we discuss these next in the context of the effects in which we are interested.

11.4 EFFECTS OF INTEREST IN THIS DESIGN

Whether the study represents a between-subjects or a within-subjects design, with two independent variables in the design there are three effects of interest: the main effect of one independent variable, the main effect of the other independent variable, and the interaction of the two independent variables. In a within-subjects design, all three of these effects are subsumed in the within-subjects variance.

11.4.1 MAIN EFFECTS

As can be seen from Table 11.1, the main effect of Vehicle is not statistically significant. We therefore conclude that driving either a sports sedan or an SUV does not result in a different number of driving errors overall.

The main effect of Drinks is statistically significant. This indicates that the number of driving errors is related to the amount of alcohol students drank. With three levels of this variable, however, we need to perform pairwise comparisons of the means to determine which are significantly different from each other. Because the strength of this effect as indexed by eta squared is so substantial (as can be seen in Table 11.1), we would probably want to follow up on this main effect even though we obtained a significant interaction.

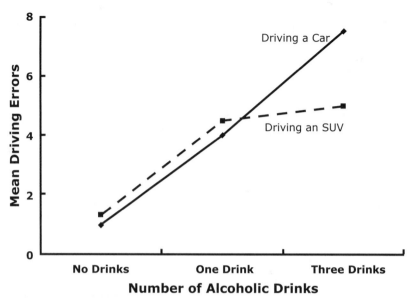

Figure 11.2   A graph of the two-way interaction.

11.4.2   TWO-WAY INTERACTION

The Vehicle × Drinks interaction is statistically significant. Thus, the particular combinations of the levels of the independent variable are more predictive of driving errors than either independent variable in isolation. The interaction is graphed in Figure 11.2. Based on visual inspection of the figure, it appears that students with no alcohol consumed have very few driving errors for both vehicles and that driving errors comparably increase for both vehicles when the students have had one drink. After three drinks, driving errors continue to increase at about the same rate when students are behind the wheel of a sports car but appear to increase only modestly (or perhaps level off) when they are driving an SUV. We need to perform tests of simple effects to fully explicate this interaction effect by determining which means are statistically different from which others.

11.5   THE ERROR TERMS IN A TWO-WAY WITHIN-SUBJECTS DESIGN

Each effect of interest in a within-subjects design is associated with its own error term. This is always true in a within-subjects design. The reason for this is because each effect interacts with the subjects effect as described in Chapter 10, and this interaction is "noise" or unpredictability in the effect across the different participants in the study. This unpredictability is taken as an error term. Because each effect can interact differently with subjects, each interaction is keyed to its respective effect.

11.5.1   THE MAIN EFFECT OF VEHICLE

The error term for the main effect of Vehicle is represented by the Vehicle × Subjects interaction, shown as Error $(A \times S)$ in the summary table.

This error component represents the errors made driving a sports car and an SUV on a student by student basis. If all of the students showed exactly the same pattern for driving the car and the SUV (averaging over the number of drinks), the functions would be completely parallel and the value of the sum of squares for this interaction would be zero.

11.5.2   THE MAIN EFFECT OF DRINKS

The error term for the main effect of Drinks is represented by the Drinks $\times$ Subjects interaction, shown as Error ($B \times S$) in the summary table. This error term focuses on the driving errors made as a function of alcohol consumed on a student by student basis. Again, if all students showed exactly the same pattern of errors (averaging across the vehicles they drove), then there would be no unpredictability and the sum of squares for the interaction would be zero.

11.5.3   THE TWO-WAY INTERACTION

The error term for the interaction follows the same logic that we just discussed for the main effects. It is really the interaction of the effect (which is in this case is the Vehicle $\times$ Drinks interaction itself) with subjects. The Vehicle $\times$ Drinks ($A \times B$) interaction is based on the cells of the design. We had laid out the matrix for the design earlier in Figure 11.1. Imagine one student's driving scores in that matrix. Those scores would reflect a particular pattern. If all of the other students had scores that fit that pattern exactly, the Vehicle $\times$ Drinks $\times$ Subjects interaction – the $A \times B \times S$ – would yield a sum of squares of zero.

11.6   COMPUTING THE OMNIBUS TWO-FACTOR WITHIN-SUBJECTS ANOVA BY HAND

The computational process for a two-factor within-subjects ANOVA is both a direct extension of the computational procedures we covered in Chapter 10 for the single-factor within-subjects design, and also parallels some of the computational procedures we encountered with the two-factor between-subjects ANOVA design of Chapter 8. We will continue with the hypothetical example introduced in this chapter. The independent variables are type of vehicle (Factor A) and amount of alcohol consumed (Factor B). The type of vehicle independent variable has two levels ($a_1 =$ car, $a_2 =$ SUV) and amount of alcohol consumed has three levels ($b_1 = 0$ drinks, $b_2 = 1$ drink, $b_3 = 3$ drinks). The dependent variable is the number of driving errors. This combination of independent variables produces a 2×3 factorial and is displayed with the raw data in the first matrix (*ABS* Matrix) of Table 11.2. The *ABS* Matrix is accompanied by the usual sums of scores and sums of squared scores, needed for future calculations. The *AB* matrix of means can be used to develop a plot of the treatment effects as in Figure 11.2.

Three additional matrices (*AB* matrix, *AS* matrix, and *BS* matrix) also need to be created from the original *ABS* matrix and can be seen

Table 11.2. Preliminary calculations for a 2×3 within-subjects ANOVA data set

(ABS matrix)

		Vehicle type (Factor A)					
		a_1 (car)			a_2 (SUV)		
		Amount of alcohol			Amount of alcohol		
	Factor B	(0)	(1)	(3)	(0)	(1)	(3)
	Subjects	b_1	b_2	b_3	b_1	b_2	b_3
	s_1	1	4	6	2	5	4
	s_2	1	5	7	1	6	7
	s_3	2	6	9	1	6	5
	s_4	2	5	10	1	4	6
	s_5	0	3	7	2	4	5
	s_6	0	1	6	1	2	3
Sum of scores ΣY:		6	24	45	8	27	30	$T = 140$
Sum of squared scores ΣY^2:		10	112	351	12	133	160
Mean $\overline{Y}$:		1.00	4.00	7.50	1.33	4.50	5.00
Standard deviation s:		0.89	1.79	1.64	0.52	1.52	1.41

AB matrix (sums)

	(0)	(1)	(3)	
	b_1	b_2	b_3	Sum
a_1 (car)	6	24	45	75
a_2 (SUV)	8	27	30	65
Sum	14	51	75	140

AS matrix

	a_1 (car)	a_2 (SUV)	Sum
s_1	11	11	22
s_2	13	14	27
s_3	17	12	29
s_4	17	11	28
s_5	10	11	21
s_6	7	6	13
Sum	75	65	140

BS matrix

	b_1 (0)	b_2 (1)	b_3 (3)	Sum
s_1	3	9	10	22
s_2	2	11	14	27
s_3	3	12	14	29
s_4	3	9	16	28
s_5	2	7	12	21
s_6	1	3	9	13
Sum	14	51	75	140

AB matrix of means

	(0) b_1	(1) b_2	(3) b_3	Marginal means
a_1 (car)	1.00	4.00	7.50	4.17
a_2 (SUV)	1.33	4.50	5.00	3.61
Marginal means	1.17	4.25	6.25	

Table 11.3. Two-factor within-subjects ANOVA summary table

Source	SS	df	MS	F
A (vehicle type)	2.78	1	2.78	1.76
B (amount of alcohol)	157.39	2	78.70	52.82*
A × B	17.06	2	8.53	25.85*
S (subject)	30.23	5	6.05	
A × S	7.88	5	1.58	
B × S	14.94	10	1.49	
A × B × S	3.28	10	0.33	
Total	233.56	35		

*$p < .05$.

in Table 11.2. These matrices are created by summing the Y scores collapsed across various treatment configurations. The reader is encouraged to examine carefully how these matrices were created before proceeding with the sum of squares computations.

11.6.1 CALCULATING SUMS OF SQUARES

There are seven sums of squares to be calculated in a two-way within-subjects ANOVA. These final values can be seen in the ANOVA summary table in Table 11.3. The preliminary calculations shown in Table 11.2 provide the ingredients for computing the following seven sums of squares:

- Sum of squares main effect of Factor A (SS_A).
- Sum of squares main effect of Factor B (SS_B).
- Sum of squares interaction effect of Factors A and B ($SS_{A \times B}$).
- Sum of squares main effects of subjects (SS_S).
- Sum of squares interaction of Factor A and subjects ($SS_{A \times S}$).
- Sum of squares interaction of Factor B and subjects ($SS_{B \times S}$).
- Sum of squares interaction of Factor A, Factor B, and subjects ($SS_{A \times B \times S}$).
- Sum of squares total (SS_T).

The formulas for these seven sums of squares and the summary of their calculations based on the data in Table 11.2 are as follows. Each sum of squares is calculated by focusing on a particular matrix in Table 11.2. The first three sums of squares (SS_A, SS_B, $SS_{A \times B}$) are derived from the AB matrix in Table 11.2.

$$SS_A = \frac{\Sigma A^2}{(b)(n)} - \frac{T^2}{(a)(b)(n)}$$

$$= \frac{(75)^2 + (65)^2}{(3)(6)} - \frac{(140)^2}{(2)(3)(6)} = \boxed{2.78} \qquad (11.1)$$

$$SS_B = \frac{\Sigma B^2}{(a)(n)} - \frac{T^2}{(a)(b)(n)}$$

$$= \frac{(14)^2 + (51)^2 + (75)^2}{(2)(6)} - \frac{(140)^2}{(2)(3)(6)} = \boxed{157.39} \qquad (11.2)$$

$$SS_{A \times B} = \frac{\Sigma AB^2}{n} - \frac{\Sigma A^2}{(b)(n)} - \frac{\Sigma B^2}{(a)(n)} + \frac{T^2}{(a)(b)(n)}$$

$$= \frac{(6)^2 + (24)^2 + (45)^2 + (8)^2 + (27)^2 + (30)^2}{6}$$

$$- \frac{(75)^2 + (65)^2}{(3)(6)} - \frac{(14)^2 + (51)^2 + (75)^2}{(2)(6)} + \frac{(140)^2}{(2)(3)(6)}$$

$$= \boxed{17.06}. \qquad (11.3)$$

The next two sums of squares are derived from the AS matrix of Table 11.2.

$$SS_S = \frac{\Sigma S^2}{(a)(b)} - \frac{T^2}{(a)(b)(n)}$$

$$= \frac{(22)^2 + (27)^2 + \cdots + (21)^2 + (13)^2}{(2)(3)} - \frac{(140)^2}{(2)(3)(6)}$$

$$= \boxed{30.23} \qquad (11.4)$$

$$SS_{A \times S} = \frac{\Sigma AS^2}{b} - \frac{\Sigma A^2}{(b)(n)} - \frac{\Sigma S^2}{(a)(b)} + \frac{T^2}{(a)(b)(n)}$$

$$= \frac{(11)^2 + (13)^2 + \cdots + (11)^2 + (6)^2}{3} - \frac{(75)^2 + (65)^2}{(3)(6)}$$

$$- \frac{(22)^2 + (27)^2 + \cdots + (21)^2 + (13)^2}{(2)(3)} + \frac{(140)^2}{(2)(3)(6)}$$

$$= \boxed{7.88}. \qquad (11.5)$$

The next sum of squares is derived from the BS matrix of Table 11.2.

$$SS_{B \times S} = \frac{\Sigma BS^2}{a} - \frac{\Sigma B^2}{(a)(n)} - \frac{\Sigma S^2}{(a)(b)} + \frac{T^2}{(a)(b)(n)}$$

$$= \frac{(3)^2 + (2)^2 + \cdots + (12)^2 + (9)^2}{2} - \frac{(14)^2 + (51)^2 + (75)^2}{(2)(6)}$$

$$- \frac{(22)^2 + (27)^2 + \cdots + (21)^2 + (13)^2}{(2)(3)} + \frac{(140)^2}{(2)(3)(6)}$$

$$= \boxed{14.94}. \qquad (11.6)$$

The next sum of squares ($SS_{A \times B \times S}$) is derived from all four of the matrices in Table 11.2.

$$SS_{A \times B \times S} = \Sigma Y^2 - \frac{\Sigma AB^2}{n} - \frac{\Sigma AS^2}{b} - \frac{\Sigma BS^2}{a} + \frac{\Sigma A^2}{(b)(n)} + \frac{\Sigma B^2}{(a)(n)}$$

$$+ \frac{\Sigma S^2}{(a)(b)} - \frac{T^2}{(a)(b)(n)}$$

$$= (1)^2 + (1)^2 + \cdots + (5)^2 + (3)^2$$

$$- \frac{(6)^2 + (24)^2 + (45)^2 + (8)^2 + (27)^2 + (30)^2}{6}$$

$$- \frac{(11)^2 + (13)^2 + \cdots + (11)^2 + (6)^2}{3}$$

$$- \frac{(3)^2 + (2)^2 + \cdots + (12)^2 + (9)^2}{2} + \frac{(75)^2 + (65)^2}{(3)(6)}$$

$$+ \frac{(14)^2 + (51)^2 + (75)^2}{(2)(6)}$$

$$+ \frac{(22)^2 + (27)^2 + \cdots + (21)^2 + (13)^2}{(2)(3)} - \frac{(140)^2}{(2)(3)(6)}$$

$$= \boxed{3.28}. \tag{11.7}$$

Lastly, from the *ABS* matrix of Table 11.2, we derive the SS_T.

$$SS_T = \Sigma Y^2 - \frac{T^2}{(a)(b)(n)}$$

$$= (1)^2 + (1)^2 + \cdots + (5)^2 + (3)^2 - \frac{(140)^2}{(2)(3)(6)}$$

$$= \boxed{233.56}. \tag{11.8}$$

All of these sums of squares have been conveniently entered into an ANOVA summary table (see Table 11.3).

11.6.2 CALCULATING DEGREES OF FREEDOM

Below are the formulas for the degrees of freedom associated with each sum of squares, and the simple computations involved based on our numerical example.

$$df_A = a - 1 = 2 - 1 = \boxed{1}$$

$$df_B = b - 1 = 3 - 1 = \boxed{2}$$

$$df_{A \times B} = (a - 1)(b - 1) = (2 - 1)(3 - 1) = (1)(2) = \boxed{2}$$

$$df_S = n - 1 = 6 - 1 = \boxed{5}$$

$$df_{A \times S} = (a - 1)(n - 1) = (2 - 1)(6 - 1) = (1)(5) = \boxed{5}$$

$$df_{B \times S} = (b - 1)(n - 1) = (3 - 1)(6 - 1) = (2)(5) = \boxed{10}$$

$$df_{A \times B \times S} = (a - 1)(b - 1)(n - 1) = (2 - 1)(3 - 1)(6 - 1)$$

$$= (1)(2)(5) = \boxed{10}$$

$$df_T = (a)(b)(n) - 1 = (2)(3)(6) - 1 = 36 - 1 = \boxed{35}.$$

11.6.3 CALCULATING MEAN SQUARES AND F RATIOS

Mean squares are calculated by dividing each sum of squares by its respective degrees of freedom. Note that we do not calculate a mean square total value. You will recall from our earlier discussion that a two-factor within-subjects design produces three F ratios. What is unique about this analysis is that each F ratio requires a separate error term for the denominator to control for the different sources of variability influencing the scores of the dependent measure under study. Thus, our three F ratios are formed with the following error terms. F_A: $MS_{A \times S}$, F_B: $MS_{B \times S}$, and $F_{A \times B}$: $MS_{A \times B \times S}$. Thus,

$$F_A = \frac{MS_A}{MS_{A \times S}} = \boxed{1.76} \tag{11.9}$$

$$F_B = \frac{MS_B}{MS_{B \times S}} = \frac{78.70}{1.49} = \boxed{52.82} \tag{11.10}$$

$$F_{A \times B} = \frac{MS_{A \times B}}{MS_{A \times B \times S}} = \frac{8.53}{0.33} = \boxed{25.85}. \tag{11.11}$$

11.6.4 EVALUATING THE F RATIOS

We test the null hypothesis for each of our three observed (calculated) F ratios by evaluating each F value with critical values of F (see Appendix C) at the following degrees of freedom: (df_{effect}, df_{error}), at a particular alpha level. Thus, our F observed for Factor A was $F_A = 1.76$ and is evaluated with (1, 5) degrees of freedom. The critical value of F at these degrees of freedom (with alpha set at .05) is 6.61. Since our F_A is less than the critical value, we do not reject the null hypothesis, and conclude that type of vehicle does not affect the error rate.

For the main effect of Factor B, our F_B was 52.82 and is evaluated with (2, 10) degrees of freedom. The critical value of F is 4.10 at the .05 level. Since our F_B is greater than the critical value, we reject the null and conclude that the amount of alcohol affects the error rate.

Lastly, we evaluate the interaction effect ($F_{A \times B} = 25.85$) with (2, 10) degrees of freedom. These happen to be the same degrees of freedom we had with F_B; hence, we reject the null hypothesis and conclude that the two independent variables (vehicle and alcohol) combine to produce a statistically significant unique joint effect. We remind the reader that because the interaction effect is statistically significant, it takes precedence over the statistically significant main effect of Factor B. We will briefly review the computational details of these analyses.

11.6.5   COMPUTING TWO-FACTOR WITHIN-SUBJECTS SIMPLE EFFECTS BY HAND

Simple Effects

As we noted in Chapter 8, a statistically significant interaction effect is typically followed by an analysis of simple effects. Recall that such an analysis is like conducting a one-way ANOVA with one of the independent variables held constant. The sums that we use to form our sums of squares come from the AB matrix of sums in Table 11.2. Thus, in the present example, the simple effects of Factor A (type of vehicle) at a particular level of Factor B (amount of alcohol), for example, b_3 (3 drinks), would have the following formulas for sums of squares, degrees of freedom, mean square, and F.

$$SS_{A \text{ at } b_3} = \frac{\Sigma(AB_{j3})^2}{n} - \frac{(B_3)^2}{(a)(n)}$$

$$df_{A \text{ at } b_3} = a - 1 \qquad\qquad (11.12)$$

$$MS_{A \text{ at } b_3} = \frac{SS_{A \text{ at } b_3}}{df_{A \text{ at } b_3}} \qquad\qquad (11.13)$$

$$F_{A \text{ at } b_3} = \frac{MS_{A \text{ at } b_3}}{MS_{A \times B \times S}}. \qquad\qquad (11.14)$$

This F is evaluated with $(df_{A \text{ at } b3}, df_{A \times B \times S})$.

Likewise, to compute the simple effects of Factor B at levels of A (e.g., level a_2), we have the following formula:

$$SS_{B \text{ at } a_2} = \frac{\Sigma(AB_{2k})^2}{n} - \frac{(A_2)^2}{(b)(n)}$$

$$df_{B \text{ at } a_2} = b - 1 \qquad\qquad (11.15)$$

$$MS_{B \text{ at } a_2} = \frac{SS_{A \text{ at } b_2}}{df_{B \text{ at } a_2}} \qquad\qquad (11.16)$$

$$F_{B \text{ at } a_2} = \frac{MS_{B \text{ at } a_2}}{MS_{A \times B \times S}}. \qquad\qquad (11.17)$$

This F is evaluated with $(df_{B \text{ at } a2}, df_{A \times B \times S})$.

Simple Comparisons

Following a statistically significant simple effect analysis that has three or more treatments, paired or simple comparisons can be conducted.

For example, because Factor B has three treatments we could examine a pairwise comparison with the following formula:

$$SS_{B_{\text{comp}} \text{ at } a_j} = \frac{(n)(\hat{\Psi}_{B \text{ at } a_j})^2}{2}. \qquad (11.18)$$

Here we compute a difference between two means (from the AB matrix of means in Table 11.2) at one of the levels of Factor A and square it. This value is multiplied by the sample size and divided by 2 because there are two means being compared:

$$MS_{B_{\text{comp}} \text{ at } a_j} = \frac{SS_{A_{\text{comp}} \text{ at } a_j}}{1} \qquad (11.19)$$

$$F_{B_{\text{comp}} \text{ at } a_j} = \frac{MS_{A_{\text{comp}} \text{ at } a_j}}{MS_{A \times B \times S}}. \qquad (11.20)$$

Main Comparisons

If the interaction effect is not statistically significant, but one or more of the main effects are statistically significant and the effect comprises three or more levels, then an analysis of the main comparisons is in order. We would begin by isolating a pairwise comparison of marginal means; in the present example, we would focus on the marginal means for Factor B (1.17, 4.25, 6.25) in the AB matrix of means in Table 11.2. Recall that we subtract one mean from another to create our difference value, which we call $\hat{\Psi}$ (read "psi hat"). The formula is as follows:

$$SS_{B_{\text{comp}}} = \frac{(a)(n)(\hat{\Psi}_B)^2}{2}$$
$$df_{B_{\text{comp}}} = 1 \qquad (11.21)$$

$$MS_{B_{\text{comp}}} = \frac{SS_{B_{\text{comp}}}}{1} \qquad (11.22)$$

$$F_{B_{\text{comp}}} = \frac{MS_{B_{\text{comp}}}}{MS_{B \times S}}. \qquad (11.23)$$

We have purposely omitted some of the computational details involved in factorial within-subjects simple effects analyses, simple comparisons, and main comparisons. However, many of the relevant computational details are covered in Chapters 7 and 8 (between-subjects designs) and can be readily applied with the present (within-subjects) formulas. For more details on these analyses, we recommend Keppel et al. (1992) and Keppel and Wickens (2004).

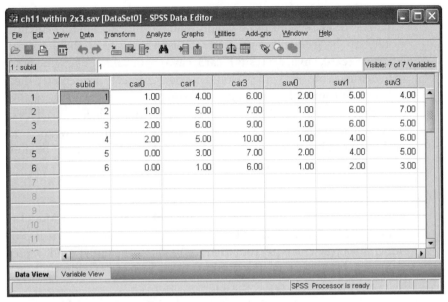

Figure 11.3 Data file for the numerical example.

11.7 PERFORMING THE OMNIBUS ANALYSIS IN SPSS

11.7.1 STRUCTURING THE DATA FILE

The data file for a two-way within-subjects design is structured in a more complex fashion than the one-way design. We show in Figure 11.3 the data file for our 2 × 3 within-subjects design example. The first column, as always, is used for our participant identification number; we have named this variable **subid**. The next six columns are devoted to the cells representing our two-way design.

The variables (columns) in the data file are ordered in a systematic manner so that they will match up with our independent variables. Recall from Figure 11.1 that the two rows represented the vehicles that the students were driving and that the three columns represented the number of drinks the students consumed prior to driving. The first cell in this 2 × 3 matrix coincides with column 2 in the data file with the variable name **car0**. It refers to students driving sports cars (the first level of **vehicle**) with no alcohol consumed (the first level of **drinks**). In our design matrix in Figure 11.1, it is the upper left cell. The next two columns are named **car1** and **car3**. These cells also refer to the first level of **vehicle** (sports car) but step us through the remaining levels of **drinks**. Thus, **car1** denotes the driving errors for sports cars with students having had one drink whereas **car3** denotes the errors for sports cars with students having had three drinks.

The next three columns in the data file step us through the second level of **Vehicle: suv0, suv1**, and **suv3** represent the cells for students driving an SUV after zero, one, or three drinks. Note that we use the same structure here as we did for the sports car level of **Vehicle**. In this structure, as we view the variables in the data file from left to right the information for **drinks** "changes" or increments more rapidly than the information for **vehicle**. It would have been equally appropriate for us to have chosen the alternative structure (**car0, suv0, car1, suv1, car3, suv3**) with **vehicle** changing faster than **drinks**. Note that while it would not be incorrect to have entered these variables in some other nonsystematic order, we strongly advise against such chaotic behavior. When we specify our within-subjects variables in the initial SPSS dialog windows, the systematic structure that we used can facilitate bringing these variables into the analysis (which must be done in a very systematic fashion).

11.7.2   STRUCTURING THE DATA ANALYSIS

From the main SPSS menu, select **Analyze → General Linear Model → Repeated Measures**. This will open the dialog window shown in Figure 11.4. As you will recall from Chapter 10, this is the initial window used by SPSS to have you name the within-subjects variable(s) in the study and to indicate the number of levels it has.

Having two within-subjects independent variables calls for defining each of them separately. Remember how we structured the data file. The variable named **vehicle** changed or incremented more slowly than the **drinks** variable. Given this data structure, the first variable we should name is **vehicle**; then we can name **drinks**.

Highlight the default name **factor1** and type in **vehicle**. Then type in the number of levels for that within-subjects variable, which in this case is 2. Finally click **Add**. We see the name **vehicle** appear in the panel to

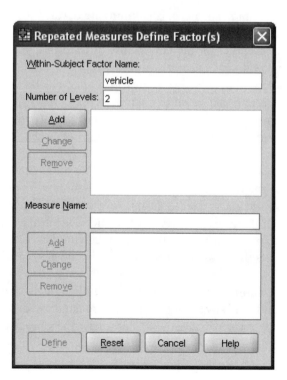

Figure 11.4 The initial **GLM Repeated Measures** dialog window.

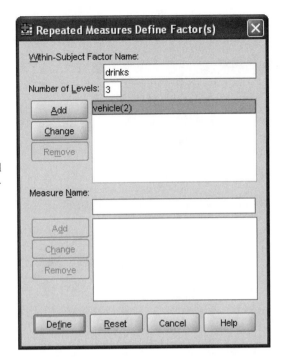

Figure 11.5 The initial window with the first variable named.

the right of the **Add** button as shown in Figure 11.5. The value of 2 in parentheses next to it marks the number of levels that we had specified.

The window is now ready for the next repeated measure to be named. We have typed **drinks** in the box for **Within-Subject Factor Name** and have specified 3 by **Number of Levels**. All of this is shown as well in Figure 11.5. We then click **Add**. The result of this typing is shown in Figure 11.6.

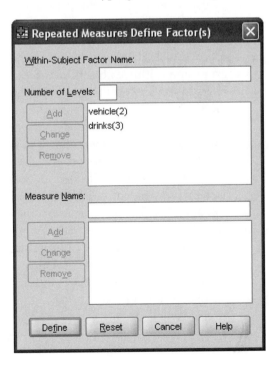

Figure 11.6 The initial window with both variables named.

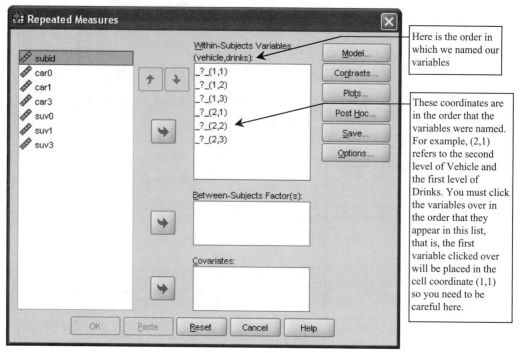

Figure 11.7 The main **GLM Repeated Measures** window reached just after clicking the **Define** pushbutton.

SPSS has registered both **vehicle** with two levels and **drinks** with three levels. Click **Define** to reach the main **GLM Repeated Measures** window.

The main dialog window for **GLM Repeated Measures** is shown in Figure 11.7. Note the number pairs in parentheses next to the lines with question marks. These are in the order that we named the variables; this order (**vehicle, drinks**) is shown at the very top of the window under to the expression **Within-Subjects Variables**. The first value in each parentheses refers to the level of the first independent variable, **vehicle** in this case; the second value in each parentheses refers to the level of the second independent variable, **drinks** in this case. SPSS requires that you define your variables before it will allow you to perform the analysis (notice that the **OK** pushbutton is not yet available).

Now we can deal with defining the variables. To define the within-subjects independent variables, we need to supply the variable (the named column in the data file) corresponding to the specified values of the levels of each. For example, the cell combination or coordinate (1,1) indicates the first level of **vehicle** (car) and the first level of **drinks** (0); the coordinate (1,2) indicates the first level of **vehicle** (car) and the second level of **drinks** (1); and the coordinate (1,3) indicates the first level of **vehicle** (car) and the third level of **drinks** (1). The last three are analogous using the second level of **vehicle** (SUV).

As we view the list of coordinates from top to bottom, the second variable (**drinks** in this case) increments more rapidly than the first variable (**vehicle** in this case). This is precisely the order of the variables in the data file, and is not a coincidence. By entering the data in this systematic manner at the outset, and by naming the variables such that the slowest incrementing variable in the data file is named first, we have made it very convenient to define our two independent variables in this window. This is because the order of the coordinates matches the order in which the variables appear in the **variables list** in the left panel of the main **GLM Repeated Measures** window.

Knowing the order in which we need to bring the variables over to the cell coordinates, we can now bring them over. The first variable brought over will be matched with the (1,1) coordinate, so you need to be very careful to use the right panel with the question marks and coordinates as your guide for the ordering of your variables. In our case the ordering in the **variable list** matches the order in which SPSS needs to be told of the order in this window, so we can highlight the list from **car0** to **suv3** and bring it over at the same time; SPSS will preserve the ordering of any highlighted set of variables clicked over. Figure 11.8 shows the levels of the variables defined.

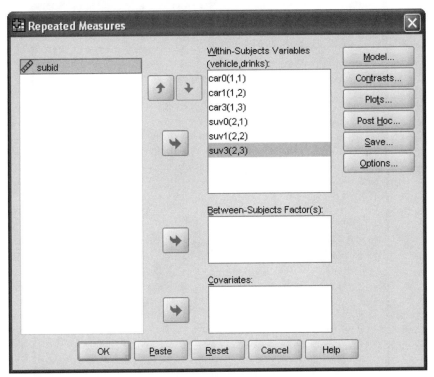

Figure 11.8 The main **GLM Repeated Measures** window after we have defined our variables.

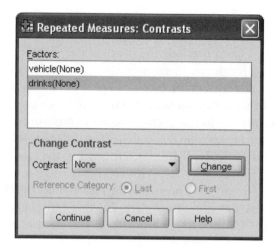

Figure 11.9 Setting **Contrasts** to **None**.

Click the **Contrasts** pushbutton to reach the window shown in Figure 11.9. SPSS uses the **Polynomial** contrast as its default. Because the repeated measure does not represent equally spaced intervals, we should not perform such an analysis. Highlight each variable in turn, select **None**

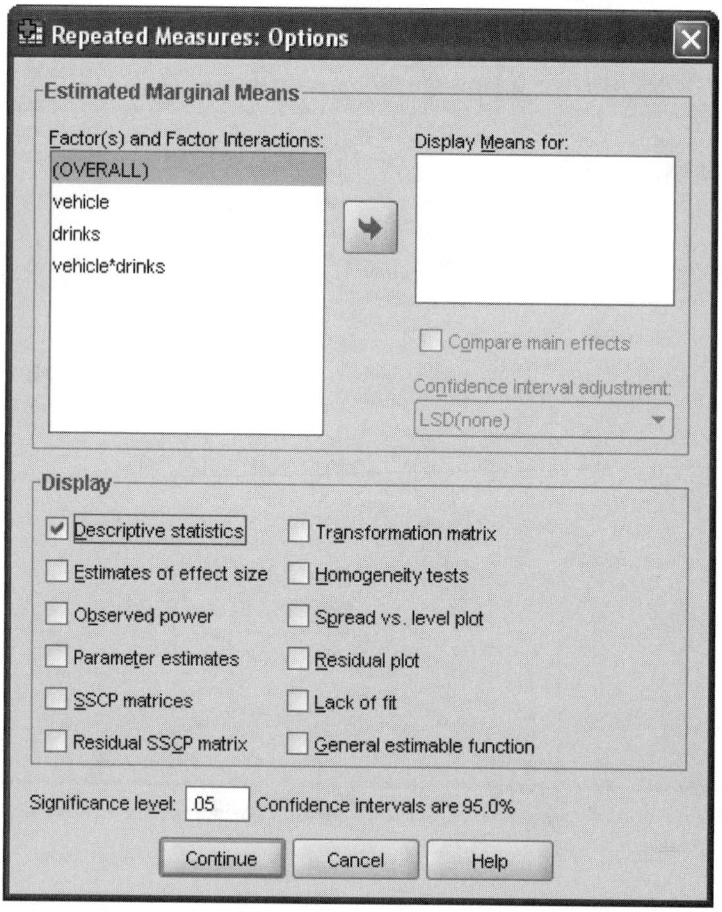

Figure 11.10 The **Options** window.

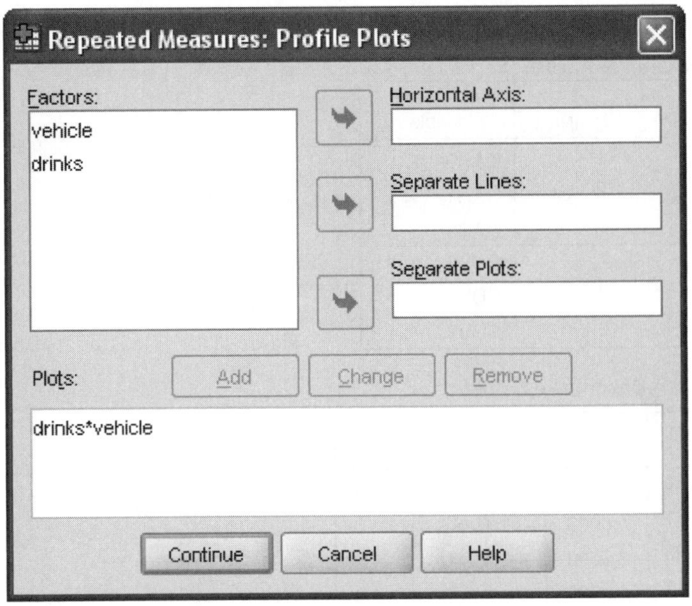

Figure 11.11   The **Plots** window.

from the drop-down menu, and click **Change**. The contrasts in parentheses next to each variable name will now say **None** as shown in Figure 11.9 (you should note that most of the recent versions of SPSS will ignore the request to run no contrasts and will perform polynomial contrasts if one of the other contrasts is not selected). Click **Continue** to return to the main dialog window.

In the bottom portion of the window labeled **Display**, check **Descriptive statistics** to output the observed means and standard deviations as shown in Figure 11.10. Click **Continue** to return to the main dialog window.

Click the **Plots** pushbutton to set up the plot for the interaction should it be found to be significant. We place **drinks** on the horizontal axis for three reasons: (a) it is conceptually more continuous than **vehicle**; (b) it has more levels than **vehicle** and thus would have produced more lines had we designated this variable for **Separate Lines**; (c) it matches with the way we drew the design matrix in Figure 11.1. We show the **Plots** window in Figure 11.11 just after clicking the **Add** pushbutton to register the plot with SPSS. Click **Continue** to return to the main dialog window and **OK** to run the omnibus analysis.

11.8   SPSS OUTPUT FOR THE OMNIBUS ANALYSIS

Figure 11.12 shows two small tables from the initial portion of the output. The **Within-Subjects Factors** table presents the levels of the repeated measures; SPSS reports the levels of each in the left portion and the actual variables corresponding to the cells in the right portion of the table.

Within-Subjects Factors

Measure: MEASURE_1

VEHICLE	DRINKS	Dependent Variable
1	1	CAR0
	2	CAR1
	3	CAR3
2	1	SUV0
	2	SUV1
	3	SUV3

Figure 11.12   Descriptive statistics.

Descriptive Statistics

	Mean	Std. Deviation	N
CAR0	1.0000	.89443	6
CAR1	4.0000	1.78885	6
CAR3	7.5000	1.64317	6
SUV0	1.3333	.51640	6
SUV1	4.5000	1.51658	6
SUV3	5.0000	1.41421	6

The **Descriptive Statistics** table presents the observed means, standard deviations, and sample sizes for each condition.

Figure 11.13 presents the results of **Mauchly's Test of Sphericity**. Recall that Mauchly's test simultaneously evaluates two assumptions: (a) that the levels of the within-subjects variable have equal variances (i.e., that there is homogeneity of variance); and (b) that the levels of the

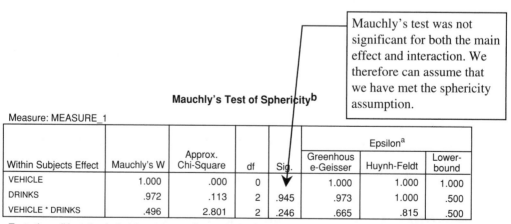

Mauchly's test was not significant for both the main effect and interaction. We therefore can assume that we have met the sphericity assumption.

Mauchly's Test of Sphericity[b]

Measure: MEASURE_1

Within Subjects Effect	Mauchly's W	Approx. Chi-Square	df	Sig.	Epsilon[a] Greenhouse-Geisser	Huynh-Feldt	Lower-bound
VEHICLE	1.000	.000	0		1.000	1.000	1.000
DRINKS	.972	.113	2	.945	.973	1.000	.500
VEHICLE * DRINKS	.496	2.801	2	.246	.665	.815	.500

Tests the null hypothesis that the error covariance matrix of the orthonormalized transformed dependent variables is proportional to an identity matrix.

a. May be used to adjust the degrees of freedom for the averaged tests of significance. Corrected tests are displayed in the Tests of Within-Subjects Effects table.

b. Design: Intercept
   Within Subjects Design: VEHICLE+DRINKS+VEHICLE*DRINKS

Figure 11.13   Mauchly's test of sphericity.

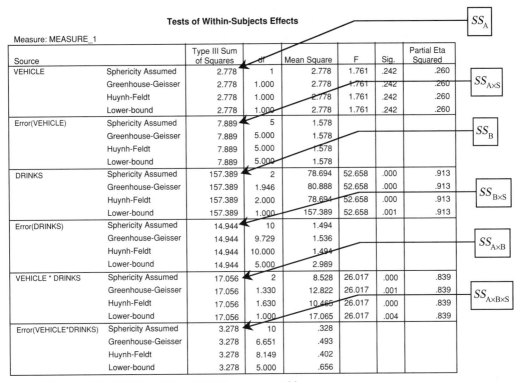

Figure 11.14 Within-subjects ANOVA summary table.

within-subjects variable are correlated to the same extent. With only two levels of **vehicle**, the test is not interpretable and SPSS does not test its significance. With three or more conditions available for evaluation with respect to **drinks** and the **vehicle∗drinks** interaction, Mauchly's sphericity statistic is computed and tested for significance; neither is statistically significant.

The summary table for the within-subjects variance is displayed in Figure 11.14. With a nonsignificant sphericity test, we can focus on the rows labeled **Sphericity Assumed**. There are three effects of interest: the two main effects and the two-way interaction. Each is evaluated by its own error term and, as we noted in Table 11.1, the main effect for **drinks** and the **vehicle∗drinks** interaction are both statistically significant.

Figure 11.15 shows the between-subjects portion of the variance. The **Intercept** concerns the full model that we do not deal with here. What

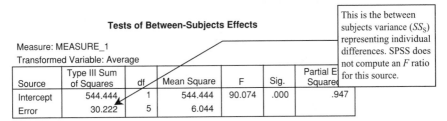

Figure 11.15 Between-subjects ANOVA summary table.

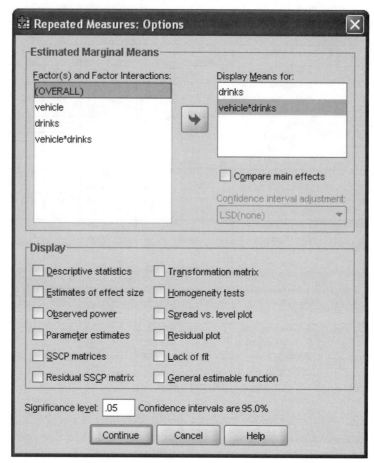

Figure 11.16   Two effects are brought into the **Estimated Margin Means** panel.

SPSS calls **Error** is the variance representing individual differences; it does not calculate an *F* ratio for this effect.

11.9   PERFORMING THE POST-ANOVA ANALYSIS IN SPSS

11.9.1   DETERMINING THE EFFECTS TO BE SIMPLIFIED

Configure the analysis as you did for the omnibus ANOVA. In the **Options** dialog window shown in Figure 11.16, bring over the effects for **drinks** and the **vehicle∗drinks** interaction (the two significant effects) to the panel labeled **Display Means for**; this will produce a line of **/EMMEANS = TABLES** syntax and tables of estimated marginal means for these effects. Do not ask for the **Descriptive statistics** – we already have obtained them and the simple effects are performed on the estimated margin means rather than the observed means in any case. Click **Continue** to reach the main dialog window and then click the **Paste** pushbutton instead of clicking **OK**. The underlying syntax will then be placed in a syntax window; locate that window on your desktop and make it active by clicking it.

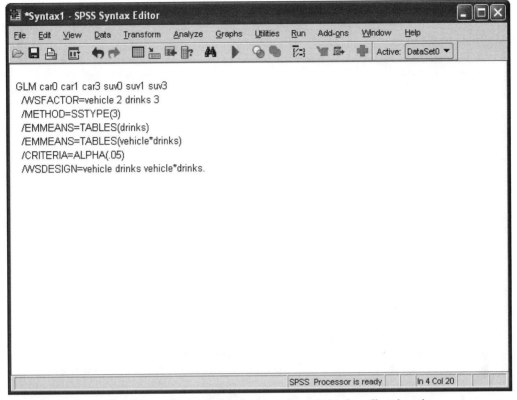

Figure 11.17 The syntax that underlies the two-way analysis after effects have been specified in **Options** window.

11.9.2 SIMPLE EFFECTS TESTING

With the syntax window active, the syntax displayed in Figure 11.17 is available. The first step in performing the simple effects testing is to duplicate the **/EMMEANS = TABLES** command line for the **vehicle*drinks** interaction. This is shown in Figure 11.18.

The next step is to add our **compare** syntax to each of these lines. The two pieces of syntax are as follows and are shown in Figure 11.19:

```
compare (vehicle) adj (bonferroni)
compare (drinks) adj (bonferroni)
```

11.9.3 MAIN EFFECT PAIRED COMPARISONS

We are going to add our **compare** syntax to the **/EMMEANS = TABLES** subcommand for the **drinks** main effect just as we did in Chapter 9. The analysis that we will generate will compare each pair of means to each other. Add the following syntax to that line:

```
compare (drinks) adj (bonferroni)
```

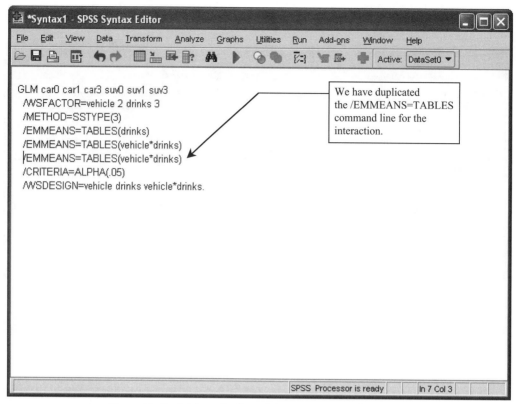

Figure 11.18 We have duplicated the /EMMEANS command for the interaction.

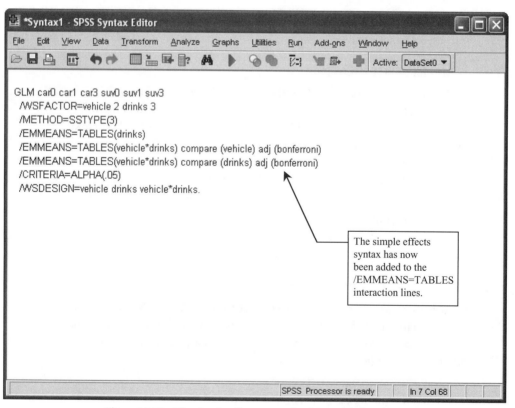

Figure 11.19 The simple effects syntax has been added to the /EMMEANS = TABLES interaction command line.

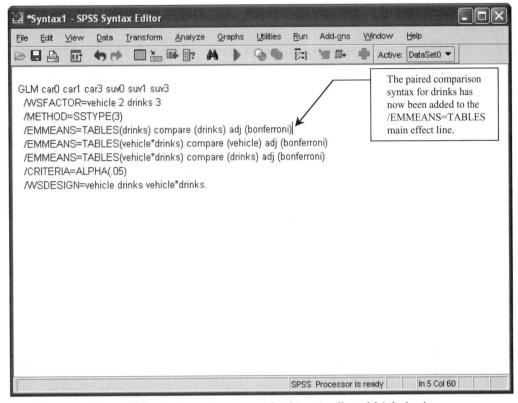

Figure 11.20   The **Paired Comparisons** syntax for the main effect of **drinks** has been added.

The syntax window with this new syntax added is shown in Figure 11.20. After completing this typing, select **Run ➔ All** from the main menu to perform the analysis.

11.10   SPSS OUTPUT FOR THE POST-ANOVA ANALYSIS

11.10.1   INTERACTION SIMPLE EFFECTS

Simple effects analyses are best grounded in the graph of the interaction. In this way, we can see exactly what is analyzed here. Use Figure 11.2 as your base for interpreting what we find here.

Figure 11.21 presents the **Pairwise Comparisons** table for **vehicle**. Note that there are three main rows, each isolating one of the levels of **drinks**. Inside each row, the first level of **vehicle** (car) is compared to the second level of **vehicle** (SUV). We are thus making the three "vertical" comparisons with respect to Figure 11.2 – we are comparing the driving errors for the two types of vehicles at each level of the number of drinks consumed by the students. For example, under the first level of **drinks** (the students have had no drinks) the difference between the means of the two **vehicle** levels is 0.333. This difference is not statistically significant. The only difference that is statistically significant, as can be seen in Figure 11.20,

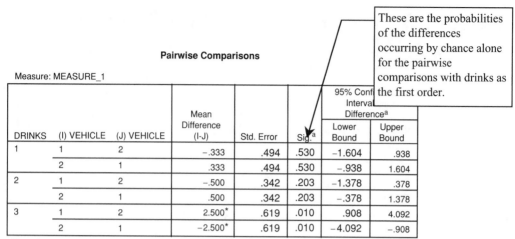

These are the probabilities of the differences occurring by chance alone for the pairwise comparisons with drinks as the first order.

Pairwise Comparisons

Measure: MEASURE_1

DRINKS	(I) VEHICLE	(J) VEHICLE	Mean Difference (I-J)	Std. Error	Sig.[a]	95% Conf Interval Difference[a] Lower Bound	Upper Bound
1	1	2	−.333	.494	.530	−1.604	.938
	2	1	.333	.494	.530	−.938	1.604
2	1	2	−.500	.342	.203	−1.378	.378
	2	1	.500	.342	.203	−.378	1.378
3	1	2	2.500*	.619	.010	.908	4.092
	2	1	−2.500*	.619	.010	−4.092	−.908

Based on estimated marginal means

*. The mean difference is significant at the .05 level.

a. Adjustment for multiple comparisons: Bonferroni.

Figure 11.21   **Drinks** structured by **vehicle** pairwise comparisons.

is at the third level of **drinks**. For students who have had three drinks, they make 2.50 more driving errors in sports cars than in SUVs.

Figure 11.22 presents the **Pairwise Comparisons** table for **drinks**. Note that there are now two main rows, each isolating one of the levels of **vehicle**. Inside each row, the three levels of **drinks** (car) are compared.

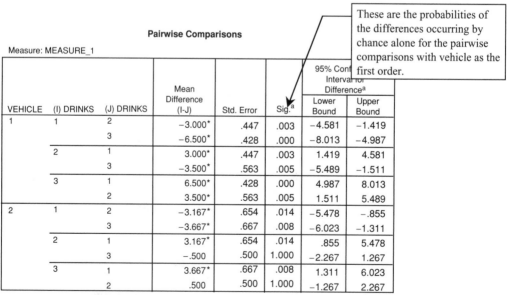

These are the probabilities of the differences occurring by chance alone for the pairwise comparisons with vehicle as the first order.

Pairwise Comparisons

Measure: MEASURE_1

VEHICLE	(I) DRINKS	(J) DRINKS	Mean Difference (I-J)	Std. Error	Sig.[a]	95% Conf Interval for Difference[a] Lower Bound	Upper Bound
1	1	2	−3.000*	.447	.003	−4.581	−1.419
		3	−6.500*	.428	.000	−8.013	−4.987
	2	1	3.000*	.447	.003	1.419	4.581
		3	−3.500*	.563	.005	−5.489	−1.511
	3	1	6.500*	.428	.000	4.987	8.013
		2	3.500*	.563	.005	1.511	5.489
2	1	2	−3.167*	.654	.014	−5.478	−.855
		3	−3.667*	.667	.008	−6.023	−1.311
	2	1	3.167*	.654	.014	.855	5.478
		3	−.500	.500	1.000	−2.267	1.267
	3	1	3.667*	.667	.008	1.311	6.023
		2	.500	.500	1.000	−1.267	2.267

Based on estimated marginal means

*. The mean difference is significant at the .05 level.

a. Adjustment for multiple comparisons: Bonferroni.

Figure 11.22   **Vehicle** structured by **drinks** pairwise comparisons.

Pairwise Comparisons

Measure: MEASURE_1

These are the probabilities of the differences occurring by chance alone for the pairwise comparisons for the main effect of drinks.

(I) DRINKS	(J) DRINKS	Mean Difference (I-J)	Std. Error	Sig.[a]	95% Confidence Interval for Difference[a]	
					Lower Bound	Upper Bound
1	2	−3.083*	.539	.007	−4.987	−1.179
	3	−5.083*	.473	.000	−6.755	−3.412
2	1	3.083*	.539	.007	1.179	4.987
	3	−2.000*	.483	.027	−3.707	−.293
3	1	5.083*	.473	.000	3.412	6.755
	2	2.000*	.483	.027	.293	3.707

Based on estimated marginal means

*. The mean difference is significant at the .05 level.

a. Adjustment for multiple comparisons: Bonferroni.

Figure 11.23 Pairwise comparison tests for the main effect of **drinks**.

We are thus making the two "horizontal" comparisons with respect to Figure 11.2 – we are comparing driving errors for the three numbers of drinks for each vehicle. It appears from the **Pairwise Comparisons** table that under the first level of **vehicle** (cars) all of the conditions differ from each other but that under the second level of **vehicle** (SUV) driving errors increase from zero drinks to one drink but do not further significantly increase after three drinks.

11.10.2   MAIN EFFECT OF DRINKS

The **Paired Comparisons** table for the main effect of **drinks** is shown in Figure 11.23. Overall, driving errors increased with increasing alcohol consumed. However, we recognize from just having discussed the interaction that this is an overgeneralization since driving errors tapered off when the students were behind the wheel of an SUV. It should therefore not be emphasized in the results section.

11.11   PERFORMING THE OMNIBUS ANALYSIS IN SAS

11.11.1   STRUCTURING THE DATA FILE

The data file in Excel format is shown in Figure 11.24. Note that it is in univariate or stacked mode as explained in Section 10.15.2. Briefly, each row can contain no more than a single score on the dependent variable. In the present 2 × 3 within-subjects design, each participant was measured six times, once under each condition. To conform to univariate structure, we must use six lines for each case, one for each condition.

To illustrate our data structure, consider the participant identified as **subid** 1 occupying lines 2 through 7 in the Excel file. Row two (the first

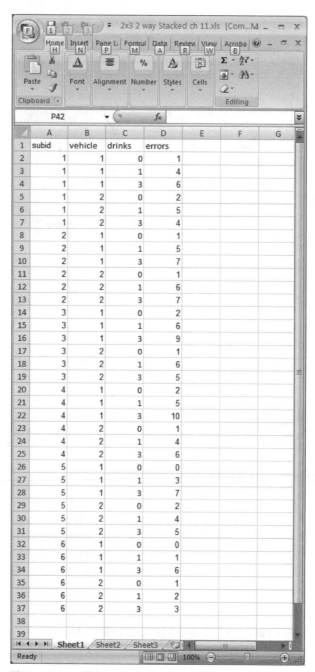

Figure 11.24   The data in an Excel spreadsheet ready to be imported.

row of data) represents the combination of **vehicle 1** and **drinks 0**; this corresponds to driving the car after consuming no alcoholic drinks. The outcome variable shows a score of 1, indicating that the individual made only one driving error on the driving course. Row three represents vehicle 1 and drinks 1, which corresponds to driving the car after consuming one alcoholic drink; as you can see, subid 1 committed four driving errors on the course under those conditions. The other four rows hold information regarding the other conditions in the study for subid 1.

Note that the data have been entered into the Excel file in a systematic fashion. The first three rows for each participant represent vehicle 1 and the last three rows represent vehicle 2. Within the each set of three rows, the second within-subjects variable drinks is listed in the order 0, 1, and 2. The pattern is the same for each subject in the data file.

11.11.2 STRUCTURING THE DATA ANALYSIS

Import the data file from Excel into a new SAS project. Then from the main menu select Analyze → ANOVA → Mixed Models. The window opens on the Task Roles tab. Drag Outcome to the icon for Dependent Variable and drag vehicle, drinks, and subid to the icon for Classification variables. The result of this is shown in Figure 11.25.

Select Fixed Effect Model in the navigation panel as shown in Figure 11.26. In the Class and quantitative variables pane, select vehicle then click the Main pushbutton; vehicle will automatically appear in the

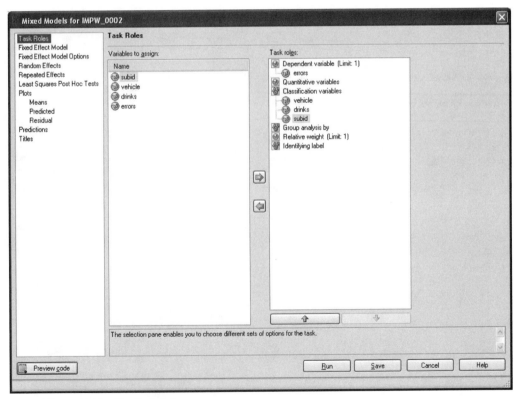

Figure 11.25 The **Task Roles** screen.

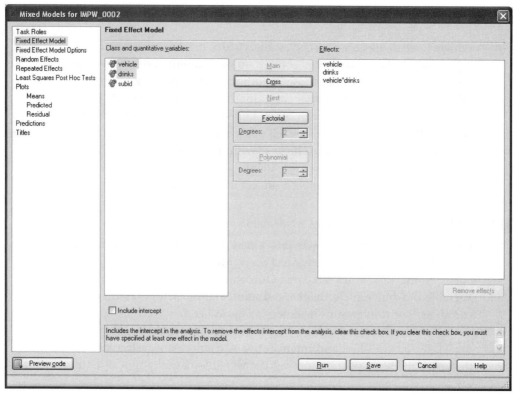

Figure 11.26 The **Fixed Effect Model** window.

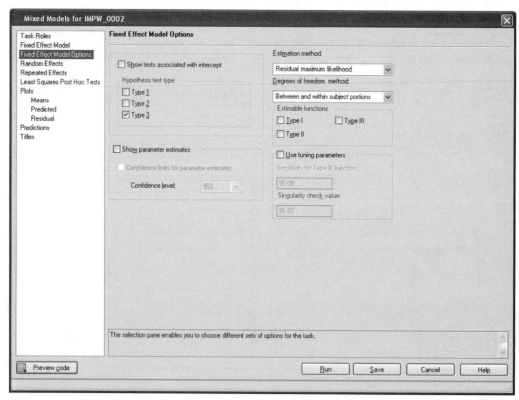

Figure 11.27 The **Fixed Effect Model Options** window.

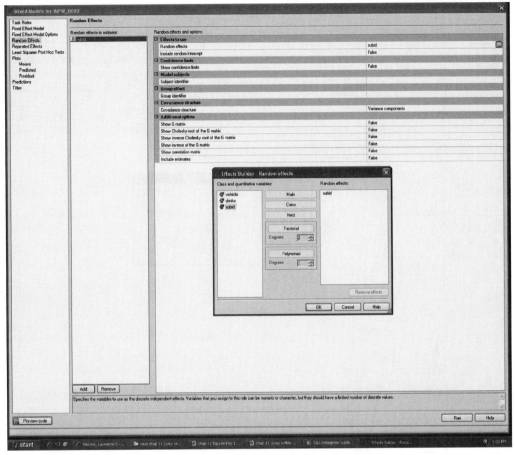

Figure 11.28   The **Effects to Use** panel in the **Random Effects** window.

Effects window. Select **drinks** then click the **Main** pushbutton; **drinks** will automatically appear in the **Effects** window. While holding the Control key down, select both **vehicle** and **drinks** then click the **Cross** pushbutton; **vehicle∗drinks** will automatically appear in the **Effects** window.

Select **Fixed Effect Model Options** in the navigation panel as shown in Figure 11.27. Check **Type 3** under **Hypothesis test type**. Then select **Residual maximum likelihood** under **Estimation method**, and **Between and within subject portions** under **Degrees of freedom method**.

Select **Random Effects** in the navigation panel as shown in Figure 11.28. Select **Random effects** under the **Effects to use** menu; a box will appear at the far right of that menu. Click on that box and the **Effects Builder – Random effects** window appears. Select **subid** and click the **Main** pushbutton; **subid** will automatically appear in the **Random effects** panel. Click the **OK** pushbutton.

Now select **Subject identifier** under the **Model subjects** menu; a box will appear at the far right of that menu. Click on that box and the **Effects Builder – Subject identifier** window appears. Select **subid** and

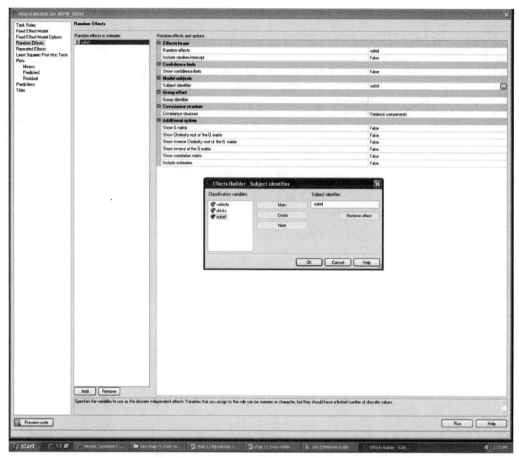

Figure 11.29   The **Model Subjects** panel in the **Random Effects** window.

click the **Main** pushbutton; **subid** will automatically appear in the **Subject identifier** panel. Select the **OK** pushbutton. This is shown in Figure 11.29.

Select **Least Squares Post Hoc Tests** in the navigation panel. Click the **Add** pushbutton at the bottom of the **Effects to estimate** pane. The screen shown in Figure 11.30 appears. Highlight **vehicle** under the **Effects to use** menu and a downward arrow will appear. Click on this arrow and a drop-down menu will appear; select **True**. Repeat this procedure for **drinks** and **drinks*vehicle**. To perform the analysis, click the **Run** pushbutton.

11.12   SAS OUTPUT FROM THE OMNIBUS ANALYSIS

The bottom portion of Figure 11.31 presents the descriptive statistics for the conditions. Because these are least squares means, we obtain the standard errors associated with the means rather than the standard deviation. The top portion presents the summary table. The F ratios produced by *SAS Enterprise Guide* are somewhat different from those produced by SPSS because of the differences in algorithms used to analyze within-subjects variables (see Section 10.17), but they both permit us to draw comparable

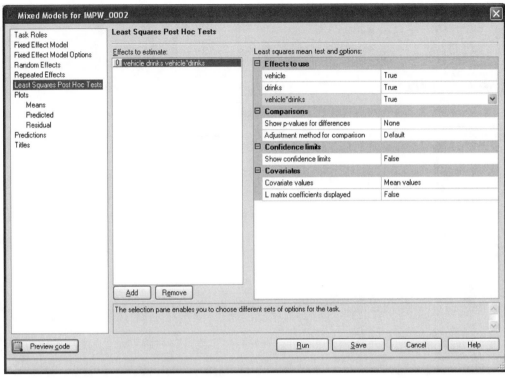

Figure 11.30 The **Least Squares Post Hoc Tests** window.

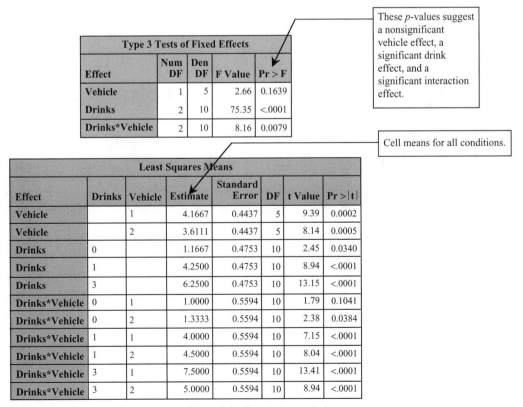

These *p*-values suggest a nonsignificant vehicle effect, a significant drink effect, and a significant interaction effect.

Type 3 Tests of Fixed Effects				
Effect	**Num DF**	**Den DF**	**F Value**	**Pr > F**
Vehicle	1	5	2.66	0.1639
Drinks	2	10	75.35	<.0001
Drinks*Vehicle	2	10	8.16	0.0079

Cell means for all conditions.

Least Squares Means							
Effect	**Drinks**	**Vehicle**	**Estimate**	**Standard Error**	**DF**	**t Value**	**Pr >|t|**
Vehicle		1	4.1667	0.4437	5	9.39	0.0002
Vehicle		2	3.6111	0.4437	5	8.14	0.0005
Drinks	0		1.1667	0.4753	10	2.45	0.0340
Drinks	1		4.2500	0.4753	10	8.94	<.0001
Drinks	3		6.2500	0.4753	10	13.15	<.0001
Drinks*Vehicle	0	1	1.0000	0.5594	10	1.79	0.1041
Drinks*Vehicle	0	2	1.3333	0.5594	10	2.38	0.0384
Drinks*Vehicle	1	1	4.0000	0.5594	10	7.15	<.0001
Drinks*Vehicle	1	2	4.5000	0.5594	10	8.04	<.0001
Drinks*Vehicle	3	1	7.5000	0.5594	10	13.41	<.0001
Drinks*Vehicle	3	2	5.0000	0.5594	10	8.94	<.0001

Figure 11.31 The summary table and the descriptive statistics.

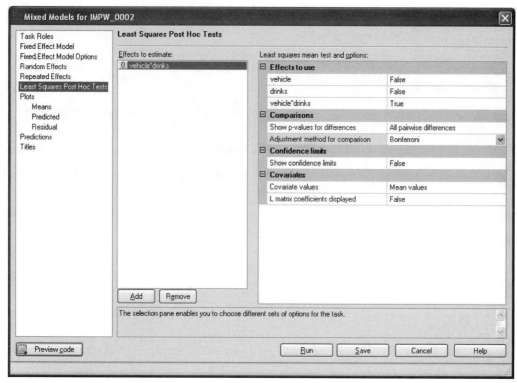

Figure 11.32   The **Least Squares Post Hoc Tests** window.

conclusions from the analysis. Thus, we determine that both **drinks** and the **drinks∗vehicle** interaction were statistically significant.

11.13   PERFORMING THE SIMPLE EFFECTS ANALYSIS IN SAS

To perform the simple effects analysis in *SAS Enterprise Guide*, configure the analysis as shown above. Then in the **Least Squares Post Hoc Tests** screen shown in Figure 11.32, in the **Effects to use** frame, set the interaction to **True** and all others to **False**. Then in the **Comparisons** frame, set **Show p-values for differences** to **All pairwise differences** and set **Adjustment method for comparison** to **Bonferroni**. Click **Run** to perform the analysis.

11.14   SAS OUTPUT FROM THE SIMPLE EFFECTS ANALYSIS

The pairwise mean comparisons are shown in Figure 11.33. The conditions that are being compared are specified in the table in the following form. The first two coded columns denote the mean of the first condition of the pair and the next two coded columns denote the second condition of the pair that is subtracted from the first. To illustrate, consider the first row in the **Differences of Least Squares Means** table. This takes the form "first condition" minus "second condition." The first condition is the

This Table presents both the unadjusted p-value and the Bonferroni adjustment

Differences of Least Squares Means										
Effect	Drinks	Vehicle	Drinks	Vehicle	Estimate	Standard Error	DF	t Value	Pr > |t|	Adjustment	Adj P
Drinks*Vehicle	0	1	0	2	−0.3333	0.5900	10	−0.56	0.5846	Bonferroni	1.0000
Drinks*Vehicle	0	1	1	1	−3.0000	0.5900	10	−5.08	0.0005	Bonferroni	0.0071
Drinks*Vehicle	0	1	1	2	−3.5000	0.5900	10	−5.93	0.0001	Bonferroni	0.0022
Drinks*Vehicle	0	1	3	1	−6.5000	0.5900	10	−11.02	<.0001	Bonferroni	<.0001
Drinks*Vehicle	0	1	3	2	−4.0000	0.5900	10	−6.78	<.0001	Bonferroni	0.0007
Drinks*Vehicle	0	2	1	1	−2.6667	0.5900	10	−4.52	0.0011	Bonferroni	0.0166
Drinks*Vehicle	0	2	1	2	−3.1667	0.5900	10	−5.37	0.0003	Bonferroni	0.0047
Drinks*Vehicle	0	2	3	1	−6.1667	0.5900	10	−10.45	<.0001	Bonferroni	<.0001
Drinks*Vehicle	0	2	3	2	−3.6667	0.5900	10	−6.21	<.0001	Bonferroni	0.0015
Drinks*Vehicle	1	1	1	2	−0.5000	0.5900	10	−0.85	0.4166	Bonferroni	1.0000
Drinks*Vehicle	1	1	3	1	−3.5000	0.5900	10	−5.93	0.0001	Bonferroni	0.0022
Drinks*Vehicle	1	1	3	2	−1.0000	0.5900	10	−1.69	0.1210	Bonferroni	1.0000
Drinks*Vehicle	1	2	3	1	−3.0000	0.5900	10	−5.08	0.0005	Bonferroni	0.0071
Drinks*Vehicle	1	2	3	2	−0.5000	0.5900	10	−0.85	0.4166	Bonferroni	1.0000
Drinks*Vehicle	3	1	3	2	2.5000	0.5900	10	4.24	0.0017	Bonferroni	0.0259

Figure 11.33 All pairwise comparisons.

one coded as **drinks** = 0 and **vehicle** = 1 (car) whose mean (see Figure 11.30) is 1.000. The second condition is the one codes as **drinks** = 0 and **vehicle** = 2 (SUV) whose mean is 1.3333. Given the expression Condition 1 − Condition 2 = 1.000 − 1.3333 = −0.3333, we see −0.3333 in the **Estimate** row. The t value is 0.56 and the Bonferroni corrected probability (labeled **Adj P**) associated with that t value for 10 df is shown as 1.0000 (SAS codes any computed Bonferroni adjusted probabilities exceeding 1 as 1.0000); we thus conclude that the mean difference is not statistically significant. Once again, the numerical values for the corrected probabilities differ from those obtained from SPSS (see Section 10.19).

11.15 PERFORMING THE POST HOC ANALYSIS IN SAS

To perform the simple effects analysis in *SAS Enterprise Guide*, configure the analysis as in Section 11.14. Then in the **Least Squares Post Hoc Tests** screen shown in Figure 11.34, in the **Effects to use** frame, set **drinks** to **True** and all others to **False**. Then in the **Comparisons** frame, set **Show p-values for differences** to **All pairwise differences** and set **Adjustment method for comparison** to **Bonferroni**. Click **Run** to perform the analysis.

11.16 SAS OUTPUT FROM THE POST HOC ANALYSIS

The pairwise mean comparisons are shown in Figure 11.35. As can be seen, all conditions differ significantly from all others.

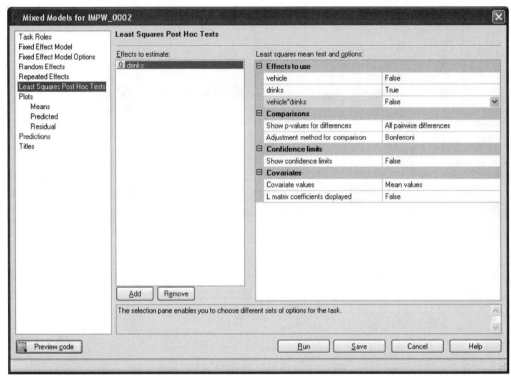

Figure 11.34 The **Least Squares Post Hoc Tests** window.

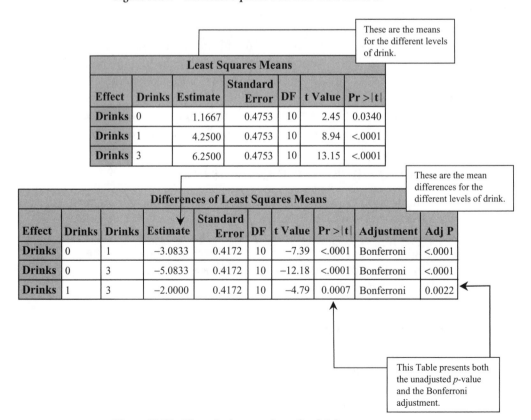

These are the means for the different levels of drink.

Least Squares Means						
Effect	Drinks	Estimate	Standard Error	DF	t Value	Pr >|t|
Drinks	0	1.1667	0.4753	10	2.45	0.0340
Drinks	1	4.2500	0.4753	10	8.94	<.0001
Drinks	3	6.2500	0.4753	10	13.15	<.0001

These are the mean differences for the different levels of drink.

Differences of Least Squares Means								
Effect	Drinks	Drinks	Estimate	Standard Error	DF	t Value	Pr >|t|	Adjustment	Adj P
Drinks	0	1	−3.0833	0.4172	10	−7.39	<.0001	Bonferroni	<.0001
Drinks	0	3	−5.0833	0.4172	10	−12.18	<.0001	Bonferroni	<.0001
Drinks	1	3	−2.0000	0.4172	10	−4.79	0.0007	Bonferroni	0.0022

This Table presents both the unadjusted *p*-value and the Bonferroni adjustment.

Figure 11.35 The paired comparisons for **drinks**.

11.17   COMMUNICATING THE RESULTS

Six students from a local community college matched for their driving experience participated in this within-subjects design. Students drove a sport car and an SUV at different times after having either zero, one, or three alcoholic drinks. The dependent variable was driving errors made. All effects met the assumption of sphericity as tested by Mauchly's procedure.

No significant main effect was found for type of vehicle, $F(1, 10) = 2.78$, $p > .05$, but a significant effect was found for number of drinks, $F(2, 10) = 157.39$, $p < .05$, $\eta^2 = .77$. However, the two-way interaction was found to be significant, $F(2, 10) = 8.53$, $p < .05$, $\eta^2 = .08$.

The graph of the interaction is presented in Figure 11.2. Simple effects analysis indicated that driving errors increased steadily as a function of alcohol consumption when they were driving a sport car. Errors made while driving an SUV increased comparably to the sport car up to one drink but leveled off from one to three drinks.

CHAPTER 11 EXERCISES

11.1. Researchers are interested in the effects of both auditory and visual stimulation on children's problem solving ability. Fifth grade children are presented with rock (a_1) and hip hop (a_2) music in rooms painted in red (b_1), white (b_2), or yellow (b_3) while solving ten anagram problems in each treatment combination. The dependent variable is the number of correct solutions. The data are as follows:

		a_1			a_2	
Subject	b_1	b_2	b_3	b_1	b_2	b_3
1	4	7	5	2	7	5
2	3	7	6	2	8	6
3	3	8	7	2	9	5
4	2	8	7	3	7	6
5	1	6	6	1	8	5

a. Conduct an ANOVA on these data by hand and with SPSS or SAS.
b. If the interaction is statistically significant, perform a simple effects analysis and multiple comparisons tests by hand and with SPSS or SAS.

11.2. Consider the hypothetical 3×3 factorial data set below. The nine treatment combinations had $n = 4$ participants receiving all treatments in a randomized order.

		a_1			a_2			a_3	
Subject	b_1	b_2	b_3	b_1	b_2	b_3	b_1	b_2	b_3
1	4	7	12	10	17	15	21	18	19
2	6	9	14	10	17	9	20	17	19
3	8	8	14	14	20	9	24	17	20
4	8	6	9	19	20	8	19	15	21

a. Conduct an ANOVA on these data by hand and with SPSS or SAS.
b. If the interaction is statistically significant, perform a simple effects analysis and multiple comparisons tests by hand and with SPSS or SAS.

11.3. Consider the following hypothetical 2×4 factorial data set. The eight treatment combinations had $n = 5$ participants receiving all treatments in a randomized order. The data are as follows:

Subject	a_1				a_2			
	b_1	b_2	b_3	b_4	b_1	b_2	b_3	b_4
1	2	2	8	3	7	2	3	9
2	3	1	8	1	7	2	1	9
3	1	1	9	3	8	2	1	7
4	2	0	9	1	8	1	1	6
5	0	2	8	3	9	1	2	9

a. Conduct an ANOVA by hand and with SPSS or SAS.
b. If the interaction is statistically significant, perform a simple effects analysis and multiple comparisons tests by hand and with SPSS or SAS.

Three-Way Within-Subjects Design

A three-way within-subjects design is one in which the levels of three within-subjects variables are combined factorially. Given that it is a within-subjects design, we still require that the participants contribute a data point to each of the separate cells (each of the unique combinations of the levels of the independent variables) of the design. The logistics of running a three-way study can be challenging but, partly because the subjects are their own controls and partly because each within-subject effect of interest is associated with its own error term (thus partitioning the total within-subjects error into separate error components), such a design has a considerable amount of statistical power. The key is to find appropriate applications of this design that are relatively immune to carry-over contamination effects.

12.1 A NUMERICAL EXAMPLE OF A THREE-WAY WITHIN-SUBJECTS DESIGN

The hypothetical data set that will serve as our example uses three time-independent variables as shown in Figure 12.1. We are engaged in market research studying the product look of powdered laundry detergent. One independent variable is the color of the powder: it is colored white, sea green, or baby blue. The second independent variable is whether or not the powder contains granules (flecks of bright shiny additions that in reality contribute nothing to the cleaning power of the detergent). The third independent variable is the color of the box: It is colored either aqua or orange.

We will assume that the manufacturer for whom this research is being conducted is less interested in directly comparing the color of the powder; thus, we have structured Figure 12.1 to separate the powder color as the primary break. Within each powder color, we present the results by the presence or absence of granules and by box color.

Six single heads of households who agreed to participate in the study are given twelve boxes of detergent corresponding to the twelve cells $(3 \times 2 \times 2 = 12)$ of the design and a detailed schedule of when to use each. The schedule randomizes use of these twelve boxes within each participant as well as between participants. Each time they clean a load

| Subid | White Powder | | | | Sea Green Powder | | | | Baby Blue Powder | | | |
| | Granules Present | | Granules Absent | | Granules Present | | Granules Absent | | Granules Present | | Granules Absent | |
	Aqua Box	Orange Box	Aqua Box	Orange Box	Aqua Box	Orange Box	Aqua Box	Orange Box	Aqua Box	Orange Box	Aqua Box	Orange Box
1	6.50	4.00	2.70	3.20	2.10	2.00	2.20	2.30	4.50	5.00	3.60	6.70
2	6.50	4.00	2.70	3.20	2.10	2.00	2.20	2.33	4.50	5.00	3.60	6.70
3	5.50	4.80	3.60	3.80	1.50	2.50	1.60	3.20	3.50	6.00	4.60	7.10
4	7.50	3.20	1.80	2.60	2.70	1.50	2.80	1.50	5.50	4.00	2.60	6.30
5	7.20	4.70	3.70	4.40	1.50	2.40	1.70	3.10	3.40	5.60	4.20	7.50
6	5.80	3.30	1.70	2.00	2.70	1.60	2.70	1.60	5.60	4.40	2.80	5.90
Means	6.50	4.00	2.70	3.20	2.10	2.00	2.20	2.33	4.50	5.00	3.57	6.70

Figure 12.1 Raw data for the numerical example.

of laundry with a designated detergent, they are to rate how satisfied they were with the outcome. The rating survey is composed of fifteen items rated on a 9-point summative response scale. For each participant, items are averaged for each condition; it is these means that are recorded in the data file and that serve as the dependent variable. Thus, overall satisfaction, the dependent variable in this study, can range between a low of 1 and a high of 9 with higher values indicating greater satisfaction.

12.2   PARTITIONING THE VARIANCE INTO ITS SOURCES

The summary table for the ANOVA (shown in Table 12.1) contains both the between and within partitions of the total variance. The portion of the summary table labeled as **Between-subjects variance** presents the between-subjects source of variance. Without a between-subjects variable in the study, the only source of these differences that we can identify here is the individual differences among the six participants. With six participants there are 5 *df*. No *F* ratio is ordinarily computed for individual differences. The portion of the table labeled **Within-subjects variance** summarizes the within-subjects partitioning of the variance. We highlight several aspects of the table in our discussion below of the effects of interest.

12.3   EFFECTS OF INTEREST IN THIS DESIGN

With three independent variables in the design there are seven effects of interest: the main effects of the three independent variables, the three two-way interactions, and the single three-way interaction. In a within-subjects design, all of these effects are subsumed in the within-subjects variance. The hierarchical structure of these effects, as discussed in the three-way between-subjects design chapter, still holds. Thus, the three-way

Table 12.1. Summary tables for within-subjects and between-subjects effects for the three-way numerical example

Within-subjects variance

Source	SS	df	MS	F	Sig.	η^2
Box color (A)	138.889	1	138.889	0.872	0.393	
$A \times S$	796.278	5	159.256			
Granules (B)	578.000	1	578.000	7.717	0.039	.028
$B \times S$	374.800	5	74.900			
Powder color (C)	9,780.333	2	4,890.167	126.579	0.000	.475
$C \times S$	386.333	10	38.633			
$A \times B$	1,720.889	1	1,720.889	25.486	0.004	.084
$A \times B \times S$	337.611	5	67.522			
$A \times C$	2,441.444	2	1,220.722	77.808	0.000	.119
$A \times C \times S$	156.889	10	15.689			
$B \times C$	2,712.333	2	1,356.167	116.243	0.000	.132
$B \times C \times S$	116.667	10	11.667			
$A \times B \times C$	677.444	2	338.722	9.394	0.005	.033
$A \times B \times C \times S$	360.556	10	36.056			
Total within subjects	2,0578.466					

Between-subjects variance

Source	SS	df	MS
Subject differences (S)	495.833	5	99.167

interaction supercedes the two-way interactions, which, in turn, supersede the main effects.

As can be seen from Table 12.1, with the exception of the main effect of box color, all of the effects of interest are statistically significant. Although it is not the strongest effect as indexed by eta squared, the significant three-way interaction informs us that satisfaction with the detergent is a function of all three independent variables; that is, satisfaction is related to the particular combinations of box color, presence or absence of granules, and color of the powder. We would therefore graph that interaction to see the interrelationships of the variables.

There are several ways to construct a graphic representation of this interaction. We opt to place all three independent variables within a single set of axes, partly because there is no pressing theoretical reason to break apart one of the variables, and partly because it seems desirable to see them together so that the manufacturer will be in a better position to decide on what the product will look like.

Placing all the data points on one axis set does have the drawback of driving the pictorial representation toward complexity, making it more difficult for readers to discern the pattern. If we are going to use just one set of axes, we want to select a type of plot that will help readers process the information. We note that all three of our variables are categorical and thus do not cry out for a line graph to convey the functional relationship

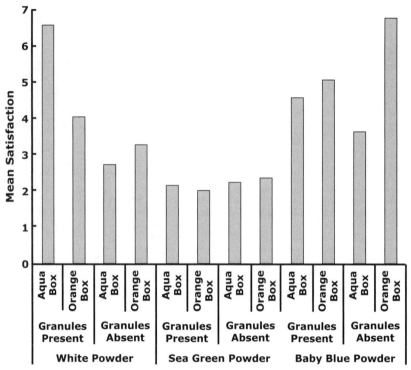

Figure 12.2   Bar graph showing the significant three-way interaction.

between the independent and dependent variables; therefore, we select a bar graph format for our figure. This also gives you the opportunity to experience a different way to visually depict an interaction from what we have presented in Chapters 8, 9, and 11.

We have plotted the interaction in Figure 12.2. The most global "break" is on the basis of powder color; thus, the first four bars represent white powder, the next four bars represent green powder, and the last four bars represent blue powder. The next most global break is based on the presence or absence of granules. That is, within each powder color, the first two bars represent the presence of granules and the last two bars represent the absence of granules. Finally, box color is considered; within each granule condition we show the aqua box and then the orange box. Thus, reading across the x axis from left to right, we can say that box color changes or increments most rapidly, granules (present or absent) increments next most rapidly, and powder color increments most slowly. This differential rate of incrementing will be represented in the data files we will construct for the analysis in SPSS and SAS.

The simple effects analyses that we will want to perform will be keyed to this plot and will reveal the fine grain details concerning mean differences. But certain global outcomes are suggested by a visual inspection of the graph. First, the green powder did not fare well at all; there is little doubt that the manufacturer will quickly decide not to pursue that option. Second, if white powder is used, it appears that it should be packaged in an aqua box with granules. Third, if blue powder is used, it seems that it might be best to package it in an orange box with no granules.

In terms of structuring our simple effects tests, it does not appear critical to directly compare the different powder colors unless further discussion with the manufacturer indicates that it is called for. As experienced consultants, we will run a subsequent analysis so that if asked to produce that information, we will have it available at the time. For the purposes of our discussion here, we will make our comparisons within each of the powder colors. Thus, we will want to (a) compare the aqua and orange box colors for each level of the granules variable for each powder color, and (b) compare the presence or absence of granules for each level of box color for each powder color. This strategy will explicitly guide us in structuring the syntax command structure in SPSS for our simple effects analyses and in determining which of the paired comparisons will be of interest in our SAS output.

12.4 THE ERROR TERMS IN A THREE-WAY WITHIN-SUBJECTS DESIGN

Each of the seven effects of interest is associated with its own error term. As we saw in Chapter 10, this is always true in a within-subjects design. Each effect interacts with the subjects effect, and this interaction represents noise or unpredictability. The most complex effect, for example, is the three-way interaction; the error term for this effect is the interaction of subjects with this three-way interaction.

12.5 COMPUTING THE OMNIBUS ANALYSIS BY HAND

As we indicated in Chapter 9, we will not be providing a computational example of any of the three-way designs (including the present three-way within-subjects design). However, we do provide the computational formulas for the three-factor within-subjects design for those students or instructors who require this level of detail (see Table 12.2).

12.6 PERFORMING THE OMNIBUS ANALYSIS IN SPSS

12.6.1 STRUCTURING THE DATA FILE

The data file for a three-way within-subjects design is structured in an analogous manner to the data file we used for our two-way example in Chapter 10. We show in Figure 12.3 the data file for our $2 \times 2 \times 3$ within-subjects design example. The first column, as always, is used for our participant identification number, and we named this variable **subid**. The next twelve columns are devoted to the cells representing our three-way design. We need twelve columns because each column represents a different combination of the levels of our independent variables and there are twelve such combinations ($2 \times 2 \times 3 = 12$).

Table 12.2. Computational formulas for the three-way within-subjects design

Source	df	SS	MS	F
A	$a-1$	$\dfrac{\Sigma A^2}{(b)(c)(n)} - \dfrac{T^2}{(a)(b)(c)(n)}$	$\dfrac{SS_A}{df_A}$	$\dfrac{MS_A}{MS_{A\times S}}$
$A \times S$	$(a-1)(n-1)$	$\dfrac{\Sigma AS^2}{(b)(c)} - \dfrac{\Sigma A^2}{(b)(c)(n)} - \dfrac{\Sigma S^2}{(a)(b)(c)} + \dfrac{T^2}{(a)(b)(c)(n)}$	$\dfrac{SS_{A\times S}}{df_{A\times S}}$	
B	$b-1$	$\dfrac{\Sigma B^2}{(a)(c)(n)} - \dfrac{T^2}{(a)(b)(c)(n)}$	$\dfrac{SS_B}{df_B}$	$\dfrac{MS_B}{MS_{B\times S}}$
$B \times S$	$(b-1)(n-1)$	$\dfrac{\Sigma BS^2}{(a)(c)} - \dfrac{\Sigma B^2}{(a)(c)(n)} - \dfrac{\Sigma S^2}{(a)(b)(c)} + \dfrac{T^2}{(a)(b)(c)(n)}$	$\dfrac{SS_{B\times S}}{df_{B\times S}}$	
C	$c-1$	$\dfrac{\Sigma C^2}{(a)(b)(n)} - \dfrac{T^2}{(a)(b)(c)(n)}$	$\dfrac{SS_C}{df_C}$	$\dfrac{MS_C}{MS_{C\times S}}$
$C \times S$	$(c-1)(n-1)$	$\dfrac{\Sigma CS^2}{(a)(b)} - \dfrac{\Sigma C^2}{(a)(b)(n)} - \dfrac{\Sigma S^2}{(a)(b)(c)} + \dfrac{T^2}{(a)(b)(c)(n)}$	$\dfrac{SS_{C\times S}}{df_{C\times S}}$	
$A \times B$	$(a-1)(b-1)$	$\dfrac{\Sigma AB^2}{(c)(n)} - \dfrac{\Sigma A^2}{(b)(c)(n)} - \dfrac{\Sigma B^2}{(a)(c)(n)} + \dfrac{T^2}{(a)(b)(c)(n)}$	$\dfrac{SS_{A\times B}}{df_{A\times B}}$	$\dfrac{MS_{A\times B}}{MS_{A\times B\times S}}$
$A \times B \times S$	$(a-1)(b-1)(n-1)$	$\dfrac{\Sigma ABS}{c} - \dfrac{\Sigma AB^2}{(c)(n)} - \dfrac{\Sigma AS^2}{(b)(c)} - \dfrac{\Sigma BS^2}{(a)(c)} + \dfrac{\Sigma S^2}{(a)(b)(c)} + \dfrac{A^2}{(b)(c)(n)} + \dfrac{B^2}{(a)(c)(n)} - \dfrac{T^2}{(a)(b)(c)(n)}$	$\dfrac{SS_{A\times B\times S}}{df_{A\times B\times S}}$	
$A \times C$	$(a-1)(c-1)$	$\dfrac{\Sigma AC^2}{(b)(n)} - \dfrac{\Sigma A^2}{(b)(c)(n)} - \dfrac{\Sigma C^2}{(a)(b)(n)} + \dfrac{T^2}{(a)(b)(c)(n)}$	$\dfrac{SS_{A\times C}}{df_{A\times C}}$	$\dfrac{MS_{A\times C}}{MS_{A\times C\times S}}$

Source	df	SS	MS	F
$A \times C \times S$	$(a-1)(c-1)(n-1)$	$\dfrac{\sum ACS^2}{b} - \dfrac{\sum AC^2}{(b)(n)} - \dfrac{\sum AS^2}{(b)(c)} - \dfrac{\sum CS^2}{(a)(b)} + \dfrac{\sum C^2}{(a)(b)(c)} + \dfrac{\sum A^2}{(b)(c)(n)} - \dfrac{\sum C^2}{(a)(b)(n)} - \dfrac{T^2}{(a)(b)(c)(n)}$	$\dfrac{SS_{A\times C\times S}}{df_{A\times C\times S}}$	
$B \times C$	$(b-1)(c-1)$	$\dfrac{\sum BC^2}{(a)(n)} - \dfrac{\sum B^2}{(a)(c)(n)} - \dfrac{\sum C^2}{(a)(b)(n)} + \dfrac{T^2}{(a)(b)(c)(n)}$	$\dfrac{SS_{B\times C}}{df_{B\times C}}$	$\dfrac{MS_{B\times C}}{MS_{B\times C\times S}}$
$B \times C \times S$	$(b-1)(c-1)(n-1)$	$\dfrac{\sum BCS^2}{a} - \dfrac{\sum BC^2}{(a)(n)} - \dfrac{\sum BS^2}{(a)(c)} - \dfrac{\sum CS^2}{(a)(b)} + \dfrac{\sum B^2}{(a)(c)(n)} + \dfrac{\sum C^2}{(a)(b)(n)} + \dfrac{\sum S^2}{(a)(b)(c)} - \dfrac{T^2}{(a)(b)(c)(n)}$	$\dfrac{SS_{B\times C\times S}}{df_{B\times C\times S}}$	
$A \times B \times C$	$(a-1)(b-1)(c-1)$	$\dfrac{\sum ABC^2}{n} - \dfrac{\sum BC^2}{(a)(n)} - \dfrac{\sum AC^2}{(b)(n)} - \dfrac{\sum AB^2}{(c)(n)} + \dfrac{\sum C^2}{(a)(b)(n)} + \dfrac{\sum B^2}{(a)(c)(n)} + \dfrac{\sum A^2}{(b)(c)(n)} - \dfrac{T^2}{(a)(b)(c)(n)}$	$\dfrac{SS_{A\times B\times C}}{df_{A\times B\times C\times S}}$	$\dfrac{MS_{A\times B\times C}}{MS_{A\times B\times C\times S}}$
$A \times B \times C \times S$	$(a-1)(b-1)(c-1)(n-1)$	$\dfrac{\sum CS^2}{(a)(b)} + \dfrac{\sum BS^2}{(a)(c)} + \dfrac{\sum AS^2}{(b)(c)} + \dfrac{\sum BC^2}{(a)(n)} + \dfrac{\sum AC^2}{(b)(n)} + \dfrac{\sum AB^2}{(c)(n)} + \dfrac{\sum C^2}{(a)(b)(n)} + \sum Y^2 + \dfrac{T^2}{(a)(b)(c)(n)} - \dfrac{\sum A^2}{(b)(c)(n)} - \dfrac{\sum B^2}{(a)(c)(n)} - \dfrac{\sum C^2}{(a)(b)(n)} - \dfrac{\sum ABC^2}{n} - \dfrac{\sum ABS^2}{c} - \dfrac{\sum ACS}{b} - \dfrac{\sum BCS}{a} - \dfrac{\sum S^2}{(a)(b)(c)}$		
Total	$(a)(b)(c)(n) - 1$	$\sum Y^2 - \dfrac{T^2}{(a)(b)(c)(n)}$		

331

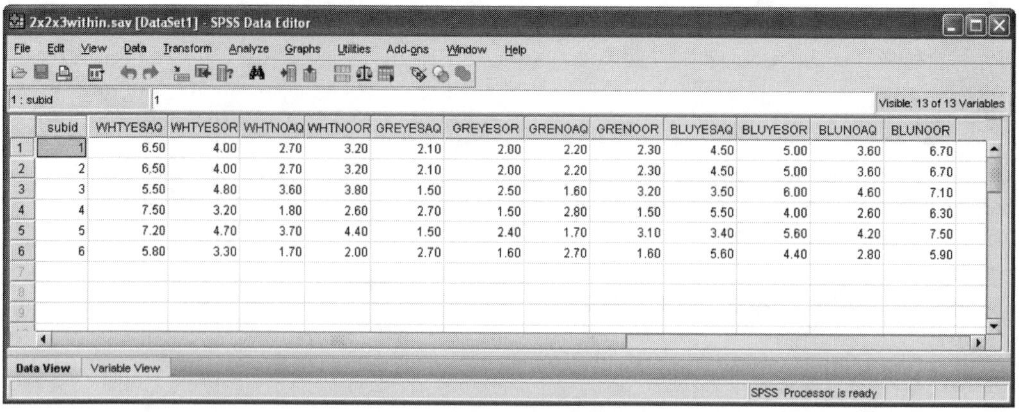

	subid	WHTYESAQ	WHTYESOR	WHTNOAQ	WHTNOOR	GREYESAQ	GREYESOR	GRENOAQ	GRENOOR	BLUYESAQ	BLUYESOR	BLUNOAQ	BLUNOOR
1	1	6.50	4.00	2.70	3.20	2.10	2.00	2.20	2.30	4.50	5.00	3.60	6.70
2	2	6.50	4.00	2.70	3.20	2.10	2.00	2.20	2.30	4.50	5.00	3.60	6.70
3	3	5.50	4.80	3.60	3.80	1.50	2.50	1.60	3.20	3.50	6.00	4.60	7.10
4	4	7.50	3.20	1.80	2.60	2.70	1.50	2.80	1.50	5.50	4.00	2.60	6.30
5	5	7.20	4.70	3.70	4.40	1.50	2.40	1.70	3.10	3.40	5.60	4.20	7.50
6	6	5.80	3.30	1.70	2.00	2.70	1.60	2.70	1.60	5.60	4.40	2.80	5.90

Figure 12.3 The data file for our three-way within-subjects example.

How the data are structured for a three-way within-subjects design (the order of the cells in the data file) should be determined by the way you will be presenting the results to your readers. This structuring, in turn, is often suggested by the theoretical framework or model within which the research was generated. If, as is sometimes the case in social and behavioral research, the study was more exploratory in nature, then the variables, while they still should be based on a systematic mapping of the independent variables, can be ordered in any structure that the researchers choose. As can be seen in Figure 12.3, we split the data most globally by powder color.

The variables (columns) in the data file shown in Figure 12.3 are ordered in a systematic manner so that they will match up with the way in which we displayed our independent variables in Figure 12.1. The names of the variables capture the levels of the three variables:

- Powder color is given in the first three letters: **WHT** for white, **GRE** for green, and **BLU** for blue.
- Presence of granules is given in the next two or three letters: **YES** for present and **NO** for absent.
- Box color is given in the final two letters: **AQ** for aqua and **OR** for orange.

In the data file, the variables are incrementing in the same manner as they are in Figures 12.1 and 12.2. Here are the specifics:

- Powder color increments most slowly – the first four columns following **subid** represent the boxes with the white powder, the next four columns represent the green powder, and the next four columns represent the blue powder.
- Within each of the four columns for each powder, **granules** increments next most slowly – the first two columns represent the presence of granules and the last two represent the absence of granules.

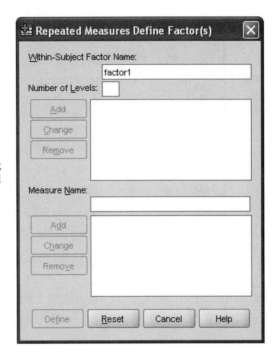

Figure 12.4 The opening window for **GLM Repeated Measures**.

- Within each level of granules we find **box color**, which increments the most rapidly – the first column represents the aqua box and the second column represents the orange box.

12.6.2 STRUCTURING THE DATA ANALYSIS

From the main SPSS menu, select **Analyze → General Linear Model → Repeated Measures**. This will open the dialog window shown in Figure 12.4. As you will recall from Chapter 10, this is the initial window used by SPSS to have you name the within-subjects variables in the study.

We will enter the variables in this **Repeated Measures Define Factor(s)** window in the order that we have incremented them in the data file. That is, we will name the variable that increments most slowly, then the one that increments next most slowly, and finally the one that increments most rapidly. We therefore highlight the default name **factor1** and type in **powcolor** and specify the **Number of Levels** for that variable as three. Click **Add** to register this within-subjects variable with SPSS. The window is now ready for the next repeated measure – the one that increments next most slowly – to be named. Type in **granules**, specify that it has two levels, and click **Add**. The last variable to be named is the one that increments most rapidly. Type in **boxcolor**, specify that it has two levels, and click **Add** one last time (see Figure 12.5). With all three within-subjects variables named, click **Define** to reach the main **GLM Repeated Measures** window.

The **GLM Repeated Measures** main dialog window is shown in Figure 12.6. Note the number triplets in parentheses next to the lines with

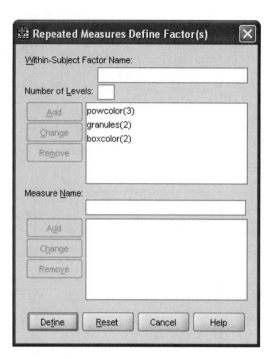

Figure 12.5   All three of the within-subjects variables have been named.

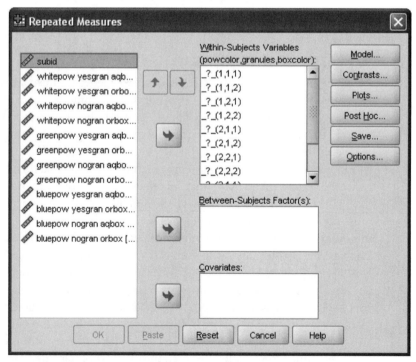

Figure 12.6   The main **GLM Repeated Measures** window.

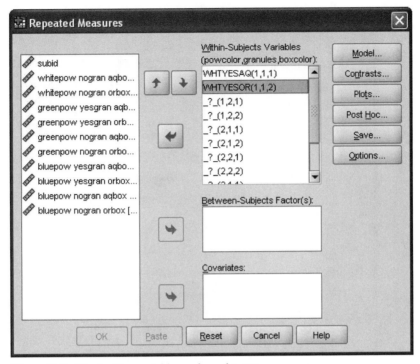

Figure 12.7 The first two variables are brought over.

question marks. These are in the order that we named the variables, which is shown at the very top of the window under the expression **Within-Subjects Variables**.

Now we can deal with defining these variables by supplying the variable (the named column in the data file) corresponding to the specified values of the levels of each. For example, the cell combination or coordinate (1,1,1) indicates the first level of **powcolor** (white), the first level of **granules** (yes, granules present), and the first level of **boxcolor** (aqua).

Knowing the order in which we need to bring the variables over to the cell coordinates, we can now bring them over. The first variable brought over will be matched with the (1,1,1) coordinate, so you need to be very careful to use the right panel with the question marks and coordinates as your guide for the ordering of your variables. We show in Figure 12.7 the first two variables that were brought over. The second variable represents the cell for white powder, granules present, and the orange box.

In our case, the ordering in the list of variables matches the order in which SPSS needs to be told of the order here (this is what we intended when we specified our variables in the initial dialog window), so we can highlight the rest of the list from **WHTNOAQ** to **BLUNOOR** and bring them over at the same time; SPSS will preserve the ordering of any highlighted set of variables clicked over. Figure 12.8 shows that all of the variables have now been defined.

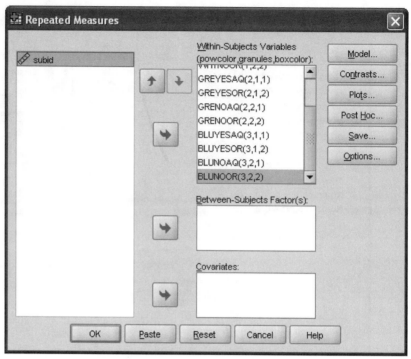

Figure 12.8   All of the variables are now defined.

Click the **Contrasts** pushbutton (we do not show this window). SPSS uses the **Polynomial** contrast as its default. Since the repeated measures do not represent equally spaced intervals, we should not perform such an analysis. Highlight each variable in turn, select **None** from the drop-down menu, and click **Change**. The contrasts in parentheses next to each variable name will now say **None**. Click **Continue** to return to the main dialog window.

Click the **Options** pushbutton to reach the dialog window shown in Figure 12.9. The top portion of the window is devoted to **Estimated Marginal Means**. Bring over all of the effects to the panel labeled **Display Means for**; this will produce a line of **/EMMEANS = TABLES** syntax and tables of estimated marginal means for these effects. In the bottom portion of the window labeled **Display**, check **Descriptive statistics** to output the observed means and standard deviations. Click **Continue** to return to the main dialog window.

Click the **Plots** pushbutton to set up a plot for the three-way interaction should it be found to be significant. We intend to ask SPSS to generate a couple of different plots, not knowing which if we will settle on once the data analysis is completed. In the first plot, we placed **boxcolor** on the horizontal axis, used separate lines for the **granules** variable, and used separate plots for **powcolor**. The separate plots for powder color correspond to the structure of our data file in which this variable is the

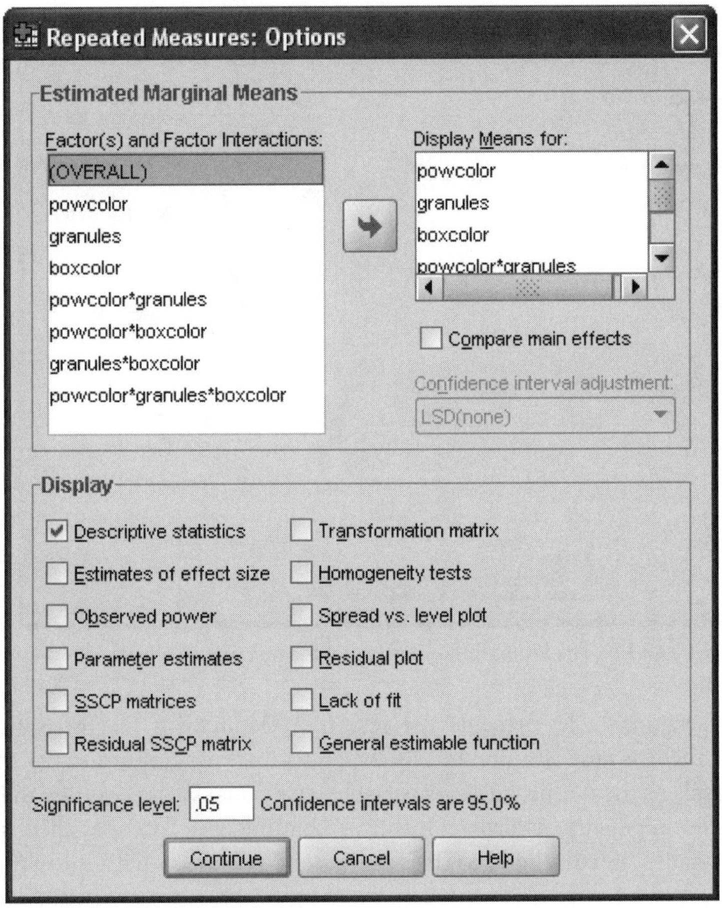

Figure 12.9   The **Options** window.

one on which the most global split is made. This is shown in Figure 12.10. Clicking **Add** changes the window to what we see in Figure 12.11. We can now specify one or more additional plots following the procedure just described. Clicking **Add** after specifying each will register it with SPSS.

We have taken you through this process in order to demonstrate what you might do in the first stage of data analysis. If the three-way interaction was not significant but one or more two-way interactions were, then we would probably want to repeat the analysis asking for those plots. As it turns out, the three-way interaction is significant but we opted to use a bar graph plot rather than these line graphs.

12.7   SPSS OUTPUT FOR THE OMNIBUS ANALYSIS

Figure 12.12 shows two tables from the initial portion of the output. The **Descriptive Statistics** table presents the observed means, standard deviations, and sample sizes for each condition.

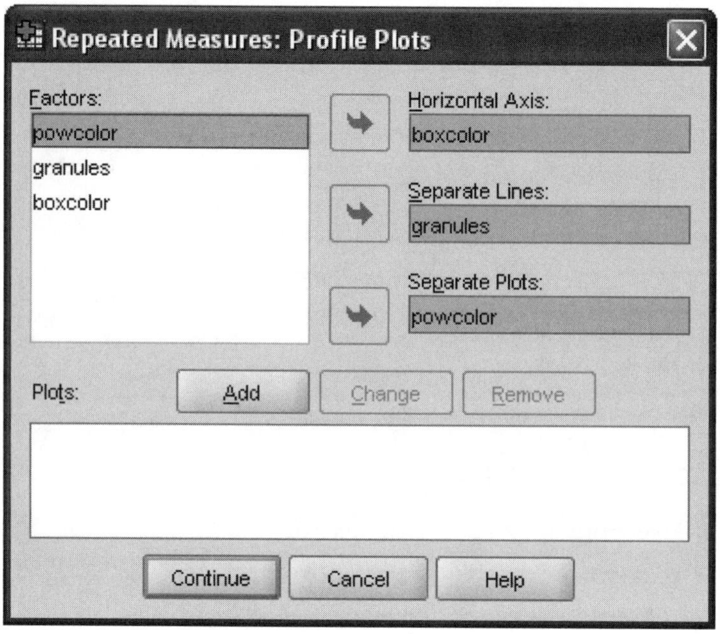

Figure 12.10 The **Plots** window setting up the first plot.

Figure 12.13 presents the results of **Mauchly's Test of Sphericity**. Mauchly's test simultaneously evaluates two assumptions: (a) that the levels of the within-subjects variable have equal variances (i.e., that there is homogeneity of variance); and (b) that the levels of the within-subjects variable are correlated to the same extent. All effects with more than two

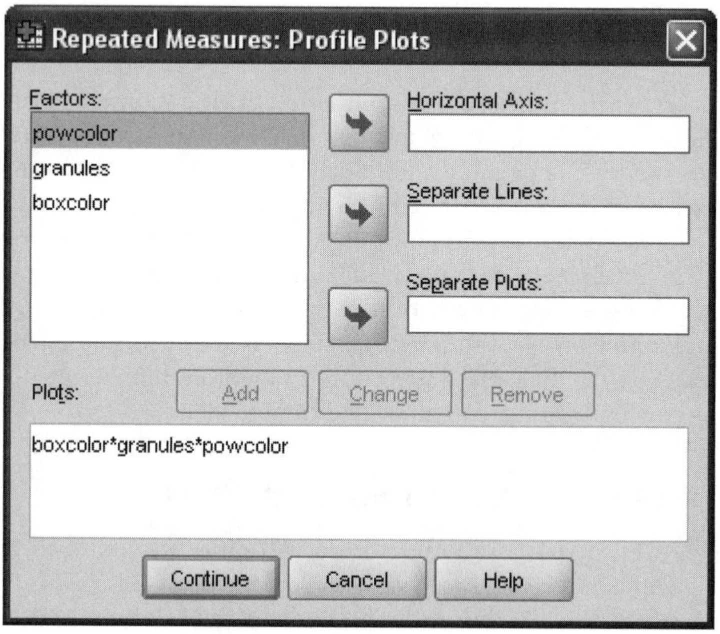

Figure 12.11 Additional plots can now be specified here.

Descriptive Statistics

	Mean	Std. Deviation	N
whitepow yesgran aqbox	6.5000	.77201	6
whitepow yesgran orbox	4.0000	.67231	6
whitepow nogran aqbox	2.7000	.85088	6
whitepow nogran orbox	3.2000	.84853	6
whitepow yesgran aqbox	2.1000	.53666	6
whitepow yesgran orbox	2.0000	.40497	6
whitepow nogran aqbox	2.2000	.49396	6
whitepow nogran orbox	2.3333	.71740	6
whitepow yesgran aqbox	4.5000	.94021	6
whitepow yesgran orbox	5.0000	.73756	6
whitepow nogran aqbox	3.5667	.77374	6
whitepow nogran orbox	6.7000	.56569	6

Figure 12.12 Descriptive statistics output.

conditions (i.e., at least 2 df) were evaluated for sphericity; none was statistically significant.

The summary table for the within-subjects variance is displayed in Figure 12.14. With a nonsignificant sphericity test, we can focus on the rows labeled **Sphericity Assumed**.

> Mauchly's test was not significant for the main and interaction effects. We therefore can assume that we have met the sphericity assumption.

Mauchly's Test of Sphericity[b]

Measure: MEASURE_1

Within Subjects Effect	Mauchly's W	Approx. Chi-Square	df	Sig.	Epsilon[a]		
					Greenhouse-Geisser	Huynh-Feldt	Lower-bound
powcolor	.253	5.499	2	.064	.572	.631	.500
granules	1.000	.000	0	.	1.000	1.000	1.000
boxcolor	1.000	.000	0	.	1.000	1.000	1.000
powcolor * granules	.733	1.245	2	.537	.789	1.000	.500
powcolor * boxcolor	.485	2.893	2	.235	.660	.805	.500
granules * boxcolor	1.000	.000	0	.	1.000	1.000	1.000
powcolor * granules* boxcolor	.494	2.818	2	.244	.664	.813	.500

Tests the null hypothesis that the error covariance matrix of the orthonormalized transformed dependent variables is proportional to an identity matrix.

a. May be used to adjust the degrees of freedom for the averaged tests of significance. Corrected tests are displayed in the Tests of Within-Subjects Effects table.

b. Design: Intercept
Within Subjects Design:
powcolor+granules+boxcolor+powcolor*granules+powcolor*boxcolor+granules*boxcolor+powcolor*granules*boxcolor

Figure 12.13 Outcome of the Mauchly's sphericity tests.

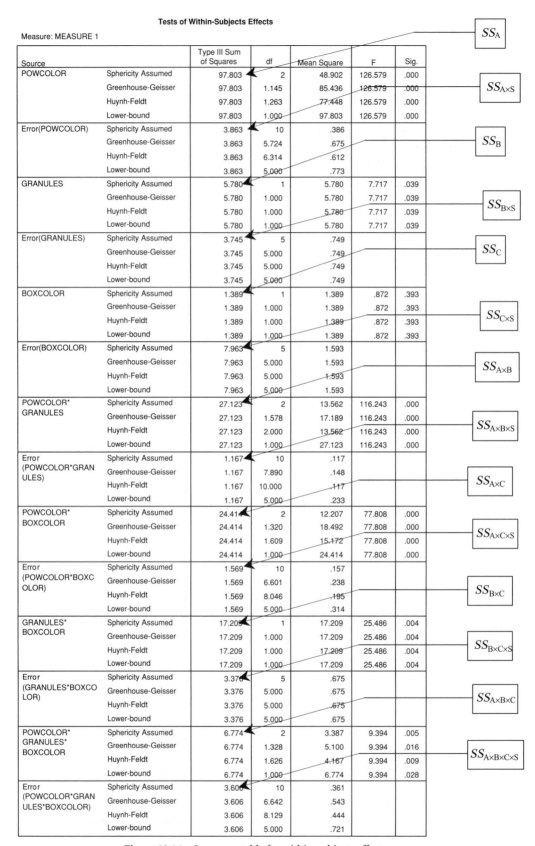

Figure 12.14 Summary table for within-subjects effects.

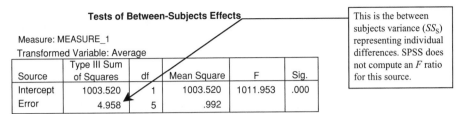

Tests of Between-Subjects Effects

Measure: MEASURE_1
Transformed Variable: Average

Source	Type III Sum of Squares	df	Mean Square	F	Sig.
Intercept	1003.520	1	1003.520	1011.953	.000
Error	4.958	5	.992		

This is the between subjects variance (SS_S) representing individual differences. SPSS does not compute an F ratio for this source.

Figure 12.15 Summary table for between-subjects effect.

There are seven effects of interest: the three main effects, the three two-way interactions, and the three-way interaction. Each is evaluated by its own error term. As we noted in Table 12.1, all of the effects except the main effect of **boxcolor** are significant.

Figure 12.15 shows the between-subjects portion of the variance. The **Intercept** concerns the full model, which we do not deal with here. What SPSS calls **Error** is the variance representing individual differences; it does not calculate an F ratio for this effect.

12.8 PERFORMING THE POST-ANOVA ANALYSIS IN SPSS

Configure the analysis as you did for the omnibus ANOVA but at the end of the process click the **Paste** pushbutton instead of clicking **OK**. The underlying syntax will then be placed in a syntax window; locate that window on your desktop and make it active by clicking it. This window is shown in Figure 12.16.

With a significant three-way interaction we need to perform our tests of simple effects. These should be based on the way we finally chose to graphically present the interaction in Figure 12.2. In that plot, we separated the bars into three groupings based on the color of the powder as the most global split (the variable that increments most slowly), presence or absence of the granules as the next most global split (the variable that increments next most slowly), and the box color as the most rapidly incrementing variable. In the set of variables enclosed by parentheses following the **/EMMEANS = TABLES** subcommand, **powcolor** therefore should be listed first, **granules** should be listed next, and **boxcolor** should be in the rightmost position.

Within each color of powder, some comparisons of the cell means are desirable. We therefore should type in two **compare** additions to the syntax, one comparing **granules** and another comparing **boxcolor**. This is shown in Figure 12.17.

Because the main effect of **powcolor** is so strong an effect and because it has three levels, it would be useful to perform a paired comparison of the means for this effect as well. This can be done in syntax using **compare** specifications on the **/EMMEANS = TABLES** subcommand for the main effect as shown as well in Figure 12.17. After entering the necessary syntax, select **Run → All** from the main menu to perform the analysis.

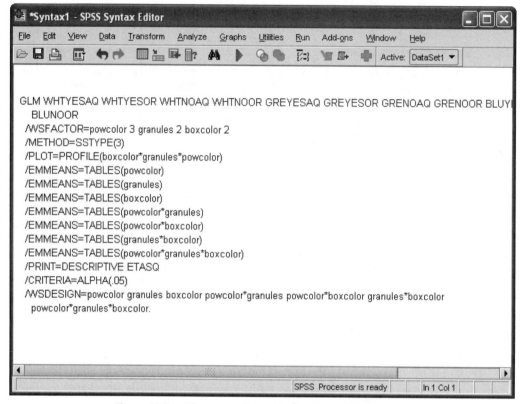

Figure 12.16 The syntax that underlies the analysis we that just performed in this screen.

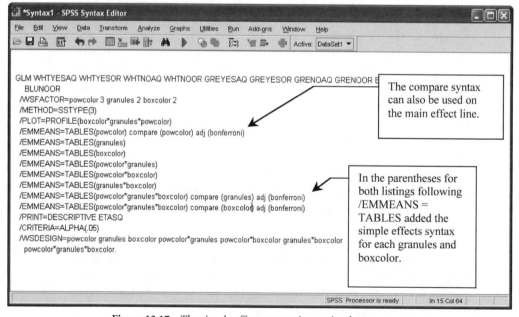

Figure 12.17 The simple effects syntax is now in place.

A. Estimates

Measure: MEASURE_1

powcolor	granules	boxcolor	Mean	Std.Error	95% Confidence Interval	
					Lower Bound	Upper Bound
1	1	1	6.500	.315	5.690	7.310
		2	4.000	.274	3.294	4.706
	2	1	2.700	.347	1.807	3.593
		2	3.200	.346	2.310	4.090
2	1	1	2.100	.219	1.537	2.663
		2	2.000	.165	1.575	2.425
	2	1	2.200	.202	1.682	2.718
		2	2.333	.293	1.580	3.086
3	1	1	4.500	.384	3.513	5.487
		2	5.000	.301	4.226	5.774
	2	1	3.567	.316	2.755	4.379
		2	6.700	.231	6.106	7.294

> These are the probabilities of the differences occurring by chance alone for the pairwise comparisons with powder color as the first order.

B. Pairwise Comparisons

Measure: MEASURE_1

powcolor	boxcolor	(I) granules	(J) granules	Mean Difference (I-J)	Std. Error	Sig.[a]	95% Confidence Interval for Difference[a]	
							Lower Bound	Upper Bound
1	1	1	2	3.800*	.497	.001	2.523	5.077
		2	1	-3.800*	.497	.001	-5.077	-2.523
	2	1	2	.800*	.139	.002	.443	1.157
		2	1	-.800*	.139	.002	-1.157	-.443
2	1	1	2	-.100*	.026	.012	-.166	-.034
		2	1	.100*	.026	.012	.034	.166
	2	1	2	-.333*	.128	.048	-.663	-.004
		2	1	.333*	.128	.048	.004	.663
3	1	1	2	.933	.695	.237	-.853	2.720
		2	1	-.933	.695	.237	-2.720	.853
	2	1	2	-1.700*	.163	.000	-2.120	-1.280
		2	1	1.700*	.163	.000	1.280	2.120

Based on estimated marginal means
*.The mean difference is significant at the .05 level.
a. Adjustment for multiple comparisons: Bonferroni.

Figure 12.18   **Powcolor** by **boxcolor** by **granules** pairwise comparisons.

12.9   SPSS OUTPUT FOR THE POST-ANOVA ANALYSIS

12.9.1   INTERACTION SIMPLE EFFECTS

Simple effects analyses are best grounded in the graph of the interaction. In this way, we can see exactly what is analyzed here. Use Figure 12.2 as your base for interpreting what we find here.

Figure 12.18 presents the **estimated marginal means**, shown as *Estimates*, matched to the **Pairwise Comparisons** table for **granules**. Note that the two tables are structured exactly the same with **granules** incrementing most rapidly and **powcolor** incrementing most slowly.

Most of the comparisons are statistically significant. Note that there are three main rows, each isolating one of the levels of **powcolor**. Inside each major row, we see separate rows for each level of **boxcolor**. Within each level of **boxcolor** the two levels of **granules** are compared. If we look at the very first comparison at the top of the table, we see that it is for

A. Estimates

Measure: MEASURE_1

powcolor	granules	boxcolor	Mean	Std. Error	95% Confidence Interval	
					Lower Bound	Upper Bound
1	1	1	6.500	.315	5.690	7.310
		2	4.000	.274	3.294	4.706
	2	1	2.700	.347	1.807	3.593
		2	3.200	.346	2.310	4.090
2	1	1	2.100	.219	1.537	2.663
		2	2.000	.165	1.575	2.425
	2	1	2.200	.202	1.682	2.718
		2	2.333	.293	1.580	3.086
3	1	1	4.500	.384	3.513	5.487
		2	5.000	.301	4.226	5.774
	2	1	3.567	.316	2.755	4.379
		2	6.700	.231	6.106	7.294

These are the probabilities of the differences occurring by chance alone for the pairwise comparisons with powder color as the first order.

B. Pairwise Comparisons

Measure: MEASURE_1

powcolor	granules	(I) boxcolor	(J) boxcolor	Mean Difference (I-J)	Std. Error	Sig.[a]	95% Confidence Difference[a]	
							Lower Bound	Upper Bound
1	1	1	2	2.500*	.465	.003	1.305	3.695
		2	1	-2.500*	.465	.003	-3.695	-1.305
	2	1	2	-.500*	.093	.003	-.739	-.261
		2	1	.500*	.093	.003	.261	.739
2	1	1	2	.100	.384	.805	-.887	1.087
		2	1	-.100	.384	.805	-1.087	.887
	2	1	2	-.133	.494	.798	-1.404	1.138
		2	1	.133	.494	.798	-1.138	1.404
3	1	1	2	-.500	.678	.494	-2.242	1.242
		2	1	.500	.678	.494	-1.242	2.242
	2	1	2	-3.133*	.158	.000	-3.541	-2.726
		2	1	3.133*	.158	.000	2.726	3.541

Based on estimated marginal means

*. The mean difference is significant at the .05 level.

a. Adjustment for multiple comparisons: Bonferroni.

Figure 12.19   **Powcolor** by **granules** by **boxcolor** pairwise comparisons.

the white powder (**powcolor 1**) in the aqua box (**boxcolor 1**) comparing the two granule conditions. The two means, shown in the **estimated marginal means** table, are 6.5 and 2.7, a difference we see in the **Pairwise Comparisons** table of 3.8 with a standard error of 0.4967. That difference is statistically significant ($p = .001$).

Figure 12.19 presents the **estimated marginal means**, shown as *Estimates*, and the **Pairwise Comparisons** table for **boxcolor**. Note here that once again the two tables are in synchrony. This time **boxcolor** increments most rapidly thanks to our reordering of the variables in parentheses following the **/EMMEANS = TABLES** subcommand. **Powcolor** still increments most slowly since it was placed first in the list.

All of the comparisons for the white powder are significant and none for the green powder are significant. For the blue powder, only the comparison for the absence of granules between box colors is significant.

A. Estimates

Measure: MEASURE_1

powcolor	Mean	Std. Error	95% Confidence Interval	
			Lower Bound	Upper Bound
1	4.100	.247	3.465	4.735
2	2.158	.011	2.131	2.185
3	4.942	.111	4.656	5.227

These are the probabilities of the differences occurring by chance alone for the pairwise comparisons for powder color.

B. Pairwise Comparisons

Measure: MEASURE_1

(I) powcolor	(J) powcolor	Mean Difference (I-J)	Std. Error	Sig.[a]	95% Confidence Interval for Difference[a]	
					Lower Bound	Upper Bound
1	2	1.942*	.241	.001	1.090	2.793
	3	-.842*	.168	.012	-1.436	-.248
2	1	-1.942*	.241	.001	-2.793	-1.090
	3	-2.783*	.102	.000	-3.142	-2.424
3	1	.842*	.168	.012	.248	1.436
	2	2.783*	.102	.000	2.424	3.142

Based on estimated marginal means

*.The mean difference is significant at the .05 level.

a.Adjustment for multiple comparisons: Bonferroni.

Figure 12.20   Paired comparisons for the main effect of powder color.

12.9.2   MAIN EFFECT OF POWCOLOR

The **estimated marginal means** and the **Paired Comparisons** table for the main effect of **powcolor** is shown in Figure 12.20. Because all means differed significantly from each other, we can order the degrees of satisfaction as follows: blue, white, and green. However, we know that this is an overgeneralization and should therefore not be emphasized in the results section.

12.10   PERFORMING THE OMNIBUS ANALYSIS IN SAS

12.10.1   STRUCTURING THE DATA FILE

The data file in an Excel spreadsheet ready to import into a *SAS Enterprise Guide* project is displayed in Figure 12.21. Note that it is structured in univariate or stacked mode as explained in Section 10.15.2. Recall that in such a form, each row can contain no more than a single score on the dependent variable. In the present three-way within-subjects design, each participant was measured twelve times, once under each condition. That is, there were three levels of **powcolor** (white, green, and blue coded 1, 2, and 3, respectively), two levels of **granules** (present or absent coded as 1 and 0), and **boxcolor** (aqua and orange coded as 1 and 2). Thus, consider the participant identified as **subid** 1 occupying lines 2 through 13 in the

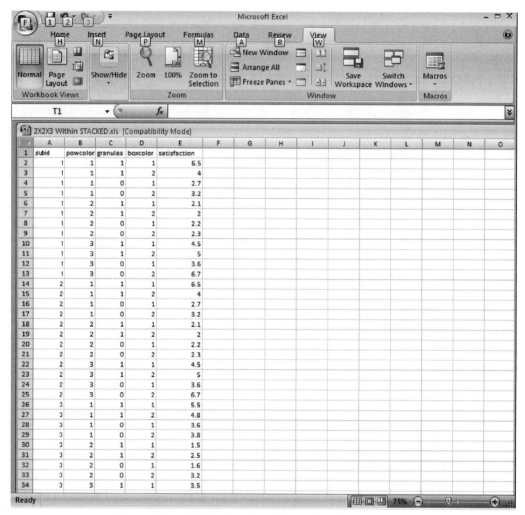

Figure 12.21 The data in an Excel file ready to import.

Excel file. Row two (the first row of data) represents the combination of **powcolor 1**, **granules 1**, and **boxcolor 1**; this corresponds to using the white powder with granules in the aqua box. The outcome variable shows a score of 6.5, indicating the level of satisfaction with that particular detergent. Row three represents **powcolor 1**, **granules 1**, and **boxcolor 2**, which corresponds to using the white powder with granules in the orange box; as you can see, that person provided a satisfaction score of 4 with that particular detergent.

Note that the data have been entered in a systematic fashion. The first four rows for each participant represent **powcolor 1** (white powder), the next four rows represent **powcolor 2** (green powder), and the last four rows represent **powcolor 3** (blue powder). Within the each set of these four rows, the second within-subjects variable **granules** is listed. With two levels of **granules**, the first pair of rows represent **granules 1** and the

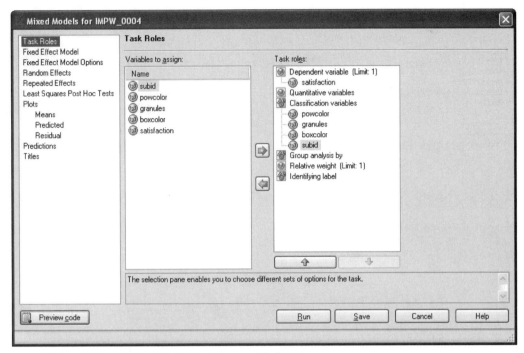

Figure 12.22 The **Task Roles** have been specified.

second pair of rows represent **granules 0**. The last within-subjects variable is **boxcolor** and will increment most rapidly down the rows. Therefore, the first row of data represents **boxcolor 1** (aqua), the next represents **boxcolor 2** (orange), the third represents **boxcolor 1** (aqua), and so on. This strategy, represented in univariate form for SAS, is equivalent to the multivariate data structure that we used for SPSS.

12.10.2 STRUCTURING THE DATA ANALYSIS

Import the data file from Excel into a new SAS project. Then, from the main menu, select **Analyze ➜ ANOVA ➜ Mixed Models**. The window opens on the **Task Roles** tab. From the **Variables to assign** panel, select **satisfaction** and drag it to the icon for **Dependent variable**. Then, one at a time, select **powcolor**, **granules**, **boxcolor**, and **subid** and drag them over to the area under **Classification variables**. The result of this is shown in Figure 12.22.

In the navigation panel, select **Fixed Effect Model** as shown in Figure 12.23. In the **Class and quantitative variables** panel, select **powcolor** and then select the **Main** pushbutton; **powcolor** will automatically appear in the **Effects** window. Repeat this procedure for **granules** and **boxcolor**. Then, while holding the Control key down, select both **powcolor** and **granules**. With both variables highlighted, click the **Factorial** pushbutton; **powcolor∗granules** will automatically appear in the **Effects** window.

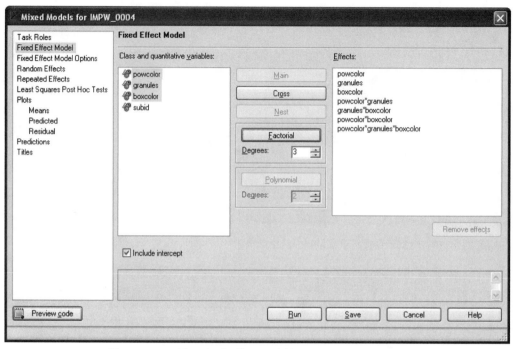

Figure 12.23   The **Fixed Effect Model** has now been specified.

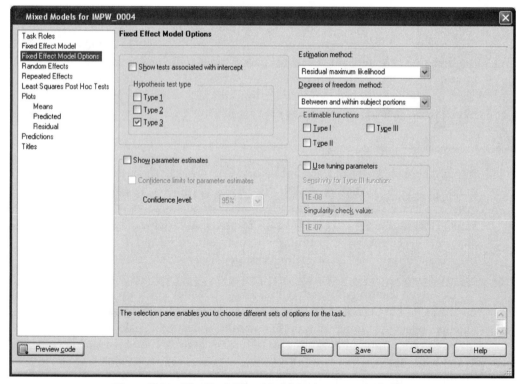

Figure 12.24   The **Fixed Effect Model Options** are specified here.

Repeat this procedure with **powcolor** and **boxcolor** and **granules** and **box-color**. Finally, while holding the Control key down, select **powcolor, gran-ules**, and **boxcolor** and then click the **Factorial** pushbutton; **powcolor∗granules∗boxcolor** will automatically appear in the **Effects** window.

Select **Fixed Effect Model Options** in the navigation panel as shown in Figure 12.24. Check **Type 3** under **Hypothesis test type**. Then select **Residual maximum likelihood** under **Estimation method**, and **Between and within subjects portions** under **Degrees of freedom method**.

Select **Random Effects** in the navigation panel as shown in Figure 12.25. When you select **Random effects** under the **Effects to use** menu, a box will appear at the far right of that menu. Click on that box and the **Effects Builder – Random effects** window appears. Select **subid** and click the **Main** pushbutton; **subid** will automatically appear in the **Random effects** panel. Click the **OK** pushbutton.

Now select **Subject identifier** under the **Model subjects** menu; a box will appear at the far right of that menu. Click on that box and the **Effects Builder – Subject identifier** window appears. Select **subid** and click the **Main** pushbutton; **subid** will automatically appear in the **Subject identifier** panel. Select the **OK** pushbutton. This is shown in Figure 12.26.

Select **Least Squares Post Hoc Tests** in the navigation panel. Click the **Add** pushbutton at the bottom of the **Effects to estimate** pane. The screen shown in Figure 12.27 appears. Highlight each of the seven effects in turn under the **Effects to use** menu and select **True** for each; this command will cause the means for each of the effects to be output. Click the **Run** pushbutton.

12.11 SAS OUTPUT FROM THE OMNIBUS ANALYSIS

The least squares means are shown in Figure 12.28. The results of the omnibus analysis are shown in Figure 12.29. These F ratios are numerically somewhat different from those produced by SPSS, although they lead us to draw the same conclusions. As can be seen, all effects except the main effect of **boxcolor** are statistically significant.

12.12 PERFORMING THE SIMPLE EFFECTS ANALYSIS IN SAS

To perform the simple effects analysis in *SAS Enterprise Guide*, configure the analysis as shown above. Then in the **Least Squares Post Hoc Tests** screen shown in Figure 12.30, in the **Effects to use** frame, set the three-way interaction to **True** and all others to **False**. Then in the **Comparisons** frame, set **Show p-values for differences** to **All pairwise differences** and set **Adjustment method for comparison** to **Bonferroni**. Click **Run** to perform the analysis.

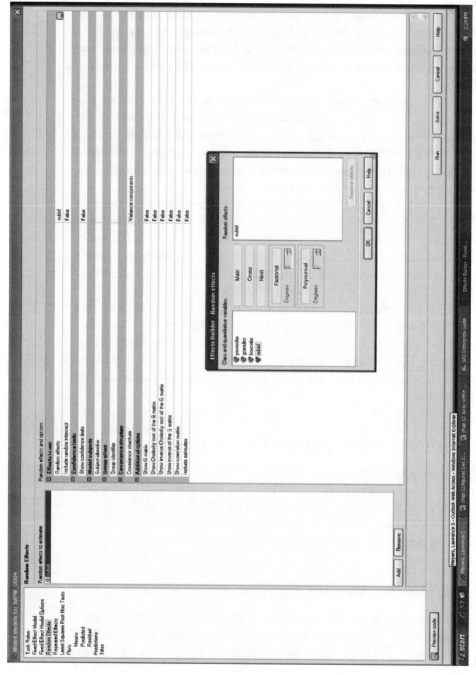

Figure 12.25 The **Random Effects** window.

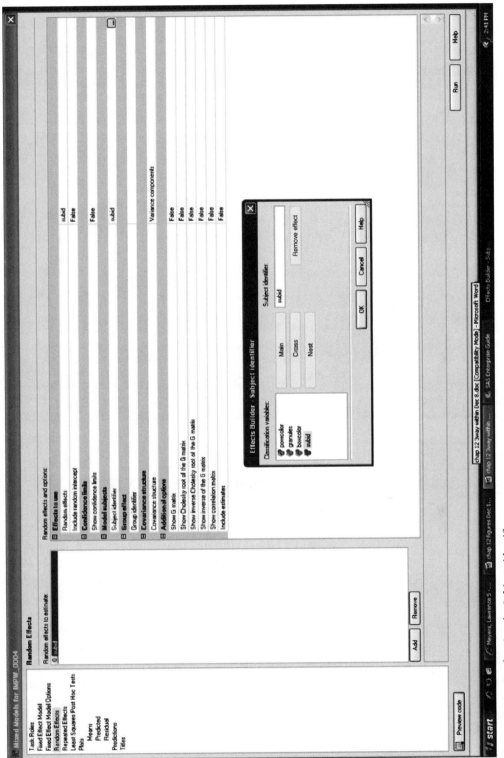

Figure 12.26 Configuring the **Subjects** identifier.

352

WITHIN-SUBJECTS DESIGNS

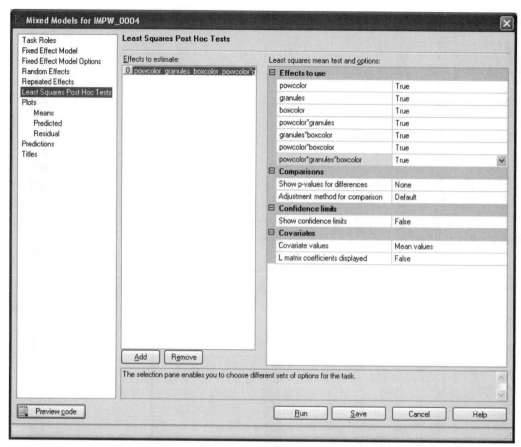

Figure 12.27   Specifying the effects for which we wish to obtain least square means.

12.13   SAS OUTPUT FROM THE SIMPLE EFFECTS ANALYSIS

The results of the simple effects tests are shown in Figure 12.31 and are read in a manner analogous to that described for the two-way interaction discussed in Section 11.14. Many of the comparisons are statistically significant, despite the large number of paired comparisons that were performed (and hence the substantial Bonferroni correction used by *SAS Enterprise Guide*). Generally, these comparisons yield similar conclusions to those we drew from the SPSS analysis.

Consider, for example, the second comparison in the table since this is the comparison we illustrated in the SPSS output for the simple effects tests. The first three coded columns represent one condition and the last three coded columns represent the other condition. We are thus comparing the mean for **powcolor 1, granules 0, boxcolor 1** (the white powder with no granules in the aqua box with a mean of 2.7) with **powcolor 1, granules 1, boxcolor 1** (the white powder with granules in the aqua box with a mean of 6.5). The t value for that comparison is -9.71, which is statistically significant ($p < .0001$). We may therefore conclude that the participants

These are the means for each cell of this $3 \times 2 \times 2$ design.

				Least Squares Means				
Effect	powcolor	granules	boxcolor	Estimate	Standard Error	DF	t Value	Pr>|t|
powcolor	1			4.1000	0.1629	10	25.16	<.0001
powcolor	2			2.1583	0.1629	10	13.25	<.0001
powcolor	3			4.9417	0.1629	10	30.33	<.0001
granules		0		3.4500	0.1420	5	24.30	<.0001
granules		1		4.0167	0.1420	5	28.29	<.0001
boxcolor			1	3.5944	0.1420	5	25.32	<.0001
boxcolor			2	3.8722	0.1420	5	27.27	<.0001
powcolor*granules	1	0		2.9500	0.2138	10	13.80	<.0001
powcolor*granules	1	1		5.2500	0.2138	10	24.56	<.0001
powcolor*granules	2	0		2.2667	0.2138	10	10.60	<.0001
powcolor*granules	2	1		2.0500	0.2138	10	9.59	<.0001
powcolor*granules	3	0		5.1333	0.2138	10	24.01	<.0001
powcolor*granules	3	1		4.7500	0.2138	10	22.22	<.0001
powcolor*granules	1		1	4.6000	0.2138	10	21.52	<.0001
powcolor*granules	1		2	3.6000	0.2138	10	16.84	<.0001
powcolor*granules	2		1	2.1500	0.2138	10	10.06	<.0001
powcolor*granules	2		2	2.1667	0.2138	10	10.13	<.0001
powcolor*granules	3		1	4.0333	0.2138	10	18.87	<.0001
powcolor*granules	3		2	5.8500	0.2138	10	27.36	<.0001
granules*boxcolor		0	1	2.8222	0.1815	5	15.55	<.0001
granules*boxcolor		0	2	4.0778	0.1815	5	22.47	<.0001
granules*boxcolor		1	1	4.3667	0.1815	5	24.06	<.0001
granules*boxcolor		1	2	3.6667	0.1815	5	20.21	<.0001
powcol*granul*boxcol	1	0	1	2.7000	0.2899	10	9.31	<.0001
powcol*granul*boxcol	1	0	2	3.2000	0.2889	10	11.04	<.0001
powcol*granul*boxcol	1	1	1	6.5000	0.2889	10	22.42	<.0001
powcol*granul*boxcol	1	1	2	4.0000	0.2889	10	13.80	<.0001
powcol*granul*boxcol	2	0	1	2.2000	0.2889	10	7.59	<.0001
powcol*granul*boxcol	2	0	2	2.3333	0.2889	10	8.05	<.0001
powcol*granul*boxcol	2	1	1	2.1000	0.2889	10	7.24	<.0001
powcol*granul*boxcol	2	1	2	2.0000	0.2889	10	6.90	<.0001
powcol*granul*boxcol	3	0	1	3.5667	0.2889	10	12.30	<.0001
powcol*granul*boxcol	3	0	2	6.7000	0.2889	10	23.11	<.0001
powcol*granul*boxcol	3	1	1	4.5000	0.2889	10	15.52	<.0001
powcol*granul*boxcol	3	1	2	5.0000	0.2889	10	17.25	<.0001

Figure 12.28 Descriptive statistics.

preferred the detergent with granules when the powder was white and when it was packaged in an aqua box.

12.14 PERFORMING THE POST HOC ANALYSIS IN SAS

To perform the simple effects analysis in *SAS Enterprise Guide*, configure the analysis as shown above. Then in the **Least Squares Post Hoc Tests**

Type 3 Tests of Fixed Effects				
Effect	Num DF	Den DF	F Value	Pr > F
powcolor	2	10	106.36	<.0001
granules	1	5	12.57	0.0165
boxcolor	1	5	3.02	0.1427
powcolor*granules	2	10	29.50	<.0001
powcolor*granules	2	10	26.55	<.0001
granules*boxcolor	1	5	37.43	0.0017
powcol*granul*boxcol	2	10	7.37	0.0108

These *p*-values suggest a nonsignificant boxcolor effect, and significant effects for all other main and interaction effects.

Figure 12.29   The results of the ANOVA.

screen shown in Figure 12.32, in the **Effects to use** frame, set **powcolor** to **True** and all others to **False**. Then in the **Comparisons** frame, set **Show p-values for differences** to **All pairwise differences** and set **Adjustment method for comparison** to **Bonferroni**. Click **Run** to perform the analysis.

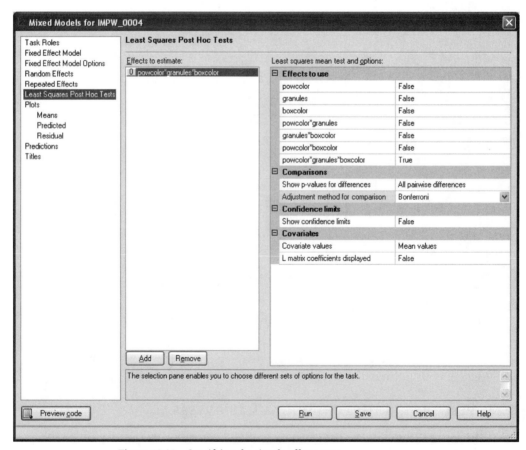

Figure 12.30   Specifying the simple effects tests.

This Table presents the unadjusted *p*-value. The Bonferroni adjustment (not shown) can be calculated by multiplying the unadjusted *p*-value by 66 (the number of all pairwise comparisons).

Differences of Least Squares Means								
Effect	powcolor	granules	boxcolor	powcolor	granules	boxcolor	t Value	Pr>|t|	Adjustment
powcol*granul*boxcol	1	0	1	1	1	2	− 1.28	0.2304	Bonferroni
powcol*granul*boxcol	1	0	1	1	1	1	− 9.71	<.0001	Bonferroni
powcol*granul*boxcol	1	0	1	1	1	2	− 3.32	0.0077	Bonferroni
powcol*granul*boxcol	1	0	1	2	0	1	1.28	0.2304	Bonferroni
powcol*granul*boxcol	1	0	1	2	0	2	0.94	0.3710	Bonferroni
powcol*granul*boxcol	1	0	1	2	1	1	1.53	0.1564	Bonferroni
powcol*granul*boxcol	1	0	1	2	1	2	1.79	0.1041	Bonferroni
powcol*granul*boxcol	1	0	1	3	0	1	− 2.21	0.0512	Bonferroni
powcol*granul*boxcol	1	0	1	3	0	2	− 10.22	<.0001	Bonferroni
powcol*granul*boxcol	1	0	1	3	1	1	− 4.60	0.0010	Bonferroni
powcol*granul*boxcol	1	0	1	3	1	2	− 5.88	0.0002	Bonferroni
powcol*granul*boxcol	1	0	2	1	1	1	− 8.43	<.0001	Bonferroni
powcol*granul*boxcol	1	0	2	1	1	2	− 2.04	0.0682	Bonferroni
powcol*granul*boxcol	1	0	2	2	0	1	2.55	0.0286	Bonferroni
powcol*granul*boxcol	1	0	2	2	0	2	2.21	0.0512	Bonferroni
powcol*granul*boxcol	1	0	2	2	1	1	2.81	0.0185	Bonferroni
powcol*granul*boxcol	1	0	2	2	1	2	3.07	0.0119	Bonferroni
powcol*granul*boxcol	1	0	2	3	0	1	− 0.94	0.3710	Bonferroni
powcol*granul*boxcol	1	0	2	3	0	2	− 8.94	<.0001	Bonferroni
powcol*granul*boxcol	1	0	2	3	1	1	− 3.32	0.0077	Bonferroni
powcol*granul*boxcol	1	0	2	3	1	2	− 4.60	0.0010	Bonferroni
powcol*granul*boxcol	1	1	1	1	1	2	6.39	<.0001	Bonferroni
powcol*granul*boxcol	1	1	1	2	0	1	10.98	<.0001	Bonferroni
powcol*granul*boxcol	1	1	1	2	0	2	10.64	<.0001	Bonferroni
powcol*granul*boxcol	1	1	1	2	1	1	11.24	<.0001	Bonferroni
powcol*granul*boxcol	1	1	1	2	1	2	11.49	<.0001	Bonferroni
powcol*granul*boxcol	1	1	1	3	0	1	7.49	<.0001	Bonferroni
powcol*granul*boxcol	1	1	1	3	0	2	− 0.51	0.6205	Bonferroni
powcol*granul*boxcol	1	1	1	3	1	1	5.11	0.0005	Bonferroni
powcol*granul*boxcol	1	1	1	3	1	2	3.83	0.0033	Bonferroni
powcol*granul*boxcol	1	1	2	2	0	1	4.60	0.0010	Bonferroni
powcol*granul*boxcol	1	1	2	2	0	2	4.26	0.0017	Bonferroni
powcol*granul*boxcol	1	1	2	2	1	1	4.85	0.0007	Bonferroni
powcol*granul*boxcol	1	1	2	2	1	2	5.11	0.0005	Bonferroni

Figure 12.31   A portion of the output from all the pairwise comparison tests.

12.15   SAS OUTPUT FROM THE POST HOC ANALYSIS

The pairwise mean comparisons are shown in Figure 12.33. As can be seen, all conditions differ significantly from all others.

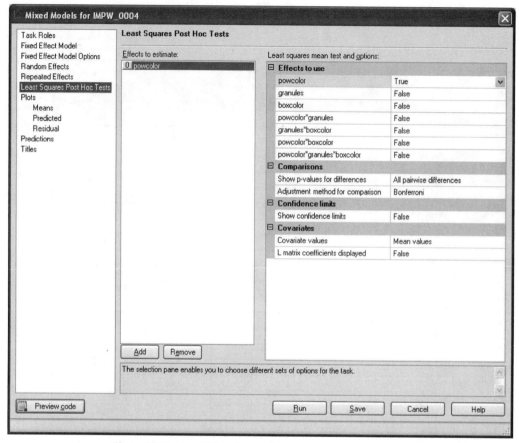

Figure 12.32   Specifying the post hoc tests.

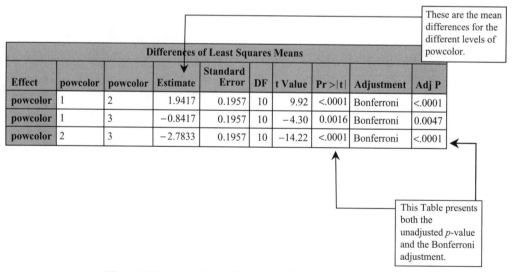

These are the mean differences for the different levels of powcolor.

Differences of Least Squares Means									
Effect	powcolor	powcolor	Estimate	Standard Error	DF	t Value	Pr > |t|	Adjustment	Adj P
powcolor	1	2	1.9417	0.1957	10	9.92	<.0001	Bonferroni	<.0001
powcolor	1	3	−0.8417	0.1957	10	−4.30	0.0016	Bonferroni	0.0047
powcolor	2	3	−2.7833	0.1957	10	−14.22	<.0001	Bonferroni	<.0001

This Table presents both the unadjusted *p*-value and the Bonferroni adjustment.

Figure 12.33   A portion of the output from the post hoc tests.

12.16 COMMUNICATING THE RESULTS

Six single heads of households participated in this within-subjects design. Individuals laundered their clothes by using aqua or orange boxes containing white, light green, or light blue powder that either did or did not contain granules. These heads of households evaluated the degree to which they were satisfied with each detergent after each load of laundry. All effects met the assumption of sphericity as tested by Mauchly's procedure.

All effects were statistically significant except for the main effect of box color. The summary tables are shown in Table 12.1. The three-way interaction was statistically significant, $F(2, 10) = 9.39$, $p < .05$, $\eta^2 = .03$. The bar graph depicting the three-way interaction is presented in Figure 12.2. Very generally, participants seemed to prefer two different combinations of the product: (a) white powder with granules packaged in an aqua box and (b) blue powder without granules packaged in an orange box.

CHAPTER 12 EXERCISES

12.1. In this experiment, participants were presented with words or non-words on a computer screen. The dependent measure was reaction time in milliseconds. The design was a $2 \times 2 \times 2$ factorial where $n = 7$ participants received all treatment combinations. Factor A was the stimulus type (a_1 = word and a_2 = nonword). Factor B was stimulus length (b_1 = 4 letters and b_2 = 8 letters). Factor C was orthographic style (c_1 = lowercase and c_2 = uppercase). The data are as follows:

	c_1				c_2			
Subject	$a_1 b_1$	$a_1 b_2$	$a_2 b_1$	$a_2 b_2$	$a_1 b_1$	$a_1 b_2$	$a_2 b_1$	$a_2 b_2$
1	1,018	1,050	1,144	1,210	1,009	1,030	1,127	1,130
2	1,100	1,140	1,149	1,220	1,040	1,070	1,140	1,145
3	1,050	1,140	1,160	1,240	1,030	1,080	1,150	1,145
4	1,040	1,090	1,170	1,250	1,044	1,085	1,160	1,166
5	1,009	1,050	1,165	1,265	1,050	1,090	1,175	1,170
6	1,027	1,070	1,170	1,245	1,055	1,095	1,160	1,155
7	1,030	1,080	1,175	1,250	1,040	1,085	1,165	1,160

   a. Conduct an ANOVA using SPSS or SAS.
   b. If the interaction(s) is(are) statistically significant, conduct simple effects analyses with SPSS or SAS.

12.2. In this study of iconic visual processing, participants are shown cartoons depicting problematic situations. The dependent measure is comprehension time in milliseconds. The design was a $2 \times 2 \times 3$ factorial, where $n = 5$ participants received all treatment combinations. Factor A was presentation length (a_1 = 500 ms and a_2 = 1,000 ms). Factor B was stimulus color ($b_1 = $ color and $b_2 = $ black and white). Factor C was cartoon theme ($c_1 = $ happy, $c_2 = $ sad, and $c_3 = $ ironic). The data are as follows:

	c_1				c_2				c_3			
Subject	$a_1 b_1$	$a_1 b_2$	$a_2 b_1$	$a_2 b_2$	$a_1 b_1$	$a_1 b_2$	$a_2 b_1$	$a_2 b_2$	$a_1 b_1$	$a_1 b_2$	$a_2 b_1$	$a_2 b_2$
1	900	920	800	820	950	930	880	901	1,010	1,040	1,050	1,060
2	980	960	820	830	940	940	890	910	1,020	1,020	1,060	1,070
3	990	970	840	840	960	940	900	920	1,025	1,030	1,070	1,080
4	960	975	830	845	970	950	860	870	1,019	1,035	1,080	1,020
5	995	990	844	850	980	960	870	880	1,029	1,018	1,050	1,050

a. Conduct an ANOVA on these data with SPSS or SAS.
b. If the interaction(s) is(are) statistically significant, conduct simple effects analyses and any necessary multiple comparisons tests.

12.3. Consider the following hypothetical $2 \times 2 \times 3$ factorial experiments where all $n = 6$ participants received all treatment combinations. The data are as follows:

	c_1				c_2				c_3			
Subject	$a_1 b_1$	$a_1 b_2$	$a_2 b_1$	$a_2 b_2$	$a_1 b_1$	$a_1 b_2$	$a_2 b_1$	$a_2 b_2$	$a_1 b_1$	$a_1 b_2$	$a_2 b_1$	$a_2 b_2$
1	10	12	13	3	15	19	18	6	20	19	18	15
2	9	14	14	6	16	17	14	5	21	20	17	18
3	7	18	15	3	18	16	12	4	22	21	20	20
4	10	10	16	6	20	15	10	3	20	22	21	14
5	11	17	18	2	21	18	15	5	20	23	22	14
6	9	11	20	4	22	20	18	3	21	20	20	10

a. Conduct an ANOVA on these data with SPSS or SAS.
b. If the interaction(s) is(are) statistically significant, conduct simple effects analyses and multiple comparisons tests.

SECTION 5

Mixed Designs

Simple Mixed Design

13.1 COMBINING BETWEEN-SUBJECTS AND WITHIN-SUBJECTS FACTORS

A mixed design is one that contains at least one between-subjects independent variable and at least one within-subjects independent variable. In a simple mixed design, there are only two independent variables, one a between-subjects factor and the other a within-subjects factor; these variables are combined factorially. The number of levels of each independent variable is not constrained by the design. Thus, we could have a 2 × 2, a 4 × 3, or even a 3 × 7 factorial design. Chapters 14 and 15 will address two complex mixed designs that contain three independent variables.

Because there are two independent variables, there are three effects of interest: the main effect of the between-subjects variable (A), the main effect of the within-subjects variable (B), and the two-way interaction ($A \times B$). Note that this is analogous to what we have seen in the two-way between-subjects and two-way within-subjects designs. Furthermore, the conceptual understanding of main effects and interactions in those designs carries forward to the simple mixed design. Main effects focus on the mean differences of the levels of each independent variable (e.g., a_1 vs. a_2) and interactions focus on whether or not the patterns of differences are parallel (e.g., a_1 vs. a_2 under b_1 compared to a_1 vs. a_2 under b_2).

The primary difference between a simple mixed design and the between-subjects and within-subjects designs is in the way that the total variance of the dependent variable is partitioned. As is true for within-subjects designs, the total variance in a mixed design is divided into between-subjects variance and within-subjects variance. The three effects of interest break out as follows:

- The main effect of A: The between-subjects variable A is subsumed in the between-subjects portion of the variance. It has its own between-subjects error term that is used in computing the F ratio associated with A.
- The main effect of B: The within-subjects variable B is subsumed in the within-subjects portion of the variance. It has its own within-subjects error term that is used in computing the F ratio associated with B. This error term is associated with the B factor and thus it

Leadership Style	Subid	Project Type	
		Simple Projects	Complex Projects
Authoritarian	1	5	0
	2	6	2
	3	7	1
	4	5	0
	5	7	2
Democratic	6	7	7
	7	5	6
	8	7	7
	9	6	7
	10	7	7
Laissez Faire	11	4	2
	12	3	3
	13	5	4
	14	3	2
	15	3	3

Figure 13.1 Data for numerical example.

is also used as an error term when computing the F ratio for the interaction effect.

- The $A \times B$ interaction: The interaction effect is subsumed in the within-subjects portion of the variance. It shares its within-subjects error term with the main effect of B.

13.2   A NUMERICAL EXAMPLE OF A SIMPLE MIXED DESIGN

Data from the hypothetical study serving as our example are shown in Figure 13.1. The fifteen cases in this study are work teams that, over the course of a month, were each given several projects to complete. Some of the projects were relatively simple whereas other projects were relatively complex. This project type variable has two levels, simple and complex. Project type is a within-subjects variable in that each work team was asked to complete both simple and complex projects.

Work teams operated under three different types of leadership style. Teams 1–5 were led by an authoritarian style leader (the leader made work decisions and directed the members of the team), Teams 6–10 experienced a democratic style of leadership (decisions were made as a group), and Teams 11–15 were led in a laissez faire style (team members were left alone to do the work). This leadership style variable has three levels, authoritarian (coded as 1), democratic (coded as 2), and laissez faire (coded as 3). It is a between-subjects variable in that each team operated under only one leadership style.

The dependent variable was an overall performance rating made by management personnel based on how well the project was done. It captured such elements as use of resources, cost, on-time delivery, and responsiveness to the initial request for the project. Ratings could vary between 0 (*not satisfactory performance*) and 7 (*performed with high quality*).

Note that the cases in this study are the teams and not individual people. This is perfectly acceptable in that the team was treated as a unit, and it was the team that was given a performance score.

Leadership Style	Project Type		Overall Mean
	Simple Projects	Complex Projects	
Authoritarian	6.00	1.00	3.50
Democratic	6.40	6.80	6.60
Laissez Faire	3.60	2.80	3.20
Overall Mean	5.33	3.53	

Figure 13.2 Cell means and marginal means for the data in our numerical example.

Figure 13.2 presents the cell means as well as the row and column means. This is a convenient way to quickly summarize the results and makes the factorial structure of the design clear. The main effect of leadership style deals with comparing the row means to each other, the main effect of project type deals with comparing the column means to each other, and the interaction effect deals with the patterns of the cell means.

13.3   EFFECTS OF INTEREST

The summary table showing the results of the analysis is presented in Table 13.1. The effects in which we are interested are analogous to the other two-way designs that we have covered in Chapters 8 and 11: two main effects and the two-way interaction. As can be seen, all three of these effects are statistically significant.

The eta squared values are shown in the last column of the summary table. Although it is customary to use the total variance as the basis for computing this statistic in between-subjects and within-subjects designs, in mixed designs it is common to keep the between-subjects and

Table 13.1. Summary table for a simple mixed factorial design

Source	SS	df	MS	F	η^2
Between subjects	85.87	14			
A (leadership style)	70.87	2	35.44	28.35*	.83
S/A	15.00	12	1.25		
Within subjects	67.50				
B (project type)	24.30	1	24.30	97.20*	.36
A × B (Leadership × Project)	40.20	2	20.10	80.40*	.60
B × S/A	3.00	12	0.25		
Total	153.37	29			

*$p < .05$.

Note: $\eta_A^2 = SS_A/SS_{S/A}$.

$\eta_B^2 = SS_B/SS_{B \times S/A}$.

$\eta_{A \times B}^2 = SS_{A \times B}/SS_{B \times S/A}$.

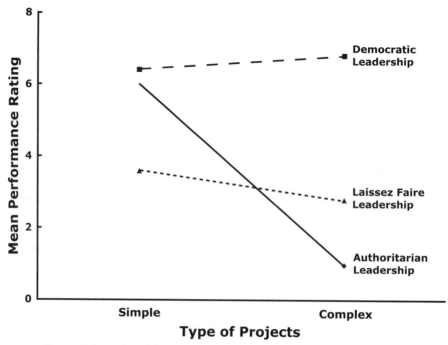

Figure 13.3   A plot of the two-way interaction.

within-subjects partitions separated in the computation. Following this tradition, we have computed the eta squared values as follows:

- Main effect of A: $SS_A \div SS$ Between Subjects
- Main effect of B: $SS_B \div SS$ Within Subjects
- Interaction $A \times B$: $SS_{A \times B} \div SS$ Within Subjects

The plot of the interaction is shown in Figure 13.3. We will be in a position to narrate these results once we have performed the simple effects analyses.

13.4   COMPUTING THE OMNIBUS ANALYSIS BY HAND

The procedures for conducting a simple mixed factorial ANOVA parallel many of the procedures covered previously in the between-subjects and within-subjects factorial designs. We begin with the basic observations or $A \times B \times$ Subjects (ABS) data matrix found in Table 13.2. Recall that in this study, the fifteen "cases" (designated s_i) are actually work teams that are split into three independent groups that operate under three types of leadership style (Factor A). This between-subjects factor has the following three levels ($a_1 =$ authoritarian, $a_2 =$ democratic, $a_3 =$ laissez faire). Factor B, project type, is a within-subjects variable with two treatment conditions ($b_1 =$ simple, $b_2 =$ complex). The dependent variable is a score from 0 to 7 reflecting overall work team performance. The scores for each subject within each (AB) treatment combination (in the ABS matrix)

Table 13.2. Preliminary calculations for a 3×2 simple mixed design factorial

ABS data matrix

	a_1 (Authoritarian)			a_2 (Democratic)			a_3 (Laissez faire)	
Subjects	(Simple) b_1	(Complex) b_2	Subjects	(Simple) b_1	(Complex) b_2	Subjects	(Simple) b_1	(Complex) b_2
s_1	5	0	s_6	7	7	s_{11}	4	2
s_2	6	2	s_7	5	6	s_{12}	3	3
s_3	7	1	s_8	7	7	s_{13}	5	4
s_4	5	0	s_9	6	7	s_{14}	3	2
s_5	7	2	s_{10}	7	7	s_{15}	3	3
ΣAB:	30	5	ΣAB:	32	34	ΣAB:	18	14
ΣY^2:	184	9	ΣY^2:	208	232	ΣY^2:	68	42
$\overline{Y}$:	6.00	1.00	$\overline{Y}$:	6.40	6.80	$\overline{Y}$:	3.60	2.80
s:	1.00	1.00	s:	0.89	0.45	s:	0.89	0.84

AB matrix (sums)

	(Simple) b_1	(Complex) b_2	Marginal sum
a_1 (Authoritarian)	30	5	35
a_2 (Democratic)	32	34	66
a_3 (Laissez faire)	18	14	32
Marginal sum	80	53	133

AS matrix

a_1 (Authoritarian)		a_2 (Democratic)		a_3 (Laissez faire)	
s_1	5	s_6	14	s_{11}	6
s_2	8	s_7	11	s_{12}	6
s_3	8	s_8	14	s_{13}	9
s_4	5	s_9	13	s_{14}	5
s_5	9	s_{10}	14	s_{15}	6
Sum	35	Sum	66	Sum	32

AB matrix (means)

Leadership style (Factor A)	Project type (Factor B)		
	(Simple) b_1	(Complex) b_2	Marginal means
a_1 (Authoritarian)	6.00	1.00	3.00
a_2 (Democratic)	6.40	6.80	6.60
a_3 (Laissez faire)	3.60	2.80	3.20
Marginal means	1.17	4.25	

have been summed and squared and means and standard deviations were computed in the usual manner. Two additional matrices are needed to provide the ingredients for the computations involved in producing sums of squares. An AB matrix of sums is generated by collapsing subjects' scores across each AB treatment combination. An AS matrix is produced by collapsing each subject's total score across the levels of the within-subjects variable (Factor B). An AB matrix of means provides the researcher with an efficient way of scrutinizing the main effect and interaction dynamics.

13.4.1 SUM OF SQUARES

There are six sums of squares that we will calculate for a simple mixed design ANOVA, all of which are produced by manipulating the scores or sums in the ABS, AB, and AS matrices.

- Sum of squares between groups (SS_A).
- Sum of squares subjects within the A treatments ($SS_{S/A}$). This provides the error component to evaluate the A main effect or between-subject portion.
- Sum of squares within groups (SS_B).
- Sum of squares A by B interaction effect ($SS_{A \times B}$).
- Sum of squares repeated measures error term ($SS_{B \times S/A}$). (Read "sum of squares B by S over A." It reflects the Factor B by subject interaction at each level of Factor A; see Keppel, 1991.) This is the error term for evaluating the B main effect and $A \times B$ interaction effect.
- Sum of squares total (SS_T).

The formulas for these six sums of squares and the summary of their calculations based on data from Table 13.1 are as follows:

$$
\begin{aligned}
SS_A &= \frac{\Sigma A^2}{(b)(n)} - \frac{T^2}{(a)(b)(n)} \\
&= \frac{(35)^2 + (66)^2 + (32)^2}{(2)(5)} - \frac{(133)^2}{(3)(2)(5)} \\
&= 660.50 - 589.63 = \boxed{70.87}
\end{aligned}
\tag{13.1}
$$

$$
\begin{aligned}
SS_{S/A} &= \frac{\Sigma AS^2}{b} - \frac{\Sigma A^2}{(b)(n)} \\
&= \frac{(5)^2 + (8)^2 + \cdots + (5)^2 + (6)^2}{2} - \frac{(35)^2 + (66)^2 + (32)^2}{(2)(5)} \\
&= 675.50 - 660.50 = \boxed{15.00}
\end{aligned}
\tag{13.2}
$$

$$SS_B = \frac{\Sigma B^2}{(a)(n)} - \frac{T^2}{(a)(b)(n)}$$
$$= \frac{(80)^2 + (53)^2}{(3)(5)} - \frac{(133)^2}{(3)(2)(5)}$$
$$= 613.93 - 589.63 = \boxed{24.30} \qquad (13.3)$$

$$SS_{A \times B} = \frac{\Sigma AB^2}{n} - \frac{\Sigma A^2}{(b)(n)} - \frac{\Sigma B^2}{(a)(n)} + \frac{T^2}{(a)(b)(n)}$$
$$= \frac{(30)^2 + (5)^2 + \cdots + (18)^2 + (14)^2}{5}$$
$$- \frac{(35)^2 + (66)^2 + (32)^2}{(2)(5)} - \frac{(80)^2 + (53)^2}{(3)(5)} + \frac{(133)^2}{(3)(2)(5)}$$
$$= 725.00 - 660.50 - 613.93 + 589.63 = \boxed{40.20} \qquad (13.4)$$

$$SS_{B \times S/A} = \Sigma Y^2 - \frac{\Sigma AB^2}{n} - \frac{\Sigma AS^2}{b} + \frac{\Sigma A^2}{(b)(n)}$$
$$= \frac{(5)^2 + (6)^2 + \cdots + (3)^2 + (3)^2 - (30)^2 + (5)^2 + \cdots + (18)^2 + (14)^2}{5}$$
$$- \frac{(5)^2 + (8)^2 + \cdots + (5)^2 + (6)^2}{2} + \frac{(35)^2 + (66)^2 + (32)^2}{(2)(5)}$$
$$= 743.00 - 725.00 - 675.50 + 660.50 = \boxed{3.00} \qquad (13.5)$$

$$SS_T = \Sigma Y^2 - \frac{T^2}{(a)(b)(n)}$$
$$= (5)^2 + (6)^2 + \cdots + (3)^2 + (3)^2 - \frac{(133)^2}{(3)(2)(5)}$$
$$= 743.00 - 589.63 = \boxed{153.37}. \qquad (13.6)$$

13.4.2 CALCULATING DEGREES OF FREEDOM

Below are the formulas for the degrees of freedom associated with each sum of squares and the simple computations involved based on our numerical example:

$$df_A = a - 1 = 3 - 1 = \boxed{2}$$
$$df_{S/A} = (a)(n-1) = (3)(5-1) = \boxed{12}$$
$$df_B = b - 1 = 2 - 1 = \boxed{1}$$
$$df_{A \times B} = (a-1)(b-1) = (2)(1) = \boxed{2}$$
$$df_{B \times S/A} = (a)(b-1)(n-1) = (3)(1)(4) = \boxed{12}$$
$$df_T = (a)(b)(n) - 1 = (3)(2)(5) - 1 = \boxed{29}.$$

13.4.3 CALCULATING MEAN SQUARES AND F RATIOS

We compute five mean squares for this analysis by dividing each sum of squares by its respective degrees of freedom. The mean square total is not calculated.

$$MS_A = \frac{SS_A}{df_A} = \frac{70.87}{20} = \boxed{35.44} \tag{13.7}$$

$$MS_{S/A} = \frac{SS_{S/A}}{df_{S/A}} = \frac{15.00}{12} = \boxed{1.25} \tag{13.8}$$

$$MS_B = \frac{SS_B}{df_B} = \frac{24.30}{1} = \boxed{24.30} \tag{13.9}$$

$$MS_{A \times B} = \frac{SS_{A \times B}}{df_{A \times B}} = \frac{40.20}{2} = \boxed{20.10} \tag{13.10}$$

$$MS_{B \times S/A} = \frac{SS_{B \times S/A}}{df_{B \times S/A}} = \frac{3.00}{12} = \boxed{0.25}. \tag{13.11}$$

Three F ratios are formed in a simple mixed design. The between-subjects factor is evaluated with the following F ratio.

$$F_A = \frac{MS_A}{MS_{S/A}} = \frac{35.44}{1.25} = \boxed{28.35}. \tag{13.12}$$

The remaining F ratios are a function of the within-subjects variability in the study and are each evaluated with $MS_{B \times S/A}$ as the error term:

$$F_B = \frac{MS_B}{MS_{B \times S/A}} = \frac{24.30}{0.25} = \boxed{97.20} \tag{13.13}$$

$$F_{A \times B} = \frac{MS_{A \times B}}{MS_{B \times S/A}} = \frac{20.10}{0.25} = \boxed{80.40}. \tag{13.14}$$

We evaluate the between-subjects main effect (F_A) with (2, 12) degrees of freedom (df_A, $df_{S/A}$). The critical value of F (see Appendix C) at $\alpha = .05$ is 3.89. Since our F_A exceeds this critical value, we reject the null hypothesis and conclude that leadership style affects group performance.

The remaining within-subjects F values are evaluated with the following degrees of freedom: $F_B(1, 12)(df_B, df_{B \times S/A})$ and $F_{A \times B}(2, 12)$ ($df_{A \times B}$, $df_{B \times S/A}$). The critical values of F at $\alpha = .05$ (see Appendix C) for F_B and $F_{A \times B}$ are 4.75 and 3.89, respectively. Both observed Fs exceed these critical values; thus we reject the null and conclude (in the case of the main effect of Factor B) that simple projects produce higher performance evaluations than do complex projects. However, this conclusion is tempered by the statistically significant interaction effect. All of these calculations are summarized in an ANOVA summary table (see Table 13.1).

Our next computational steps would be to examine the main comparisons of our between-subjects variable (Factor A) and our within-subjects variable (Factor B) and explore the nature of the statistically significant $A \times B$ interaction through simple effects analysis of Factor A at b_k or Factor B at a_j. The procedures for the between-subjects main comparisons are straightforward and follow the same procedures and formulas we reviewed in Chapter 8 on between-subjects factorial analyses of main effects. The only difference is that the error term becomes $MS_{S/A}$.

The hand computations for the within-subjects main comparison and the simple effects analyses require the computation of separate error terms and are somewhat tedious to do and will not be covered here. The interested reader is encouraged to review Keppel (1991, Chapter 5) and Keppel and Wickens (2004, Chapter 19) for numerical examples. Keppel et al. (1992, Chapter 12) offers a more "simplified" and computationally more elegant approach using $MS_{w.cell}$ (the average of the variances of all treatment combinations) as the error term when conducting simple effects analyses of A at b_k and simple comparisons of A at b_k. This latter approach is recommended only for students who are just beginning to learn hand computational procedures of simple mixed designs.

13.5 PERFORMING THE OMNIBUS ANALYSIS IN SPSS

13.5.1 STRUCTURING THE DATA FILE

The data file for our 3×2 simple mixed design example is shown in Figure 13.4. The first column, as always, is used for our participant identification number; we have named this variable **subid**. The next column holds our between-subjects variable of leadership style. Between-subjects variables take arbitrary numeric codes; here we have coded authoritarian, democratic, and laissez faire as 1, 2, and 3, respectively.

The last two columns in the data file represent our repeated measure of project type. Under each level of this within-subjects variable we have recorded the performance evaluation of management.

13.5.2 STRUCTURING THE DATA ANALYSIS

From the main SPSS menu, select **Analyze** → **General Linear Model** → **Repeated Measures**. You select **Repeated Measures** for the following reason. If there are two or more cases in the study and at least one of the independent variables is a within-subjects variable, then we must partition the total variance of the dependent variable into a between-subjects portion and a within-subjects portion. This partitioning is accomplished by the **Repeated Measures** module of the **General Linear Model** procedure in SPSS.

Selecting this path will open the dialog window shown in Figure 13.5. As you will recall from Chapters 10, 11, and 12, this is the initial window used by SPSS to have you name the within-subjects variable(s) in the

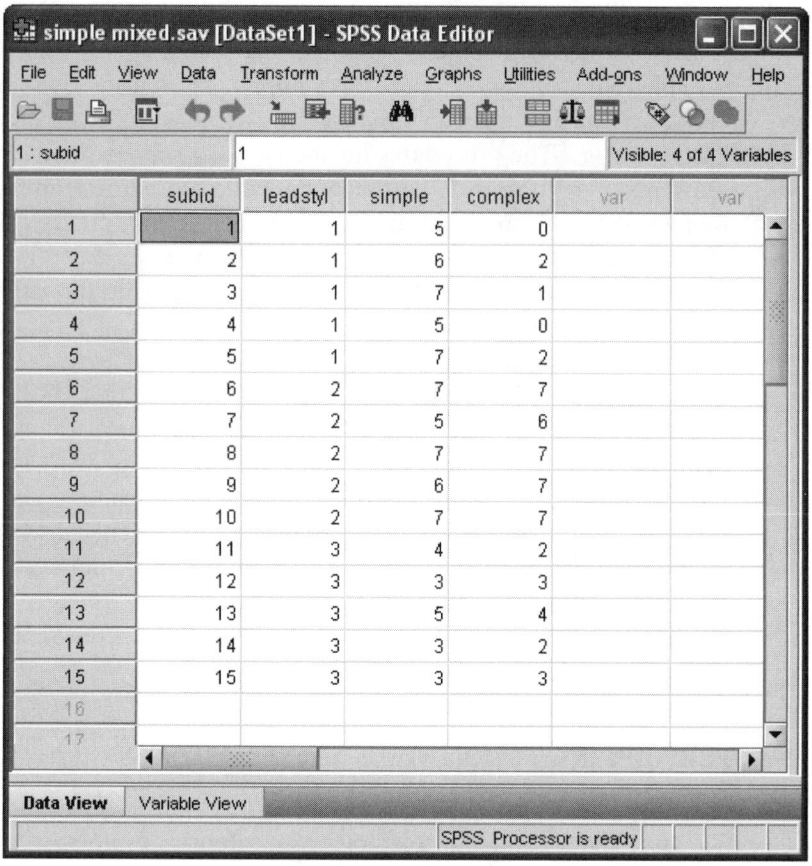

Figure 13.4   The data file for our numerical example.

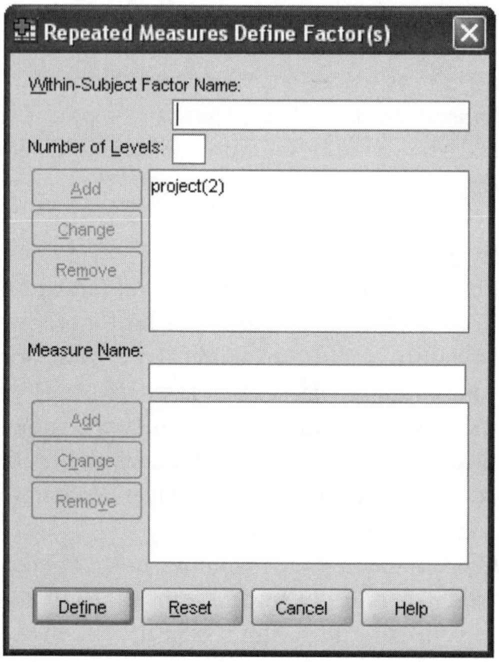

Figure 13.5 Initial dialog window for **GLM Repeated Measures**.

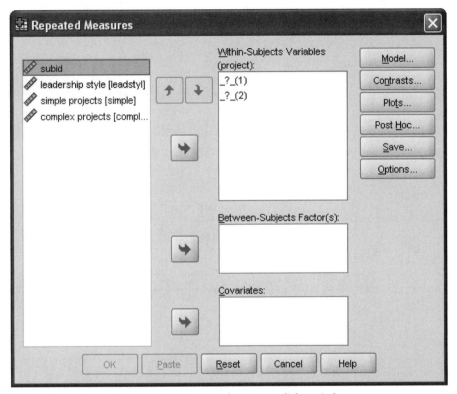

Figure 13.6   The main **GLM Repeated Measures** dialog window.

study. We have specified the variable name **project** and indicated that it has two levels.

Clicking **Define** in the **Define Factor(s)** window brings us to the **GLM Repeated Measures** main dialog window shown in Figure 13.6. We place **simple** and **complex** in slots 1 and 2, respectively, of the **Within-Subjects Variables** panel and **leadstyl** in the panel for **Between-Subjects Factor(s)** as shown in Figure 13.7.

Select the **Options** pushbutton to reach the **Options** dialog window. Bring over the three effects of interest to the **Display Means** panel and check **Descriptive statistics** and **Homogeneity tests**. This is shown in Figure 13.8. Click **Continue** to return to the main dialog window.

In the **Plots** dialog box, click over **project** to the **Horizontal Axis** because it is the within-subjects variable; we generally want the within-subjects variable on the x axis of the plot. Click over **leaderstyl** to the **Separate Lines** panel. We have shown this in Figure 13.9. This will produce a plot with three lines (one for each leadership style), each representing a different set of cases. We can then trace how each has performed across the two kinds of projects. Click **Add** to register your plot with SPSS and click **Continue** to return to the main dialog window. Then click **OK** to perform the omnibus analysis.

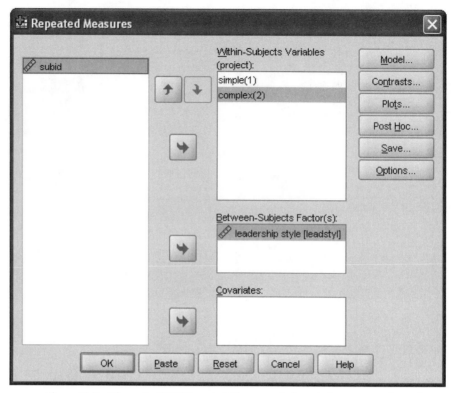

Figure 13.7   The main **GLM Repeated Measures** dialog window now configured.

13.6   SPSS OUTPUT OF THE OMNIBUS ANALYSIS

The results of Box's test, shown in Figure 13.10, is not statistically significant and indicates that the assumption of the equality of the covariance matrices cannot be rejected. Mauchly's test of sphericity, the results of which are shown in Figure 13.11, cannot be performed with only two levels of the within-subjects variable. Levene's test of error variance equality, shown in Figure 13.12, was not statistically significant; we therefore cannot reject that assumption.

The within-subjects summary table is presented in Figure 13.13 and the between-subjects summary table is presented in Figure 13.14. From these two tables we can see that all three effects of interest are statistically significant. Because the two-way interaction is statistically significant, we now proceed to our simple effects analyses.

13.7   PERFORMING THE POST-ANOVA ANALYSIS IN SPSS

13.7.1   POST HOC TUKEY TEST

Configure the analysis as you did for the omnibus ANOVA except that now you can request a Tukey test for **leadstyl** in the **Post Hoc** window. This window is shown in Figure 13.15. At the end of the process click the **Paste** pushbutton instead of clicking **OK**. The underlying syntax will then be placed in a syntax window; locate that window on your desktop and make it active by clicking it. This window is shown in Figure 13.16.

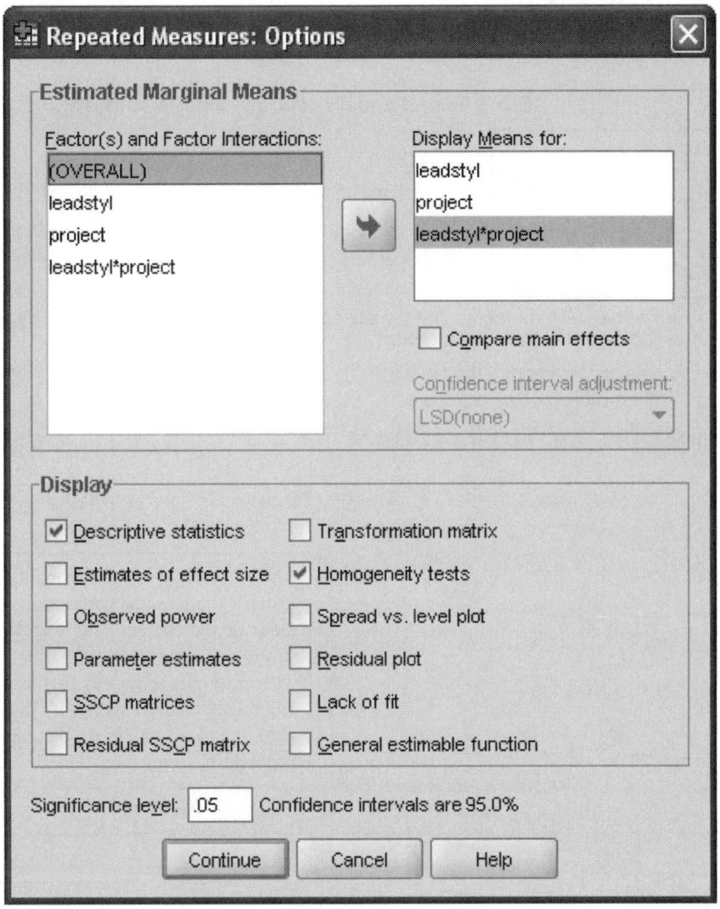

Figure 13.8 The **Options** dialog window.

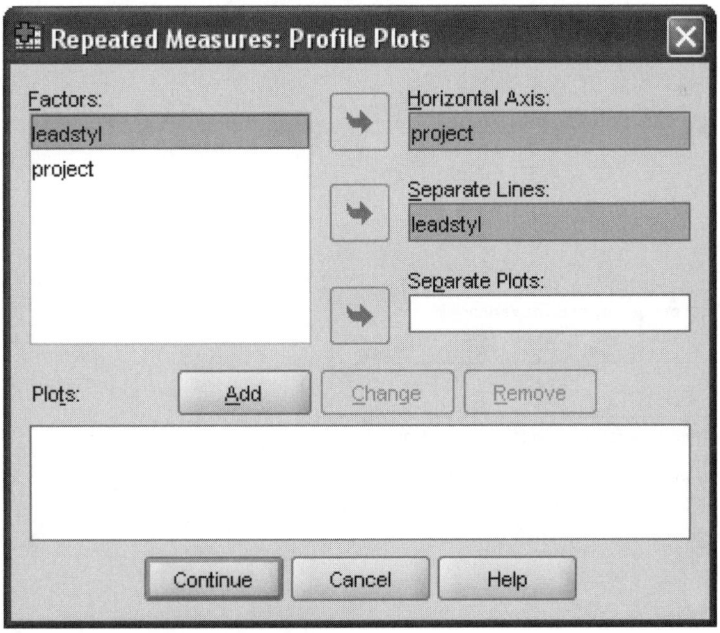

Figure 13.9 The **Plots** dialog window.

Box's Test of Equality of Covariance Matrices[a]

Box's M	5.613
F	.709
df1	6
df2	3588.923
Sig.	.643

Tests the null hypothesis that the observed covariance matrices of the dependent variables are equal across groups.

a. Design: Intercept + leadstyl
Within Subjects Design: project

Figure 13.10 Box's M test.

Mauchly's Test assesses equality of group variances and correlations. Because there is only one correlation, it cannot be compared to any other. Therefore, the sphericity assumption cannot be tested. This is why there is zero df and why there is no computed probability (Sig.).

Mauchly's Test of Sphericity[b]

Measure: MEASURE_1

					Epsilon[a]		
Within Subjects Effect	Mauchly's W	Approx. Chi-Square	df	Sig.	Greenhouse e-Geisser	Huynh-Feldt	Lower-bound
project	1.000	.000	0	.	1.000	1.000	1.000

Tests the null hypothesis that the error covariance matrix of the orthonormalized transformed dependent variables is proportional to an identity matrix.

a. May be used to adjust the degrees of freedom for the averaged tests of significance. Corrected tests are displayed in the Tests of Within-Subjects Effects table.

b. Design: Intercept+leadstyl
Within Subjects Design: project

Figure 13.11 Mauchly's test.

The equal variances assumption is not violated, probabilities (Sig.) are > .05.

Levene's Test of Equality of Error Variances[a]

	F	df1	df2	Sig.
simple projects	.063	2	12	.939
complex project	1.948	2	12	.185

Tests the null hypothesis that the error variance of the dependent variable is equal across groups.

a. Design: Intercept+leadstyl
Within Subjects Design: project

Figure 13.12 Levene's test.

Tests of Within-Subjects Effects

Measure: MEASURE_1

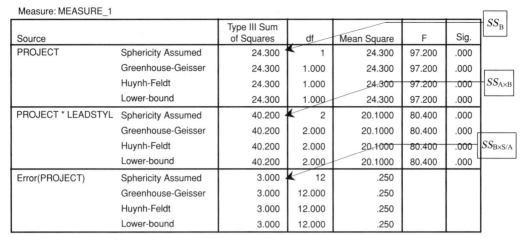

Source		Type III Sum of Squares	df	Mean Square	F	Sig.	
PROJECT	Sphericity Assumed	24.300	1	24.300	97.200	.000	SS_B
	Greenhouse-Geisser	24.300	1.000	24.300	97.200	.000	
	Huynh-Feldt	24.300	1.000	24.300	97.200	.000	$SS_{A\times B}$
	Lower-bound	24.300	1.000	24.300	97.200	.000	
PROJECT * LEADSTYL	Sphericity Assumed	40.200	2	20.1000	80.400	.000	
	Greenhouse-Geisser	40.200	2.000	20.1000	80.400	.000	
	Huynh-Feldt	40.200	2.000	20.1000	80.400	.000	$SS_{B\times S/A}$
	Lower-bound	40.200	2.000	20.1000	80.400	.000	
Error(PROJECT)	Sphericity Assumed	3.000	12	.250			
	Greenhouse-Geisser	3.000	12.000	.250			
	Huynh-Feldt	3.000	12.000	.250			
	Lower-bound	3.000	12.000	.250			

Figure 13.13   The within-subjects summary table.

Tests of Between-Subjects Effects

Measure: MEASURE_1

Transformed Variable: Average

Source	Type III Sum of Squares	df	Mean Square	F	Sig.	
Intercept	589.633	1	589.633	471.707	.000	SS_A
LEADSTYL	70.867	2	35.433	28.347	.000	
Error	15.000	12	1.250			$SS_{S/A}$

Figure 13.14   The between-subjects summary table.

Figure 13.15   Requesting a post hoc test for **leadstyl**.

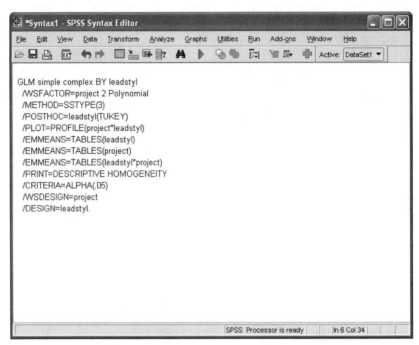

Figure 13.16   The syntax resulting from our point-and-click process.

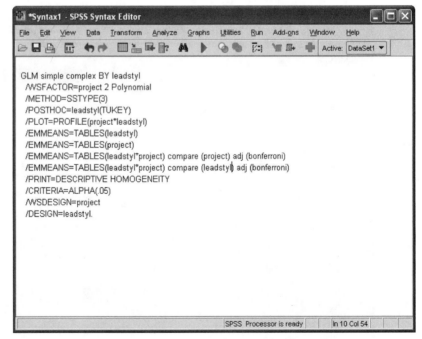

Figure 13.17   The syntax after we have added our simple effects syntax.

A. Estimates

Measure: MEASURE_1

leadership style	project	Mean	Std. Error	95% Confidence Interval	
				Lower Bound	Upper Bound
authoritarian	1	6.000	.416	5.093	6.907
	2	1.000	.356	.225	1.775
democratic	1	6.400	.416	5.493	7.307
	2	6.800	.356	6.025	7.575
lassez faire	1	3.600	.416	2.693	4.507
	2	2.800	.356	2.025	3.575

This Table presents significant differences between the two project levels for each level of leadership style.

B. Pairwise Comparisons

Measure: MEASURE_1

leadership style	(I)project	(J)project	Mean Difference (I-J)	Std. Error	Sig.[a]	95% Confidence Interval Difference[a]	
						Lower Bound	Upper Bound
authoritarian	1	2	5.000 *	.316	.000	4.311	5.689
	2	1	-5.000 *	.316	.000	-5.689	-4.311
democratic	1	2	-.400	.316	.230	-1.089	.289
	2	1	.400	.316	.230	-.289	1.089
lassez faire	1	2	.800 *	.316	.026	.111	1.489
	2	1	-.800 *	.316	.026	-1.489	-.111

Based on estimated marginal means

*. The mean difference is significant at the .05 level.

a. Adjustment for multiple comparisons: Bonferroni.

Figure 13.18 The multiple comparisons for project type.

13.7.2 SIMPLE EFFECTS TESTS

We have added our simple effects syntax as shown in Figure 13.17 by duplicating the /EMMEANS = TABLES subcommand containing the two-way interaction and adding our **compare** syntax to each. Select **Run → All** from the main menu.

13.8 OUTPUT FOR THE POST-ANOVA ANALYSIS IN SPSS

The simple effects output is best viewed when looking at the plot of the results in Figure 13.3. In this way, you can relate the statistical outcomes to a visual depiction of the means. Figure 13.18 presents the results of the first set of simple effects comparing levels of **project**. We are thus comparing **simple** and **complex** for each style of leadership; with respect to Figure 13.3, we are determining if each line significantly slants with respect to the *x* axis.

Our results indicate that there is a statistically significant difference for the authoritarian and laissez faire styles but not for the democratic

A. Estimates

Measure: MEASURE_1

leadership style	project	Mean	Std. Error	95% Confidence Interval	
				Lower Bound	Upper Bound
authoritarian	1	6.000	.416	5.093	6.907
	2	1.000	.356	.225	1.775
democratic	1	6.400	.416	5.493	7.307
	2	6.800	.356	6.025	7.575
lassez faire	1	3.600	.416	2.693	4.507
	2	2.800	.356	2.025	3.575

> Pairwise comparisons with project as the first order.

B. Pairwise Comparisons

Measure: MEASURE_1

project	(I)leadership style	(J)leadership style	Mean Difference (I-J)	Std. Error	Sig.[a]	95% Confidence Interval Difference[a]	
						Lower Bound	Upper Bound
1	authoritarian	democratic	-.400	.589	1.000	-2.037	1.237
		lassez faire	2.400*	.589	.005	.763	4.037
	democratic	authoritarian	.400	.589	1.000	-1.237	2.037
		lassez faire	2.800*	.589	.001	1.163	4.437
	lassez faire	authoritarian	-2.400*	.589	.005	-4.037	-.763
		democratic	-2.800*	.589	.001	-4.437	-1.163
2	authoritarian	democratic	-5.800*	.503	.000	-7.199	-4.401
		lassez faire	-1.800*	.503	.011	-3.199	-.401
	democratic	authoritarian	5.800*	.503	.000	4.401	7.199
		lassez faire	4.000*	.503	.000	2.601	5.399
	lassez faire	authoritarian	1.800*	.503	.011	.401	3.199
		democratic	-4.000*	.503	.000	-5.399	-2.601

Based on estimated marginal means

*. The mean difference is significant at the .05 level.

a. Adjustment for multiple comparisons: Bonferroni.

Figure 13.19   The multiple comparisons for leadership style.

style. Interfacing with our graph, we can say that both laissez faire and authoritarian styles result in poorer performance when teams are engaged in complex projects compared to simple ones. We can also say that it makes no difference what kind of projects are tackled by teams with a democratic leadership style.

Figure 13.19 presents the results of the second set of simple effects comparing levels of **leadstyl**. We are thus comparing the three styles of leadership for **simple** projects and we are comparing the three styles of leadership for **complex** projects; with respect to Figure 13.3, we are determining if the three data points at **simple** differ significantly and we are determining if the three data points at **complex** differ significantly.

Our results indicate that under **project** level 1 (**simple** projects), authoritarian and democratic styles are equally high but both result in better performance ratings than the laissez faire group. Under **project** 2 (complex projects), all groups differ significantly from all others.

Although we performed the Tukey post hoc test on **leadstyl** (shown in Figure 13.20), it should now be clear that the interaction effects are

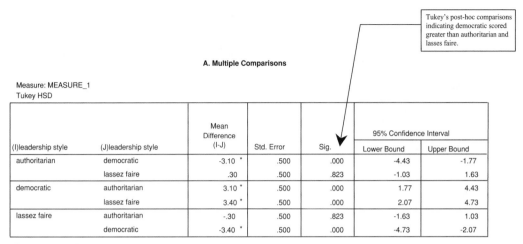

A. Multiple Comparisons

Measure: MEASURE_1
Tukey HSD

(I)leadership style	(J)leadership style	Mean Difference (I-J)	Std. Error	Sig.	95% Confidence Interval	
					Lower Bound	Upper Bound
authoritarian	democratic	-3.10 *	.500	.000	-4.43	-1.77
	lassez faire	.30	.500	.823	-1.03	1.63
democratic	authoritarian	3.10 *	.500	.000	1.77	4.43
	lassez faire	3.40 *	.500	.000	2.07	4.73
lassez faire	authoritarian	-.30	.500	.823	-1.63	1.03
	democratic	-3.40 *	.500	.000	-4.73	-2.07

Based on observed means.
   *. The mean difference is significant at the .05 level.

Tukey's post-hoc comparisons indicating democratic scored greater than authoritarian and lasses faire.

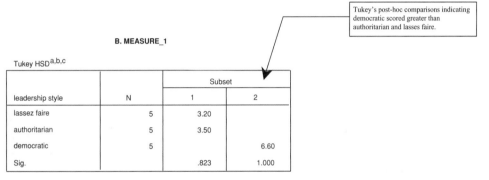

B. MEASURE_1

Tukey HSD[a,b,c]

leadership style	N	Subset	
		1	2
lassez faire	5	3.20	
authoritarian	5	3.50	
democratic	5		6.60
Sig.		.823	1.000

Means for groups in homogeneous subsets are displayed

Based on Type III Sum of Squares

The error term is Mean Square(Error) = .625.

   a. Uses Harmonic Mean Sample Size = 5.000.

   b. The group sizes are unequal. The harmonic

      of the group sizes is used. Type I error I

      not guaranteed.

   c. Alpha = .05.

Tukey's post-hoc comparisons indicating democratic scored greater than authoritarian and lasses faire.

Figure 13.20   The Tukey results for leadership style.

sufficiently compelling that to address style of leadership without considering the nature of the projects on which the teams are working would oversimplify the results. Thus, we will not address this main effect at all in our narrative of the results.

13.9   PERFORMING THE OMNIBUS ANALYSIS IN SAS

13.9.1   STRUCTURING THE DATA FILE

The data file in an Excel spreadsheet ready to import into a *SAS Enterprise Guide* project is displayed in Figure 13.21. It is structured in univariate or stacked mode as explained in Section 10.15.2, and is similar to those we

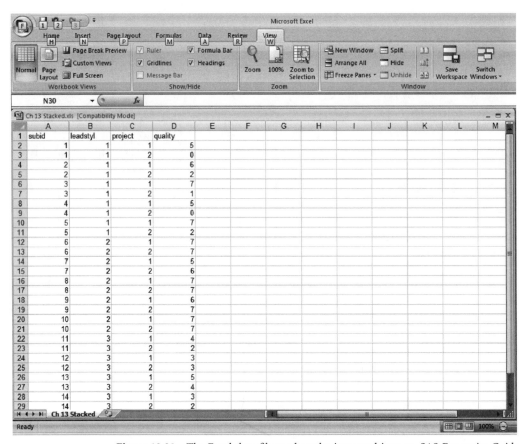

Figure 13.21 The Excel data file ready to be imported into an *SAS Enterprise Guide* project.

have seen for the examples in Chapter 11 and 12. The **leadstyl** and **project** independent variables show the codes for the levels corresponding to the **quality** dependent variable for each case. Note that by examining the data file we can determine that **leadstyl** is a between-subjects variable because each subject is associated with only one level of that variable. By the same reasoning, we can determine that **project** is a within-subjects variable because each subject has values on the dependent variable for both levels of **project**. To illustrate how to read the data file, we can see that the team identified as **subid 1** is under leadership style 1 (authoritarian). That team scored an average of five on quality for simple projects (coded as 1) and an average of zero on quality for complex projects (coded as 2).

13.9.2   STRUCTURING THE DATA ANALYSIS

Import the data file from Excel into a new SAS project. Then from the main menu select **Analyze → ANOVA → Mixed Models**. The window opens on the **Task Roles** tab. From the **Variables to assign** panel, select **quality** and drag it to the icon for **Dependent variable**. Then one at a time

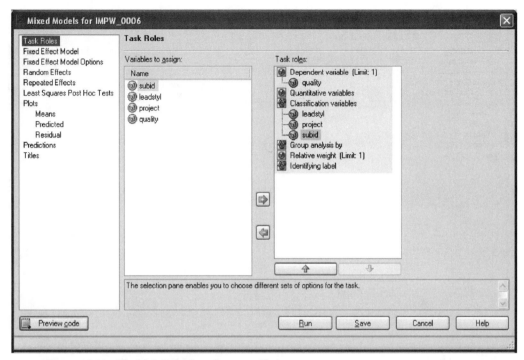

Figure 13.22 The **Task Roles** have been specified.

select **leadstyl, project**, and **subid** and drag them over to the area under **Classification variables**. The result of this is shown in Figure 13.22.

In the navigation panel, select **Fixed Effect Model** as shown in Figure 13.23. In the **Class and quantitative variables** panel, select **leadstyl** and then select the **Main** pushbutton; **leadstyl** will automatically appear in the **Effects** window. Repeat this procedure for **project**. Then while holding the Control key down, select both **leadstyl** and **project**. With both variables highlighted, click the **Factorial** pushbutton; **leadstyl*project** will automatically appear in the **Effects** window.

Select **Fixed Effect Model Options** in the navigation panel as shown in Figure 13.24. Check **Type 3** under **Hypothesis test type**. Then select **Residual maximum likelihood** under **Estimation method**, and **Between and within subject portions** under **Degrees of freedom method**.

Select **Random Effects** in the navigation panel as shown in Figure 13.25. Select **Random effects** under the **Effects to use** menu; a box will appear at the far right of that menu. Click on that box and the **Effects Builder – Random effects** window appears. Select **subid** and click the **Main** pushbutton; **subid** will automatically appear in the **Random effects** panel. Click the **OK** pushbutton.

Now select **Subject identifier** under the **Model subjects** menu; a box will appear at the far right of that menu. Click on that box and the **Effects Builder – Subject identifier** window appears. Select **subid** and click the **Main** pushbutton; **subid** will automatically appear in the **Subject identifier** panel. Select the **OK** pushbutton. This is shown in Figure 13.26.

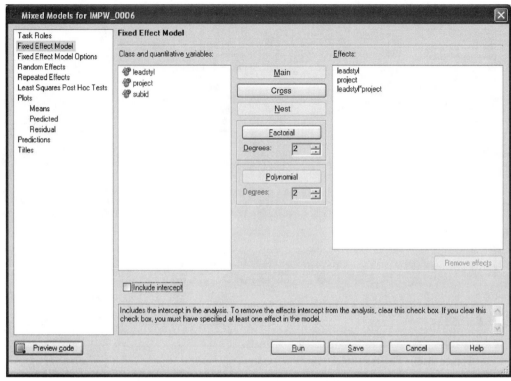

Figure 13.23   The **Fixed Effect Model** has now been specified.

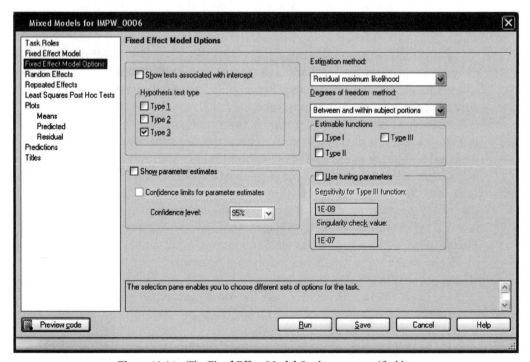

Figure 13.24   The **Fixed Effect Model Options** are specified here.

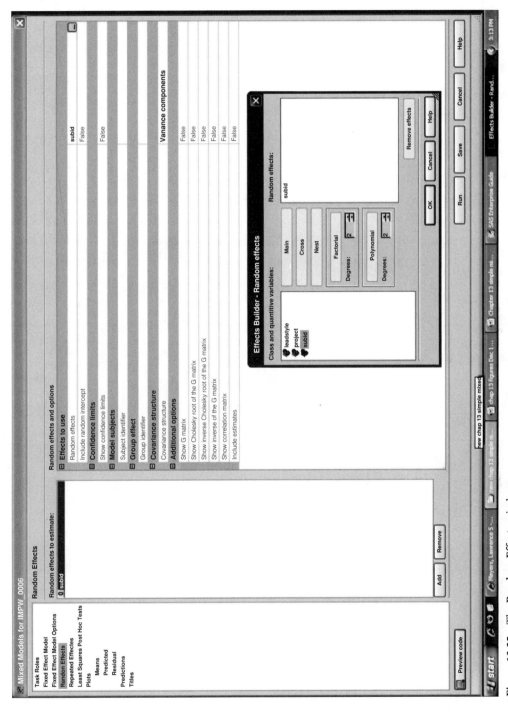

Figure 13.25   The Random Effects window.

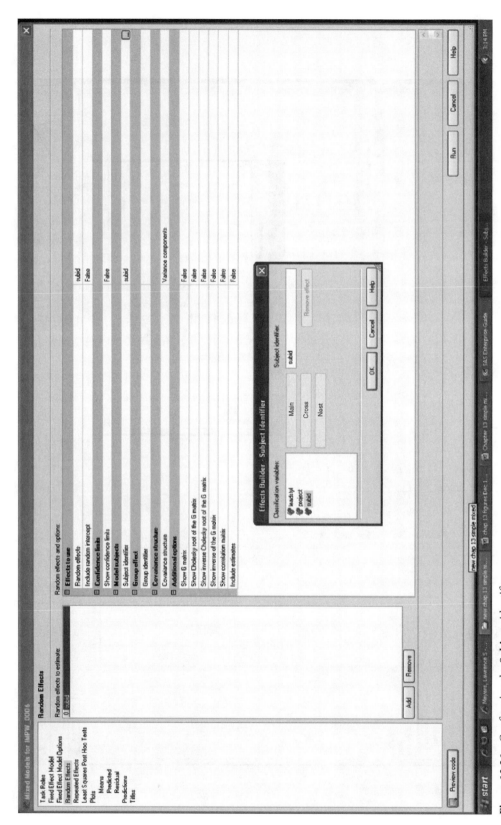

Figure 13.26 Configuring the **Subjects** identifier.

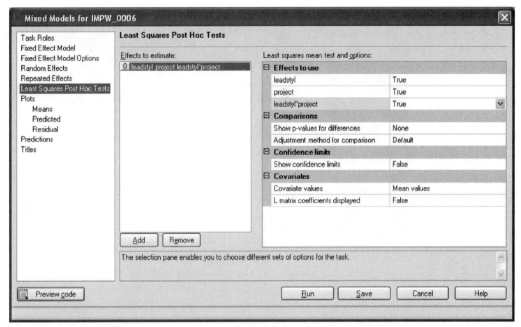

Figure 13.27 Specifying the effects for which we wish to obtain least square means.

Select **Least Squares Post Hoc Tests** in the navigation panel. Click the **Add** pushbutton at the bottom of the **Effects to estimate** pane. The screen shown in Figure 13.27 appears. Highlight each of the three effects in turn under the **Effects to use** menu and select **True** for each; this command will cause the means for each of the effects to be output. Click the **Run** pushbutton.

13.10 SAS OUTPUT OF THE OMNIBUS ANALYSIS

The least squares means are shown in Figure 13.28. The results of the omnibus analysis are shown in Figure 13.29. These *F* ratios are the same as those produced by SPSS. As can be seen, all of the effects are statistically significant.

13.11 PERFORMING THE SIMPLE EFFECTS ANALYSIS IN SAS

To perform the simple effects analysis in *SAS Enterprise Guide*, configure the analysis as shown above. Then in the **Least Squares Post Hoc Tests** screen shown in Figure 13.30, in the **Effects to use** frame, set the two-way interaction to **True** and all others to **False**. Then in the **Comparisons** frame, set **Show p-values for differences** to **All pairwise differences** and set **Adjustment method for comparison** to **Bonferroni**. Click **Run** to perform the analysis.

Cell means for all
eleven conditions.

Least Squares Means							
Effect	**leadstyl**	**project**	**Estimate**	**Standard Error**	**DF**	**t Value**	**Pr>|t|**
leadstyl	1		3.5000	0.3536	12	9.90	<.0001
leadstyl	2		6.6000	0.3536	12	18.67	<.0001
leadstyl	3		3.2000	0.3536	12	9.05	<.0001
project		1	5.3333	0.2236	12	23.85	<.0001
project		2	3.5333	0.2236	12	15.80	<.0001
leadstyl*project	1	1	6.0000	0.3873	12	15.49	<.0001
leadstyl*project	1	2	1.0000	0.3873	12	2.58	0.0240
leadstyl*project	2	1	6.4000	0.3873	12	16.52	<.0001
leadstyl*project	2	2	6.8000	0.3873	12	17.56	<.0001
leadstyl*project	3	1	3.6000	0.3873	12	9.30	<.0001
leadstyl*project	3	2	2.8000	0.3873	12	7.23	<.0001

Figure 13.28 The least squares means for all of the effects.

These *p*-values suggest statistical
significances for both main
effects and interaction effect.

Type 3 Tests of Fixed Effects				
Effect	**Num DF**	**Den DF**	**F Value**	**Pr>F**
leadstyl	2	12	28.35	<.0001
project	1	12	97.20	<.0001
leadstyl*project	2	12	80.40	<.0001

Figure 13.29 The results of the ANOVA.

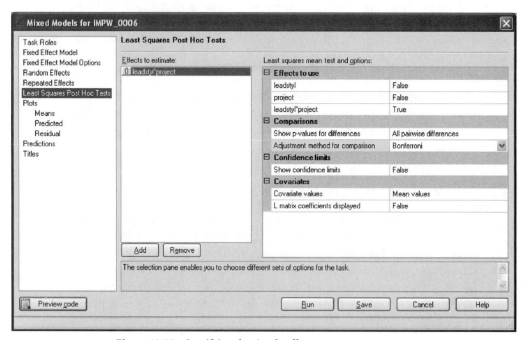

Figure 13.30 Specifying the simple effects tests.

> This Table presents both the unadjusted p-value and the Bonferroni adjustment.

					Differences of Least Squares Means						
Effect	leadstyl	project	leadstyl	project	Estimate	Standard Error	DF	t Value	Pr >ltl	Adjustment	Adj P
leadstyl*project	1	1	1	2	5.0000	0.3162	12	15.81	<.0001	Bonferroni	<.0001
leadstyl*project	1	1	2	1	−0.4000	0.5477	12	−0.73	0.4792	Bonferroni	1.0000
leadstyl*project	1	1	2	2	−0.8000	0.5477	12	−1.46	0.1698	Bonferroni	1.0000
leadstyl*project	1	1	3	1	2.4000	0.5477	12	4.38	0.0009	Bonferroni	0.0134
leadstyl*project	1	1	3	2	3.2000	0.5477	12	5.84	<.0001	Bonferroni	0.0012
leadstyl*project	1	2	2	1	−5.4000	0.5477	12	−9.86	<.0001	Bonferroni	<.0001
leadstyl*project	1	2	2	2	−5.8000	0.5477	12	−10.59	<.0001	Bonferroni	<.0001
leadstyl*project	1	2	3	1	−2.6000	0.5477	12	−4.75	0.0005	Bonferroni	0.0071
leadstyl*project	1	2	3	2	−1.8000	0.5477	12	−3.29	0.0065	Bonferroni	0.0975
leadstyl*project	2	1	2	2	−0.4000	0.3162	12	−1.26	0.2299	Bonferroni	1.0000
leadstyl*project	2	1	3	1	2.8000	0.5477	12	5.11	0.0003	Bonferroni	0.0038
leadstyl*project	2	1	3	2	3.6000	0.5477	12	6.57	<.0001	Bonferroni	0.0004
leadstyl*project	2	2	3	2	3.2000	0.5477	12	5.84	<.0001	Bonferroni	0.0012
leadstyl*project	2	2	3	2	4.0000	0.5477	12	7.30	<.0001	Bonferroni	0.0001
leadstyl*project	3	1	3	2	0.8000	0.3162	12	2.53	0.0264	Bonferroni	0.3964

Figure 13.31 The output from the simple effects tests.

13.12 SAS OUTPUT FROM THE SIMPLE EFFECTS ANALYSIS

The results of the simple effects tests are shown in Figure 13.31 and are read in a manner analogous to that described for the two-way interaction discussed in Section 11.14. To illustrate this, consider for example the first comparison in the table. The first two coded columns represent one condition and the last two coded columns represent the other condition. We are thus comparing the mean for **leadstyl 1, project 1** (authoritarian leadership style for simple projects with a mean of 6.0) with **leadstyl 1, project 2** (authoritarian leadership style for complex projects with a mean of 1.0). The t value for that comparison is 15.81, which is statistically significant ($p < .0001$). We may therefore conclude that the teams under authoritarian leadership produced significantly higher quality simple projects than complex projects.

13.13 PERFORMING THE POST HOC ANALYSIS IN SAS

To perform the simple effects analysis in *SAS Enterprise Guide*, configure the analysis as shown in Section 13.11. Then in the **Least Squares Post Hoc Tests** screen shown in Figure 13.32, in the **Effects to use** frame, set **leadstyl** to **True** and all others to **False**. Then in the **Comparisons** frame, set **Show p-values for differences** to **All pairwise differences** and set **Adjustment method for comparison** to **Bonferroni**. Click **Run** to perform the analysis.

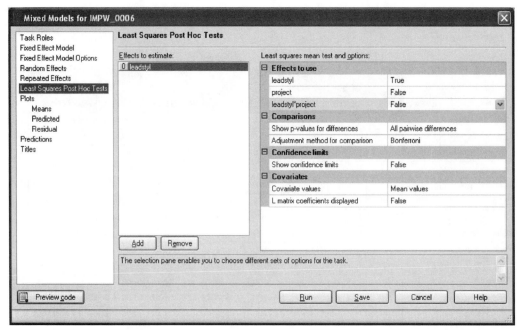

Figure 13.32 Specifying the post hoc tests.

13.14 SAS OUTPUT FROM THE POST HOC ANALYSIS

The pairwise mean comparisons together with the least squares means are shown in Figure 13.33. As we indicated in discussing the SPSS output, this main effect is completely subsumed by the interaction effect and will therefore not be described.

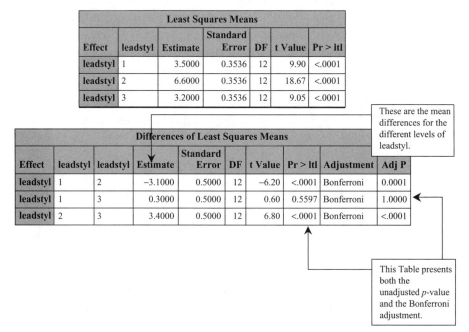

Figure 13.33 A portion of the output from the post hoc tests.

13.15 COMMUNICATING THE RESULTS

Fifteen teams worked on both simple and complex projects over the course of a month. Five teams each were headed by a leader using one of the three following leadership styles: authoritarian, democratic, or laissez faire. Performance of the teams on the projects was judged by management using a 0 (*not satisfactory*) to 7 (*high quality*) rating scale. All effects met the equality of error variances as tested by Levene's procedure and of equality of covariance matrices as tested by Box's M test.

The main effects for leadership style, $F(2, 12) = 28.35$, $p < .05$, and project type, $F(1, 12) = 97.20$, $p < .05$, were both statistically significant but can be best understood in the context of the significant Leadership Style $\times$ Project Type interaction, $F(2, 12) = 80.40$, $p < .05$, which accounted for 70 percent of the within-subjects variance.

The graph of the interaction is presented in Figure 13.3. Simple effects analysis indicated that the democratic leadership style was very effective with both types of projects. Authoritarian leadership was equally effective as democratic leadership with simple projects but was the least effective with complex projects. Laissez faire leadership was of intermediate effectiveness for both types of projects, although somewhat less so with complex projects than with simple projects.

CHAPTER 13 EXERCISES

13.1. Assume that we are interested in the effects of music on recall. Participants ($n = 7$) are presented with lists of twenty-five words as a within-subjects factor (five-minute study) under the following type of music conditions (b_1 = heavy metal, b_2 = jazz, and b_3 = soft rock). The between-subjects factor was participant gender (a_1 = male and a_2 = female). Type of music was counterbalanced across all participants. The dependent measure was the total number of words recalled. The data are as follows:

	a_1				a_2		
Subject	b_1	b_2	b_3	Subject	b_1	b_2	b_3
s_1	10	15	19	s_8	11	16	21
s_2	12	16	19	s_9	12	16	22
s_3	11	17	20	s_{10}	13	15	22
s_4	10	15	20	s_{11}	13	15	23
s_5	13	18	19	s_{12}	13	14	24
s_6	10	15	18	s_{13}	12	15	21
s_7	9	12	17	s_{14}	11	17	20

a. Conduct an ANOVA on these data by hand and with SPSS or SAS.
b. If the interaction is statistically significant, conduct simple effects and multiple comparisons analyses by hand and with SPSS or SAS.

13.2. Assume that we are interested in the effects of client–therapist ethnic/racial match on client service satisfaction. Factor A, the between-subjects factor, is ethnic/racial match ($a_1 =$ match and $a_2 =$ no match). Factor B, the within-subjects factor, consists of the first four treatment sessions. The dependent variable, service satisfaction, was a single question that $n = 5$ participants rated each session on, using a $1 = poor$ to $10 = excellent$ rating of client perceived satisfaction with service. The data are as follows:

		a_1					a_2		
Subject	b_1	b_2	b_3	b_4	Subject	b_1	b_2	b_3	b_4
s_1	5	6	7	8	s_6	2	2	4	5
s_2	5	8	9	10	s_7	3	6	7	8
s_3	6	7	8	9	s_8	4	4	4	5
s_4	7	7	8	8	s_9	5	5	6	6
s_5	9	9	9	10	s_{10}	5	6	7	8

a. Conduct an ANOVA on these data by hand and with SPSS or SAS.
b. If the interaction is statistically significant, conduct simple effects analyses and multiple comparisons tests by hand and with SPSS or SAS.

13.3. Consider the following hypothetical 3×3 factorial data set. Factor A is the between-subjects factor and Factor B is the within-subjects factor.

		a_1				a_2				a_3	
Subject	b_1	b_2	b_3	Subject	b_1	b_2	b_3	Subject	b_1	b_2	b_3
s_1	5	10	15	s_6	2	0	3	s_{11}	7	8	9
s_2	6	8	12	s_7	5	1	2	s_{12}	9	10	11
s_3	7	14	18	s_8	4	6	3	s_{13}	14	15	16
s_4	6	6	12	s_9	2	6	7	s_{14}	17	18	19
s_5	9	10	11	s_{10}	4	4	2	s_{15}	15	20	20

a. Conduct an ANOVA on these data by hand and with SPSS or SAS.
b. If the interaction is statistically significant, conduct simple effects analyses and multiple comparisons tests by hand and with SPSS or SAS.

Complex Mixed Design: Two Between-Subjects Factors and One Within-Subjects Factor

14.1 COMBINING BETWEEN- AND WITHIN-SUBJECTS FACTORS

As discussed in Chapter 13, a mixed design is one that contains at least one between-subjects independent variable and at least one within-subjects independent variable. Simple mixed designs have only two independent variables and so, by definition, must have one of each type of variable. Complex mixed designs contain at least three independent variables. The three-way complex mixed designs presented in this chapter and in Chapter 15 must (by definition) have two of one type and one of the other type of factor. In this chapter, we will focus on the design with two between-subjects factors and one within-subjects factor; Chapter 15 will present the design with one between-subjects factor and two within-subjects factors.

14.2 A NUMERICAL EXAMPLE OF A COMPLEX MIXED DESIGN

College students who signed up for a research study read vignettes in which their romantic partner was described as being attracted to another person; this attraction was depicted as being either emotional attraction with no physical component or physical attraction with no emotional component. Because this type-of-attraction variable was intended to be a within-subjects variable, students read both vignettes. After reading each vignette, the students completed a short inventory evaluating their feelings of jealousy; the response to this inventory served as the dependent variable. For the purposes of this hypothetical example, assume that no effects concerning the order of reading these vignettes was obtained; thus, we will present the results without considering the vignette ordering factor.

This within-subjects type of attraction variable was factorially combined with the following two between-subjects factors: (a) the students were either teenagers (either eighteen or nineteen years of age) or were in their middle twenties (either twenty-five or twenty-six years of age); (b) the students were either female or male.

Data from this hypothetical study of jealousy serving as our example are shown in Figure 14.1. Students with identification code numbers

Age Group	Gender	Subid	Type of Attraction	
			Emotional Attraction	Physical Attraction
Teen	Female	1	3	5
		2	2	4
		3	2	3
		4	4	5
		5	3	4
	Male	6	0	2
		7	1	1
		8	1	2
		9	0	2
		10	1	1
Mid 20s	Female	11	6	7
		12	5	4
		13	5	6
		14	6	5
		15	7	6
	Male	16	2	5
		17	1	4
		18	3	6
		19	1	4
		20	2	5

Figure 14.1   Data for numerical example.

of 1–5 are female teenagers, those with identification code numbers of 6–10 are male teenagers, those with identification code numbers of 11–15 are females in their middle twenties, and those with identification code numbers of 16–20 are males in their middle twenties. Their respective jealously scores are shown under emotional attraction and physical attraction; ratings were made on a seven-point response scale with higher values indicating greater amounts of expressed jealousy.

Figure 14.2 shows the cell means for our numerical example. These cell means tie directly to the highest order interaction in the design, in this case the three-way interaction. If this three-way interaction is statistically significant, it is these means that we will want to plot and that will serve as the basis of the simple effects testing; if the three-way interaction is not statistically significant, then we may need to collapse some of these cells to deal with one of the two-way interactions or the main effects.

Age Group	Gender	Type of Attraction	
		Emotional Attraction	Physical Attraction
Teen	Female	2.80	4.20
	Male	0.60	1.60
Mid 20s	Female	5.80	5.60
	Male	1.80	4.80

Figure 14.2   Cell means for the data in our numerical example.

Table 14.1. Summary table of the ANOVA

Source	SS	df	MS	F	η^2
Between subjects	122.60				
Age (A)	48.40	1	48.40	46.65^c	$.395^a$
Gender (B)	57.60	1	57.60	55.52^c	$.470^b$
Age × Gender ($A \times B$)	0.00	1	0.00	0.00	
Error$_{Between}$ (S/AB)	16.60	16	1.04		
Within subjects	35.00				
Attract (C)	16.90	1	16.90	54.08^c	$.483^b$
Age × Attract ($A \times C$)	0.10	1	0.10	0.32	
Gender × Attract ($B \times C$)	4.90	1	4.90	15.68^c	$.140^b$
Age × Gender × Attract ($A \times B \times C$)	8.10	1	8.10	25.92^c	$.231^b$
Error$_{Within}$ ($C \times S/AB$)	5.00	16	0.31		

[a] η^2 is computed based on the total between-subjects portion of the variance.
[b] η^2 is computed based on the total within-subjects portion of the variance.
[c] $p < .05$.

14.3   EFFECTS OF INTEREST

The summary table showing the results of the analysis is presented in Table 14.1. The effects of interest are the same as in any three-way design: three main effects, three two-way interactions, and one three-way interaction. As can be seen, all three main effects, the Gender × Attract interaction, and the three-way interaction are all statistically significant.

The eta squared values are shown in the last column of the summary table. As we did in Chapter 13, we based these computations on the separate (sub)total variances of the between-subjects and within-subjects partitions. Thus, we have computed the eta squared values as follows:

- Main effect of A: $SS_A \div SS_A + SS_B + SS_{A \times B} + SS_{S/AB}$
- Main effect of B: $SS_B \div SS_A + SS_B + SS_{A \times B} + SS_{S/AB}$
- Main effect of C: $SS_C \div SS_C + SS_{A \times C} + SS_{B \times C} + SS_{A \times B \times C} + SS_{C \times S/AB}$
- Interaction $B \times C$: $SS_{B \times C} \div SS_C + SS_{A \times C} + SS_{B \times C} + SS_{A \times B \times C} + SS_{C \times S/AB}$
- Interaction $A \times B \times C$: $SS_{A \times B \times C} \div SS_C + SS_{A \times C} + SS_{B \times C} + SS_{A \times B \times C} + SS_{C \times S/AB}$.

The plot of the interaction is shown in Figure 14.3. We will narrate these results once we have performed the simple effects analyses.

14.4   COMPUTING THE OMNIBUS COMPLEX MIXED DESIGN BY HAND

As we have done with previous three-factor designs in this book, we will provide the reader with the necessary computational formulas should they choose to do these analyses by hand (see Table 14.2).

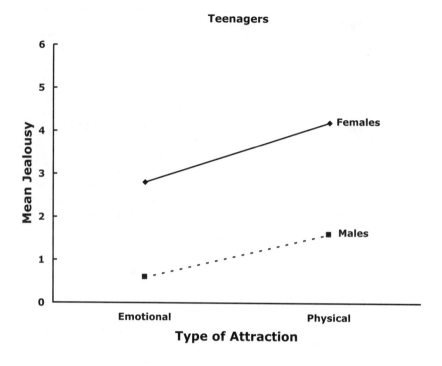

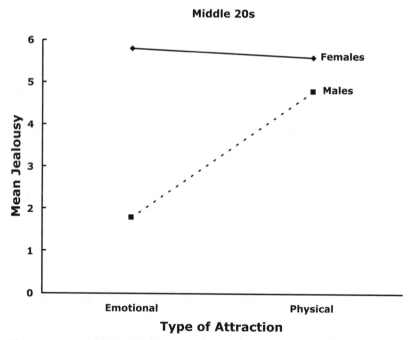

Figure 14.3   A plot of the three-way interaction.

Table 14.2. Computational formulas for the complex mixed design of two between-subjects factors and one within-subjects factor

Source	df	SS	MS	F
Between subjects				
A	$a-1$	$\dfrac{\sum A^2}{(b)(c)(n)} - \dfrac{T^2}{(a)(b)(c)(n)}$	$\dfrac{SS_A}{df_A}$	$\dfrac{MS_A}{MS_{S/AB}}$
B	$b-1$	$\dfrac{\sum B^2}{(a)(c)(n)} - \dfrac{T^2}{(a)(b)(c)(n)}$	$\dfrac{SS_B}{df_B}$	$\dfrac{MS_B}{MS_{S/AB}}$
$A \times B$	$(a-1)(b-1)$	$\dfrac{\sum AB^2}{(c)(n)} - \dfrac{\sum A^2}{(b)(c)(n)} - \dfrac{\sum B^2}{(a)(c)(n)} + \dfrac{T^2}{(a)(b)(c)(n)}$	$\dfrac{SS_{A\times B}}{df_{A\times B}}$	$\dfrac{MS_{A\times B}}{MS_{S/AB}}$
S/AB	$(a)(b)(n-1)$	$\dfrac{\sum S^2}{c} - \dfrac{\sum AB^2}{(c)(n)}$	$\dfrac{SS_{S/AB}}{df_{S/AB}}$	
Within subjects				
C	$c-1$	$\dfrac{\sum C^2}{(a)(b)(n)} - \dfrac{T^2}{(a)(b)(c)(n)}$	$\dfrac{SS_C}{df_C}$	$\dfrac{MS_C}{MS_{C\times S/AB}}$
$A \times C$	$(a-1)(c-1)$	$\dfrac{\sum AC^2}{(b)(n)} - \dfrac{\sum A^2}{(b)(c)(n)} - \dfrac{\sum C^2}{(a)(b)(n)} + \dfrac{T^2}{(a)(b)(c)(n)}$	$\dfrac{SS_{A\times C}}{df_{A\times C}}$	$\dfrac{MS_{A\times C}}{MS_{C\times S/AB}}$
$B \times C$	$(b-1)(c-1)$	$\dfrac{\sum BC^2}{(a)(n)} - \dfrac{\sum B^2}{(a)(c)(n)} - \dfrac{\sum C^2}{(a)(b)(n)} + \dfrac{T^2}{(a)(b)(c)(n)}$	$\dfrac{SS_{B\times C}}{df_{B\times C}}$	$\dfrac{MS_{B\times C}}{MS_{C\times S/AB}}$
$A \times B \times C$	$(a-1)(b-1)(c-1)$	$\dfrac{\sum ABC^2}{n} - \dfrac{\sum AC^2}{(b)(n)} - \dfrac{\sum BC^2}{(a)(n)} + \dfrac{\sum C^2}{(a)(b)(n)} - \dfrac{\sum AB^2}{(c)(n)} + \dfrac{\sum A^2}{(b)(c)(n)} + \dfrac{\sum B^2}{(a)(c)(n)} - \dfrac{T^2}{(a)(b)(c)(n)}$	$\dfrac{SS_{A\times B\times C}}{df_{A\times B\times C}}$	$\dfrac{MS_{A\times B\times C}}{MS_{C\times S/AB}}$
$C \times S/AB$	$(a)(b)(n-1)(c-1)$	$\sum Y^2 + \dfrac{\sum AB^2}{(c)(n)} - \dfrac{\sum ABC^2}{n} - \dfrac{\sum S^2}{c}$	$\dfrac{SS_{C\times S/AB}}{df_{C\times S/AB}}$	
Total	$(a)(b)(c)(n)-1$	$\sum Y^2 - \dfrac{T^2}{(a)(b)(c)(n)}$		

The present complex mixed design has the following sources of both between-subjects variance: main effects of factors A, B, and their interaction, $A \times B$, which are all evaluated with the following error term, S/AB. Conversely, the within-subjects component consists of the main effect of Factor C, and the interaction effects of $A \times C$, $B \times C$, and $A \times B \times C$, which are all evaluated with the $C \times S/AB$ error term. Further discussion of this topic can be found in Keppel (1991) and Keppel and Wickens (2004).

14.5 PERFORMING THE OMNIBUS ANALYSIS IN SPSS

14.5.1 STRUCTURING THE DATA FILE

The data file for our $2 \times 2 \times 2$ complex mixed-design example is shown in Figure 14.4. The first column, as always, is used for our participant

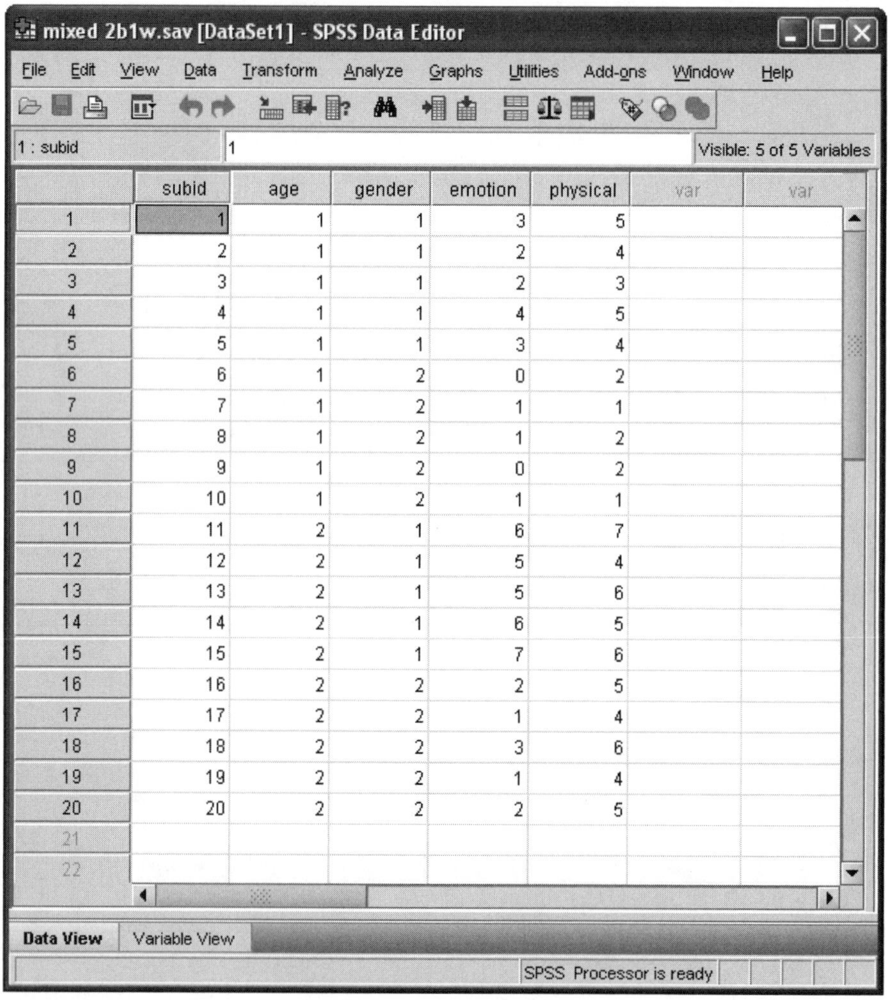

	subid	age	gender	emotion	physical	var	var
1	1	1	1	3	5		
2	2	1	1	2	4		
3	3	1	1	2	3		
4	4	1	1	4	5		
5	5	1	1	3	4		
6	6	1	2	0	2		
7	7	1	2	1	1		
8	8	1	2	1	2		
9	9	1	2	0	2		
10	10	1	2	1	1		
11	11	2	1	6	7		
12	12	2	1	5	4		
13	13	2	1	5	6		
14	14	2	1	6	5		
15	15	2	1	7	6		
16	16	2	2	2	5		
17	17	2	2	1	4		
18	18	2	2	3	6		
19	19	2	2	1	4		
20	20	2	2	2	5		

Figure 14.4 The data file for our numerical example.

identification number; we have named this variable **subid**. The next two columns represent our between-subjects variables. The first variable listed is **age**; teenagers are coded as 1 and those subjects in their middle twenties are coded as 2; **gender** is the next variable with females coded as 1 and males as 2. The last two variables represent the two levels of the within-subjects variable of attraction type; the first level of attraction type is named **emotion** and the second is named **physical**. Under each you find the values on the dependent variable (amount of reported jealousy).

14.5.2 STRUCTURING THE DATA ANALYSIS

From the main SPSS menu, select **Analyze ➜ General Linear Model ➜ Repeated Measures**. You selected **Repeated Measures** for the following reason. If there are two or more cases in the study and at least one of the independent variables is a within-subjects variable, then we must partition the total variance of the dependent variable into a between-subjects portion and a within-subjects portion. This partitioning is accomplished by the **Repeated Measures** module of the **General Linear Model** procedure in SPSS.

Selecting this path will open the dialog window shown in Figure 14.5. As you will recall from Chapters 11, 12, and 13, this is the initial window used by SPSS to have you name the within-subjects variable(s) in the study. In this dialog window, we have specified the variable name **attract** and indicated that it has two levels.

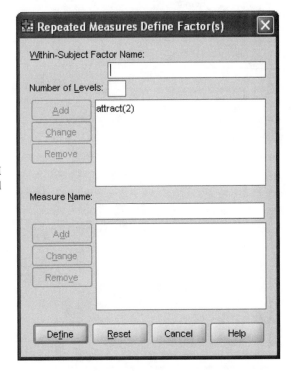

Figure 14.5 The **GLM Repeated Measures** initial **Define Factors** windows.

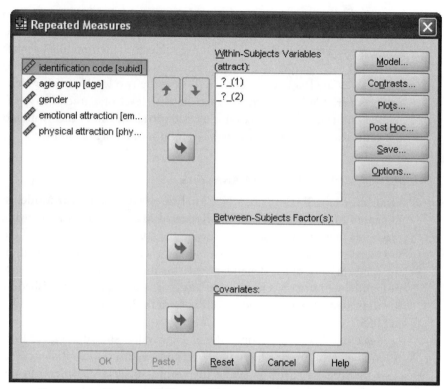

Figure 14.6   The **GLM Repeated Measures** main dialog window.

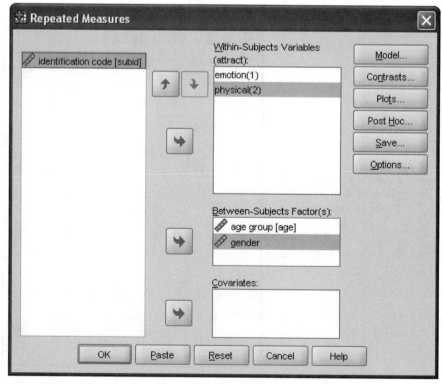

Figure 14.7   The **GLM Repeated Measures** main dialog window with the within-subjects factor defined and the between-subjects factors specified.

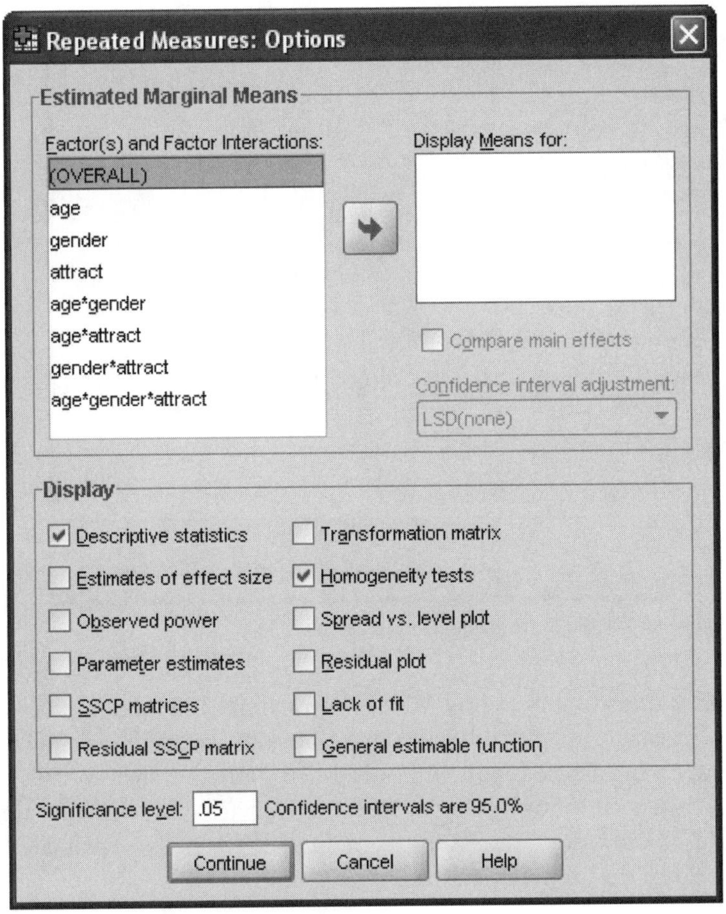

Figure 14.8 The **Options** dialog window.

Clicking **Define** in the **Define Factor(s)** window brings us to the **GLM Repeated Measures** main dialog window of shown in Figure 14.6. We place **emotion** and **physical** in slots 1 and 2, respectively, of the **Within-Subjects Variables** panel and **age** and **gender** in the panel for **Between-Subjects Factor(s)** as shown in Figure 14.7.

Select the **Options** pushbutton to reach the **Options** dialog window. With seven effects of interest here, we will wait until we perform the post-ANOVA tests to determine for which effects we want to generate tables of estimated marginal means. For this omnibus analysis, we simply check **Descriptive statistics** and **Homogeneity tests**. This is shown in Figure 14.8. Click **Continue** to return to the main dialog window.

In the **Plots** dialog window, click over **attract** to the **Horizontal Axis** because it is the within-subjects variable in that we generally want the within-subjects variable on the *x* axis. Place **age** in the panel for **Separate Plots;** this will result in one plot for the teenagers and a second plot for the students in their middle twenties. Click over **gender** to the **Separate Lines** panel; this will give us one line for females and another line for

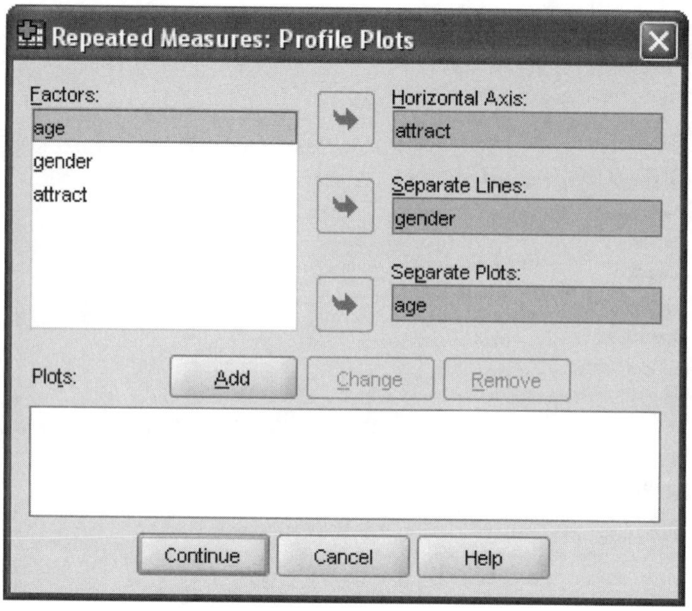

Figure 14.9   The **Plots** dialog window setup.

males in each of our plots. We have shown this in Figure 14.9. Click **Add** to register your plot with SPSS, the outcome of which is shown in the lower panel labeled **Plots** in Figure 14.10. Note that the general structure of the specification is **Horizontal Axis∗Separate Lines∗Separate Plots**. Click **Continue** to return to the main dialog window. Then click **OK** to perform the omnibus analysis.

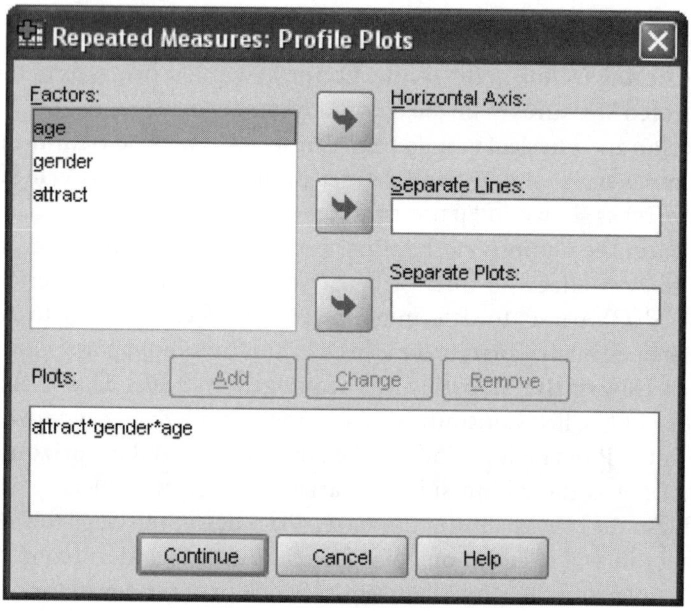

Figure 14.10   The **Plots** dialog window with the plot specified.

Box's Test of Equality of Covariance Matrices[a]

Box's M	7.961
F	1.005
df1	6
df2	3.589E3
Sig.	.420

Tests the null hypothesis that the observed covariance matrices of the dependent variables are equal across groups.

a. Design: Intercept + age + gender + age * gender
Within Subjects Design: attract

Figure 14.11 Box's test results.

14.6   SPSS OUTPUT OF THE OMNIBUS ANALYSIS

The results of Box's test, shown in Figure 14.11, is not statistically significant; this indicates that the assumption of the equality of the covariance matrices cannot be rejected. Mauchly's Test of Sphericity, the results of which are shown in Figure 14.12, cannot be performed with only two levels of the within-subjects variable. Levene's test of error variance equality was not statistically significant (see Figure 14.13); we therefore cannot reject the assumption of homogeneity of variance.

The within-subjects summary table is presented in Figure 14.14. All of the effects except for the Attract × Age interaction are statistically significant. The three-way interaction of Attract × Age × Gender is significant: We know that the combinations of all three independent variables affect performance on the dependent variable. Because this is highest order interaction, this is where we need to concentrate our simple effects analyses.

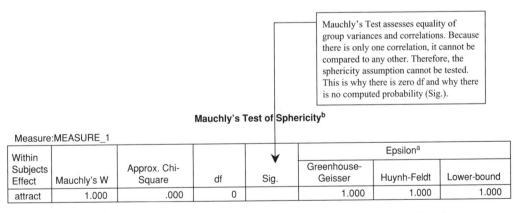

Mauchly's Test assesses equality of group variances and correlations. Because there is only one correlation, it cannot be compared to any other. Therefore, the sphericity assumption cannot be tested. This is why there is zero df and why there is no computed probability (Sig.).

Mauchly's Test of Sphericity[b]

Measure:MEASURE_1

Within Subjects Effect	Mauchly's W	Approx. Chi-Square	df	Sig.	Epsilon[a]		
					Greenhouse-Geisser	Huynh-Feldt	Lower-bound
attract	1.000	.000	0		1.000	1.000	1.000

Tests the null hypothesis that the error covariance matrix of the orthonormalized transformed dependent variables is proportional to an identity matrix.

a. May be used to adjust the degrees of freedom for the averaged test of significance. Corrected test are displayed in the Test of Within-Subjects Effects table.

b. Design: Intercept + age + gender + age * gender
Within Subjects Design: attract

Figure 14.12   Results of Mauchly's sphericity test.

> The equal variances assumptions are not violated; probabilities (Sigs.) are > .05.

Levene's Test of Equality of Error Variances[a]

	F	df1	df2	Sig.
emotional attraction	.222	3	16	.880
physical attraction	.756	3	16	.535

Tests the null hypothesis that the error variance of the dependent variable is equal across groups.

a. Design: Intercept + age + gender + age * gender
Within Subjects Design: attract

Figure 14.13   Results of Levene's sphericity test.

We present the between-subjects summary table in Figure 14.15. From this table, we see that both main effects were statistically significant. However, the three-way interaction supercedes these effects, and we may now proceed to our simple effects analyses of the three-way interaction.

14.7   PERFORMING THE POST-ANOVA ANALYSIS IN SPSS

Configure the analysis as you did for the omnibus ANOVA except that now in the **Options** dialog window we will click over the three-way interaction

Tests of Within-Subjects Effects

Measure: MEASURE_1

Source		Type III Sum of Squares	df	Mean Square	F	Sig.
ATTRACT	Sphericity Assumed	16.900	1	16.900	54.080	.000
	Greenhouse-Geisser	16.900	1.000	16.900	54.080	.000
	Huynh-Feldt	16.900	1.000	16.900	54.080	.000
	Lower-bound	16.900	1.000	16.900	54.080	.000
ATTRACT * AGE	Sphericity Assumed	1.000E-01	1	1.000E-01	.320	.579
	Greenhouse-Geisser	1.000E-01	1.000	1.000E-01	.320	.579
	Huynh-Feldt	1.000E-01	1.000	1.000E-01	.320	.579
	Lower-bound	1.000E-01	1.000	1.000E-01	.320	.579
ATTRACT * GENDER	Sphericity Assumed	4.900	1	4.900	15.680	.001
	Greenhouse-Geisser	4.900	1.000	4.900	15.680	.001
	Huynh-Feldt	4.900	1.000	4.900	15.680	.001
	Lower-bound	4.900	1.000	4.900	15.680	.001
ATTRACT * AGE * GENDER	Sphericity Assumed	8.100	1	8.100	25.920	.000
	Greenhouse-Geisser	8.100	1.000	8.100	25.920	.000
	Huynh-Feldt	8.100	1.000	8.100	25.920	.000
	Lower-bound	8.100	1.000	8.100	25.920	.000
Error(ATTRACT)	Sphericity Assumed	5.000	16	.313		
	Greenhouse-Geisser	5.000	16.000	.313		
	Huynh-Feldt	5.000	16.000	.313		
	Lower-bound	5.000	16.000	.313		

SS_C

$SS_{A \times C}$

$SS_{B \times C}$

$SS_{A \times B \times C}$

$SS_{C \times S/AB}$

Figure 14.14   The within-subjects summary table.

Tests of Between-Subjects Effects

Measure: MEASURE_1

Transformed Variable: Average

Source	Type III Sum of Squares	df	Mean Square	F	Sig.	
Intercept	462.400	1	462.400	445.687	.000	SS_A
AGE	48.400	1	48.400	46.651	.000	SS_B
GENDER	57.600	1	57.600	55.518	.000	$SS_{A \times B}$
AGE * GENDER	.000	1	.000	.000	1.000	
Error	16.600	16	16.600			$SS_{S/AB}$

These are the *p*-values for the main effects and interaction effect.

Figure 14.15   The between-subjects summary table.

term as shown in Figure 14.16. Furthermore, because we have already run the descriptives, homogeneity, and plotting portions of our analysis, there is no need to duplicate these again.

At the end of this short process click the **Paste** pushbutton instead of clicking **OK**. The underlying syntax will then be placed in a syntax window;

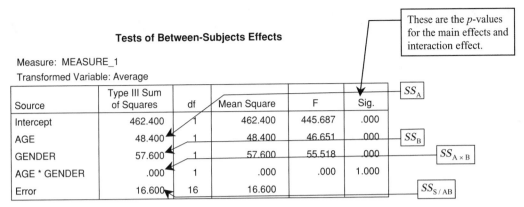

Figure 14.16   The **Options** dialog window.

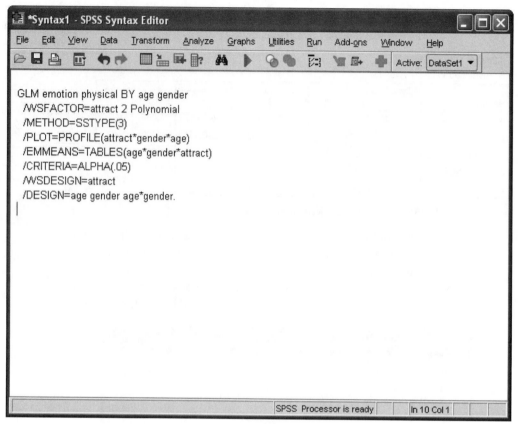

Figure 14.17   The syntax underlying the analysis.

locate that window on your desktop and make it active by clicking it. This window is shown in Figure 14.17.

We have added our simple effects syntax as shown in Figure 14.18 by duplicating the **/EMMEANS = TABLES** subcommand containing the two-way interaction and adding our **compare** syntax to each. Select **Run → All** from the main menu.

14.8   SPSS OUTPUT OF THE OMNIBUS ANALYSIS

The simple effects output is best viewed when looking at the plot of the results in Figure 14.3. In this way, you can relate the statistical outcomes to a visual depiction of the means. Figure 14.19 presents the results of the first set of simple effects comparing levels of **attract**. We are thus comparing **emotion** and **physical** for females and then for males, first for the teenagers and then for those in their middle twenties; with respect to Figure 14.3, we are determining if each line significantly slants with respect to the *x* axis.

Our results indicate that there are statistically significant differences all of the comparisons except for the females in their middle twenties.

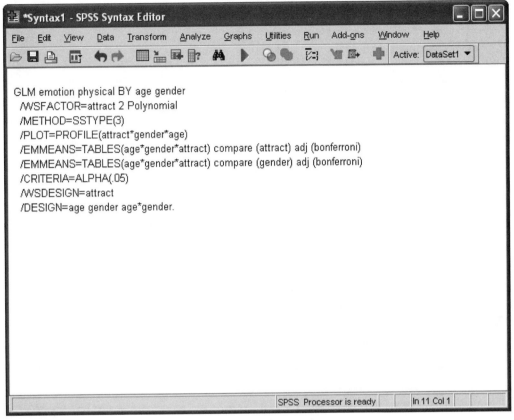

Figure 14.18 The syntax including the simple effects analyses.

Examining the lines one by one in Figure 14.3, we can derive the following summary:

- For the teenagers, both the females and males reported significantly more jealousy when their partners were physically attracted to someone else than when their partners were emotionally attracted to someone else.
- For the students in their middle twenties, males reported significantly more jealousy when their partners were physically attracted to someone else than when their partners were emotionally attracted to someone else; for females, there was no difference between the two jealousy conditions.

Figure 14.20 presents the results of the simple effects comparing levels of **gender**. We are thus comparing females to males for emotional attraction and we are comparing females to males for physical attraction separately for the two age groups; with respect to Figure 14.3, for each plot we are determining if the two data points at **emotional** attraction differ significantly and we are determining if the two data points at **physical** attraction differ significantly.

Estimates

Measure: MEASURE_1

age group	gender	attract	Mean	Std. Error	95% Confidence Interval	
					Lower Bound	Upper Bound
teen	female	1	2.800	.346	2.066	3.534
		2	4.200	.387	3.379	5.021
	male	1	.600	.346	-.134	1.334
		2	1.600	.387	.779	2.421
mid 20s	female	1	5.800	.346	5.066	6.534
		2	5.600	.387	4.779	6.421
	male	1	1.800	.346	1.066	2.534
		2	4.800	.387	3.979	5.621

> These are the probabilities of the differences occurring by chance alone for the pairwise comparisons with age as first order.

Pairwise Comparisons

Measure: MEASURE_1

age group	gender	(I) attract	(J) attract	Mean Difference (I-J)	Std. Error	Sig.[a]	95% Confidence Interval for Difference[a]	
							Lower Bound	Upper Bound
teen	female	1	2	-1.400*	.354	.001	-2.149	-.651
		2	1	1.400*	.354	.001	.651	2.149
	male	1	2	-1.000*	.354	.012	-1.749	-.251
		2	1	1.000*	.354	.012	.251	1.749
mid 20s	female	1	2	.200	.354	.579	-.549	.949
		2	1	-.200	.354	.579	-.949	.549
	male	1	2	-3.000*	.354	.000	-3.749	-2.251
		2	1	3.000*	.354	.000	2.251	3.749

Based on estimated marginal means

*. The mean difference is significant at the .05 level.

a. Adjustment for multiple comparisons: Bonferroni.

Figure 14.19   Comparing the two levels of **attract.**

Our results indicate that all pairs of data points differ except the female-to-male comparison for students in their middle twenties. Examining the comparisons one by one in Figure 14.3, we can derive the following summary:

- For the teenagers, females reported higher levels of jealousy than males when their partners were emotionally as well as physically attracted to someone else.
- For the students in their middle twenties, females reported higher levels of jealousy than males when their partners were emotionally as well as physically attracted to someone else; females and males did not differ after reading vignettes in which their partners were physically attracted to someone else.

14.9   PERFORMING THE OMNIBUS ANALYSIS IN SAS

14.9.1   STRUCTURING THE DATA FILE

The data file in an Excel spreadsheet ready to import into an *SAS Enterprise Guide* project is displayed in Figure 14.21. It is structured in univariate

Estimates

Measure: MEASURE_1

age group	gender	attract	Mean	Std. Error	95% Confidence Interval	
					Lower Bound	Upper Bound
teen	female	1	2.800	.346	2.066	3.534
		2	4.200	.387	3.379	5.021
	male	1	.600	.346	-.134	1.334
		2	1.600	.387	.779	2.421
mid 20s	female	1	5.800	.346	5.066	6.534
		2	5.600	.387	4.779	6.421
	male	1	1.800	.346	1.066	2.534
		2	4.800	.387	3.979	5.621

> These are the probabilities of the differences occurring by chance alone for the pairwise comparisons with age as first order.

Pairwise Comparisons

Measure: MEASURE_1

age group	attract	(I)gender	(J)gender	Mean Difference (I-J)	Std. Error	Sig.[a]	95% Confidence Interval for Difference[a]	
							Lower Bound	Upper Bound
teen	1	female	male	2.200*	.490	.000	1.161	3.239
		male	female	-2.200*	.490	.000	-3.239	-1.161
	2	female	male	2.600*	.548	.000	1.439	3.761
		male	female	-2.600*	.548	.000	-3.761	-1.439
mid 20s	1	female	male	4.000*	.490	.000	2.961	5.039
		male	female	-4.000*	.490	.000	-5.039	-2.961
	2	female	male	.800	.548	.163	-.361	1.961
		male	female	-.800	.548	.163	-1.961	.361

Based on estimated marginal means

*.The mean difference is significant at the .05 level.

a. Adjustment for multiple comparisons: Bonferroni.

Figure 14.20 Comparing the two levels of **gender.**

or stacked mode as explained in Section 10.15.2, and is similar to those we have seen for the examples in Chapters 10–13. The **age, gender,** and **attract** independent variables show the codes for the levels corresponding to the **jealousy** dependent variable for each case. Note that by examining the data file we can determine that **age** and **gender** are between-subjects variables because each subject is associated with only one level of each of those variables. By the same reasoning, we can determine that **attract** is a within-subjects variable because each subject has values on the dependent variable for both levels of **attract.** To illustrate how to read the data file, the individual identified as **subid 1** whose record is shown in Row 2 (the first row of data) is coded as **age 1** (teen) and **gender 1** (female). That person provided a jealousy rating of 3 for **attract 1** (emotional) and a jealousy rating of 5 for **attract 2** (physical).

14.9.2 STRUCTURING THE DATA ANALYSIS

Import the data file from Excel into a new SAS project. Then from the main menu select **Analyze ➜ ANOVA ➜ Mixed Models.** The window

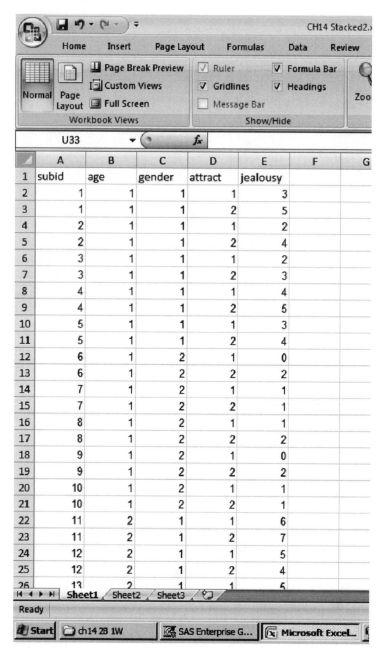

Figure 14.21 The Excel data file ready to be imported into an *SAS Enterprise Guide* project.

opens on the **Task Roles** tab. From the **Variables to assign** panel, select **quality** and drag it to the icon for **Dependent variable**. Then one at a time select **age, gender, attract**, and **subid** and drag them over to the area under **Classification variables**. The result of this is shown in Figure 14.22.

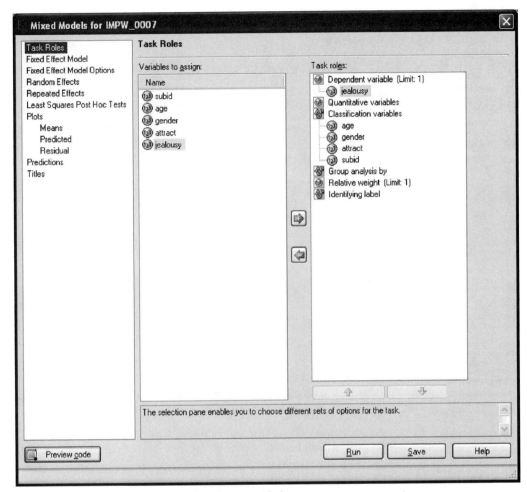

Figure 14.22 The **Task Roles** have been specified.

In the navigation panel, select **Fixed Effect Model** as shown in Figure 14.23. In the **Class and quantitative variables** panel, select **age** and then select the **Main** pushbutton; **age** will automatically appear in the **Effects** window. Repeat this procedure for **gender** and **attract**. Then while holding the Control key down, select both **age** and **gender**. With both variables highlighted, click the **Factorial** pushbutton; **age*gender** will automatically appear in the **Effects** window. Repeat this process for the remaining two-way interactions as well as the three-way interaction. All of the effects that need to be specified are shown in Figure 14.23.

Select **Fixed Effect Model Options** in the navigation panel as shown in Figure 14.24. Check **Type 3** under **Hypothesis test type**. Then select **Residual maximum likelihood** under **Estimation method**, and **Between and within subject portions** under **Degrees of freedom method**.

Select **Random Effects** in the navigation panel as shown in Figure 14.25. Select **Random effects** under the **Effects to use** menu; a box will appear at the far right of that menu. Click on that box and the **Effects**

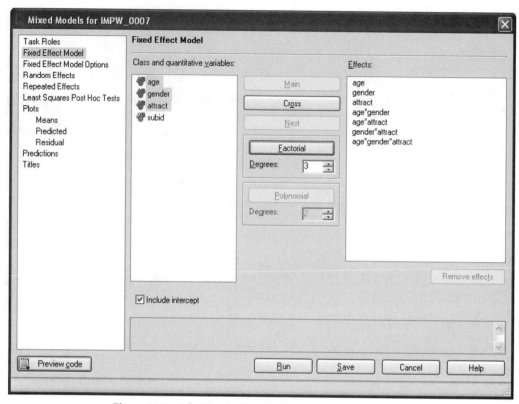

Figure 14.23 The **Fixed Effect Model** has now been specified.

Builder – Random effects window appears. Select **subid** and click the **Main** pushbutton; **subid** will automatically appear in the **Random effects** panel. Click the **OK** pushbutton.

Now select **Subject identifier** under the **Model subjects** menu; a box will appear at the far right of that menu. Click on that box and the **Effects Builder – Subject identifier** window appears. Select **subid** and click the **Main** pushbutton; **subid** will automatically appear in the **Subject identifier** panel. Select the **OK** pushbutton. This is shown in Figure 14.26.

Select **Least Squares Post Hoc Tests** in the navigation panel. Click the **Add** pushbutton at the bottom of the **Effects to estimate** pane. The screen shown in Figure 14.27 appears. Highlight each of the seven effects in turn under the **Effects to use** menu and select **True** for each; this command will cause the means for each of the effects to be output. Click the **Run** pushbutton.

14.10   SAS OUTPUT OF THE OMNIBUS ANALYSIS

The least squares means are shown in Figure 14.28. The results of the omnibus analysis are shown in Figure 14.29. These F ratios are the same as those produced by SPSS. As can be seen, most of the effects are statistically significant.

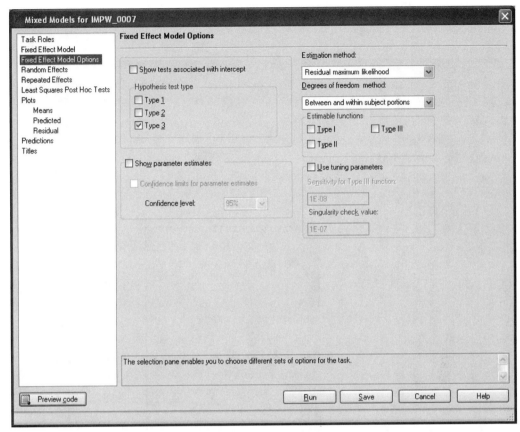

Figure 14.24 The **Fixed Effect Model Options** are specified here.

14.11 PERFORMING THE SIMPLE EFFECTS ANALYSIS IN SAS

To perform the simple effects analysis in *SAS Enterprise Guide*, configure the analysis as shown in Section 14.9.2. Then in the **Least Squares Post Hoc Tests** screen shown in Figure 14.30, in the **Effects to use** frame, set the three-way interaction to **True** and all others to **False**. Then in the **Comparisons** frame, set **Show p-values for differences** to **All pairwise differences** and set **Adjustment method for comparison** to **Bonferroni**. Click **Run** to perform the analysis.

14.12 SAS OUTPUT FROM THE SIMPLE EFFECTS ANALYSIS

A portion of the results of the simple effects tests are shown in Figure 14.31, and is read in a manner analogous to how we described this process in Chapters 12–13. To illustrate this, consider for example the first comparison in the table. The first three coded columns represent one condition and the last three coded columns represent the other condition. We are thus comparing the mean for **age 1, gender 1, attract 1** (teens who were female rating their jealousy for emotional attraction with a mean of 2.8) with **age 1, gender 1, attract 2** (teens who were female rating their jealousy

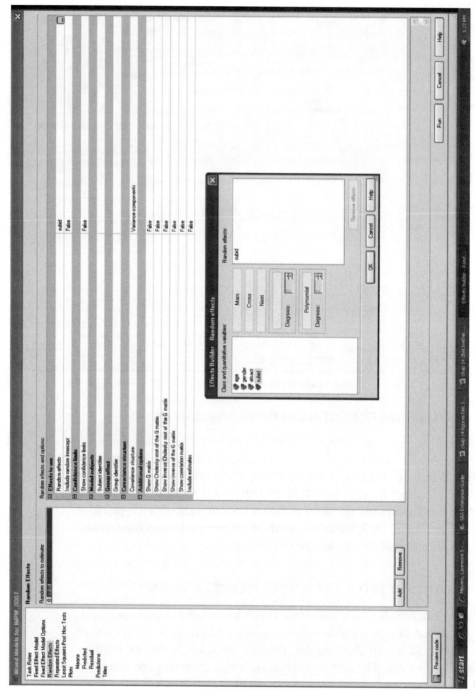

Figure 14.25 The **Random Effects** window.

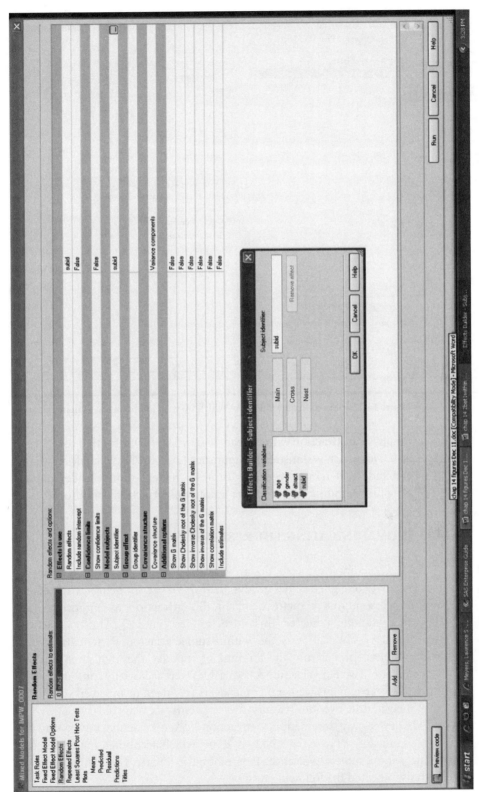

Figure 14.26 Configuring the **Subject** identifier.

413

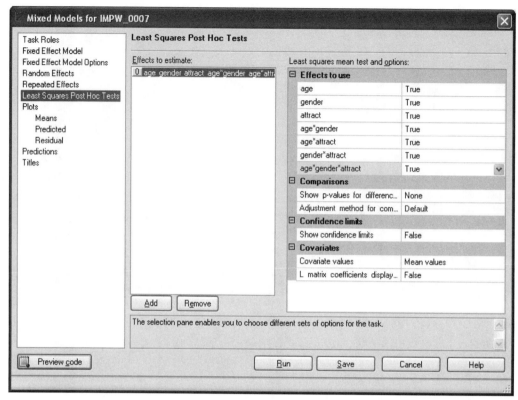

Figure 14.27 Specifying the effects for which we wish to obtain least square means.

for physical attraction with a mean of 4.2). The t value for that comparison is -3.96, which is statistically significant ($p < .0011$). Therefore, we may conclude that female teenagers express greater jealousy under physical attraction conditions than under emotional attraction conditions.

14.13 COMMUNICATING THE RESULTS

A $2 \times 2 \times 2$ complex mixed design was used to assess jealousy. Participants read two vignettes describing their partners as being attracted to another person; one vignette depicted this attraction as emotional and the other vignette depicted this attraction as physical. Thus, attraction (emotional or physical) was the within-subjects factor. Participants rated as the dependent measure the amount of jealousy they felt after reading each vignette. Two between-subjects variables were factorially combined with attraction: participants were either teenagers (eighteen or nineteen years old) or in their middle twenties (twenty-five or twenty-six years old).

The three-way interaction (Attraction × Age × Gender) was statistically significant, $F(1, 16) = 25.92$, $p < .05$, which accounted for 23 percent of the within-subjects variance. It is plotted in Figure 14.3. Simple effects analysis indicated the following results.

For teenagers, all participants reported more jealousy when their partners were physically attracted to someone else than when their partners were emotionally attracted to another. Females reported greater levels of jealousy than males.

For the students in their middle twenties, males were more jealous in the physical attraction condition than in the emotional attraction condition. Females did not differ in jealousy across the two conditions; they did report more jealousy when their partners were emotionally attracted to another but were not significantly different from the males under the condition of their partners being physically attracted to someone else.

Cell means for the 28 conditions.

				Least Squares Means				
Effect	age	gender	attract	Estimate	Standard Error	DF	t Value	Pr >|t|
age	1			2.3000	0.2278	16	10.10	<.0001
age	2			4.5000	0.2278	16	19.76	<.0001
gender		1		4.6000	0.2278	16	20.20	<.0001
gender		2		2.2000	0.2278	16	9.66	<.0001
attract			1	2.7500	0.1837	16	14.97	<.0001
attract			2	4.0500	0.1837	16	22.05	<.0001
age*gender	1	1		3.5000	0.3221	16	10.87	<.0001
age*gender	1	2		1.1000	0.3221	16	3.42	0.0035
age*gender	2	1		5.7000	0.3221	16	17.70	<.0001
age*gender	2	2		3.3000	0.3221	16	10.20	<.0001
age*attract	1		1	1.7000	0.2598	16	6.54	<.0001
age*attract	1		2	2.9000	0.2598	16	11.16	<.0001
age*attract	2		1	3.8000	0.2598	16	14.63	<.0001
age*attract	2		2	5.2000	0.2598	16	20.01	<.0001
gender*attract		1	1	4.3000	0.2598	16	16.55	<.0001
gender*attract		1	2	4.9000	0.2598	16	18.86	<.0001
gender*attract		2	1	1.2000	0.2598	16	4.62	0.0003
gender*attract		2	2	3.2000	0.2598	16	12.32	<.0001
age*gender*attract	1	1	1	2.8000	0.3674	16	7.62	<.0001
age*gender*attract	1	1	2	4.2000	0.3674	16	11.43	<.0001
age*gender*attract	1	2	1	0.6000	0.3674	16	1.63	0.1220
age*gender*attract	1	2	2	1.6000	0.3674	16	4.35	0.0005
age*gender*attract	2	1	1	5.8000	0.3674	16	15.79	<.0001
age*gender*attract	2	1	2	5.6000	0.3674	16	15.24	<.0001
age*gender*attract	2	2	1	1.8000	0.3674	16	4.90	0.0002
age*gender*attract	2	2	2	4.8000	0.3674	16	13.06	<.0001

Figure 14.28   Portion of the least squares means for all of the effects.

These are the *p*-values
for the main effects and
interaction effects.

Type 3 Tests of Fixed Effects				
Effect	Num DF	Den DF	F Value	Pr > F
age	1	16	46.65	<.0001
gender	1	16	55.52	<.0001
attract	1	16	54.08	<.0001
age*gender	1	16	0.00	1.0000
age*attract	1	16	0.32	0.5795
gender*attract	1	16	15.68	0.0011
age*gender*attract	1	16	25.92	0.0001

Figure 14.29   The results of the ANOVA.

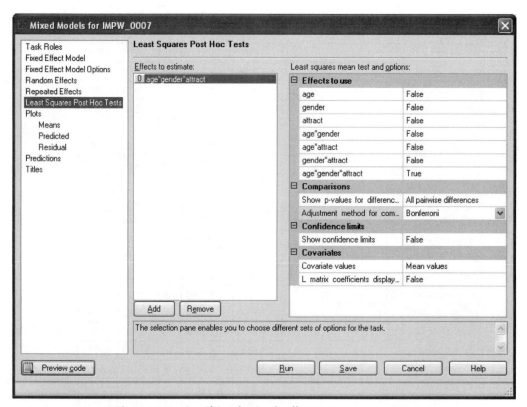

Figure 14.30   Specifying the simple effects tests.

This Table presents the unadjusted *p*-value. The Bonferroni adjustment (not shown) can be calculated by multiplying the unadjusted *p*-value by 28 (the number of all pairwise comparisons).

Differences of Least Squares Means										
Effect	age	gender	attract	age	gender	attract	Estimate	Standard Error	DF	t Value	Pr > ltl
age*gender*attract	1	1	1	1	1	2	−1.4000	0.3536	16	−3.96	0.0011
age*gender*attract	1	1	1	1	2	1	2.2000	0.5196	16	4.23	0.0006
age*gender*attract	1	1	1	1	2	2	1.2000	0.5196	16	2.31	0.0346
age*gender*attract	1	1	1	2	1	1	−3.0000	0.5196	16	−5.77	<.0001
age*gender*attract	1	1	1	2	1	2	−2.8000	0.5196	16	−5.39	<.0001
age*gender*attract	1	1	1	2	2	1	1.0000	0.5196	16	1.92	0.0723
age*gender*attract	1	1	1	2	2	2	−2.0000	0.5196	16	−3.85	0.0014
age*gender*attract	1	1	2	1	2	1	3.6000	0.5196	16	6.93	<.0001
age*gender*attract	1	1	2	1	2	2	2.6000	0.5196	16	5.00	0.0001
age*gender*attract	1	1	2	2	1	1	−1.6000	0.5196	16	−3.08	0.0072
age*gender*attract	1	1	2	2	1	2	−1.4000	0.5196	16	−2.69	0.0160
age*gender*attract	1	1	2	2	2	1	2.4000	0.5196	16	4.62	0.0003
age*gender*attract	1	1	2	2	2	2	−0.6000	0.5196	16	−1.15	0.2652
age*gender*attract	1	2	1	1	2	2	−1.0000	0.3536	16	−2.83	0.0121
age*gender*attract	1	2	1	2	1	1	−5.2000	0.5196	16	−10.01	<.0001
age*gender*attract	1	2	1	2	1	2	−5.0000	0.5196	16	−9.62	<.0001
age*gender*attract	1	2	1	2	2	1	−1.2000	0.5196	16	−2.31	0.0346
age*gender*attract	1	2	1	2	2	2	−4.2000	0.5196	16	−8.08	<.0001
age*gender*attract	1	2	2	2	1	1	−4.2000	0.5196	16	−8.08	<.0001
age*gender*attract	1	2	2	2	1	2	−4.0000	0.5196	16	−7.70	<.0001
age*gender*attract	1	2	2	2	2	1	−0.2000	0.5196	16	−0.38	0.7054
age*gender*attract	1	2	2	2	2	2	−3.2000	0.5196	16	−6.16	<.0001
age*gender*attract	2	1	1	2	1	2	−0.2000	0.3536	16	0.57	0.5795
age*gender*attract	2	1	1	2	2	1	4.0000	0.5196	16	7.70	<.0001
age*gender*attract	2	1	1	2	2	2	1.0000	0.5196	16	1.92	0.0723
age*gender*attract	2	1	2	2	2	1	3.8000	0.5196	16	7.31	<.0001
age*gender*attract	2	1	2	2	2	2	0.8000	0.5196	16	1.54	0.1432
age*gender*attract	2	2	1	2	2	2	−3.0000	0.5196	16	−8.49	<.0001

Figure 14.31 A portion of the output from the simple effects tests.

CHAPTER 14 EXERCISES

14.1. Assume we are interested in the effects of client–therapist ethnic/racial match, service location, and service satisfaction across treatment sessions for community mental health clients. Hence, the between-subjects factors are Factor A, client–therapist, ethnic/racial match (a_1 = match and a_2 = no match) and Factor B, service location (b_1 = in clinic and b_2 = at home). Assume that for Factor A, clients were held constant, but their therapists across sessions were varied in order to produce the match/no match manipulation. The first three treatment sessions for the n = 3 participants serve as the within-subjects factor. Rated service satisfaction ($1 = poor$ to $10 = excellent$) served as the dependent measure. The data are as follows:

			Sessions		
		Subject	c_1	c_2	c_3
		s_1	7	8	9
	b_1	s_2	7	8	9
a_1		s_3	6	9	10
		s_4	7	8	9
	b_2	s_5	7	8	9
		s_6	8	8	10
		s_7	3	3	5
	b_1	s_8	3	4	4
a_2		s_9	3	3	5
		s_{10}	1	4	2
	b_2	s_{11}	2	4	2
		s_{12}	2	3	2

a. Conduct an ANOVA on these data with SPSS or SAS.

b. If the interaction(s) is(are) statistically significant, conduct simple effects analyses and multiple comparisons tests with SPSS or SAS.

14.2. Suppose the management of a large hotel chain was interested in evaluating the effect of type of view (Factor A) and bed size (Factor B) on customer perception of quality over a one-week period. The between-subjects factors were Factor A, type of view (a_1 = ocean and a_2 = parking lot) and Factor B, bed size (a_1 = king and b_2 = queen). The within-subjects factor was a block of five contiguous days that n = 3 customers stayed at the flagship hotel. The dependent measure was an overall customer satisfaction question ($1 = extremely\ unsatisfied$ to $10 = extremely\ satisfied$). The data are as follows:

			Days				
		Subject	c_1	c_2	c_3	c_4	c_5
		s_1	8	9	9	9	10
	b_1	s_2	8	8	8	9	9
a_1		s_3	9	9	9	9	9
		s_4	7	6	5	5	5
	b_2	s_5	7	8	8	8	8
		s_6	6	6	6	6	7
		s_7	2	2	2	2	3
	b_1	s_8	3	4	4	4	4
a_2		s_9	4	4	5	5	5
		s_{10}	1	1	1	1	1
	b_2	s_{11}	2	2	2	2	2
		s_{12}	3	3	2	2	1

a. Conduct an ANOVA on these data with SPSS or SAS.

b. If the interaction(s) is(are) statistically significant, conduct simple effects analyses and multiple comparisons tests with SPSS or SAS.

14.3. Assume the following $2 \times 2 \times 3$ factorial data set with Factor C as the repeated or within-subjects factor. The data are as follows:

		Subject	c_1	c_2	c_3
		s_1	18	14	10
	b_1	s_2	15	13	10
a_1		s_3	18	12	9
		s_4	17	10	11
	b_2	s_5	20	10	10
		s_6	19	9	7
		s_7	10	5	2
	b_1	s_8	11	3	2
a_2		s_9	12	5	4
		s_{10}	13	8	4
	b_2	s_{11}	14	5	5
		s_{12}	15	7	6

a. Conduct an ANOVA on these data with SPSS or SAS.

b. If the interaction(s) is(are) statistically significant, conduct simple effects analyses and multiple comparisons tests with SPSS or SAS.

Complex Mixed Design: One Between-Subjects Factor and Two Within-Subjects Factors

15.1 A NUMERICAL EXAMPLE OF A COMPLEX MIXED DESIGN

The complex mixed design that we consider in this chapter contains one between-subjects factor (the gender of the participants) and two within-subjects factors (the color of toys and the type of toy). Over the course of several days, eight-year-old children were brought into a room containing toys. On some occasions, the toys were of the hands-on type (e.g., balls, beads, building blocks); sometimes these toys were yellow and sometimes they were blue. On other occasions, the toys were of the pretend type (e.g., dolls, stuffed animals, action figures, dress-up clothes); sometimes these toys were yellow and sometimes they were blue.

The children were placed in the room for sixty seconds each time. While they were in the room, the children were allowed to do anything they liked (within the bounds of reason and safety). Observers recorded the number of seconds each child interacted with toys of a particular category (e.g., yellow hands-on toys). For each child, the average number of seconds of such interaction comprised the dependent measure.

The data for this hypothetical study are shown in Figure 15.1. There were five girls (**subid** 1 through 5) and five boys (**subid** 6 through 10). Girl 3, for example, spent an average of seven seconds playing with yellow hands-on toys, twelve seconds playing with yellow pretend toys, three seconds playing with blue hands-on toys, and four seconds playing with blue pretend toys. Because the children were measured under both the hands-on and the pretend toy type condition, this manipulation represents a within-subjects (repeated measures) factor. Furthermore, because the children were measured under both the yellow and blue toy conditions, this second manipulation also represents a second within-subjects variable. When dealing with multiple repeated measures factors, we must record the values of the dependent variable for each combination of the within-subjects factors. Hence, each of the last four columns in Figure 15.1 captures the play time for each of the four combinations.

Figure 15.2 presents the means of the conditions for the girls and boys. Generally, the boys are registering more play time than the girls when playing with the blue toys, but the pattern is different for the yellow toys.

Gender	Subid	Yellow Colored Toys		Blue Colored Toys	
		Hands-On	Pretend	Hands-On	Pretend
Girls	1	5	6	0	2
	2	9	20	0	7
	3	7	12	3	4
	4	4	10	5	5
	5	6	15	2	3
Boys	6	18	4	9	7
	7	12	8	6	5
	8	8	7	8	9
	9	10	3	7	8
	10	14	6	8	11

Figure 15.1   Data for the numerical example.

Looking at the yellow toys, we see that the boys spend more time than girls with the hands-on toys but the girls spend more time than boys with the pretend toys. Given that all three factors seem to differentially affect performance depending on their combination, it is likely that we will obtain a significant three-way interaction.

15.2   EFFECTS OF INTEREST

The effects of interest in this complex mixed design are the same as in any three-way design, including the three-way complex mixed design described in Chapter 14: three main effects (gender, color, and toy type), three two-way interactions (Gender × Color, Gender × Toy Type, and Color × Toy Type), and one three-way interaction (Gender × Color × Toy Type). What makes this design especially interesting is that it contains two repeated measures. With more than a single within-subjects factor in the study, we have multiple within-subjects error terms; each of these error terms is keyed to a within-subjects effect just as they are for fully within-subjects designs as discussed in Chapters 11 and 12. This can be seen in the summary table presented in Table 15.1 that shows the results of the analysis.

With only one between-subjects factor (Factor *A*: gender), the between-subjects portion of the variance is partitioned into the effect

Gender	Yellow Colored Toys		Blue Colored Toys	
	Hands-On	Pretend	Hands-On	Pretend
Girls	6.20	12.60	2.00	4.20
Boys	12.40	5.60	7.60	8.00

Figure 15.2   Cell means for the numerical example.

Table 15.1. Summary ANOVA table

Source	SS	df	MS	F	η^2
Between subjects	135.53				
Gender (A)	46.23	1	46.23	4.14	
S/A	89.30	8	11.16		
Within subjects	607.28				
Color (B)	140.63	1	140.63	15.82[c]	.232[b]
Gender × Color $(A \times B)$	65.03	1	65.03	7.32[c]	
$B \times S/A$	71.10	8	8.89		
Toy type (C)	3.03	1	3.03	0.36	
Gender × Toy Type $(A \times C)$	140.63	1	140.63	16.52[c]	.232[b]
$C \times S/A$	68.10	8	8.51		
Color × Toy Type $(B \times C)$	5.63	1	5.63	1.41	
Gender × Color × Toy Type $(A \times B \times C)$	81.23	1	81.23	20.37[c]	.134[b]
$B \times C \times S/A$	31.90	8	3.99		

[a] η^2 is computed based on the total between-subjects portion of the variance.
[b] η^2 is computed based on the total within-subjects portion of the variance.
[c] $p < .05$.

of the between-subjects independent variable and an error term (S/A). The within-subjects effects have specialized error terms keyed to the particular within-subjects "effect." There are three "pure" within-subjects effects in this design: Factor B: color, Factor C: toy type, and the $(B \times C)$ Color × Toy Type interaction. Each of these effects is associated with its own error term as shown in the summary table in Table 15.1.

These pure within-subjects effects not only stand by themselves – each of them also interacts with gender, the between-subjects factor. The error term for the pure within-subjects effect contained in the interaction is used in the computation of the F ratio. The specifics follow here:

- Color is one of the pure within-subjects effects and is associated with its own error term $(B \times S/A)$. We can also evaluate its interaction with gender (A). This results in the $(A \times B)$ Gender × Color two-way interaction; the error term for this interaction is the error term associated with color $(B \times S/A)$.
- Toy type (C) is another of the pure within-subjects effects and is associated with its own error term $(C \times S/A)$. We can also evaluate its interaction with gender. This results in the $(A \times C)$ Gender × Toy Type two-way interaction; the error term for this interaction $(C \times S/A)$ is the error term associated with toy type (C).
- The $(B \times C)$ Color × Toy Type interaction is the third of the pure within-subjects effects and is associated with its own error term $(B \times C \times S/A)$. We can also evaluate its interaction with gender. This results in the $(A \times B \times C)$ Gender × Color × Toy Type three-way interaction; the error term for this three-way interaction $(B \times C \times S/A)$ is the error term associated with $(B \times C)$ Color × Toy Type two-way interaction.

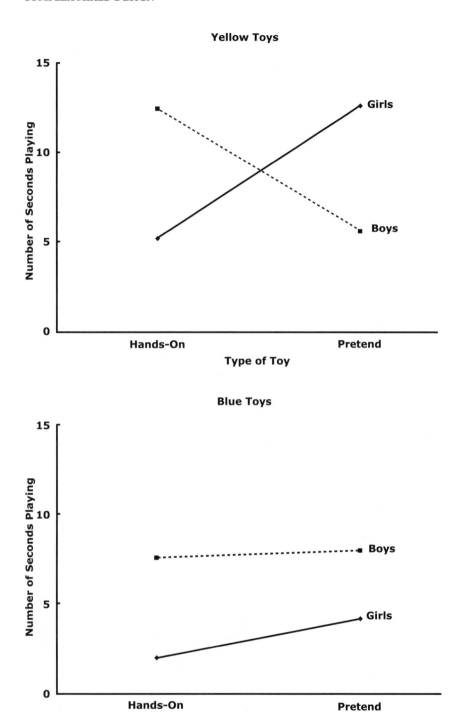

Figure 15.3   Plot of the three-way interaction.

The eta squared values are shown in the last column of the summary table. As we did in Chapter 14, we based these computations on the separate (sub)total variances of the between-subjects and within-subjects partitions. Thus, we have computed the eta squared values as follows:

- Main effect of B: $SS_B \div SS_{\text{Within Subjects}}$
- Interaction $A \times C$: $SS_{A \times C} \div SS_{\text{Within Subjects}}$
- Interaction $A \times B \times C$: $SS_{A \times B \times C} \div SS_{\text{Within Subjects}}$

The plot of the interaction is shown in Figure 15.3. We will narrate these results once we have performed the simple effects analyses.

15.3   COMPUTING THE OMNIBUS COMPLEX MIXED DESIGN BY HAND

As we have done with previous three-factor designs in this book, we will provide you, the reader, with the necessary computational formulas should you choose to do these analyses by hand (see Table 15.2).

From our previous discussions within this chapter, the present complex mixed design has the following sources of between-subjects variance: main effects of Factor A, which is evaluated with the following error term, S/A. Conversely, the within-subjects component is more complex and consists of the main effect of Factor B and the interaction effect of $A \times B$, both of which are evaluated with the $B \times S/A$ error term. Likewise, the main effect of Factor C and the $A \times C$ interaction effect are evaluated by the $C \times S/A$ error term. Finally, the $B \times C$ and the $A \times B \times C$ interaction effects are evaluated with the $B \times C \times S/A$ error term. Further discussion of this topic can be found in Keppel (1991) and Keppel and Wickens (2004).

15.4   PERFORMING THE OMNIBUS ANALYSIS IN SPSS

15.4.1   STRUCTURING THE DATA FILE

The data file for our three-way, complex mixed design example is shown in Figure 15.4. The first column, as always, is used for our participant identification number; we have named this variable **subid**. The next two columns represent our between-subjects variable of **gender** with females coded as 1 and males as 2. The last four variables represent the combinations of the within-subjects variables. In the data file, the variable that increments most rapidly is **toytype**. That is, the third and fourth columns both relate to yellow toys; the third column, named **yelhand**, represents the yellow hands-on toys and the fourth column, named **yelpret**, represents the yellow pretend toys. Columns five and six both relate to the blue toys; **bluhand** represents the blue hands-on toys and **blupret** represents the blue pretend toys. In each of these last four columns, the value of the dependent variable (the number of seconds playing with the toys) is recorded.

Table 15.2. Computational formulas for the complex mixed design: One between-subjects factor and two within-subjects factors

Source	df	SS	MS	F
Between subjects				
A	$a-1$	$\dfrac{\Sigma A^2}{(b)(c)(n)} - \dfrac{T^2}{(a)(b)(c)(n)}$	$\dfrac{SS_A}{df_A}$	$\dfrac{MS_A}{MS_{S/A}}$
S/A	$(a)(n-1)$	$\dfrac{\Sigma S^2}{(b)(c)} - \dfrac{\Sigma A^2}{(b)(c)(n)}$	$\dfrac{SS_{S/A}}{df_{S/A}}$	
Within subjects				
B	$b-1$	$\dfrac{\Sigma B^2}{(a)(c)(n)} - \dfrac{T^2}{(a)(b)(c)(n)}$	$\dfrac{SS_B}{df_B}$	$\dfrac{MS_B}{MS_{B\times S/A}}$
$A \times B$	$(a-1)(b-1)$	$\dfrac{\Sigma AB^2}{(c)(n)} - \dfrac{\Sigma A^2}{(b)(c)(n)} - \dfrac{\Sigma B^2}{(a)(c)(n)} + \dfrac{T^2}{(a)(b)(c)(n)}$	$\dfrac{SS_{A\times B}}{df_{A\times B}}$	$\dfrac{MS_{A\times B}}{MS_{B\times S/A}}$
$B \times S/A$	$(a)(n-1)(b-1)$	$\dfrac{\Sigma BS^2}{c} + \dfrac{\Sigma A^2}{(b)(c)(n)} - \dfrac{\Sigma AB^2}{(c)(n)} - \dfrac{\Sigma S^2}{(b)(c)}$	$\dfrac{SS_{B\times S/A}}{df_{B\times S/A}}$	
C	$c-1$	$\dfrac{\Sigma C^2}{(a)(b)(n)} - \dfrac{T^2}{(a)(b)(c)(n)}$	$\dfrac{SS_C}{df_C}$	$\dfrac{MS_C}{MS_{C\times S/A}}$
$A \times C$	$(a-1)(c-1)$	$\dfrac{\Sigma AC^2}{(b)(n)} - \dfrac{\Sigma A^2}{(b)(c)(n)} - \dfrac{\Sigma C^2}{(a)(b)(n)} + \dfrac{T^2}{(a)(b)(c)(n)}$	$\dfrac{SS_{A\times C}}{df_{A\times C}}$	$\dfrac{MS_{A\times C}}{MS_{C\times S/A}}$
$C \times S/A$	$(a)(n-1)(c-1)$	$\dfrac{\Sigma CS^2}{b} + \dfrac{\Sigma A^2}{(b)(c)(n)} - \dfrac{\Sigma AC^2}{(b)(n)} - \dfrac{\Sigma S^2}{(b)(c)}$	$\dfrac{SS_{C\times S/A}}{df_{C\times S/A}}$	
$B \times C$	$(b-1)(c-1)$	$\dfrac{\Sigma BC^2}{(a)(n)} - \dfrac{\Sigma B^2}{(a)(c)(n)} - \dfrac{\Sigma C^2}{(a)(b)(n)} + \dfrac{T^2}{(a)(b)(c)(n)}$	$\dfrac{SS_{B\times C}}{df_{B\times C}}$	$\dfrac{MS_{B\times C}}{MS_{B\times C\times S/A}}$
$A \times B \times C$	$(a-1)(b-1)(c-1)$	$\dfrac{\Sigma ABC^2}{n} + \dfrac{\Sigma BC^2}{(a)(n)} + \dfrac{\Sigma AC^2}{(b)(n)} + \dfrac{\Sigma C^2}{(a)(b)(n)} + \dfrac{\Sigma B^2}{(a)(c)(n)} + \dfrac{\Sigma AB^2}{(c)(n)} + \dfrac{\Sigma A^2}{(b)(c)(n)} - \dfrac{T^2}{(a)(b)(c)(n)}$	$\dfrac{SS_{A\times B\times C}}{df_{A\times B\times C}}$	$\dfrac{MS_{A\times B\times C}}{MS_{B\times C\times S/A}}$
$B \times C \times S/A$	$a(n-1)(b-1)((c-1)$	$\Sigma Y^2 + \dfrac{\Sigma AB^2}{(c)(n)} + \dfrac{\Sigma AC^2}{(b)(n)} + \dfrac{\Sigma S^2}{(b)(c)} + \dfrac{\Sigma CS^2}{b} - \dfrac{\Sigma ABC^2}{n} - \dfrac{\Sigma BS^2}{c} - \dfrac{\Sigma A^2}{(b)(c)(n)}$	$\dfrac{SS_{B\times C\times S/A}}{df_{B\times C\times S/A}}$	
Total	$(a)(b)(c)(n)-1$	$\Sigma Y^2 - \dfrac{T^2}{(a)(b)(c)(n)}$		

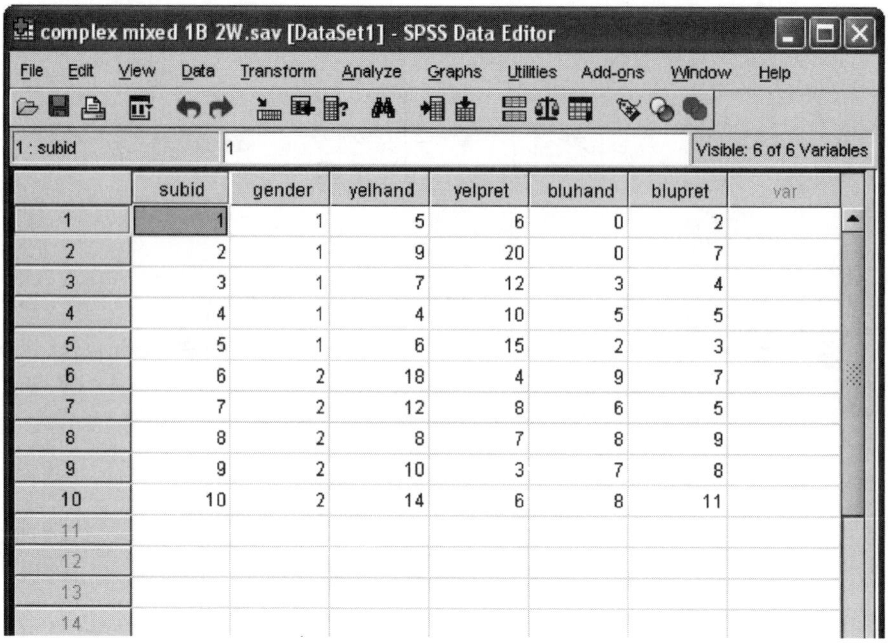

	subid	gender	yelhand	yelpret	bluhand	blupret	var
1	1	1	5	6	0	2	
2	2	1	9	20	0	7	
3	3	1	7	12	3	4	
4	4	1	4	10	5	5	
5	5	1	6	15	2	3	
6	6	2	18	4	9	7	
7	7	2	12	8	6	5	
8	8	2	8	7	8	9	
9	9	2	10	3	7	8	
10	10	2	14	6	8	11	
11							
12							
13							
14							

Figure 15.4 The data file for our numerical example.

15.4.2 STRUCTURING THE DATA ANALYSIS

From the main SPSS menu select **Analyze → General Linear Model → Repeated Measures**. You selected **Repeated Measures** for the following reason. If there are two or more cases in the study and at least one of the independent variables is a within-subjects variable, then we must partition the total variance of the dependent variable into a between-subjects portion and a within-subjects portion. This partitioning is accomplished by the **Repeated Measures** module of the **General Linear Model** procedure in SPSS.

Selecting this path will open the dialog window shown in Figure 15.5. As you will recall from Chapters 12–14, this is the initial window used by SPSS to have you name the within-subjects variable(s) in the study. There are two such factors in our example: **color** and **toytype**. In the data file (as noted in section 15.4.1), reading the columns from left to right, **color** increments most slowly and **toytype** increments most rapidly. We will use this information in identifying the within-subjects factors in the **Repeated Measures Define Factor(s)** window. That is, we will first identify the slowest incrementing variable, **color** in this case, and we will then identify the more quickly incrementing variable, **toytype** in this case. The results of this are shown in Figure 15.5.

Clicking **Define** in the **Define Factor(s)** window brings us to the main dialog window of **GLM Repeated Measures** shown in Figure 15.6. There are four slots that need to be specified in the panel to the right of the variable list. With two levels for each of our two repeated measures, that yields four variables; hence, our count of variables matches the number of slots shown in the window (this is a very good thing).

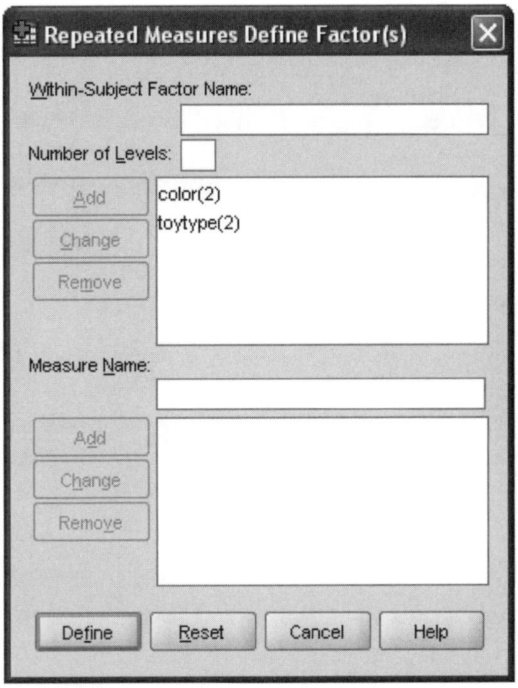

Figure 15.5 The **GLM Repeated Measures** initial **Define Factor(s)** window.

Figure 15.6 The **GLM Repeated Measures** main dialog window.

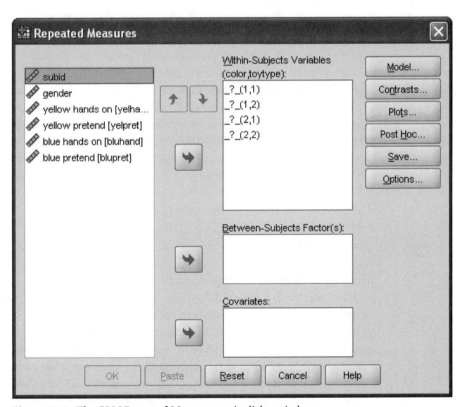

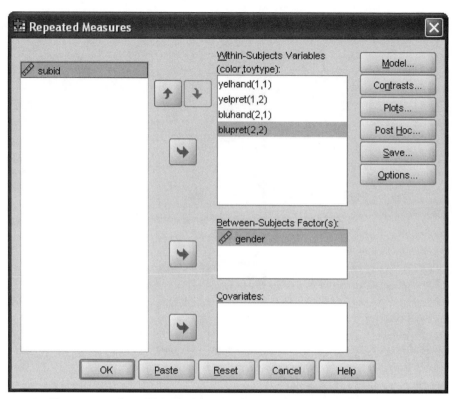

Figure 15.7   The **GLM Repeated Measures** main dialog window.

Note that near the top of the window under **Within-Subjects Variables** we see the expression (**color, toytype**). This is the order in which we identified our within-subjects factors and gives us guidance for filling in the slots. Recall that color incremented most slowly and was identified first. The codes in the panel, reading down, have the first code changing most slowly. This first code corresponds to **color** and the first two slots are reserved for the **color** coded as 1, that is, yellow.

The first slot is reserved for the variable with the code 1,1. We know that the first value stands for yellow. The second value is the code for **toytype**, and a code of 1 refers to hands-on toys. Thus, the first slot is reserved for **yelhand**. In an analogous manner, the second slot can be understood as referring to **yelpret**, the third slot refers to **bluhand**, and the fourth slot refers to **blupret**. These variables are brought over to the panel as shown in Figure 15.7, where we also see that **gender** has been dragged into the **Between-Subjects Factor(s)** panel.

Select the **Options** pushbutton to reach the **Options** dialog window. With seven effects of interest here, we will wait until we perform the post-ANOVA tests to determine for which effects we want to generate tables of estimated marginal means. For this omnibus analysis, we simply check **Descriptive statistics** and **Homogeneity tests**. This is shown in Figure 15.8. Click **Continue** to return to the main dialog window.

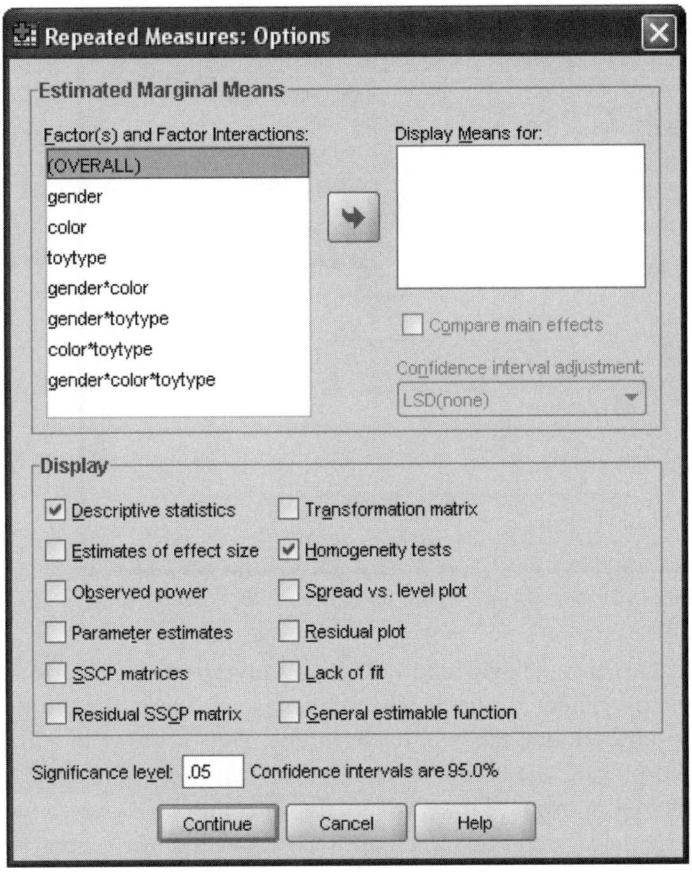

Figure 15.8 The **Options** window.

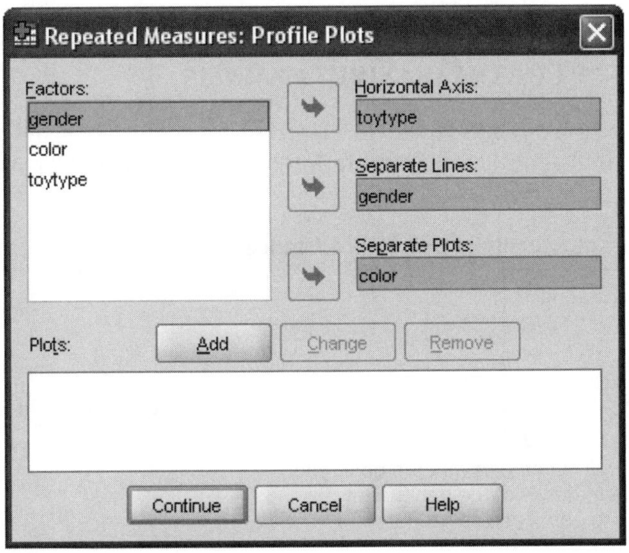

Figure 15.9 The setup for the plot.

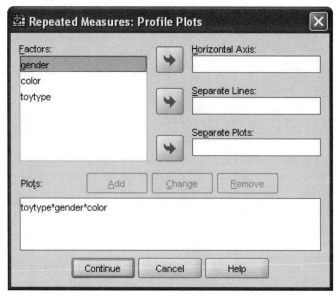

Figure 15.10   The specification for the plot.

In the **Plots** dialog window, click over **toytype** to the **Horizontal Axis** panel and **gender** to the **Separate Lines** panel. Place **color** in the panel for **Separate Plots;** this will result in one plot for the yellow toys and a second plot for the blue toys. We have shown this in Figure 15.9. Click **Add** to register your plot with SPSS, the outcome of which is shown in the lower panel labeled **Plots** in Figure 15.10. Note that the general structure of the specification is **Horizontal Axis∗Separate Lines∗Separate Plots**. Click **Continue** to return to the main dialog window. Then click **OK** to perform the omnibus analysis.

15.5   SPSS OUTPUT OF THE OMNIBUS ANALYSIS

The results of Box's test, shown in Figure 15.11, is not statistically significant; this indicates that the assumption of the equality of the covariance

Box's Test of Equality of Covariance Matrices[a]

Box's M	17.981
F	.773
df1	10
df2	305.976
Sig.	.655

Figure 15.11   The results of Box's M test.

Tests the null hypothesis that the observed covariance matrices of the dependent variables are equal across groups.

a. Design: Intercept + GENDER
   Within Subjects Design:
   COLOR + TOYTYPE + COLOR*TOYTYPE

<table>
<tr><td>Mauchly's Test assesses equality of group variances and correlations. Because there is only one correlation for each within variable, they cannot be compared to any other. Therefore, the sphericity assumption cannot be tested. This is why there are zero df and why there are no computed probabilities (Sig.).</td></tr>
</table>

Mauchly's Test of Sphericity[b]

Measure: MEASURE_1

Within Subjects Effect	Mauchly's W	Approx. Chi-Square	df	Sig.	Epsilon[a]		
					Greenhouse e-Geisser	Huynh-Feldt	Lower-bound
COLOR	1.000	.000	0	.	1.000	1.000	1.000
TOYTYPE	1.000	.000	0	.	1.000	1.000	1.000
COLOR*TOYTYPE	1.000	.000	0	.	1.000	1.000	1.000

Tests the null hypothesis that the error covariance matrix of the orthonormalized transformed dependent variables is proportional to an identity matrix.

  a. May be used to adjust the degrees of freedom for the averaged tests of significance. Corrected tests are displayed in the Tests of Within-Subjects Effects table.

  b. Design: Intercept + GENDER
    Within Subjects Design: COLOR + TOYTYPE + COLOR*TOYTYPE

Figure 15.12   The results of Mauchly's test.

matrices cannot be rejected. Mauchly's Test of Sphericity, the results of which are shown in Figure 15.12, cannot be performed with only two levels of the within-subjects variable. Levene's test of error variance equality was not statistically significant (see Figure 15.13) for any of the within-subjects factors; we therefore cannot reject the homogeneity of variance assumption for our repeated measures.

The within-subjects summary table is presented in Figure 15.14. We see from this table that the main effects of **color**, the two-way interaction of **Gender∗Toytype**, and the three-way interaction of **Gender∗Color∗Toytype** are statistically significant. With the highest order interaction (the three-way) being significant, our simple effects analyses need to be focused at this level.

<table>
<tr><td>The equal variances assumptions are not violated; probabilities (Sigs.) are > .05</td></tr>
</table>

Levene's Test of Equality of Error Variances[a]

	F	df1	df2	Sig.
YELHAND yellow hands on	1.871	1	8	.208
YELPRET yellow pretend	2.679	1	8	.140
BLUHAND blue hands on	1.588	1	8	.243
BLUPRET blue pretend	.044	1	8	.839

Tests the null hypothesis that the error variance of the dependent variable is equal across groups.

  a. Design: Intercept + GENDER
    Within Subjects Design:
    COLOR + TOYTYPE + COLOR*TOYTYPE

Figure 15.13   The results of Levene's test.

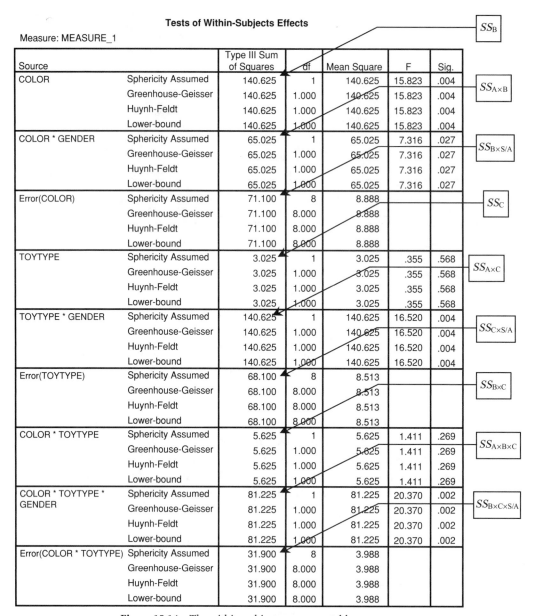

Tests of Within-Subjects Effects

Measure: MEASURE_1

Source		Type III Sum of Squares	df	Mean Square	F	Sig.
COLOR	Sphericity Assumed	140.625	1	140.625	15.823	.004
	Greenhouse-Geisser	140.625	1.000	140.625	15.823	.004
	Huynh-Feldt	140.625	1.000	140.625	15.823	.004
	Lower-bound	140.625	1.000	140.625	15.823	.004
COLOR * GENDER	Sphericity Assumed	65.025	1	65.025	7.316	.027
	Greenhouse-Geisser	65.025	1.000	65.025	7.316	.027
	Huynh-Feldt	65.025	1.000	65.025	7.316	.027
	Lower-bound	65.025	1.000	65.025	7.316	.027
Error(COLOR)	Sphericity Assumed	71.100	8	8.888		
	Greenhouse-Geisser	71.100	8.000	8.888		
	Huynh-Feldt	71.100	8.000	8.888		
	Lower-bound	71.100	8.000	8.888		
TOYTYPE	Sphericity Assumed	3.025	1	3.025	.355	.568
	Greenhouse-Geisser	3.025	1.000	3.025	.355	.568
	Huynh-Feldt	3.025	1.000	3.025	.355	.568
	Lower-bound	3.025	1.000	3.025	.355	.568
TOYTYPE * GENDER	Sphericity Assumed	140.625	1	140.625	16.520	.004
	Greenhouse-Geisser	140.625	1.000	140.625	16.520	.004
	Huynh-Feldt	140.625	1.000	140.625	16.520	.004
	Lower-bound	140.625	1.000	140.625	16.520	.004
Error(TOYTYPE)	Sphericity Assumed	68.100	8	8.513		
	Greenhouse-Geisser	68.100	8.000	8.513		
	Huynh-Feldt	68.100	8.000	8.513		
	Lower-bound	68.100	8.000	8.513		
COLOR * TOYTYPE	Sphericity Assumed	5.625	1	5.625	1.411	.269
	Greenhouse-Geisser	5.625	1.000	5.625	1.411	.269
	Huynh-Feldt	5.625	1.000	5.625	1.411	.269
	Lower-bound	5.625	1.000	5.625	1.411	.269
COLOR * TOYTYPE * GENDER	Sphericity Assumed	81.225	1	81.225	20.370	.002
	Greenhouse-Geisser	81.225	1.000	81.225	20.370	.002
	Huynh-Feldt	81.225	1.000	81.225	20.370	.002
	Lower-bound	81.225	1.000	81.225	20.370	.002
Error(COLOR * TOYTYPE)	Sphericity Assumed	31.900	8	3.988		
	Greenhouse-Geisser	31.900	8.000	3.988		
	Huynh-Feldt	31.900	8.000	3.988		
	Lower-bound	31.900	8.000	3.988		

The right-side annotations point to: SS_B, $SS_{A \times B}$, $SS_{B \times S/A}$, SS_C, $SS_{A \times C}$, $SS_{C \times S/A}$, $SS_{B \times C}$, $SS_{A \times B \times C}$, $SS_{B \times C \times S/A}$.

Figure 15.14   The within-subjects summary table.

We present the between-subjects summary table in Figure 15.15. From this table, we see that the main effect of gender is not statistically significant.

15.6   PERFORMING THE POST-ANOVA ANALYSIS IN SPSS

Configure the analysis as you did for the omnibus ANOVA except that now in the **Options** dialog window we will click over the three-way interaction

Tests of Between-Subjects Effects

Measure: MEASURE_1
Transformed Variable: Average

Source	Type III Sum of Squares	df	Mean Square	F	Sig.
Intercept	2146.225	1	2146.225	192.271	.000
GENDER	46.225	1	46.225	4.141	.076
Error	89.300	8	11.163		

SS_A

This is the *p*-value for the gender main effect.

$SS_{S/A}$

Figure 15.15   The between-subjects summary table.

term as shown in Figure 15.16. Furthermore, since we have already run the descriptives, homogeneity, and plotting portions of our analysis, there is no need to run these.

At the end of this short process click the **Paste** pushbutton instead of clicking **OK**. The underlying syntax will then be placed in a syntax window; locate that window on your desktop and make it active by clicking it. This window is shown in Figure 15.17.

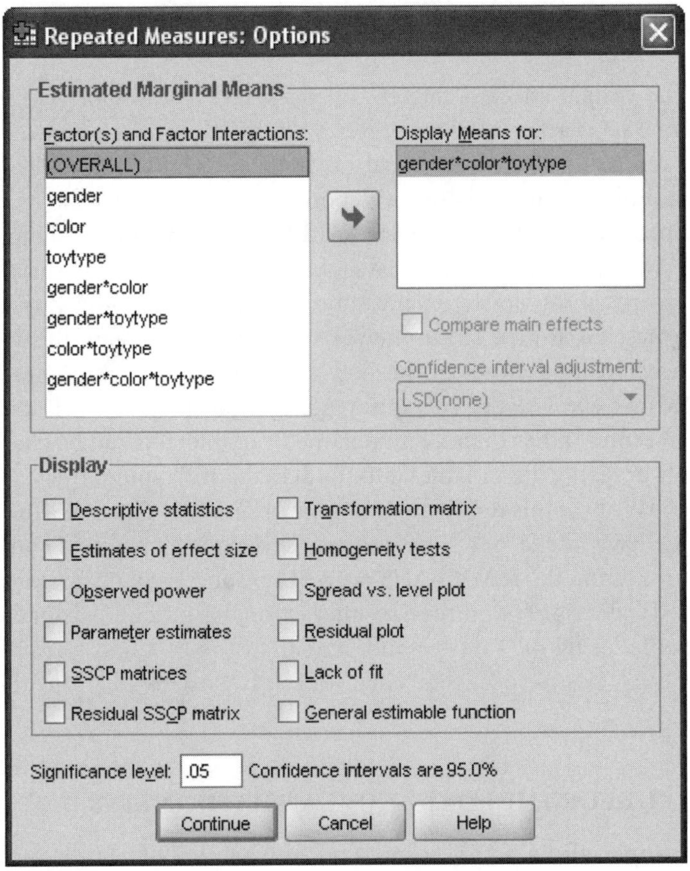

Figure 15.16   The **Options** window prepared for the simple effects analyses.

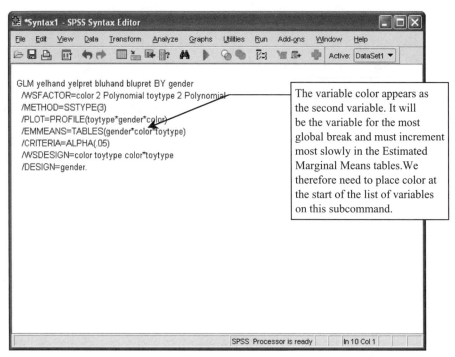

Figure 15.17   The syntax underlying the analysis.

The simple effects analyses will be keyed to the plots we show in Figure 15.3. Each plot shows the results for a particular toy color, and thus represents the most global of divisions of the factors. This most global breakdown needs to correspond to the order that the factors are specified on the **/EMMEANS = TABLES** subcommand containing the three-way interaction. The most global break would be the variable that increments most slowly on the subcommand, that is, the variable that is furthest toward the left in the parentheses. Therefore, in this instance, we need to have **color** as the first variable in the parentheses on the **/EMMEANS = TABLES** subcommand. We note in Figure 15.17 that **color** is the second rather than the first named variable; this can be changed by simply dragging the variables into (or deleting and typing) their necessary order. We show this modification in the ordering in Figure 15.18.

We have added our simple effects syntax as shown in Figure 15.19 by duplicating the **/EMMEANS = TABLES** subcommand containing the revised ordering of our three-way interaction. We have added our **compare** syntax focusing on **toytype** and **gender** to the first and second of our duplicate lines, respectively. Now we select **Run → All** from the main menu.

15.7   OUTPUT FOR THE POST-ANOVA ANALYSIS IN SPSS

The simple effects output is best viewed when looking at the plot of the results in Figure 15.3. In this way, you can relate the statistical outcomes

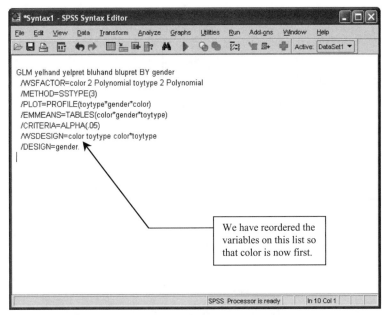

Figure 15.18 We have reordered the variables on the **/EMMEANS = TABLES** subcommand so that **color** is the first named variable in parentheses.

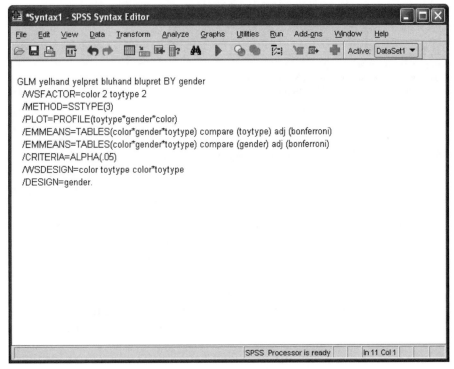

Figure 15.19 The syntax for the simple effects analyses.

A. Estimates

Measure: MEASURE_1

COLOR	GENDER	TOYTYPE	Mean	Std. Error	95% Confidence Interval Lower Bound	Upper Bound
1	1	1	6.200	1.360	3.063	9.337
		2	12.600	1.792	8.468	16.732
	2	1	12.400	1.360	9.263	15.537
		2	5.600	1.792	1.468	9.732
2	1	1	2.000	.762	.244	3.756
		2	4.200	.933	2.049	6.351
	2	1	7.600	.762	5.844	9.356
		2	8.000	.933	5.849	10.151

> These are the probabilities of the differences occurring by chance alone for the pairwise comparisons with color as first order.

B. Pairwise Comparisons

Measure: MEASURE_1

COLOR	GENDER	(I) TOYTYPE	(J) TOYTYPE	Mean Difference (I-J)	Std. Error	Sig.[a]	95% Confidence Interval for Difference[a] Lower Bound	Upper Bound
1	1	1	2	−6.400*	1.962	.011	−10.925	−1.875
		2	1	6.400*	1.962	.011	1.875	10.925
	2	1	2	6.800*	1.962	.008	2.275	11.325
		2	1	−6.800*	1.962	.008	−11.325	−2.275
2	1	1	2	−2.200	1.072	.074	−4.673	.273
		2	1	2.200	1.072	.074	−.273	4.673
	2	1	2	−.400	1.072	.719	−2.873	2.073
		2	1	.400	1.072	.719	−2.073	2.873

Based on estimated marginal means
*. The mean difference is significant at the .05 level.
a. Adjustment for multiple comparisons: Bonferroni.

Figure 15.20   Pairwise comparisons.

to a visual depiction of the means. Figure 15.20 presents the results of the first set of simple effects comparing levels of **toytype**. We are thus comparing the amount of time the children spent playing with **hands-on** toys versus the time they spent playing with **pretend** toys. This comparison is performed separately for females and then for males, first for the yellow toys and then for the blue toys; with respect to Figure 15.3, we are determining if each line significantly slants with respect to the *x* axis.

Our results indicate that only one pair of means differed significantly. There was a difference in play time between the **hands-on** and **pretend** toys for **gender 1** (girls) for **color 1** (yellow). Examining the comparisons one by one in Figure 15.3, we can derive the following summary:

- For the yellow toys, girls played with pretend toys more than they played with the hands-on toys; boys played a comparable amount of time with both types of toys.

- For the blue toys, neither sex demonstrated a preference for either type of toy.

Figure 15.21 presents the results of the simple effects comparing levels of **gender**. We are thus comparing females to males for the time spent playing with **hands-on** toys and we are comparing females to males for the time spent playing with **pretend** toys; with respect to Figure 15.3, for each plot we are determining if the two data points at **hands-on** differ significantly, and we are determining if the two data points at **pretend** differ significantly.

Our results indicate that most of the pairs of data points differ except for the female-to-male comparison for the yellow pretend toys. Examining

A. Estimates

Measure: MEASURE_1

COLOR	GENDER	TOYTYPE	Mean	Std. Error	Lower Bound	Upper Bound
1	1	1	6.200	1.360	3.063	9.337
		2	12.600	1.792	8.468	16.732
	2	1	12.400	1.360	9.263	15.537
		2	5.600	1.792	1.468	9.732
2	1	1	2.000	.762	.244	3.756
		2	4.200	.933	2.049	6.351
	2	1	7.600	.762	5.844	9.356
		2	8.000	.933	5.849	10.151

95% Confidence Interval for columns Lower Bound / Upper Bound.

These are the probabilities of the differences occurring by chance alone for the pairwise comparisons with color as first order.

B. Pairwise Comparisons

Measure: MEASURE_1

COLOR	TOYTYPE	(I) GENDER	(J) GENDER	Mean Difference (I-J)	Std. Error	Sig.[a]	Lower Bound	Upper Bound
1	1	1	2	−6.200*	1.924	.012	−10.636	−1.764
		2	1	6.200*	1.924	.012	1.764	10.636
	2	1	2	7.000*	2.534	.025	1.157	12.843
		2	1	−7.000*	2.534	.025	−12.843	−1.157
2	1	1	2	−5.600*	1.077	.001	−8.084	−3.116
		2	1	5.600*	1.077	.001	3.116	8.084
	2	1	2	−3.800*	1.319	.020	−6.842	−.758
		2	1	3.800*	1.319	.020	.758	6.842

Based on estimated marginal means
*. The mean difference is significant at the .05 level.
a. Adjustment for multiple comparisons: Bonferroni.

Figure 15.21 Pairwise comparisons.

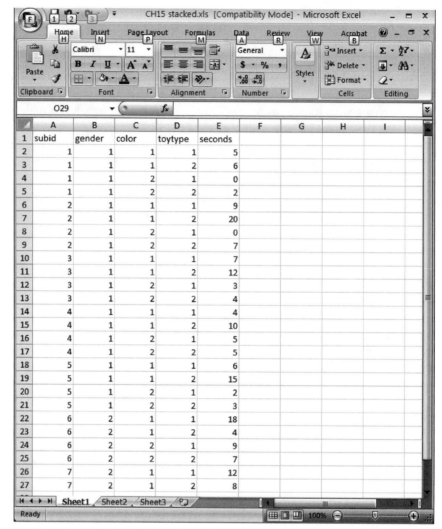

Figure 15.22 The Excel data file ready to be imported into an *SAS Enterprise Guide* project.

the comparisons one by one in Figure 15.3, we can derive the following summary:

- For the yellow toys, boys played with the hands-on toys significantly more than girls; there was no difference in the time the girls and boys spent playing with pretend toys.
- For the blue toys, boys spent more time with both hands-on and with pretend toys than the girls.

15.8 PERFORMING THE OMNIBUS ANALYSIS IN SAS

15.8.1 STRUCTURING THE DATA FILE

A portion of the data file in an Excel spreadsheet ready to import into an *SAS Enterprise Guide* project is displayed in Figure 15.22. It is structured

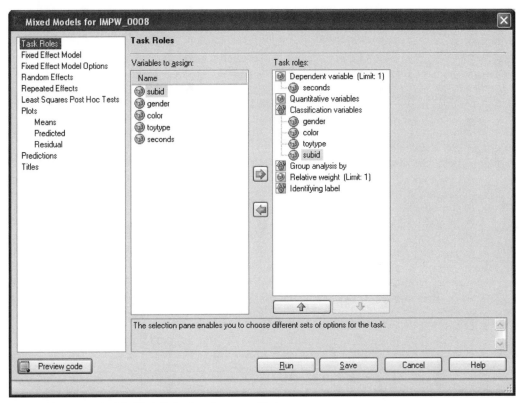

Figure 15.23 The **Task Roles** have been specified.

in univariate or stacked mode as explained in Section 10.15.2, and is similar to those we have seen for the examples in Chapters 12–14. The **gender**, **color**, and **toytype** independent variables show the codes for the levels corresponding to the **seconds** dependent variable (the time the child played with the toys) for each case. Note that by examining the data file we can determine that **gender** is a between-subjects variable because each subject is associated with only one level of that variable. By the same reasoning, we can determine that **color and toytype** are within-subjects variables because each subject has values on the dependent variable for all levels of those two variables. To illustrate how to read the data file, the case identified as **subid 1** whose record is shown in Row 2 (the first row of data) is coded as **gender 1** (female). That child played with toys coded as **color 1** (yellow) and **toytype 1** (hands-on) for five seconds. As another example, the case identified as **subid 6** whose record is shown in Row 25 is coded as **gender 2** (male). That child played with toys coded as **color 2** (blue) and **toytype 2** (pretend) for seven seconds.

15.8.2 STRUCTURING THE DATA ANALYSIS

Import the data file from Excel into a new SAS project. Then from the main menu select **Analyze ➔ ANOVA ➔ Mixed Models**. The window

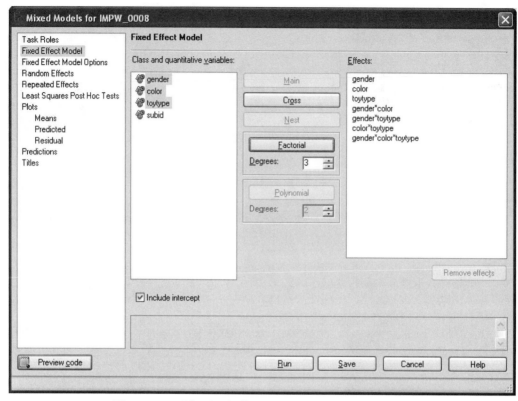

Figure 15.24   The **Fixed Effect Model** has now been specified.

opens on the **Task Roles** tab. From the **Variables to assign** panel, select **quality** and drag it to the icon for **Dependent variable**. Then one at a time select **gender**, **color**, **toytype**, and **subid** and drag them over to the area under **Classification variables**. The result of this is shown in Figure 15.23.

In the navigation panel, select **Fixed Effect Model** as shown in Figure 15.24. In the **Class and quantitative variables** panel, select **gender** and then select the **Main** pushbutton; **gender** will automatically appear in the **Effects** window. Repeat this procedure for **color** and **toytype**. Then while holding the Control key down, select both **gender** and **color**. With both variables highlighted, click the **Factorial** pushbutton; **age∗gender** will automatically appear in the **Effects** window. Repeat this process for the remaining two-way interactions as well as the three-way interaction. All of the effects that need to be specified are shown in Figure 15.24.

Select **Fixed Effect Model Options** in the navigation panel as shown in Figure 15.25. Check **Type 3** under **Hypothesis test type**. Then select **Residual maximum likelihood** under **Estimation method**, and **Between and within subject portions** under **Degrees of freedom method**.

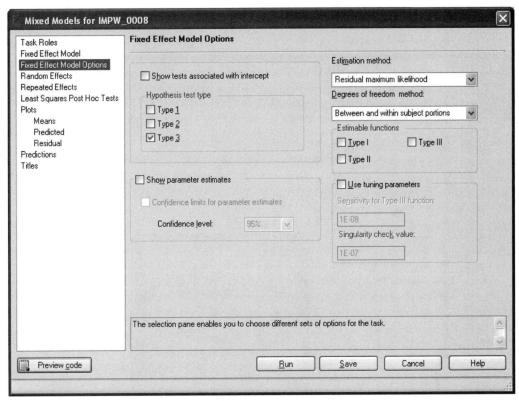

Figure 15.25   The **Fixed Effect Model Options** are specified here.

Select **Random Effects** in the navigation panel as shown in Figure 15.26. Select **Random effects** under the **Effects to use** menu; a box will appear at the far right of that menu. Click on that box and the **Effects Builder – Random effects** window appears. Select **subid** and click the **Main** pushbutton; **subid** will automatically appear in the **Random effects** panel. Click the **OK** pushbutton.

Now select **Subject identifier** under the **Model subjects** menu; a box will appear at the far right of that menu. Click on that box and the **Effects Builder – Subject identifier** window appears. Select **subid** and click the **Main** pushbutton; **subid** will automatically appear in the **Subject identifier** panel. Select the **OK** pushbutton. This is shown in Figure 15.27.

Select **Least Squares Post Hoc Tests** in the navigation panel. Click the **Add** pushbutton at the bottom of the **Effects to estimate** pane. The screen shown in Figure 15.28 appears. Highlight each of the seven effects in turn under the **Effects to use** menu and select **True** for each; this command will cause the means for each of the effects to be output. Click the **Run** pushbutton.

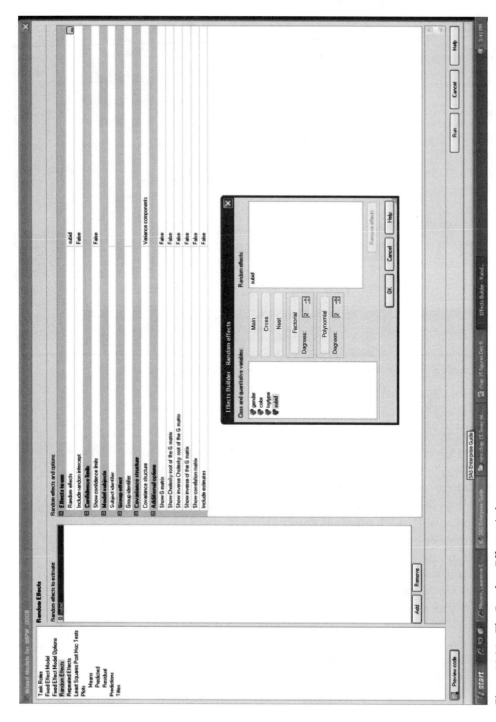

Figure 15.26   The **Random Effects** window.

442

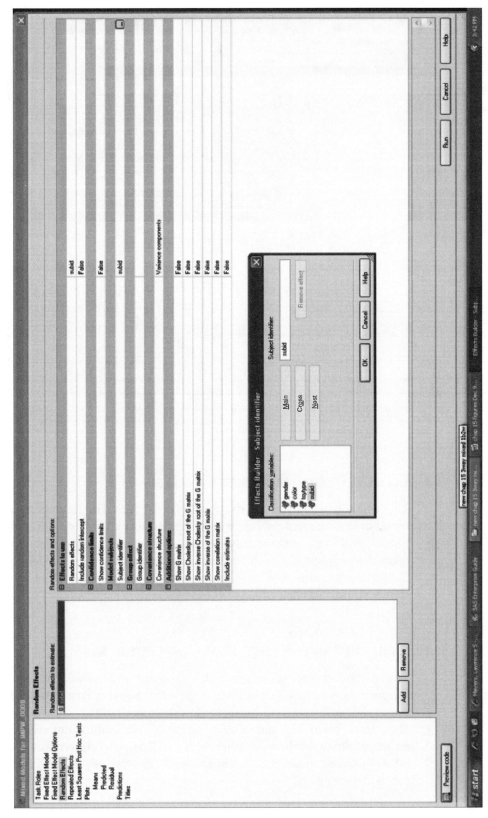

Figure 15.27 Configuring the **Subjects** identifier.

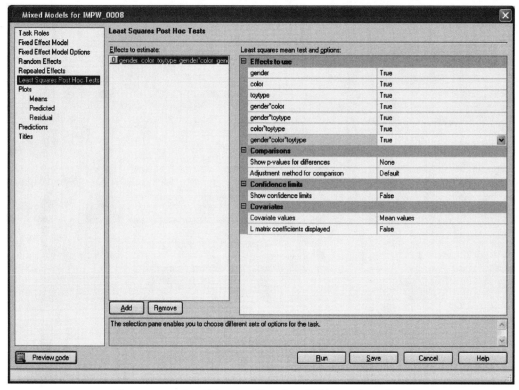

Figure 15.28   Specifying the effects for which we wish to obtain least square means.

15.9   SAS OUTPUT OF THE OMNIBUS ANALYSIS

The least squares means are shown in Figure 15.29. The results of the omnibus analysis are shown in Figure 15.30. The F ratios obtained through SAS are similar to but not exactly the same as those produced by SPSS (see Section 10.17) because of the two within-subjects factors in the design. Examining the summary table informs us that most of the effects are statistically significant.

15.10   PERFORMING THE POST-ANOVA ANALYSIS IN SAS

To perform the simple effects analysis in *SAS Enterprise Guide*, configure the analysis as shown above. Then in the **Least Squares Post Hoc Tests** screen shown in Figure 15.31, in the **Effects to use** frame, set the three-way interaction to **True** and all others to **False**. Then in the **Comparisons** frame, set **Show p-values for differences** to **All pairwise differences** and set **Adjustment method for comparison** to **Bonferroni**. Click **Run** to perform the analysis.

Cell means for
all conditions.

| | | | | | Least Squares Means | | | | |
Effect	gender	color	toytype	Estimate	Standard Error	DF	t Value	Pr>|t|
gender	1			6.2500	0.7471	8	8.37	<.0001
gender	2			8.4000	0.7471	8	11.24	<.0001
color		1		9.2000	0.6762	8	13.60	<.0001
color		2		5.4500	0.6762	8	8.06	<.0001
toytype			1	7.0500	0.6762	8	10.43	<.0001
toytype			2	7.6000	0.6762	8	11.24	<.0001
gender*color	1	1		9.4000	0.9563	8	9.83	<.0001
gender*color	1	2		3.1000	0.9563	8	3.24	0.0119
gender*color	2	1		9.0000	0.9563	8	9.41	<.0001
gender*color	2	2		7.8000	0.9563	8	8.16	<.0001
gender*toytype	1		1	4.1000	0.9563	8	4.29	0.0027
gender*toytype	1		2	8.4000	0.9563	8	8.78	<.0001
gender*toytype	2		1	10.0000	0.9563	8	10.46	<.0001
gender*toytype	2		2	6.8000	0.9563	8	7.11	0.001
color*toytype		1	1	9.3000	0.9021	8	10.31	<.0001
color*toytype		1	2	9.1000	0.9021	8	10.09	<.0001
color*toytype		2	1	4.8000	0.9021	8	5.32	0.0007
color*toytype		2	2	6.1000	0.9021	8	6.76	0.0001
gender*color*toytype	1	1	1	6.2000	1.2757	8	4.86	0.0013
gender*color*toytype	1	1	2	12.6000	1.2757	8	9.88	<.0001
gender*color*toytype	1	2	1	2.0000	1.2757	8	1.57	0.1556
gender*color*toytype	1	2	2	4.2000	1.2757	8	3.29	0.0110
gender*color*toytype	2	1	1	12.4000	1.2757	8	9.72	<.0001
gender*color*toytype	2	1	2	5.6000	1.2757	8	4.39	0.0023
gender*color*toytype	2	2	1	7.6000	1.2757	8	5.96	0.0003
gender*color*toytype	2	2	2	8.0000	1.2757	8	6.27	0.0002

Figure 15.29   The least squares means for all of the effects.

15.11   OUTPUT FOR THE POST-ANOVA ANALYSIS IN SAS

A portion of the results of the simple effects tests are shown in Figures 15.32 and 15.33. The conditions being compared are read in a manner analogous to how we described this process in the earlier chapters. To illustrate this, consider, for example, the eleventh comparison in the tables. The first three coded columns represent one condition and the last three coded columns represent the other condition. We are thus comparing

Type 3 Tests of Fixed Effects				
Effect	Num DF	Den DF	F Value	Pr > F
gender	1	8	4.14	0.0763
color	1	8	19.73	0.0022
toytype	1	8	0.42	0.5331
gender*color	1	8	9.12	0.0166
gender*toytype	1	8	19.73	0.0022
color*toytype	1	8	0.79	0.4003
gender*color*toytype	1	8	11.39	0.0097

These are the *p*-values for the main effects and interaction effects.

Figure 15.30   The results of the ANOVA.

the mean for **gender 1, color 1, toytype 2** (girls who were playing with yellow pretend toys) with **gender 2, color 1, toytype 2** (boys who were playing with yellow pretend toys). The girls played with these toys an average of 12.6 seconds and the boys played with these toys an average of 5.6 seconds. The uncorrected probability associated with the Student's *t* value of 3.88, shown in the last column of Figure 15.32, indicates that

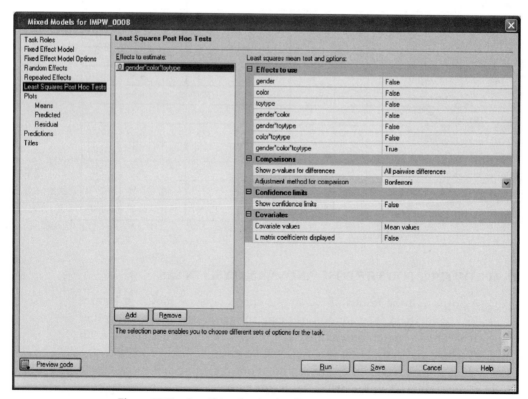

Figure 15.31   Specifying the simple effects tests.

This Table presents the unadjusted *p*-value.

Differences of Least Squares Means											
Effect	gender	color	toytype	gender	color	toytype	Estimate	Standard Error	DF	t Value	Pr>|t|
gender*color*toytype	1	1	1	1	1	2	−6.4000	1.6887	8	−3.79	0.0053
gender*color*toytype	1	1	1	1	2	1	4.2000	1.6887	8	2.49	0.0377
gender*color*toytype	1	1	1	1	2	2	2.0000	1.6887	8	1.18	0.2703
gender*color*toytype	1	1	1	2	1	1	−6.2000	1.8042	8	−3.44	0.0089
gender*color*toytype	1	1	1	2	1	2	0.6000	1.8042	8	0.33	0.7480
gender*color*toytype	1	1	1	2	2	1	−1.4000	1.8042	8	−0.78	0.4601
gender*color*toytype	1	1	1	2	2	2	−1.8000	1.8042	8	−1.00	0.3476
gender*color*toytype	1	1	2	1	2	1	10.6000	1.6887	8	6.28	0.0002
gender*color*toytype	1	1	2	1	2	2	8.4000	1.6887	8	4.97	0.0011
gender*color*toytype	1	1	2	2	1	1	0.2000	1.8042	8	0.11	0.9145
gender*color*toytype	1	1	2	2	1	2	7.0000	1.8042	8	3.88	0.0047
gender*color*toytype	1	1	2	2	2	1	5.0000	1.8042	8	2.77	0.0242
gender*color*toytype	1	1	2	2	2	2	4.6000	1.8042	8	2.55	0.0342
gender*color*toytype	1	2	1	2	2	2	−2.2000	1.6887	8	−1.30	0.2289
gender*color*toytype	1	2	1	2	1	1	−10.4000	1.8042	8	−5.76	0.0004
gender*color*toytype	1	2	1	2	1	2	−3.6000	1.8042	8	−2.00	0.0811
gender*color*toytype	1	2	1	2	2	1	−6.0000	1.8042	8	−3.10	0.0146
gender*color*toytype	1	2	1	2	2	2	−6.0000	1.8042	8	−3.33	0.0105
gender*color*toytype	1	2	2	2	1	1	−8.2000	1.8042	8	−4.55	0.0019
gender*color*toytype	1	2	2	2	1	2	−1.4000	1.8042	8	−0.78	0.4601
gender*color*toytype	1	2	2	2	2	1	−3.4000	1.8042	8	−1.88	0.0962
gender*color*toytype	1	2	2	2	2	2	−3.8000	1.8042	8	−2.11	0.0683
gender*color*toytype	2	1	1	2	1	2	6.8000	1.6887	8	4.03	0.0038
gender*color*toytype	2	1	1	2	2	1	4.8000	1.6887	8	2.84	0.0217
gender*color*toytype	2	1	1	2	2	2	4.4000	1.6887	8	2.61	0.0313
gender*color*toytype	2	1	2	2	2	1	−2.0000	1.6887	8	−1.18	0.2703
gender*color*toytype	2	1	2	2	2	2	−2.4000	1.6887	8	−1.42	0.1930
gender*color*toytype	2	2	1	2	2	2	−0.4000	1.6887	8	−0.24	0.8187

Figure 15.32 Uncorrected tests of significance between means.

this 7-second difference is statistically significant ($p = .0047$). However, the corresponding SAS Bonferroni adjusted probability value shown in Figure 15.33 ($p = .1309$) is not statistically significant.

It is worthwhile noting that the SPSS output showed that its Bonferroni adjusted probability value presented in Figure 15.21 ($p = .025$) is statistically significant. This discrepancy results from different algorithms used in computing effects for repeated measures designs (see Section 10.19).

> This Table presents the Bonferroni adjusted *p*-values.

Differences of Least Squares Means								
Effect	gender	color	toytype	gender	color	toytype	Adjustment	Adj P
gender*color*toytype	1	1	1	1	1	2	Bonferroni	0.1487
gender*color*toytype	1	1	1	1	2	1	Bonferroni	1.0000
gender*color*toytype	1	1	1	1	2	2	Bonferroni	1.0000
gender*color*toytype	1	1	1	2	1	1	Bonferroni	0.2483
gender*color*toytype	1	1	1	2	1	2	Bonferroni	1.0000
gender*color*toytype	1	1	1	2	2	1	Bonferroni	1.0000
gender*color*toytype	1	1	1	2	2	2	Bonferroni	1.0000
gender*color*toytype	1	1	2	1	2	1	Bonferroni	0.0067
gender*color*toytype	1	1	2	1	2	2	Bonferroni	0.0304
gender*color*toytype	1	1	2	2	1	1	Bonferroni	1.0000
gender*color*toytype	1	1	2	2	1	2	Bonferroni	0.1309
gender*color*toytype	1	1	2	2	2	1	Bonferroni	0.6789
gender*color*toytype	1	1	2	2	2	2	Bonferroni	0.9574
gender*color*toytype	1	2	1	2	2	2	Bonferroni	1.0000
gender*color*toytype	1	2	1	2	1	1	Bonferroni	0.0118
gender*color*toytype	1	2	1	2	1	2	Bonferroni	1.0000
gender*color*toytype	1	2	1	2	2	1	Bonferroni	0.4082
gender*color*toytype	1	2	1	2	2	2	Bonferroni	0.2927
gender*color*toytype	1	2	2	2	1	1	Bonferroni	0.0528
gender*color*toytype	1	2	2	2	1	2	Bonferroni	1.0000
gender*color*toytype	1	2	2	2	2	1	Bonferroni	1.0000
gender*color*toytype	1	2	2	2	2	2	Bonferroni	1.0000
gender*color*toytype	2	1	1	2	1	2	Bonferroni	0.1065
gender*color*toytype	2	1	1	2	2	1	Bonferroni	0.6084
gender*color*toytype	2	1	1	2	2	2	Bonferroni	0.8777
gender*color*toytype	2	1	2	2	2	1	Bonferroni	1.0000
gender*color*toytype	2	1	2	2	2	2	Bonferroni	1.0000
gender*color*toytype	2	2	1	2	2	2	Bonferroni	1.0000

Figure 15.33 Bonferroni corrected tests of significance between means.

Although SAS and SPSS frequently yield repeated measures adjusted probabilities that lead researchers to the same conclusions, this is one of those times when they do not. In this instance, we would be inclined to abide by the SPSS results, since the total number of paired comparisons (which SAS uses as its count to perform the Bonferroni correction) is more than we intended to deal with when we configured our simple effects strategy.

15.12   COMMUNICATING THE RESULTS

A $2 \times 2 \times 2$ complex mixed design was used to assess children's toy preferences. Children eight years of age were brought into a room for sixty seconds. The room contained toys, some of which were colored yellow and others of which were colored blue. Furthermore, some of the toys were of a hands-on type (e.g., balls, building blocks) whereas others were pretend toys (e.g., stuffed animals, action figures). These two factors (color and toy type) constituted the within-subjects variables in the design. The between-subjects variable was the gender of the child, and the dependent measure was the number of seconds the children played with each kind and color of toy.

The three-way interaction (Gender × Color × Toy Type) was statistically significant, $F(1, 8) = 38.03$, $p < .05$, which accounted for 7 percent of the within-subjects variance. It is plotted in Figure 15.3. Simple effects analysis indicated the following results.

For the yellow toys, girls played with pretend toys more than they played with the hands-on toys; boys played a comparable amount of time with both types of toys. Furthermore, boys played with the hands-on toys significantly more than girls; there was no difference in the time the girls and boys spent playing with pretend toys.

For the blue toys, neither sex demonstrated a preference for either type of toy. Furthermore, boys spent more time with both hands-on and with pretend toys than the girls.

CHAPTER 15 EXERCISES

15.1. Assume we are interested in the effects of gender, viewing anger management videos, and participation in brief psychotherapy on an anger response scale. The between-subjects factor is gender ($a_1 =$ male and $a_2 =$ female). Male and female clients, suffering from anger management problems, are provided with psychotherapy ($b_1 =$ group and $b_2 =$ individual). Both treatments are given in a randomized order over the course of three weekly sessions (Factor C). The dependent measure is a five-item Anger Response scale with $1 = $ *not at all angry* to $4 = $ *very angry*. The five items are summed, such that individual scores range from 5 to 20. Data were collected at the end of each weekly session. The data are as follows:

	Subject	b_1	b_2	b_1	b_2	b_1	b_2
		c_1		c_2		c_3	
a_1	s_1	18	14	12	10	9	12
	s_2	19	13	11	10	9	11
	s_3	19	12	10	11	10	10
	s_4	18	11	14	10	8	9
a_2	s_5	15	13	8	9	5	5
	s_6	16	12	9	8	5	9
	s_7	16	11	10	7	8	5
	s_8	14	10	9	7	8	5

a. Conduct an ANOVA on these data with SPSS or SAS.
b. If the interaction(s) is(are) statistically significant, conduct simple effects analyses and multiple comparisons tests with SPSS or SAS.

15.2. Assume the management of a large hotel chain was interested in evaluating the effect of type of view (Factor A) and time of day of evaluation (Factor B) on customer satisfaction over a four-day period (Factor C). The between-subjects factor was Factor A, type of view ($a_1 =$ ocean and $a_2 =$ parking lot). The within-subjects factor were Factor B, time of day ($b_1 =$ early morning and $b_2 =$ early afternoon), and Factor C, a block of four contiguous days that the $n = 3$ customers spent at the hotel. The dependent measure was the rating on an overall satisfaction question ($1 = $ *extremely unsatisfied* to 10 *extremely satisfied*). The data are as follows:

	Subject	b_1	b_2	b_1	b_2	b_1	b_2	b_1	b_2
		c_1		c_2		c_3		c_4	
a_1	s_1	8	8	8	8	9	8	9	8
	s_2	9	8	8	7	8	9	10	9
	s_3	8	8	8	8	9	8	10	8
a_2	s_4	3	2	2	2	3	2	1	1
	s_5	4	2	2	1	1	1	1	2
	s_6	3	2	2	1	1	1	1	1

a. Conduct an ANOVA using SPSS or SAS.
b. If the interaction(s) is(are) statistically significant, conduct simple effects analyses and multiple comparisons tests with SPSS or SAS.

15.3. Assume the following $2 \times 3 \times 3$ factorial data set with Factor A as the between-subjects factor and Factors B and C as the within-subjects factors, and $n = 4$ participants per treatment combination.

	Subject	b_1	b_2	b_3	b_1	b_2	b_3	b_1	b_2	b_3
		c_1			c_2			c_3		
a_1	s_1	5	1	2	10	9	7	15	13	15
	s_2	4	6	2	9	9	9	16	12	15
	s_3	3	5	3	8	8	8	17	18	17
	s_4	2	4	1	7	8	9	18	14	17
a_2	s_5	1	5	1	4	5	6	10	11	13
	s_6	2	4	1	3	4	7	10	11	13
	s_7	3	3	2	3	5	5	9	12	12
	s_8	4	2	2	2	5	6	10	11	12

a. Conduct an ANOVA on these data using SPSS or SAS.
b. If the interaction(s) is(are) statistically significant, conduct simple effects analyses and multiple comparisons tests with SPSS or SAS.

SECTION 6

Advanced Topics

CHAPTER SIXTEEN

Analysis of Covariance

16.1 EXPERIMENTAL AND STATISTICAL CONTROL

The designs we have discussed to this point represent techniques of experimental control in order to reduce error variance in a research study (Kirk, 1995). It is also possible to reduce error variance by employing statistical control using analysis of covariance (ANCOVA). Statistical control is used when we know how participants stand (we know their score) on an additional variable that was not part of or could not readily be incorporated into the experimental design. This additional variable is brought into the data analysis by treating it as a *covariate*. Although covariates can be categorical or quantitative, we will limit our discussion and data analysis procedures to quantitatively measured covariates.

The covariate represents a potential source of variance that has not been experimentally controlled but could covary with the dependent variable and could therefore confound our interpretation of the results. By identifying that variable as a covariate and collecting measures of it on the study participants, it is possible to "remove" or "neutralize" its effect on the dependent variable (by statistical procedures described in this chapter) prior to analyzing the effects of the independent variable on the dependent variable.

16.2 A SIMPLE ILLUSTRATION OF COVARIANCE

Consider a test of math skills that we might administer to school children in a particular grade. We wish to compare girls and boys in their math problem solving skills. The test is composed of word problems that are each presented in a few sentences (e.g., "A train leaves from one city traveling east at 45 miles per hour. A second train leaves from another city traveling west at 65 miles per hour . . . "). Our dependent variable is the number of problems correctly answered in the hour that we designate for testing.

Framed as we have just described it, we would say that this study represents a two-group between-subjects design; the independent variable is the gender of the child and the dependent variable is the score on the math test. If we obtained a statistically significant *F* ratio, we would be inclined to conclude that the girls and boys differed in the ability to solve math word problems; if the *F* ratio was not statistically significant – although we cannot accept the null hypothesis (we can only fail to reject it) – we might be inclined to entertain the possibility that the girls and boys did not differ in their ability to solve math word problems.

Now consider this complication. Performance on math word problems may be affected by factors other than math ability. Among such other factors might be those concerning how verbally proficient the children are, how motivated they are to do well on the test, their ability to concentrate during the test, and a host of other factors. The dependent variable, number of correctly answered problems, results from the compilation of everything we just listed, including the actual focus of the study: math word problem ability.

ANCOVA is one way to investigate the effects of one or more of these other concomitant (Kirk, 1995) or potentially confounding variables. Maxwell and Delaney (2000, p. 356) express it very well:

The logic of ANCOVA is to address the conditional question of Would the groups have been different on the ... [dependent measure] ... if they had been equivalent on the covariate?

Applying Maxwell and Delaney's question to our illustration, we could ask this: To what extent might we have obtained a gender difference had the groups been equivalent in their verbal ability and/or their motivation and/or their concentration and so on? Had we thought of these issues prior to the study, we might have been able to collect information on one or more of them just prior to the time of the data collection. These variables could then have been used as covariates in our analysis.

16.3   THE EFFECT OF A COVARIATE ON GROUP DIFFERENCES

In our illustration, there were two possible outcomes on the one-way ANOVA: the *F* ratio could have been statistically significant or not. When one or more covariates are included in the analysis – transforming the design into an ANCOVA – we have four possible outcome scenarios, only one of which will materialize. Let's use verbal ability as the single covariate in an ANCOVA design to simplify our discussion; the following are then the four possible alternative outcome scenarios.

- An ANOVA results in the researchers determining that the groups are significantly different; when the dependent variable of number of math word problems solved is "statistically adjusted" to take verbal ability into consideration by using it as a covariate, an ANCOVA still shows the groups to be significantly different. Even here, however, the strength of the effect of the independent variable (on the dependent variable) might be similar or quite different in the two analyses.
- An ANOVA results in the researchers determining that the groups are significantly different; when the dependent variable of number

of math word problems solved is statistically adjusted to take verbal ability into consideration by using it as a covariate, an ANCOVA now indicates that the groups do not differ significantly on the dependent variable.

- An ANOVA results in the researchers determining that the groups are not significantly different; when the dependent variable of number of math word problems solved is statistically adjusted to take verbal ability into consideration by using it as a covariate, an ANCOVA still shows that the groups do not differ significantly on the dependent variable.

- An ANOVA results in the researchers determining that the groups are not significantly different; when the dependent variable of number of math word problems solved is statistically adjusted to take verbal ability into consideration by using it as a covariate, an ANCOVA now shows the groups to be significantly different.

These four alternative outcomes are all possible. In two of them, the researchers draw the same conclusion from the analysis (the groups differ significantly, the groups do not differ significantly). But in two of them the ANCOVA reverses the conclusion that would have been drawn from the ANOVA, and these circumstances are much more striking.

In one of these reversal-of-outcome circumstances, the researchers conclude, based on the ANOVA, that there is a difference between girls and boys in math word problem ability. However, when verbal ability is statistically controlled by using ANCOVA, that apparent difference no longer manifests itself. Thus, the difference in problem solving attributed to gender in the ANOVA was now explained by verbal ability in the ANCOVA. When the independent variable of gender was finally evaluated in the ANCOVA, it was unable to account for a statistically significant portion of the variance in math score remaining after the effects of verbal ability were first accounted for.

In the other reversal-of-outcome circumstance, the researchers conclude, based on the ANOVA, that girls and boys perform comparably in math word problem ability. However, when verbal ability is statistically controlled by using ANCOVA, a statistically significant difference between the genders is obtained. Here, verbal ability can be thought of as masking or compensating for the effect of gender. When verbal ability is statistically controlled by removing the variance in math scores associated with it, the difference between girls and boys can be detected.

16.4 THE PROCESS OF PERFORMING ANCOVA

The key to understanding ANCOVA is not fundamentally different from the key to understanding ANOVA: We attempt to explain the variance of the dependent variable on the basis of the effects of the other variables in the design. In ANOVA, these other effects are the main effects of the independent variables and their interactions. In ANCOVA, we still have effects of the independent variables and their interactions, but now we also have the effect of the covariate (we will limit our discussion to the situation where we have a single covariate). As is true for the effects of the independent variables and their interactions, we now also use the covariate to explain variance of the dependent variable.

There are three steps that are involved in performing an ANCOVA: (a) using the covariate to predict the dependent variable, (b) adjusting the dependent variable to remove the effects of the covariate, and (c) performing an ANOVA on the adjusted dependent variable scores.

16.4.1   PREDICTING THE DEPENDENT VARIABLE FROM THE COVARIATE

We use a covariate in the design because we have hypothesized that it may bear a relationship to (covary with) the dependent variable. We thus want to "remove" its effect before we evaluate the effect of our independent variable.

To say that the covariate may be related to the dependent variable is to say that the covariate may be *correlated* to the dependent variable. To the extent that two variables are correlated, we may *predict* the values of one from the values of the other. Prediction is evaluated by means of a linear regression procedure, which results in a prediction equation often referred to as a *linear regression model.*

In this first major step of ANCOVA, we use the covariate as a predictor of the dependent variable without taking into account group membership, that is, without involving the independent variable. Hence, this analysis is performed on the sample as a whole; cases in the analysis are pooled regardless of their group. Because our focus is on the covariate and not on the independent variable, the covariate is thus given the first opportunity to account for the variance of the dependent variable; the independent variable is not yet involved in the analysis.

If the covariate and the dependent variable are correlated, the linear regression model (using the covariate as the predictor) will account for a significant amount of the variance of the dependent variable. The strength of this relationship can be indexed by a *squared multiple correlation coefficient* computed by dividing the sum of squares associated with the regression model by the sum of squares of the total variance of the dependent variable; it is ordinarily shown in the ANCOVA summary table as an R squared value.

16.4.2   ADJUSTING THE DEPENDENT VARIABLE VALUES

The regression model uses the scores on the covariate of all participants to predict their scores on the dependent variable. At the completion of the regression procedure, each case in the data file has a predicted dependent variable score. These predicted values will be different from the original scores for most if not all of the cases in that they reflect only what we are able to predict from the covariate. Using the variables in our earlier illustration, we would have predicted the number of math problems solved based only on a knowledge of each child's verbal ability score. These predicted values therefore represent (they have "used up") all of the information that we have available to us in the verbal ability variable; that is, verbal ability has done all that it can do for us in predicting or explaining the variance in math scores.

There are several ways to express the idea that the predicted number of math problems solved reflects the absence of their verbal ability component:

- The effect verbal ability has been statistically nullified.
- The effect verbal ability has been statistically controlled.
- The effect verbal ability has been statistically removed.
- The effect verbal ability has been statistically partialled out.
- The effect verbal ability has been statistically equated in the groups.

You will likely see one or more of these expressions used in the literature when the authors are describing a covariance procedure.

The predicted values from the linear regression procedure can be viewed as scores on the dependent measures that have been adjusted by removing the effects of the covariate from them. These values are referred to as *adjusted values* of the dependent variable in that they no longer contain information related to the covariate. Both the originally observed as well as these predicted or adjusted values of the dependent variable exist side by side during the process of computing the ANCOVA; SPSS and SAS allow the predicted values to be saved to the data file for future reference.

16.4.3   PERFORMING AN ANOVA ON THE ADJUSTED DEPENDENT VARIABLE

In the final stage of ANCOVA, we use as the dependent variable not the scores on the observed dependent variable but rather the adjusted values of the dependent variable that were generated from the linear regression analysis. In terms of our illustration, in the ANCOVA we evaluate the effect of gender on the adjusted number of math problems solved. Because the effect of verbal ability is not reflected in (has been statistically removed from) these adjusted values, any differences between the groups can be more confidently attributed to the independent variable.

16.4.4   EXAMINING MEAN DIFFERENCES IN ANCOVA

A simple way of thinking about ANCOVA is that it is just an ANOVA performed on adjusted rather than the original or observed scores on the dependent variable. Because the original scores are adjusted, the groups are compared on the adjusted scores. Therefore, if a statistically significant *F* ratio is obtained for the independent variable in an ANCOVA, it means that *the groups differ on the adjusted means*, which may be quite different from a comparison of the means based on the observed dependent variable scores. If we wish to perform a multiple comparisons test to examine the mean differences of the groups, this test must be performed on the *adjusted* and not the observed means. Furthermore, if we wish to present the group means on which the comparison was based, we must report the *adjusted* rather than the observed means.

16.5 ASSUMPTIONS OF ANCOVA

ANCOVA is subject to all of the assumptions underlying ANOVA. These are as follows:

- Normal distribution of the dependent variable.
- Independence of variance estimates.
- Homogeneity of variance.
- Random sampling.

These assumptions have been discussed in Chapter 5. In addition, there are two additional assumptions that are important to meet when performing an ANCOVA:

- Linearity of regression.
- Homogeneity of regression.

16.5.1 LINEARITY OF REGRESSION

We discussed in Section 16.4.1 that, based on the sample as a whole, the scores on the covariate are used in a linear regression procedure to predict the scores of the dependent variable. In order to properly interpret the results of the regression procedure, it is assumed that the relationship between the two variables is linear. Technically, the linear regression procedure evaluates the predictability of the dependent measure based on a linear model incorporating the covariate; if the dependent variable and the covariate are not related linearly (even if they are strongly related in a more complex way), the linear regression procedure will return an outcome of "no viable prediction."

The most common way to determine if the data meet this linearity assumption is to graph the data in a scatterplot. The y axis of such a plot represents the dependent variable and the x axis represents the covariate. Each data point represents the coordinate of these two for each case. For example, consider the simplified data set shown in the first four columns in Table 16.1.

Although there are two different groups represented in the data file as shown in the first column of Table 16.1, the linearity of regression assumption is evaluated on the sample as a whole. Thus, regardless of group membership, Case 1 scored a 1 on the dependent variable and a 1 on the covariate, Case 2 scored a 2 on the dependent variable and a 5 on the covariate, and so on.

Table 16.1. A small hypothetical data set

Group membership code	Case number	Score on dependent variable	Score on covariate	Adjusted score on dependent variable
1	1	1	1	0.54610
1	2	2	5	2.29078
1	3	3	7	3.16312
2	4	5	6	5.31560
2	5	6	8	6.18794
2	6	8	11	7.49645
Mean	4.167	6.33	4.167	

The scatterplot for these data is shown in Figure 16.1. As can be seen in the plot, it appears that the line of best fit for these data points is a straight line. This is confirmed by noting that the Pearson correlation of the dependent variable and the covariate (the Pearson r assesses the degree to which variables are linearly related) is 0.904. Thus, we would conclude that these data meet the linearity of regression assumption.

To give you a sense of what adjusted scores look like, we present the adjusted dependent variable values in the last column of Table 16.1. These were generated by the SPSS GLM procedure (by saving the predicted values in the ANCOVA analysis) and are shown to five decimal points to reinforce the idea that these values are statistically produced.

Based on the straight line fit through the data points of Figure 16.1 (the linear model), given a value on the covariate of, say, 8, we would

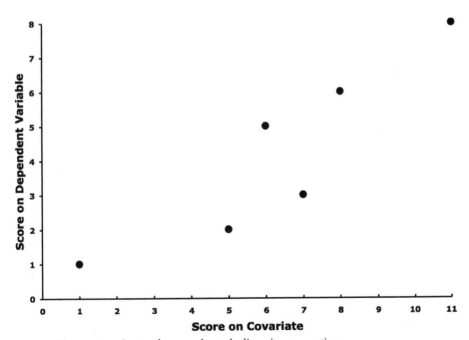

Figure 16.1 Scatterplot to evaluate the linearity assumption.

Table 16.2. Observed and adjusted group means for the data shown in Table 16.1

Group membership code	Mean of observed dependent variable	Mean of adjusted dependent variable
1	2.00	2.827
2	6.33	5.461

predict a value on the dependent variable of 6.18794. The observed value on the dependent variable is 6 and so the model overpredicted the score of Case 5 by a small margin. Not all of the scores were overpredicted; the model underpredicted the observed values twice (Cases 1 and 6).

The last row of Table 16.1 presents the means for the dependent variable, the covariate, and the adjusted dependent variable. Note that the adjusted overall mean is the same as the observed overall mean because the center of the distribution of dependent values has not shifted.

The observed and adjusted means of the groups are presented in Table 16.2. The observed means for Groups 1 and 2 are 2.00 and 6.33, respectively. Note that the adjusted group means are different from the observed means. The adjusted mean of Group 1 is 2.827, and the adjusted mean of Group 2 is 5.461. We should note that these adjusted group means are not the simple average of the adjusted scores for the cases in the respective groups; rather, they are the adjusted means for the two groups at the mean value of the covariate (which is 6.33). In an ANCOVA, it is these adjusted means, not the observed means, that are being evaluated by the F ratio associated with the group effect.

16.5.2 HOMOGENEITY OF REGRESSION

We have just seen that the assumption of linearity of regression is tested on the sample as a whole typically by visually inspecting a scatterplot of the data. In contrast, the assumption of homogeneity of regression focuses on the individual groups and is typically evaluated by performing a statistical analysis.

Every linear regression model has a value for the slope of the line; the slope indexes how steeply the function rises or falls with respect to the x axis. The slope of interest in testing the assumption of homogeneity of regression is associated with the covariate as it predicts the dependent variable, as we described in Section 16.5.1. What is of interest here is not the results based on the sample as a whole; instead, we examine each group separately. *Homogeneity of regression assumes that the slope of the regression line is the same for each group.* When the slopes of the regression models for the individual groups are significantly different, that is, when the slope for one group differs significantly from the slope of at least one other group, then the assumption of homogeneity of regression has been violated.

The way in which this assumption is statistically evaluated makes use of a concept that we have discussed from Chapter 8 on: the presence of an interaction effect. We have seen many times in Chapters 8–15 that an interaction is obtained when the lines for the levels of an independent variable are not parallel. Another way to express the idea of the lines not being parallel is to say that the slopes of the lines are different.

Thus, the way in which we test the homogeneity of regression assumption is by setting up an analysis containing the interaction of the experimental effect(s) of the independent variable(s) and the covariate. This can be done relatively easily in SPSS and SAS. In our simplified example study shown in Table 16.1, where we have only one independent variable, there is only one experimental effect, namely, the main effect of group. Keeping with our example, if the Group × Covariate interaction effect is not statistically significant, then we presume that the slopes are comparable and that the assumption of homogeneity of regression is satisfied; if the Group × Covariate interaction effect is statistically significant, then we presume that the slopes are not parallel and that the assumption of homogeneity of regression is violated.

When the assumption of homogeneity of regression is not met, researchers have some options, three of which are briefly mentioned here. First, there are some nonparametric ANCOVA procedures that may be used; a sample of these is described by Bonate (2000) and Maxwell, Delaney, and O'Callaghan (1993). Second, with moderate to pronounced departures from the assumption of homogeneity of regression, researchers can engage in a relatively more complex analysis that assesses the treatment effect(s) as a function of the value of the covariate; this approach is discussed by Maxwell and Delaney (2000). Third, although it is not necessarily the preferred strategy, researchers can proceed with the ANCOVA, recognizing that with mild to moderate violations of the assumption (a) the analysis will be able to withstand such violations and (b) the effects of the violation will be in the conservative direction (Maxwell & Delaney, 2000).

16.6   NUMERICAL EXAMPLE OF A ONE-WAY ANCOVA

16.6.1   A BRIEF DESCRIPTION OF OUR EXAMPLE

In our hypothetical numerical example, researchers have administered a test of math word problems to thirty-six school children in a particular grade who were exposed to one of three different math training programs during the school year. Children in Group 1 (Traditional) learned math under the traditional curriculum in place for decades in the school district. Children in Group 2 (Freeform) were allowed to interact with the material in whatever way they chose during the math sessions; the teacher was

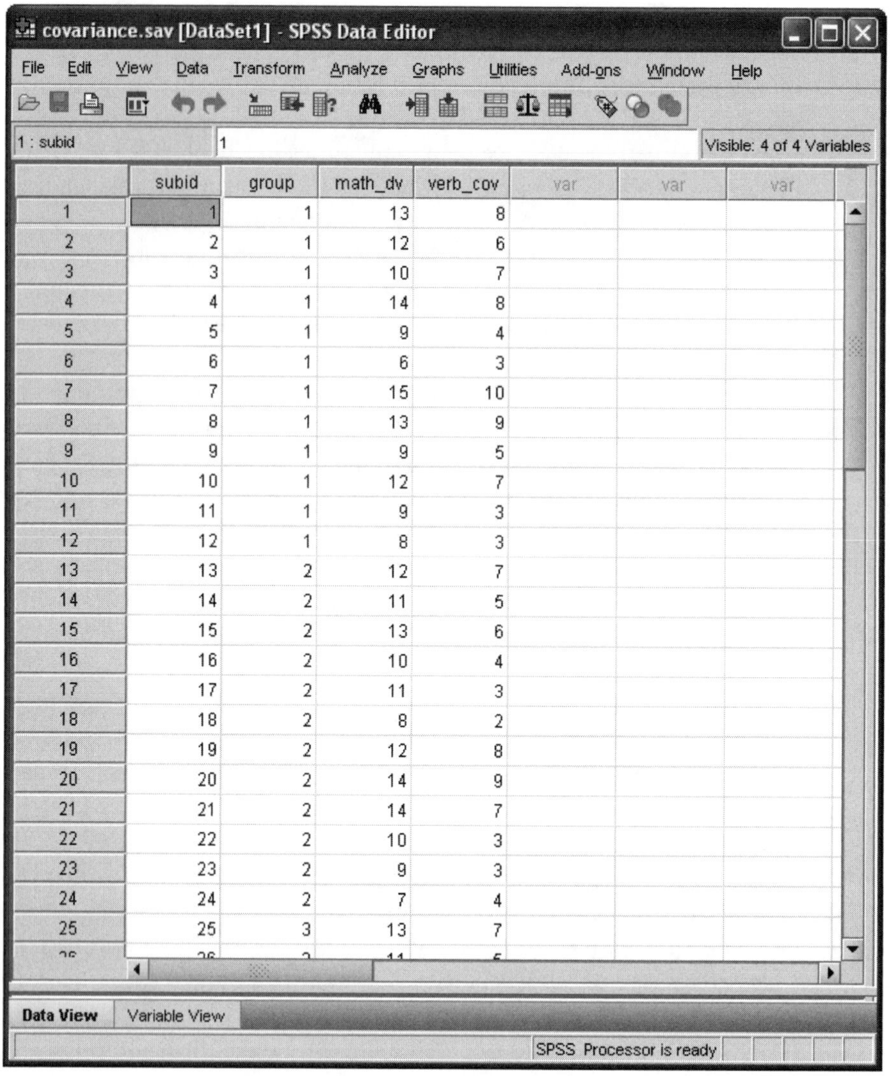

	subid	group	math_dv	verb_cov
1	1	1	13	8
2	2	1	12	6
3	3	1	10	7
4	4	1	14	8
5	5	1	9	4
6	6	1	6	3
7	7	1	15	10
8	8	1	13	9
9	9	1	9	5
10	10	1	12	7
11	11	1	9	3
12	12	1	8	3
13	13	2	12	7
14	14	2	11	5
15	15	2	13	6
16	16	2	10	4
17	17	2	11	3
18	18	2	8	2
19	19	2	12	8
20	20	2	14	9
21	21	2	14	7
22	22	2	10	3
23	23	2	9	3
24	24	2	7	4
25	25	3	13	7

Figure 16.2 The data file for our numerical example.

always available as a resource. Children in Group 3 (Software) learned math through math training software developed by a team of faculty and students from the local university; again, the teacher was available as a resource. The dependent variable was the number of math word problems correctly answered in the hour that was allocated for testing. Hypothesizing that verbal ability should be used as a covariate in this study, the researchers administered a test of verbal proficiency to the children a few days prior to the math test.

The SPSS data file for this study is shown in Figure 16.2. The variable named **subid** is our participant identification code. The **group** variable represents the type of educational program to which the children were exposed; the codes of 1, 2, and 3 stand for Traditional, Freeform, and

Software, respectively. The variable named **math_dv** indicates the number of math word problems solved correctly; we have used the letters *dv* in the ending of the name to help you remember that this is the dependent variable. The variable named **verb_cov** indicates the score on the verbal proficiency test (higher scores indicate greater verbal proficiency); we have used the letters *cov* in the ending of the name to help you remember that this variable is used as the covariate.

16.6.2   OUR DATA ANALYSIS STRATEGY

We will perform the data analysis for this numerical example by using SPSS and then by using SAS, foregoing the hand calculations for the sake of space. The analysis will proceed as follows:

- First, we will run the simplified ANOVA design, using **group** as the independent variable and **math_dv** as the dependent variable. This will establish a context for the covariance analysis to show us what we would have obtained without the covariate in the analysis.
- Second, in preparation for the covariance analysis, we will evaluate the data to ensure that the assumptions of linearity and homogeneity of regression are met.
- Third, we will perform the ANCOVA, using **group** as the independent variable, **math_dv** as the dependent variable, and **verb_cov** as the covariate.

16.7   PERFORMING THE ANOVA IN SPSS

From the main SPSS menu select **Analyze** ➜ **General Linear Model** ➜ **Univariate**. Selecting this path will open the dialog window shown in Figure 16.3. We have configured it with **group** as the fixed factor and **math_dv** as the dependent variable. Because the only information of interest in this analysis is the main effect of **group**, we click **OK** to run the basic ANOVA.

The summary table for the ANOVA is shown in Figure 16.4. As can be seen from the output, the *F* ratio associated with **group** is not statistically significant. Based only on this information, it would ordinarily be concluded that the educational training programs do not differ in their effectiveness of teaching math word problem solving; that is, there are no significant differences between the means of the three groups.

16.8   EVALUATING THE ANCOVA ASSUMPTIONS IN SPSS

16.8.1   LINEARITY OF REGRESSION

The first assumption we will evaluate is the linearity of regression. The goal here is to view the scatterplot of the covariate (verbal proficiency) and the dependent variable (number of math problems correctly solved).

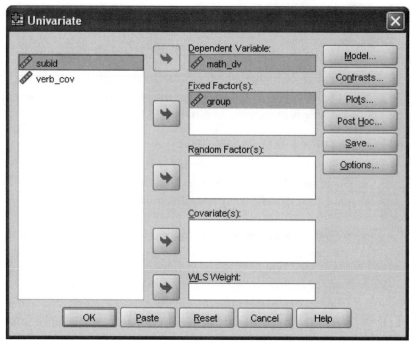

Figure 16.3 The main **GLM Univariate** dialog window configured for ANOVA.

If we determine that the relationship between these two variables is linear, then the assumption of linearity of regression will have been satisfied.

From the main SPSS menu select **Graphs → Legacy Dialogs → Interactive → Scatterplot**. Selecting this path will open the dialog window shown in Figure 16.5. The window opens on the **Assign Variables** tab. Drag **math_dv** to the **y Axis** panel, which is located on the vertical arrow (always place the dependent variable on the *y* axis). Then drag **verb_cov** to the **x Axis** panel, which is located on the vertical arrow.

Select the **Fit** tab. Under the **Method** drop-down menu, the default shows **None** (see Figure 16.6). Choose **Regression** as shown in Figure 16.7.

Tests of Between-Subjects Effects

Dependent Variable: MATH_DV

Source	Type III Sum of Squares	df	Mean Square	F	Sig.
Corrected Model	1.722[a]	2	.861	.168	.846
Intercept	4378.028	1	4378.028	853.618	.000
GROUP	1.722	2	.861	.168	.846
Error	169.250	33	5.129		
Total	4549.000	36			
Corrected Total	170.972	35			

a. R Squared = .010 (Adjusted R Squared = −.050)

The effect of group (representing the training variable) is not statistically significant in this ANOVA.

Figure 16.4 The summary table for the ANOVA.

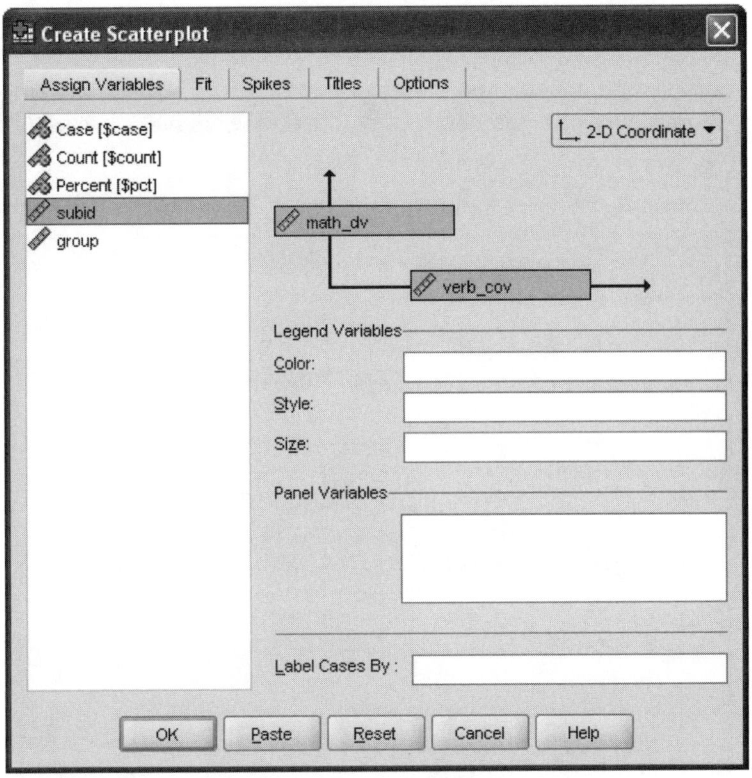

Figure 16.5 The **Create Scatterplot** dialog window.

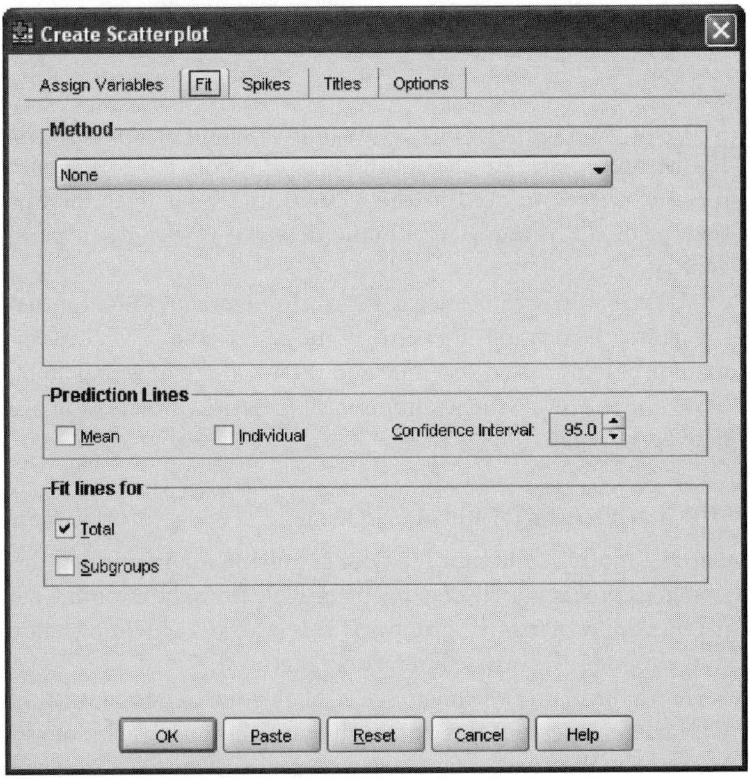

Figure 16.6 The default **Method** is **None**.

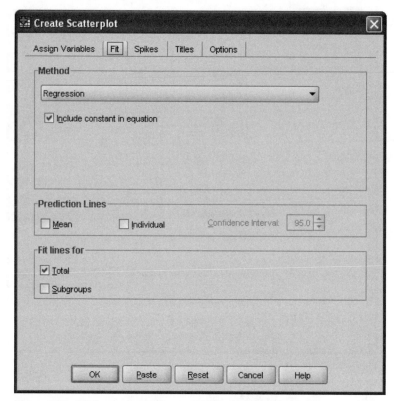

Figure 16.7   We have selected **Regression** on the **Method** menu.

Make sure that the checkbox for **Include constant in equation** is checked. Furthermore, because the linearity assumption is tested on the sample as a whole, we need to check **Total** under **Fit lines for** toward the bottom of the window (this is the default). Click **OK** to perform the analysis.

The resulting scatterplot is shown in Figure 16.8. As can be seen in the figure, the relationship between the dependent variable of number of math problems solved correctly and the covariate of verbal ability seems to be linear. Hence, the assumption of linearity of regression appears to have been met.

16.8.2   HOMOGENEITY OF REGRESSION

The assumption of homogeneity of regression maintains that the regression lines predicting the dependent variable from the covariate have comparable slopes across the groups. This is tested by obtaining a nonsignificant Groups × Covariate interaction effect.

From the main SPSS menu select **Analyze → General Linear Model → Univariate.** Selecting this path will open the dialog window shown in Figure 16.9. We have configured it with **group** as the fixed factor, **math_dv** as the dependent variable, and **verb_cov** as the covariate.

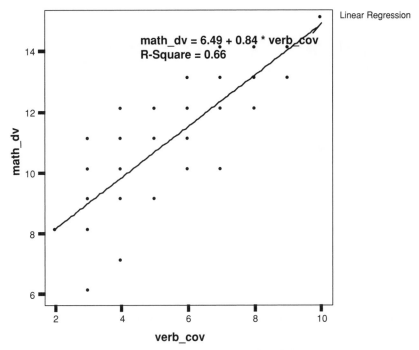

Figure 16.8   The scatterplot of the dependent variable and the covariate.

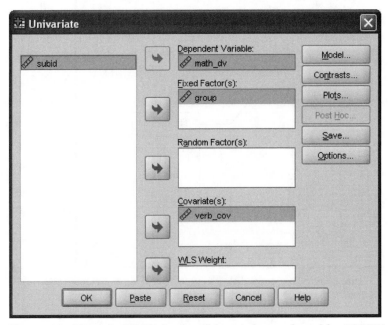

Figure 16.9   The main **GLM Univariate** dialog window configured for ANCOVA.

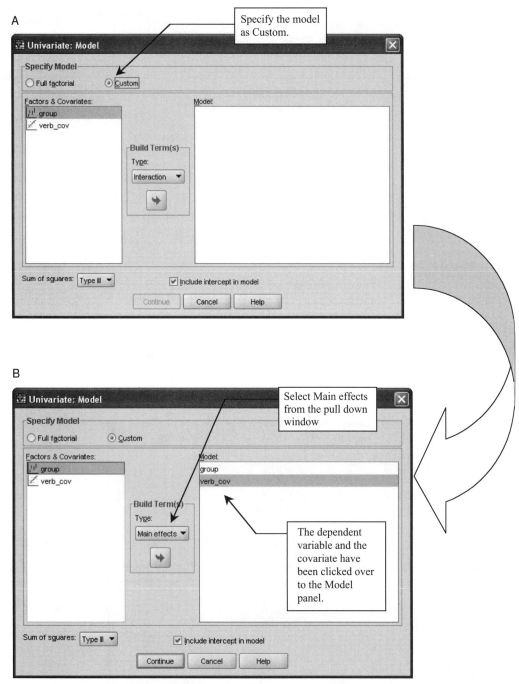

Figure 16.10 The **Model** dialog window, specifying A, **Custom** model, and B, **Main effects**.

Select the **Model** pushbutton to reach the dialog screen shown in Figure 16.10A. Click **Custom** in the **Specify Model** area and select **Main effects** from the pull-down menu under **Build Terms**. Now select **group** and **verb_cov** in the **Factors & Covariates** panel and click them over to the **Model** panel. This is shown in Figure 16.10B.

A

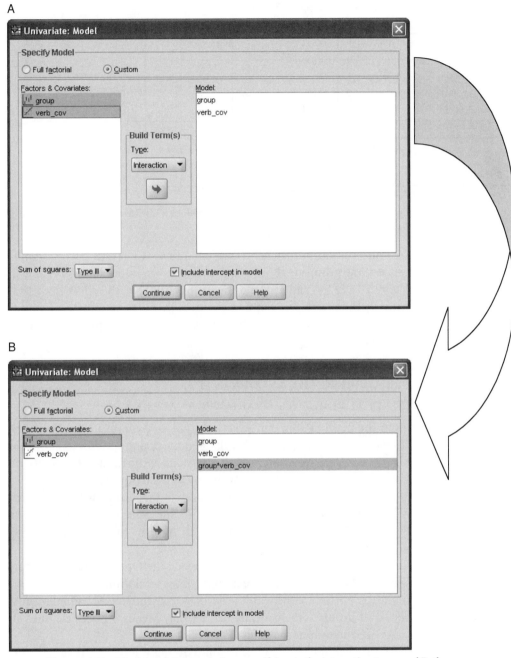

B

Figure 16.11 The **Model** dialog window, specifying A, the interaction term, and B, the **group∗verb_cov** interaction.

Now select **Interaction** from the pull-down menu (replacing the **Main effects** choice). Click on both **group** and **verb_cov** by holding down the control button while clicking them one at a time. Clicking this set over to the **Model** panel specifies the **group∗verb_cov** interaction as shown in Figure 16.11B. Click **Continue** to return to the main GLM window and click **OK** to run the analysis.

Tests of Between-Subjects Effects

Dependent Variable: MATH_DV

Source	Type III Sum of Squares	df	Mean Square	F	Sig.
Corrected Model	127.219[a]	5	25.444	17.446	.000
Intercept	153.214	1	153.214	105.052	.000
GROUP	5.156	2	2.578	1.768	.188
VERB_COV	102.362	1	102.362	70.185	.000
GROUP * VERB_COV	1.598	2	.799	.548	.584
Error	43.754	30	1.458		
Total	4549.000	36			
Corrected Total	170.972	35			

The group × verb_cov interaction effect is not statistically significant.

a. R Squared = .744 (Adjusted R Squared = .701)

Figure 16.12 The summary table for our custom model testing the homogeneity of regression assumption.

The only output in which we are interested is the test of significance of the **group*verb_cov** interaction shown in the summary table in Figure 16.12. As can be seen in the summary table, the effect is not statistically significant. We can thus presume that the assumption of homogeneity of regression has not been violated and we can proceed with the ANCOVA.

16.9 PERFORMING THE ANCOVA IN SPSS

16.9.1 STRUCTURING THE COVARIANCE ANALYSIS

From the main SPSS menu select **Analyze ➜ General Linear Model ➜ Univariate.** Selecting this path will open the dialog window as shown in Figure 16.9. We have configured it with **group** as the fixed factor, **math_dv** as the dependent variable, and **verb_cov** as the covariate.

Select the **Model** pushbutton to reach the dialog window shown in Figure 16.10. Click **Full factorial** in the **Specify Model** area (see Figure 16.13) and click **Continue** to return to the main dialog window.

Select the **Options** pushbutton to reach the **Options** dialog window shown in Figure 16.14. We are going to do several things in this window. In the top portion of the window devoted to **Estimated Marginal Means**, click over **group** to the panel named **Display Means for** because this will cause SPSS to output the adjusted group means.

Then click the checkbox for **Compare main effects** and select **Bonferroni** from the drop-down menu for **Confidence interval adjustment** as also shown in Figure 16.14. We are here anticipating the possibility of obtaining a significant main effect of **group** and want to perform our multiple comparisons tests. Because the ANCOVA assesses the differences between the adjusted means of the groups, we cannot use the **Post Hoc** dialog window; those post hoc tests are performed on the observed means. By checking the **Compare main effects** option we will be using a

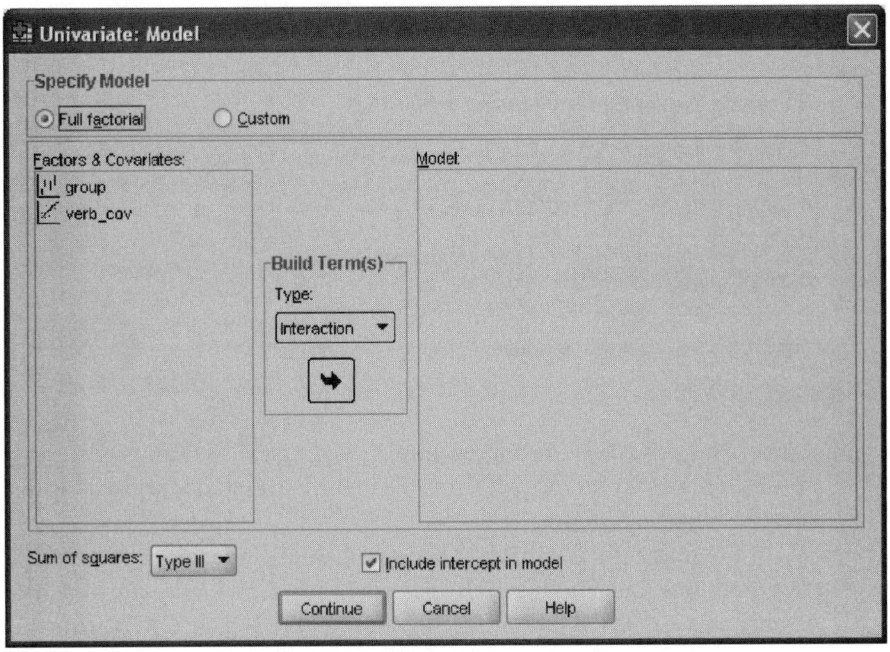

Figure 16.13　The **Univariate: Model** window set to full factorial.

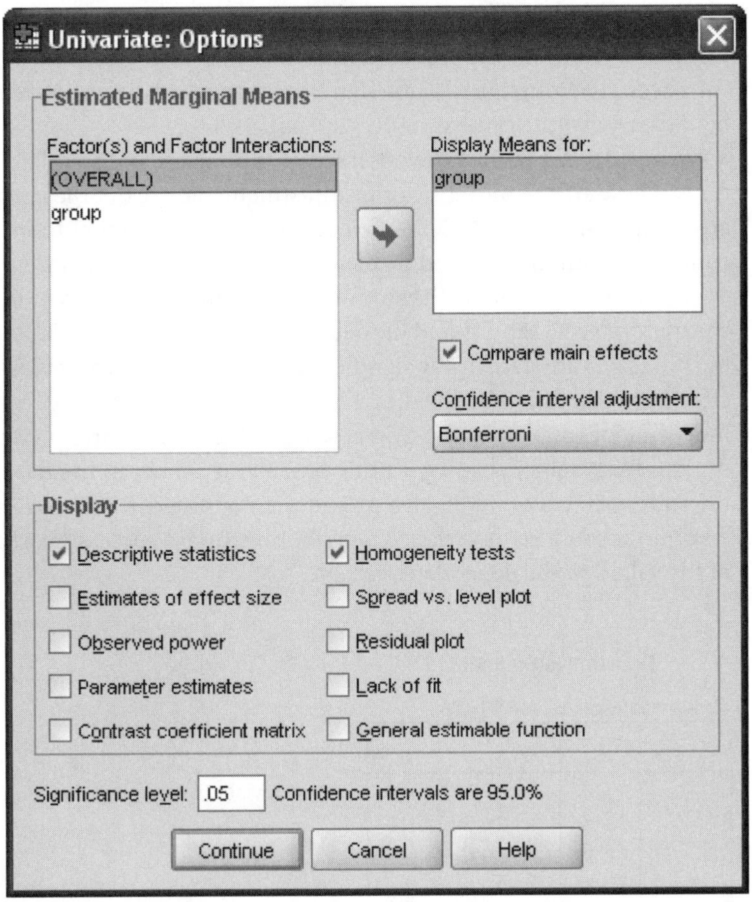

Figure 16.14　The **Options** dialog window.

Levene's Test of Equality of Error Variances[a]

Dependent Variable: MATH_DV

F	df1	df2	Sig.
.105	2	33	.901

> The equal variances assumption is not violated, probabilities (Sigs.) are > .05

Tests the null hypothesis that the error variance of the dependent variable is equal across groups.

a. Design: Intercept+VERB_COV+GROUP

Figure 16.15   Levene's test results.

Bonferroni adjusted t-test procedure to compare the adjusted group means. This is why this option is located in the **Estimated Marginal Means** area.

In the lower portion of the **Options** dialog window, select **Descriptive statistics** and **Homogeneity tests** as shown in Figure 16.14. Click **Continue** to reach the main dialog window and click **OK** to run the analysis.

16.9.2   THE OUTPUT OF THE COVARIANCE ANALYSIS

Figure 16.15 shows the results of Levene's test of the equality of error variances. The outcome is not significant, and so we cannot reject the hypothesis that the error variances are equal.

Figure 16.16 presents the observed group means and standard deviations. These are the means prior to adjustment and are not the means that are compared in the ANCOVA. We show them here so that they may be contrasted with the adjusted means we will show in a moment.

Figure 16.17 contains the summary table for the ANCOVA. Both the covariate of verbal ability and the effect of the independent variable (training program) are statistically significant. The full eta squared value for the effects are computed by dividing the sum of squares for the effect of the independent variable by the **Corrected Total** sum of squares presented in the summary table, yielding a value of .735 (as shown in the footnote to the summary table). SPSS also provides an adjusted R squared value to take into account error variance contributing toward successful prediction of the values of the dependent variable.

Descriptive Statistics

Dependent Variable: MATH_DV

GROUP	Mean	Std. Deviation	N
1	10.83	2.725	12
2	10.92	2.234	12
3	11.33	1.723	12
Total	11.03	2.210	36

Figure 16.16   Observed means and standard deviations for the groups.

The outcome of this ANCOVA leads the researchers to very different conclusions than the outcome of the ANOVA we performed earlier. Based on the ANOVA it would be concluded that the training programs did not result in different levels of performance on the math word problem test. The ANCOVA overrules that conclusion; by statistically controlling for verbal ability, we find that there is some difference between the training programs that is statistically significant.

The top portion of Figure 16.18 presents the adjusted (estimated marginal) means for the three groups. Because these are predicted values, the most appropriate indication of their stability is the standard error associated with these predictions. In reporting these means, we report the standard errors rather than the standard deviations. Note that these predicted (adjusted) group means are different than the observed means shown in Figure 16.16. The *F* ratio for the **group** effect evaluated the differences of these adjusted means.

The bottom portion of Figure 16.18 provides the results of the pairwise comparisons of the adjusted means. Based on these results, it appears that Group 1 (Traditional) performed significantly lower on the math test than Groups 2 (Freeform) and 3 (Software) when verbal ability was controlled; these latter two groups appeared to perform comparably.

16.10   PERFORMING THE ANOVA IN SAS

The data file that we use to perform our ANCOVA in *SAS Enterprise Guide* is identical to what we used in SPSS. It is shown in Figure 16.19 in an Excel file just prior to being imported into SAS.

We first perform an ANOVA without taking the covariate into consideration. From the main menu select **Analyze → ANOVA → Linear Models.** As shown in Figure 16.20, specify **math_dv** as the **Dependent variable** and **group** as the **Classification variable.**

Select **Model** in the navigation panel on the far left of the screen. Highlight **group** and click the **Main** pushbutton in the middle of the window. This action will place **group** into the **Effects** panel as shown in Figure 16.21.

Select **Model Options** in the navigation panel. Select only **Type III** as shown in Figure 16.22.

We also wish to obtain the observed means for the groups. This can be accomplished by selecting the **Arithmetic** portion of the **Post Hoc Tests** window in the navigation panel shown in Figure 16.23. Click **Add** to show the effects in the model and to display the panels in the **Options for means tests.** Under **Class effects to use** (the first panel on the right portion of the window), set the **group** effect to **True** (it is initially set to **False** – double-clicking it displays the **True/False** menu from which you select **True**). Click **Run** to perform the analysis.

The output of the ANOVA is shown in Figure 16.24. This matches the results produced by SPSS.

These are the *p*-values for the main effects and covariate effect.

Tests of Between-Subjects Effects

Dependent Variable: MATH_DV

Source	Type III Sum of Squares	df	Mean Square	F	Sig.
Corrected Model	125.620[a]	3	41.873	29.546	.000
Intercept	169.433	1	169.433	119.551	.000
VERB_COV	123.898	1	123.898	87.422	.000
GROUP	13.029	2	6.515	4.597	.018
Error	45.352	32	1.417		
Total	4549.000	36			
Corrected Total	170.972	35			

a. R Squared = .735 (Adjusted R Squared = .710)

Figure 16.17   The summary table for the ANCOVA.

Estimates

Dependent Variable: MATH_DV

GROUP	Mean	Std. Error	95% Confidence Interval Lower Bound	95% Confidence Interval Upper Bound
1	10.202[a]	.350	9.488	10.915
2	11.195[a]	.345	10.492	11.897
3	11.687[a]	.346	10.983	12.391

a. Evaluated at covariates appeared in the model: VERB_COV = 5.39.

These are the probabilities of the differences occurring by chance alone for the pairwise comparisons.

Pairwise Comparisons

Dependent Variable: MATH_DV

(I) GROUP	(J) GROUP	Mean Difference (I-J)	Std. Error	Sig.[a]	95% Confidence Interval for Difference[a] Lower Bound	95% Confidence Interval for Difference[a] Upper Bound
1	2	−.993	.496	.161	−2.245	.260
	3	−1.485*	.497	.016	−2.742	−.229
2	1	.993	.496	.161	−.260	2.245
	3	−.492	.486	.956	−1.720	.736
3	1	1.485*	.497	.016	.229	2.742
	2	.492	.486	.956	−.736	1.720

Based on estimated marginal means

*. The mean difference is significant at the .05 level.

a. Adjustment for multiple comparisons: Bonferroni.

Figure 16.18   Bonferroni pairwise comparisons.

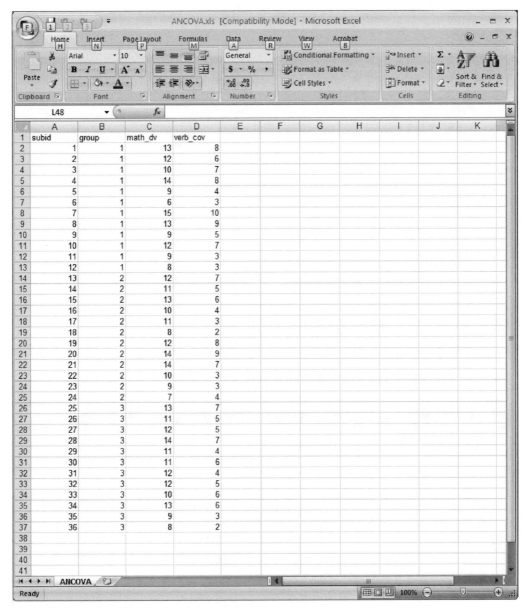

Figure 16.19   The data file equivalent to what we used in SPSS.

16.11   EVALUATING THE ANCOVA ASSUMPTIONS IN SAS

16.11.1   LINEARITY OF REGRESSION

The assumption that the dependent variable and the covariate are related in a linear manner is most easily assessed by examining the scatterplot. If you have not already done so, import the data file from Excel into an *SAS Enterprise Guide* project. From the main menu select **Graph ➔ Scatter Plot**. This brings you to the screen shown in Figure 16.25. Select **2D Scatter Plot**.

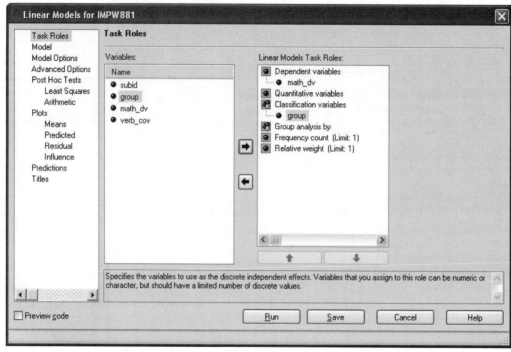

Figure 16.20   Task roles for the ANOVA.

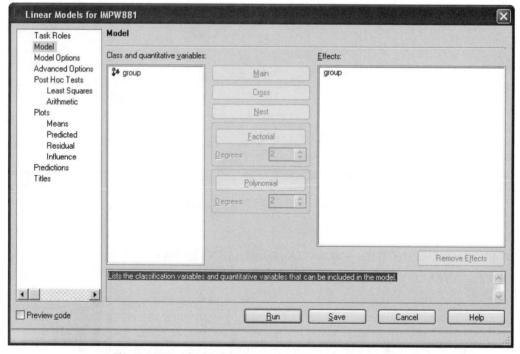

Figure 16.21   The **Model** screen.

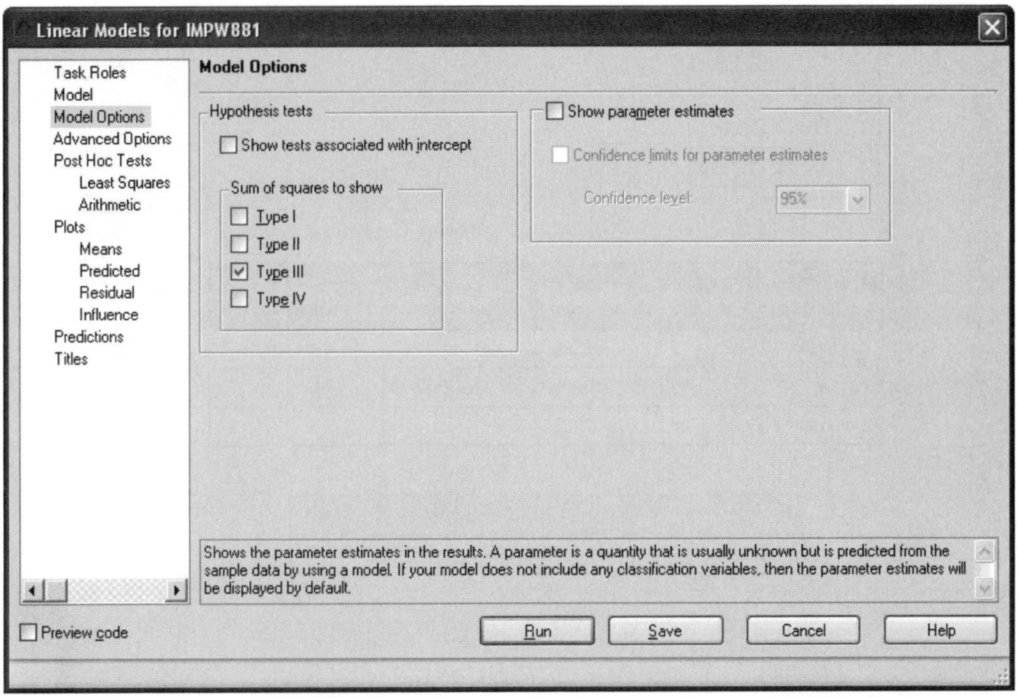

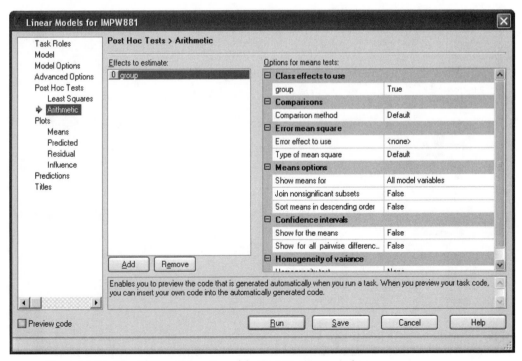

Figure 16.22 The **Model Options** screen.

Figure 16.23 The **Arithmetic** portion of the **Post Hoc Tests** window.

group	math_dv LSMEAN
1	10.8333333
2	10.9166667
3	11.3333333

The effect of group (representing the training variable) is not statistically significant in this ANOVA.

Source	DF	Sum of Squares	Mean Square	F Value	Pr > F
Model	2	1.7222222	0.8611111	0.17	0.8462
Error	33	169.2500000	5.1287879		
Corrected Total	35	170.9722222			

R-Square	Coeff Var	Root MSE	math_dv Mean
0.010073	20.53617	2.264683	11.02778

Source	DF	Type III SS	Mean Square	F Value	Pr > F
group	2	1.72222222	0.86111111	0.17	0.8462

Figure 16.24   Output of the ANOVA.

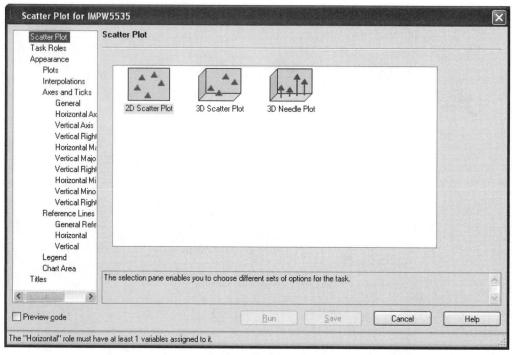

Figure 16.25   Selecting the type of scatterplot.

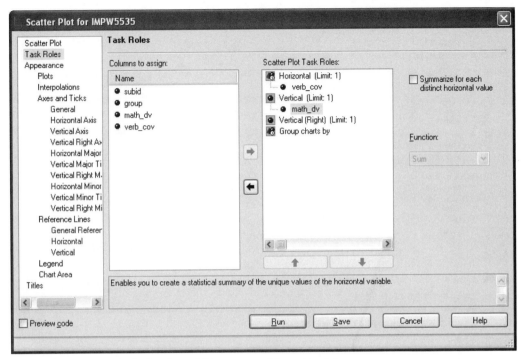

Figure 16.26   Assigning the variables to axes of the scatterplot.

Select **Task Roles** in the navigation panel to the left of the window to display the screen shown in Figure 16.26. Drag **math_dv** to the icon for **Vertical** and drag **verb_cov** to the icon for **Horizontal**.

Next, select **Plots** in the navigation panel to configure the scatterplot. As shown in Figure 16.27, we have selected a black color for our output; you can select other options if you wish. Click **Run** to perform the procedure.

The scatterplot is presented in Figure 16.28. It can be determined from the plot that the covariate and the dependent variable are related to each other in a linear manner. Therefore, the data appear to meet the linearity of regression assumption.

16.11.2   HOMOGENEITY OF REGRESSION

To evaluate the assumption of homogeneity of regression in *SAS Enterprise Guide*, we configure the analysis similar to how we ran the ANOVA described above. From the main menu select **Analyze → ANOVA → Linear Models**. As shown in Figure 16.29, specify **math_dv** as the **Dependent variable** and **group** as the **Classification variable**. Now, however, we will also specify **verb_cov** as the **Quantitative variable** (SAS makes explicit the idea that it treats covariates as quantitative rather than categorical variables).

Select **Model** in the navigation panel on the far left of the screen. Highlight **verb_cov** and click the **Main** pushbutton in the middle of the window. This action will place **verb_cov** in the **Effects** panel as shown

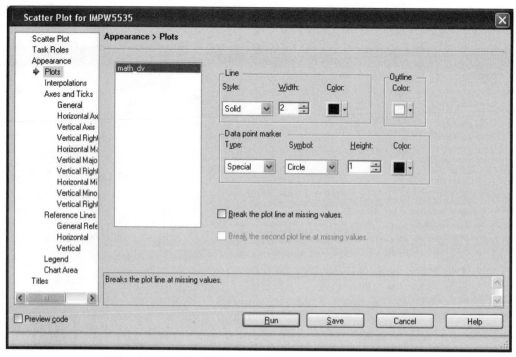

Figure 16.27   The **Plots** screen.

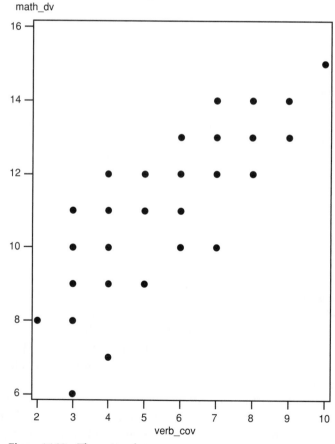

Figure 16.28   The scatterplot output.

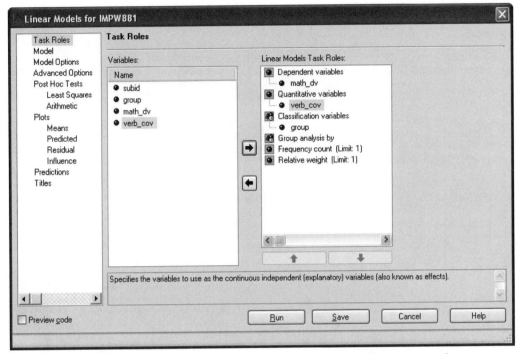

Figure 16.29 **Task Roles** are specified for testing the assumption of homogeneity of regression.

in Figure 16.30. Do the same for **group.** Next, highlight **verb_cov** and, while holding down the Control key, highlight **group.** Click the **Cross** pushbutton to place the **verb_cov*group** interaction in the **Effects** panel.

Select **Model Options** in the navigation panel. Select only Type III as shown in Figure 16.31. Then click **Run** to perform the analysis.

The output of the analysis is shown in Figure 16.32. This matches the output obtained from SPSS. The interaction of the independent variable and covariate is not statistically significant, allowing us to conclude that the data do not violate the assumption of homogeneity of regression.

16.12   PERFORMING THE ANCOVA IN SAS

16.12.1   STRUCTURING THE COVARIANCE ANALYSIS

To perform the omnibus ANCOVA, we configure the analysis similar to how we ran the test for homogeneity of regression described above. From the main menu, select **Analyze → ANOVA → Linear Models**. As shown in Figure 16.33, specify **math_dv** as the **Dependent variable, group** as the **Classification variable,** and **verb_cov** as the **Quantitative variable**.

In the **Model** window, click over only the two main effects (see Figure 16.34). In the **Model Options** window, specify only **Type III** (not shown).

We also wish to obtain the means for the groups. Because ANCOVA analyzes the adjusted means, we can no longer refer to the observed means

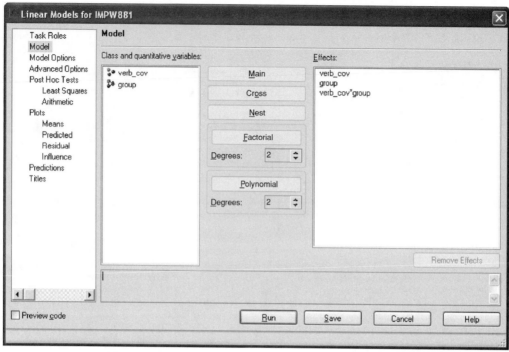

Figure 16.30   The **Model** is now specified.

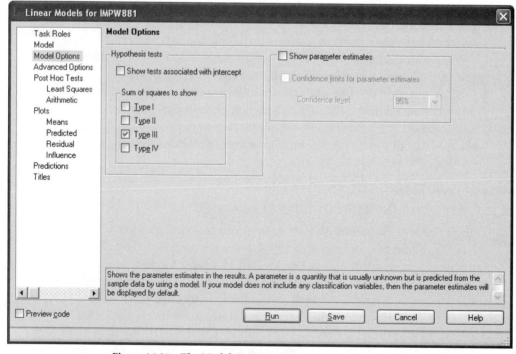

Figure 16.31   The **Model Options** screen.

Source	DF	Type III SS	Mean Square	F Value	Pr > F
verb_cov	1	102.3616413	102.3616413	70.19	<.0001
group	2	5.1560536	2.5780268	1.77	0.1881
verb_cov*group	2	1.5983137	0.7991569	0.55	0.5838

The group × verb_cov interaction effect is not statistically significant.

Figure 16.32 The output of the analysis.

as these were not the means that were evaluated. Instead, we need to obtain the adjusted means. SAS calls these means "least squares means." We can have them displayed in the output by selecting the **Least Squares** portion of the **Post Hoc Tests** window in the navigation panel shown in Figure 16.35. Click **Add** to show the effects in the model and to display the panels in the **Options for means tests.** Under **Class effects to use** (the first panel on the right portion of the window), set the **group** effect to **True** (it is initially set to **False** – double-clicking it displays the **True/False** menu from which you select **True**).

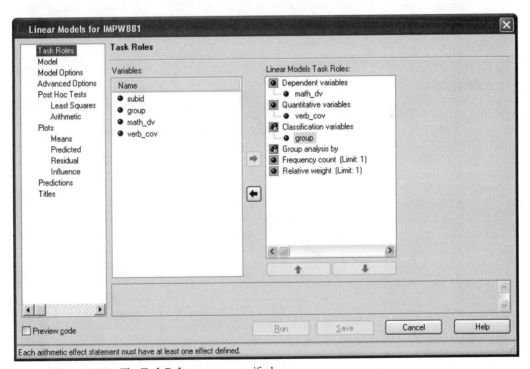

Figure 16.33 The **Task Roles** are now specified.

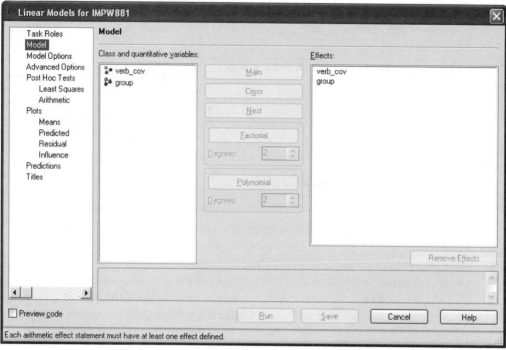

Figure 16.34 The **Model** is specified.

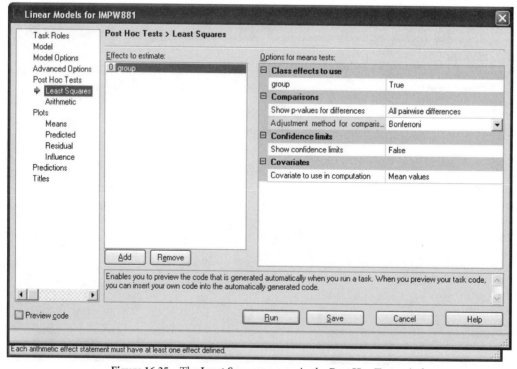

Figure 16.35 The **Least Squares** screen in the **Post Hoc Tests** window.

These are the *p*-values for the main effects and covariate effect.

Source	DF	Type III SS	Mean Square	F Value	Pr > F
verb_cov	1	123.8980812	123.8980812	87.42	<.0001
group	2	13.0293005	6.5146503	4.60	0.0176

group	math_dv LSMEAN	LSMEAN Number
1	10.2018446	1
2	11.1945217	2
3	11.6869670	3

Least Squares Means for effect group Pr > |t| for HO: LSMean(i)=LSMean(j) Dependent Variable: math_dv			
i/j	1	2	3
1		0.1612	0.0161
2	0.1612		0.9558
3	0.0161	0.9558	

These are the *p*-values for the pairwise comparisons.

Figure 16.36 The adjusted pairwise comparisons.

To make the analysis complete, we will specify our pairwise mean comparisons as well, rather than waiting to view the omnibus analysis results and then go back and run our post-ANOVA mean comparisons. Under the **Comparisons** panel, set **Show p-values for differences** to **All pairwise differences** and set **Adjustment method for comparisons** to **Bonferroni**. Click **Run** to perform the analysis.

16.12.2 THE OUTPUT OF THE COVARIANCE ANALYSIS

The output of the ANOVA is shown in Figure 16.36. This matches the results produced by SPSS.

16.13 COMMUNICATING THE RESULTS

A one-way between-subjects covariance design was used to assess the effectiveness of three different instructional programs to teach math. Thirty-six eighth grade children, twelve children per group, were exposed to either a traditional, freeform, or computer-based instructional program at the end of which they were tested on math word problems. The dependent variable was the number of math problems solved during the hour of testing. To statistically control for the math scores the degree to which verbal proficiency might affect performance, a test of verbal skill was administered to the children earlier in the week that the math test was given; the verbal proficiency score was then used as a covariate in the analysis.

The ANCOVA yielded a significant effect for both the covariate, $F(1, 32) = 87.42$, $p < .05$, and the instructional program independent variable, $F(2, 32) = 6.52$, $p < .05$; this latter effect accounted for just under 8 percent of the total variance. Bonferroni corrected multiple comparison tests indicated that the computer-based (adjusted $M = 11.69$, $SE = 0.35$, 95 percent $CI = 10.98 - 12.39$) and freeform (adjusted $M = 11.20$, $SE = 0.35$, 95 percent $CI = 10.49 - 11.90$) programs resulted in higher math word problem performance than the tradition instructional program (adjusted $M = 10.20$, $SE = 0.35$, 95 percent $CI = 9.49 - 10.92$) when controlling for the level of verbal proficiency.

CHAPTER 16 EXERCISES

16.1. Assume we are interested in the effects of teaching method, Factor A (a_1 = lecture, a_2 = PowerPoint presentation, and a_3 = group discussion), on the teaching of an Introductory Statistics course. The three independent groups of students ($n = 6$) are all measured at the end of the semester on a comprehensive (100-point) final exam. Further assume that students' statistical aptitude (X) was measured with 50 multiple-choice questions prior to the start of the experiment to be used as a covariate in the analysis. The data are as follows:

Teaching method (A)

Lecture		PowerPoint		Group discussion	
a_1	X	a_2	X	a_3	X
88	25	70	24	60	23
80	28	71	22	64	23
87	30	73	22	66	27
90	20	80	24	69	22
95	24	75	26	70	25
92	26	70	20	61	24

a. Conduct an ANCOVA and any necessary multiple comparisons tests using SPSS or SAS.

b. If appropriate, perform multiple comparisons to determine which pair or pairs of group means are significantly different.

16.2. Assume we are interested in the effect of treatment type (Factor A: a_1 = brief, a_2 = cognitive-behavioral, and a_3 = psychoanalytic) on community mental health clients' ($n = 5$) GAF at the end of eight weeks of treatment (Time 2). The covariate for this study would be the clients' GAF scores at the start of treatment (Time 1). The data are as follows:

Treatment type (A)

	Brief		Cognitive-behavioral		Psychoanalytic	
Time 1	Time 2	Time 1	Time 2	Time 1	Time 2
40	70	45	75	50	55
45	72	45	75	50	55
42	68	50	70	45	50
35	70	55	70	40	50
40	70	40	69	40	45

a. Conduct an ANCOVA and any necessary multiple comparisons tests on these data using SPSS or SAS.

b. If appropriate, perform multiple comparisons to determine which pair or pairs of group means are significantly different.

16.3. Assume we want to examine the effectiveness of three therapeutic interventions (Factor A: $a_1 =$ group, $a_2 =$ individual, and $a_3 =$ family) with Latino community mental health clients. The dependent variable is client service satisfaction ($1 =$ *very unsatisfied* to $10 =$ *very satisfied*). Client acculturation status (x) ($1 =$ *unacculturated* to $5 =$ *acculturated*) was measured for each client ($n = 5$) prior to treatment, and served as the covariate. The data are as follows:

Therapeutic intervention (A)

Group		Individual		Family	
a_1	x	a_2	x	a_3	x
5	2	6	3	9	2
4	2	7	2	10	1
3	1	6	1	10	4
2	1	5	1	8	5
1	1	5	2	8	2

a. Conduct an ANCOVA and any necessary multiple comparisons tests with SPSS or SAS.

b. If appropriate, perform multiple comparisons to determine which pair or pairs of group means are significantly different.

CHAPTER SEVENTEEN

Advanced Topics in Analysis of Variance

The purpose of this final chapter is to provide a brief introduction to some selected topics related to experimental design and ANOVA procedures. These topics include interaction comparisons, random and fixed factors, nested factors, Latin squares, unequal sample sizes, and multivariate analysis of variance. Because complete coverage of these topics requires at least a separate chapter per topic, which is beyond the scope of the present book, our coverage will be somewhat cursory; sources that may be consulted for further information are provided in connection with each topic.

17.1 INTERACTION COMPARISONS

17.1.1 SIMPLE EFFECTS ANALYSES

As we indicated in Chapter 8, most researchers explore a statistically significant $A \times B$ interaction effect by conducting simple effects analyses. The simple effects strategy that we have used throughout this book was to perform pairwise comparisons using t tests directly following the omnibus ANOVA that yielded a statistically significant interaction effect. An alternative but similar strategy with three or more levels of one of the independent variables can be illustrated by considering the means displayed in Figure 17.1. This 3×2 factorial was originally presented in Figure 8.2. In this alternative but similar strategy, we focus on one level of one of the independent variables at a time. In Figure 17.1 we have outlined the means of the females to highlight one focus, and would repeat this focus with the males. To implement this strategy, we would do the following:

1. Perform a one-way ANOVA for the females comparing the means of type of residence.
2. If the F ratio for type of residence is statistically significant, we would then perform either planned or unplanned comparisons to determine the locus of the interaction effect.
3. We would then repeat this procedure for the males.

These two strategies for conducting simple effects analyses, direct pairwise t tests and one-way ANOVAs followed by multiple comparisons tests, are conceptually very similar in that they both immediately decompose

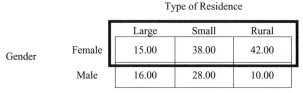

Figure 17.1 Simple effects strategy.

the interaction effect into a direct comparison of means. The use of interaction contrasts exemplifies a somewhat different strategy to decompose the omnibus interaction effect.

17.1.2 INTERACTION CONTRASTS

Interaction components or *contrasts* examines statistically significant interaction effects by decomposing complex factorial designs into a series of smaller factorial analyses rather than moving directly to a comparison of means. Such an approach encapsulates the factorial quality of the analysis, as opposed to the more means-oriented nature of simple effects analyses.

The design displayed in Figure 17.2 illustrates the strategy of performing interaction contrasts. Here we have focused on a subset of the conditions in the full design by "reducing" or simplifying the full factorial into a 2×2, eliminating (for the moment) the two rural conditions from the analysis. This smaller factorial arrangement allows us to focus on the degree of loneliness of females and males and living in large and small cities.

This new $A \times B$ interaction effect (with the reduced factorial) addresses a different question than did the original $A \times B$ interaction. The original interaction effect asked individuals from large, small, or rural population centers whether their rated loneliness varied as a function of gender. The interaction contrast poses a more focused question: Is the difference in loneliness between large and small population centers the same for women and men? If the 2×2 interaction effect was statistically

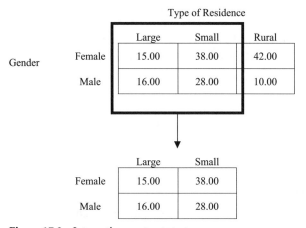

Figure 17.2 Interaction contrast strategy.

significant, we would then perform one-way ANOVAs to further explicate this simple interaction. That is, we would compare the loneliness means of large and small communities for females, and then repeat that analysis for males. Because there would be only two groups involved in any one analysis, a statistically significant F ratio would indicate that the two means were reliably different.

As Keppel et al. (1992) note, "most large factorial experiments can be profitably transformed into a number of interaction contrasts, each one focusing on a different aspect of the $A \times B$ interaction" (p. 307). For example, in the present study, our full 3×2 factorial can be broken into three simpler 2×2 analyses: A (female, male) $\times B$ (large, small), A (female, male) $\times B$ (small, rural), and A (female, male) $\times B$ (large, rural). Each of these minifactorial designs provides a separate piece of the interaction mosaic puzzle. Under a strategy of performing the 2×2 interaction contrasts, we would then perform one-way ANOVAs following up on any statistically significant interaction effects to determine where the group differences were.

The computational steps involved in calculating the necessary sums of squares, degrees of freedom, mean squares, and F ratios are beyond the scope of the present chapter, but are fairly straightforward and are introduced in Keppel et al. (1992). For more advanced and comprehensive treatment of this topic, see Boik (1979), Keppel (1991, Chapter 12), Keppel and Wickens (2004, Chapter 13), and Winer et al. (1992, Chapter 8).

17.2 FIXED AND RANDOM FACTORS

Throughout this text, we have been focusing on various ANOVA designs that can be construed as *completely randomized designs*; that is, equal numbers of participants (n) are randomly assigned to the various treatments or treatment combinations. These designs are typically assessed with the within-groups mean square ($MS_{S/A}$ or $MS_{S/AB}$, etc.) as the error term in the F ratio. The use of such an error term is based on a statistical model called the fixed effects model. This model is sometimes contrasted with other statistical models, less often used in the social and behavioral sciences, called random effects and mixed effects models. We briefly mentioned in Section 10.16 that *SAS Enterprise Guide* treated the subject identification variable as a random effect; here we expand somewhat on that discussion.

The distinction among fixed effects, random effects, and mixed effects models is based on how a researcher identifies the levels of the independent variables in an experiment or research project. We consider each in turn.

17.2.1 FIXED EFFECTS

Fixed effects or *factors* use levels of the independent variable that are selected purposefully, rationally, and systematically. For example, fixed factor independent variables could be client dysfunction level (severe, moderate, minor), word concreteness (high, medium, low), or population of the area in which people reside (large city, small town, rural community). Each of these examples depicts independent variables whose levels could be easily replicated in a subsequent study.

17.2.2 RANDOM EFFECTS

Random effects or *factors* use levels of the independent variables that have been randomly and unsystematically selected from the overall population of all possible independent variable levels. For example, a researcher defines an independent variable as the number of therapy sessions completed by clients. Assume that the clients in the population to be studied had between 1 and 100 therapy sessions. Because of resource availability, the researchers wish to study therapy progress on four different amounts of therapy. Under a random effect model, the researchers would randomly select those four different amounts of therapy (e.g., 10, 27, 44, and 82 sessions completed) and designate them as the levels of the independent variable. Because the four sessions are randomly chosen, amount of therapy sessions is defined as a random factor in the data analysis. In a subsequent replication study, a different random selection might net the following four levels: 6, 16, 55, and 60.

As another example, assume that individuals participating in a study are recruited from a larger population of all students taking an introductory psychology course at a given university. But selection of those who agree to participate is not under the control of the researcher. From the researcher's perspective, those who participate in the study represent an unpredictable (haphazard or perhaps quasirandom) subset of the population. Only a certain subset actually volunteered, although anyone in the population (at least theoretically) could have participated. By this reasoning, participants can be viewed as a random effect, and this is why *SAS Enterprise Guide* elected to treat the subjects as a random effect in the repeated measures analysis model.

17.2.3 MIXED EFFECTS

A *mixed effects model* contains at least one independent variable that represents a fixed effect and at least one independent variable that represents a random effect. Note that although we require at least two independent variables to qualify for a mixed effects model, there is no constraint on (a) whether those variables must be between-subjects or within-subjects variables, and (b) which represents the fixed effect and which represents the random effect.

17.2.4 GENERALIZABILITY ISSUES

The main advantage of employing a random effects model over a fixed effect model lies in the degree to which a researcher may generalize the results of the study. With a fixed effects model, the researchers can generalize the results of the study only to the specific levels of the independent variables used in the study. For most situations in the social and behavioral sciences, this constraint on generalizability is not a major detriment to using fixed factor designs. This is true because many independent variables exhaust the possible pool of alternative levels, and investigators typically choose levels that they deem to be representative of the independent variable (Keppel, 1973, 1991).

With a random effects model a researcher can generalize the results to many more, perhaps even all possible, levels of the independent variable. The rationale upon which this rests is analogous to why we prefer to randomly sample participants in a research study. Recall that *SAS Enterprise Guide* treats subjects in a repeated measures or mixed design as a random effect, and for very good reason. When we are able to obtain a random sample it means, all else equal, that every member of the population had the same chances of participating. Under such a circumstance, any one subset from this population is presumed to react to the treatment in a manner similar to any other subset. This, in turn, allows us to generalize our findings based on the obtained sample to other members of the population.

The same reasoning can be applied to other independent variables that are treated, or can be conceived, as representing a random effect. If the levels of a variable have been randomly selected from a set or population of levels, then we should be able to generalize our findings to other members of that population. Thompson (2006, pp. 345–346) expresses the idea as follows:

Logically, if the random sampling of participants generates data that support generalization to a larger field of participants, why could we not randomly sample levels . . . from a wider population of potential levels, and thereby achieve generalization beyond the levels . . . actually used in the study? . . . A random effect presumes a representative sample of levels from the more numerous potential levels on the way, along with interest in generalizing from the sampled levels to the population of all possible levels.

Keppel (1991, p. 486) makes the case as follows:

Generalizations based on fixed effects are restricted to the specific levels or conditions actually included in the experiment. On the other hand, generalizations based on random effects may be extended beyond those levels included in the experiment to the population or pool of levels from which they were randomly selected.

We should note that the generalizability of the results to all possible levels of the independent variable should not be carried out in a blind and uninformed manner (not that Thompson and Keppel have suggested this), but our inferences must be tempered by both common sense and critical thinking. Random sampling is a theoretical ideal but falls short when applied in the empirical world. We can approximate random sampling with very long series (10,000 rolls of an unbiased die should result in

virtually the same number of 1s, 2s, 3s, 4s, 5s, and 6s), but the shorter
the run or the smaller the sample, the less we are able to truly sample
randomly from any population (6 rolls of an unbiased die are unlikely to
yield one of each possible value). In light of this, consider two scenarios,
which, although quite unlikely, could occur under random sampling:

- We randomly sample 30 college students of the 600 students enrolled
  in introductory psychology courses at a given university. When we
  look at the sample, we note that we have only male students; that is,
  females (representing at least half of the enrolled students) were not
  sampled.
- Clients in participating mental health settings have received 100
  therapy sessions. We randomly sample four different amounts of
  therapy sessions to examine treatment progress. When we look at
  the sampled sessions, we note that we have sampled only sessions
  1, 2, 3, and 4; that is, therapy sessions between 5 and 100 were not
  sampled.

In principle, students and therapy sessions in the above two examples represent random effects. And in the statistical analysis we would
treat them as random effects (if we actually chose not to resample before
collecting any data). But how comfortable would researchers be in generalizing their findings to introductory psychology students in general or to
therapy outcome, in general? Most researchers would feel very uncomfort-
able doing so and almost certainly would not choose to make any strong
inferences based on their data collection – they would rightfully wonder
the extent to which their findings would hold, respectively for each study,
for females and for clients who have more than a very few sessions of therapy. Now, these are admittedly extreme examples, but we hope to have
made the point that, while we may follow a given experimental procedure
(randomly select levels of a variable), it is necessary to critically evaluate
the results of what we have done before overgeneralizing the findings from
any one study.

Most social and behavioral science research operates from a fixed effect
model perspective, because we often do not randomly select the levels of
our independent variables. The decision of which statistical model to use
is somewhat arbitrary, as Keppel and Wickens (2004) note:

There is some latitude in which model we assign to a particular factor, although
the choice should be made in a way that is consistent with the goals of the
research and will be accepted by those to whom it is to be presented. (p. 533)

The computational procedures involved in conducting a fixed effects
model versus a random effects model are comparable with one notable
exception. Selecting the appropriate error terms for random effects and
mixed effects models is more complicated than for the fixed effects models
that we have been discussing throughout this book. Interested readers are
encouraged to review Clark (1973), Coleman (1964), Keppel (1973, 1991),

Keppel and Wickens (2004), and Kirk (1995), for more information on this topic.

17.3 NESTED DESIGNS

The factorial designs used throughout this book have purposefully been conceptualized to include all possible combinations of the levels of the independent variables. These designs are known as *crossed* or *completely crossed factorial designs* and allow researchers to examine the interaction among the independent variables or factors. Occasionally researchers are unable or do not wish to employ a full factorial design. One instance of a design that is not a full factorial is the *nested* or *hierarchical design*. In a nested design the levels of one factor do not occur at all levels of another factor.

As an example of a nested design, consider the hypothetical crossed and nested designs depicted in Figure 17.3. In the crossed design shown in Figure 17.3A, Factor A has three levels. Assume that it represents type of treatment (brief psychotherapy, cognitive-behavioral, psychoanalytic). This variable is completely crossed with Factor B, which could be the community mental health center location (Location 1, 2, 3, 4, 5, and 6) at which the data were collected. Each level of treatment (a) appears once with each level of treatment (b), producing $(a)(b) = (3)(6) = 18$ treatment combinations.

In the nested design shown in Figure 17.3B, Factor B (location) is nested within Factor A (type of treatment), or B/A. Here, locations b_1

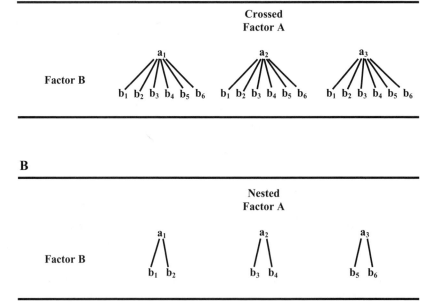

Figure 17.3 Example of a completely crossed design, A, and a nested design, B.

and b_2 only appear with a_1 (brief psychotherapy), locations b_3 and b_4 only appear with a_2 (cognitive-behavioral), and locations b_5 and b_6 only appear with a_3 (psychoanalytic).

Nested designs are commonly found in educational and animal research, where subjects form small groups or blocks (e.g., classrooms or cages) and the particular treatment is given to all members of the block. In such a case, the classroom or cage is said to be nested in the particular treatment being studied.

Keppel and Wickens (2004) note two important considerations concerning nested designs. First, in a nested design it is not possible to investigate the $A \times B$ interaction effect because not all of the levels of Factor B occur under all of the levels of Factor A. Second, nested factors are typically random factors and therefore require a different error term in computing the F ratio than fixed factors. For more comprehensive discussions of nested designs, we recommend Honeck, Kibler, and Sugar (1983), Keppel (1991), Keppel and Wickens (2004), and Winer et al. (1991).

17.4 LATIN SQUARES

The term *Latin square* is said to be derived from an ancient word puzzle game that focused on placing Latin letters in a matrix such that each letter appeared only once in each row and column (Kirk, 1995). In practice, Latin squares have two fundamental uses (Cardinal & Aitken, 2006). The first use is to employ a Latin square as a research method or experimental design technique to ensure that a nuisance variable (i.e., an uncontrolled or incidental factor) such as time of day or type of experimenter is not confounded with the treatment(s) under study. The second use controls for the possible confounding influence of these nuisance variable(s) by incorporating this information into the statistical analysis through an adjustment of the subsequent error term.

Figure 17.4 displays a simple 4×4 Latin square, which we will use to illustrate the design. Each Latin letter in the square denotes a separate treatment condition (i.e., $A = a_1$, $B = a_2$, $C = a_3$, $D = a_4$). Notice how the square is configured such that every letter appears only once in each row and column. This Latin square provides a researcher with a "roadmap" for *counterbalancing* (or systematically balancing) the effects of a nuisance variable in a within-subjects or repeated measures design. For example, suppose that we are working with four adults in examining the effects of type of music on subsequent recall. Participants listened to either rock, pop, classical, or jazz music during four, five-minute study

Figure 17.4   Example of a 4×4 Latin square design.

A	B	C	D
B	D	A	C
C	A	D	B
D	C	B	A

sessions, followed by a free recall test after each encoding (study) session. In order to control for the possible confounding effects of the presentation order of the musical stimuli, we give each participant a presentation sequence depicted by each row of the Latin square in Figure 17.4. Hence, one participant receives the order rock, pop, classical, jazz, a second participant receives the order pop, jazz, rock, classical, and so on. Such an arrangement neutralizes the potentially confounding effect of stimulus presentation order, by spreading it evenly across all participants.

The statistical analyses associated with Latin square designs and their derivatives can be quite complex. We recommend the interested reader to consult Anderson (2001), Keppel (1991), Keppel and Wickens (2004), and Winer et al. (1991).

17.5 UNEQUAL SAMPLE SIZE

Most of the analyses we have covered throughout this book, particularly the computations, have been predicated on the expectation of equal numbers of participants assigned to each treatment condition. When participants contribute equally to each treatment or treatment combination, statistical assumption violations (which were covered in Chapter 5) are less likely to occur. Thus, it makes good research sense to strive for equal treatment group sample sizes. Unfortunately, during the course of our research practice we will encounter unequal sample sizes as a result of the loss of participants from our study, or as a result of the realities of conducting real-world archival or field research.

Anderson (2001) offers five possible causes for unequal sample sizes, all of which can affect how we conduct an ANOVA. The critical issue common to all five potential causes is whether the unequal samples are independent or related to the treatment itself.

- Accidental participant loss: Participants are lost as a result of scheduling errors or equipment malfunctions, unrelated to the treatment. Such loss does not bias the individual treatment means.
- Planned unequal samples: Additional or fewer participants are randomly assigned to treatments. This can occur with specialized control groups, or unusual, rare subpopulations that have unique histories or qualities. Such differential sample sizes should not bias the observed treatment means.
- Treatment-related participant loss: Animals may die or become incapacitated as a result of medical intervention. Task demands may encourage participant fatigue, demoralization, and subsequent dropout. Mean differences assessed with ANOVA procedures, under these circumstances, may be biased and should be noted.
- Unknown participant loss: For the most part, participant loss or attrition is ultimately an experimental or research design issue. Tighter controls, clearer instructions, more realistic or relevant task

demands, and adequate pilot research should reduce participant attrition. Such loss could be random across treatment levels or the loss could be disproportionate across conditions. It is important that researchers pay attention to where their losses are occurring.

- Unequal samples caused by demographic variables: The relative frequency of subpopulations (e.g., racial/ethnic, socioeconomic groups) will almost always produce unequal samples when these populations are sampled. Comparison of the means of these subpopulations may therefore become distorted because of the inequality of group sizes. For example, the precision of the sample mean as an estimate of the population mean (e.g., standard error of the mean) is a function of sample size, and different sample sizes will therefore be associated with different levels of precision. One way to overcome this bias is to conduct a weighted means ANOVA (or analysis of weighted means). This approach is accomplished by weighting each mean in proportion to its sample size. Researchers employ a special average called a "harmonic mean," which is computed by dividing the number of groups by the sum of the reciprocals of the group sample sizes:

$$\frac{a}{\Sigma \left(\frac{1}{n_i} \right)}. \tag{17.1}$$

This harmonic mean is then used to produce the sums of squares for the weighted means ANOVA (Keppel, 1991).

For more discussion of this topic, as it relates to ANOVA procedures, see Anderson, (2001), Keppel (1991), Keppel and Wickens (2004), and Tabachnick and Fidell (2001, 2007).

17.6 MULTIVARIATE ANALYSIS OF VARIANCE (MANOVA)

To this point, the focus of the present book has been to assess the effects of the independent variable(s) on a single dependent variable. In the context of ANOVA, this is referred to as *univariate* ANOVA. Sometimes research questions can be framed to include the simultaneous assessment of two or more dependent variables. Such an approach is characterized as *multivariate* ANOVA (MANOVA).

In a MANOVA design, the dependent variables are combined into a weighted linear composite or *variate*. The weights are created to maximally differentiate between the groups or levels of the independent variable(s). As in ANOVA, researchers employing MANOVA designs can examine the effects of a single independent variable (i.e., a one-way MANOVA) or two or more independent variables (i.e., two-way or *k*-way MANOVA). The difference between the univariate and multivariate approaches is that researchers using MANOVA are concerned with assessing the effects of the independent variable(s) on the combined dependent variate – a multivariate effect.

We have identified elsewhere (Meyers et al., 2006) several benefits or advantages to using MANOVA designs. We can briefly note some of them here.

- Multiple dependent measures provide the design with useful redundancy and greater conceptual clarity.
- MANOVA reduces Type I error rate (falsely rejecting the null hypothesis) by avoiding separate univariate F tests that may inflate the operational alpha level.
- MANOVA considers the intercorrelation of dependent measures by an examination of the variance–covariance matrices.
- MANOVA can pinpoint group differences that sometimes become masked at the univariate level of analysis.

While the topic of MANOVA is certainly beyond the scope of the present book, we encourage the interested reader to explore the following publications: Huberty and Morris (1989), Hummel and Sligo (1971), Meyers et al. (2006), Stevens (2002), and Weinfurt (1995).

Appendixes

Primer on SPSS

A.1 HISTORICAL OVERVIEW

The SPSS Web site provides a comprehensive history of the company and the software they have produced. Here is a synopsis of that information. SPSS, an acronym for Statistical Package for the Social Sciences, is a set of statistical analysis procedures housed together within a large application. The software was developed in 1968 by two doctoral students at Stanford (Norman H. Nie and Dale H. Brent) and one recent Stanford graduate (Hadlai Hull). Nie and Hull brought the software to the University of Chicago and published the first manual in 1970. They incorporated the company in Illinois in 1975.

SPSS began to be used on mainframe computers, the only kind of computer available in many organizations such as universities. A user would sit down at a computer terminal and type SPSS syntax (the programming language) onto computer cards via a card-reader machine, and in the early 1980s, onto a blank screen. The syntax would look like word strings but conformed to rules that enabled the software to perform the statistical analysis that was specified by the syntax.

SPSS was first marketed for personal computers in 1984. The revolutionary difference in that marketplace was the use of a graphical user interface (abbreviated GUI but thought of by most people today as using a mouse to point and click) to make selections from dialog screens. These selections were translated "behind the scenes" to SPSS syntax, but the syntax was not directly presented on the screen to the human user. However, the code could be viewed at any time by a click of the mouse. Over the years, SPSS has put a great deal of effort into the continuing development of both its software and its GUI over the last quarter century, releasing new versions on a very frequent basis. The application has been available on both Windows and Mac platforms for many years.

A.2 DIFFERENT KINDS OF FILES AND THEIR EXTENSIONS

In working with any commercial software application, whether it is a word processing program or SPSS, you will have occasion to deal with different kinds of files, some of which are specific to that software. Files that you have created have the "regular" names that you have assigned to them, but they also have "extensions." These extensions take the form of **period-letter-letter-letter**. The three letters after the period identify the type of file or document it is. The following is a list of some of the different kinds of files and their extensions that you may be working with in the context of processing the results of a research study using SPSS:

- **.doc**: a file saved in Microsoft Word format (SPSS does not open these types of files).
- **.xls**: a file saved in Microsoft Excel format (SPSS can import data contained in these files).
- **.txt**: a text-only file containing only alphanumeric characters (SPSS can import data contained in these files).
- **.sav**: a saved SPSS data file (contains data and some information about the variables in the data file; such files are unique to SPSS).
- **.spo**: a saved SPSS output file (the results of your analysis are contained in such files; such files are unique to SPSS).
- **.sps**: a saved SPSS syntax file (text structured in specific ways that instructs SPSS to transform or analyze the data; such files are unique to SPSS).
- **.pdf**: a file saved as a portable document format; such documents are faithful mirrors of their originals, including pictures and fonts.

Especially when you are looking at SPSS files, it is very important that you can see these extensions. The reason for this is that you may be using the same regular name for data, output, and even syntax files (if you use them). For example, the way you tell the files apart from your study of self-esteem (e.g., **esteem.sav**, **esteem.spo**, and **esteem.sps**) is to note their file extensions.

Depending on the settings on your computer, you may or may not be able to see these file extensions. To make sure that these file extensions are visible, you should do the following. In Windows, set your preferences to show the extensions on your files using this route: **My Computer → Tools → Folder Options → View → Hidden files and folders → Hide extensions for known files**. Make sure that the **Hide extensions** checkbox is *not* checked (the default is the checkmark, which suppresses displaying the extensions). On a Mac, just check **Append file extension** on the **Save As** dialog box every time you save a file (it is usually the default). In either platform, you will now be able to determine in your navigation what type of file you are intending to select.

A.3   OPENING SPSS

SPSS is ordinarily stored in a folder on the primary internal drive of your computer. In Windows, it should be found in the **Program Files** folder; on a Mac, it should be stored in the **Applications** folder. Clicking on its icon will open the application. It is very convenient to place a *shortcut* (as it is called in Windows) or an *alias* (Mac terminology) for SPSS on your desktop to avoid navigating through your folder hierarchy each time you want to open SPSS.

A.4   SAVING SPSS FILES

A.4.1   GENERAL ADVICE

You should assume the following will happen to you. After putting in several hours of work on your computer, and not having saved your work during that time, the application (e.g., Word, SPSS) locks up or your computer crashes. As a result, you lose all of the work that you invested. To avoid this problem, our general advice is to save your work often to both your internal hard drive and to external media, such as a USB flash drive. If you get in the habit of saving, it will become much less of a chore.

When you save, choose a name that will quickly tell you enough about the file to identify it for you. For example, **length of study data** is preferable to **my data** or **data1**.

A.4.2   SAVING TO SPSS

When an SPSS window is the active window on your desktop (i.e., you can interact with it in some way such as typing characters, marking checkboxes, clicking on a pushbutton, or even closing the window), there will be a main menu bar at the top of the screen (see Section A.10.1).

- If the window is new (not yet saved) then to save it you may select **File ➜ Save As** to reach a dialog window that will ask you to name the file and indicate where you want to save it.
- If the window contains an already saved file that you have modified and you wish to replace the older version with the newer one, then select **File ➜ Save**.
- If the window contains an already saved file that you have modified and you wish to preserve the old file but also retain the updated or changed file, then to save it you may select **File ➜ Save As** and give it a different name than the older file.

Probably the most important type of file to save is your data file. With access to your data, you can always re-create if necessary statistical analyses that were performed in the past. If you are typing in data, save often by using the same data file name so that you replace the older version with the updated one. Once the data file is complete and error free, save it again. If you make any changes to it in the course of performing analyses, you can either save it under the same name (thus replacing the older version) or use the **Save As** function to save it under a different name (thus preserving the older version).

A.5   SETTING PREFERENCES

SPSS allows you to set a wide range of preferences to control the manner in which it displays information. As you become more familiar with the way

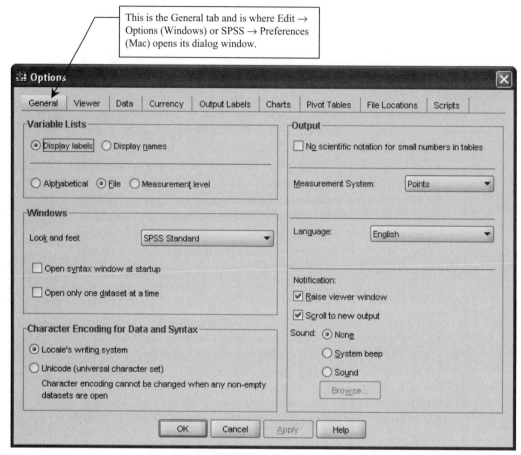

This is the General tab and is where Edit →
Options (Windows) or SPSS → Preferences
(Mac) opens its dialog window.

Figure A1   Setting preferences on the **General** tab in SPSS.

the application works, you might want to change some of our suggested
preferences. Here are some of our recommendations for starting you on
this process.

Open SPSS. On a Windows platform select from the main menu **Edit →**
Options. On a Mac select from the main menu **SPSS → Preferences**. In
either case, you should get the screen for the **General** tab that looks similar
to Figure A1.

A.5.1   GENERAL TAB

The **General** window is what opens when you interact with SPSS preferences. On that **General** tab, you want to check the following: **Display**
labels and **File** in the **Variable Lists** panel. The **File** display shows the variables in the order that they appear in the data file. This is useful if you are
the person who actually built the data file but can be confusing if you are
viewing the variables for the very first time based on a data file constructed
by someone else. SPSS provides the option of ordering the variables alphabetically; if you wish, you can instruct SPSS to display the **Variable Lists** in

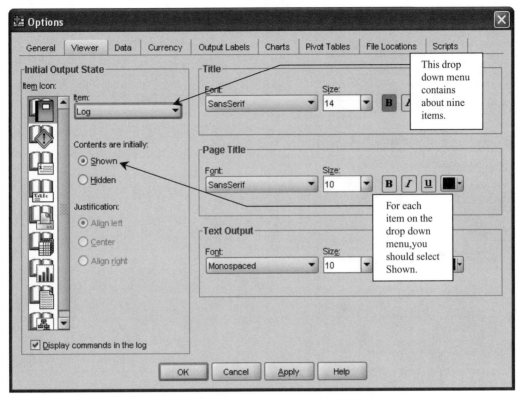

Figure A2   Setting preferences on the **Viewer** tab in SPSS.

Alphabetical order. Keep all other default values provided by SPSS. When you have finished, click on the tab labeled **Viewer**.

A.5.2   VIEWER TAB

Figure A2 shows the window that is displayed on the **Viewer** tab. Our concern is with the left panel labeled **Initial Output State**. In the drop-down menu where **Log** is presently shown are about nine items, including **Warnings**, **Notes**, **Title** and so forth. For each item, make sure that the choice of **Contents are initially** is set to **Shown**. You can leave the **Justification** choices at the SPSS defaults.

A.5.3   OUTPUT LABELS TAB

Click on the **Output Labels** tab. Focus on the bottom set of choices in the **Pivot Table Labeling** area. SPSS places the output of statistical analyses in neatly organized tables called *pivot tables*. To make these tables as readable as possible when you are viewing the output, select **Names and Labels** on the drop-down menu for **Variables in labels shown as**, and select **Values and Labels** on the drop-down menu for **Variable values in labels shown as**. This is illustrated in Figure A3. Then click **OK** to register all of these preferences with SPSS. You may see an output confirming that

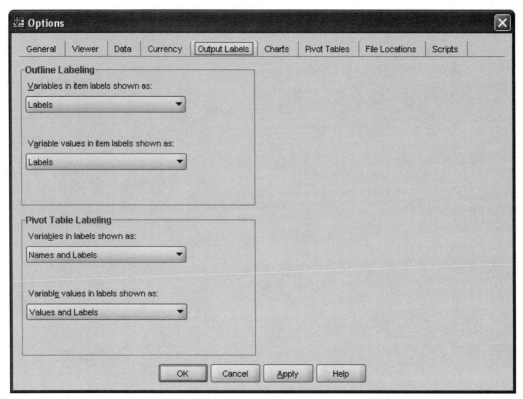

Figure A3   Setting preferences on the **Output Labels** tab in SPSS.

SPSS acknowledges your changed settings; if so, simply delete that output. Then open a new data file as described in Section A.3.

A.6   CREATING NEW DATA FILES IN SPSS

At some point early in your work with SPSS you will be creating data files. Probably the most common way that you will do this, especially if the size of the data set is not unwieldy, is to enter the data directly into an SPSS data file (you eventually reach a point in either the number of variables or the number of cases where you might want the alternative data recording strategy as described in Section A.9). Open the SPSS application. From the main SPSS menu select **File → New → Data**. This will present you with a blank data file as shown in Figure A4.

There are two ways that data files can be viewed:

- **Variable View** shows the information on the variables in the data file. Each row represents a single variable; each column deals with a different piece of information about the variable.
- **Data View** shows the actual data values. Each case occupies a row and each variable occupies a column. This is the view SPSS provides when you first open a data file.

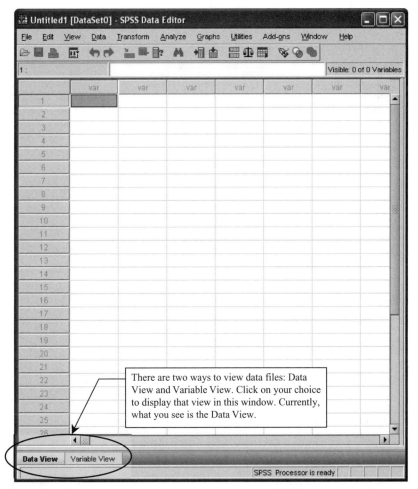

There are two ways to view data files: Data View and Variable View. Click on your choice to display that view in this window. Currently, what you see is the Data View.

Figure A4 This is the window that opens when you ask SPSS for a new data file.

A.7   VARIABLE VIEW OF THE DATA FILE

We recommend that when you construct a data file for a new data set you first click on **Variable View** at the bottom left portion of the data file window (see Figure A5) to type in the names of your variables and to record some information about them in SPSS. Once you have recorded this information, you can switch to **Data View** and enter your data.

A.7.1   VARIABLE NAMES

In the **Variable View** shown in Figure A5, we have typed in the name of four variables: **subid**, which is our identification code for each participant; **gender**, which is the gender of the participant; **group**, which is the group to which the participants have been randomly assigned (this is the independent variable); and **score**, which is the numerical value of the dependent variable. SPSS has filled in the rest of the information with its defaults.

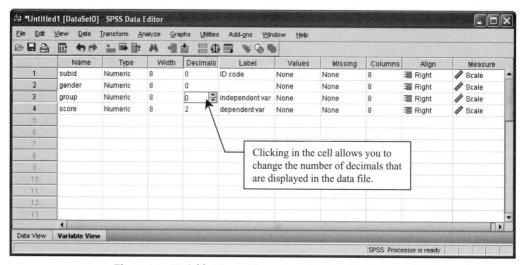

Figure A5   **Variable View** with number of decimals being specified.

A.7.2   DECIMAL SPECIFICATION

In Figure A5, we are in the process of changing the decimal specification. For **subid**, **gender**, and **group**, we use whole number codes rather than decimal values. By clicking inside of the cell, you are able to access the up and down arrows to change the value shown; you can also just double click on the old value and type the new value in its place.

A.7.3   VALUE LABELS

We recommend that you deal only with numerical data in your data file if you are just starting to use SPSS. Thus, categorical variables need to be assigned arbitrary number codes. For example, our **gender** variable will have codes of 1, 2, and 9 corresponding to the following information:

- 1 = Female participant
- 2 = Male participant
- 9 = Missing information

The wording assigned to these codes is accomplished in just a few steps as follows:

- Click in the **Values** cell for **gender** to produce a little "dialog box icon." This is shown in Figure A6.1.
- Clicking on the little box icon produces a dialog box as shown in Figure A6.2. Type the numerical value in the **Value** panel, type the appropriate **Value label**, and click **Add** to register the value with SPSS. In Figure A6.3, we have typed in the information for **female** and have clicked **Add**.
- Type the information for the remaining codes. After clicking the last **Add**, the window should resemble Figure A6.4. Click **OK** to return to the **Variable View** of the data file.
- **Save** your data file.

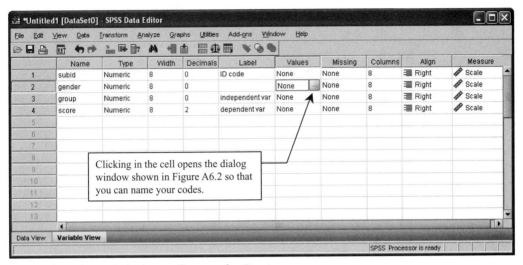

Figure A6.1   Assigning wording to codes: Step 1.

When you have completed providing value labels for all of the arbitrary codes used for those variables containing such arbitrary codes, you will be able to refer to them at any time to refresh your memory of how you coded certain categories. For example, some period of time after the data have been entered, you may need to enter the data for another couple of cases. You see a **gender** variable in the data file but may not recall if females were coded as 1 or 2; referring back to your **Value Labels** will allow you to select the proper code for each case. You can also view these labels from **Utilities** on the main menu (see Section A.12).

These **Value Labels** are just that – labels – for the numbers in your data file. SPSS uses these as "literals" or "strings of letters" without assigning meaning to the labels. For example, you indicated that the letters

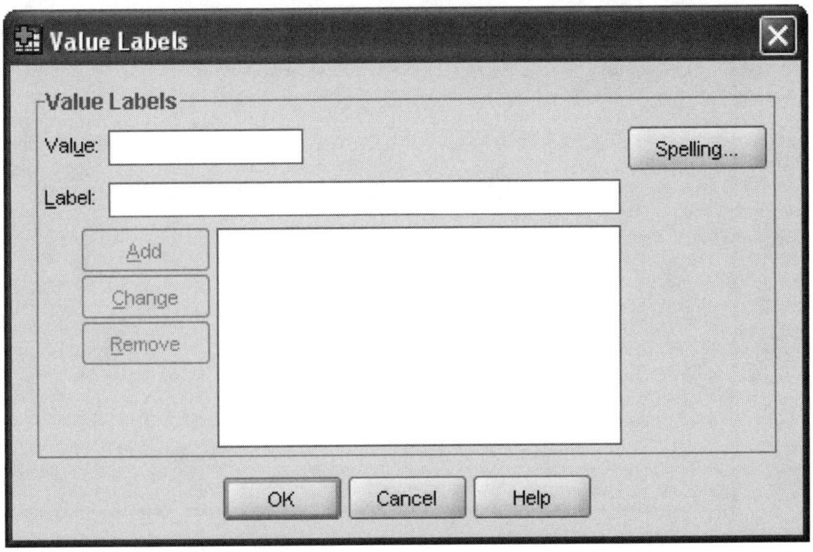

Figure A6.2   Assigning wording to codes: Step 2.

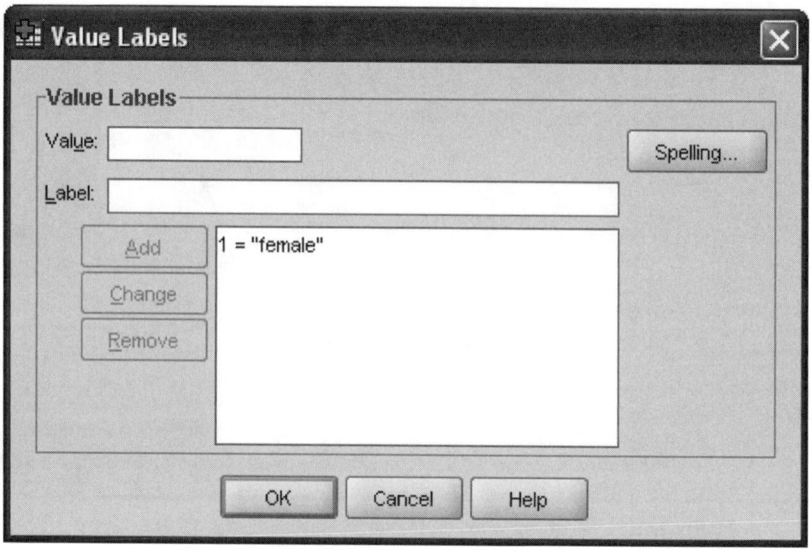

Figure A6.3   Assigning wording to codes: Step 3.

m-i-s-s-i-n-g were to be associated with the value of 9 on the gender variables. This is a great reminder for you, but SPSS treats the label simply as a string of letters that could have been anything; thus, just because you labeled the value of 9 as missing does not in any manner actually instruct SPSS to treat the value 9 as a missing value – that is why there is a variable characteristic of **Missing Values**.

A.7.4   VALID AND MISSING VALUES

Data entered into the data file are defined by SPSS as being in one of two states: the data can be either *missing* or *valid*. Missing values are treated by

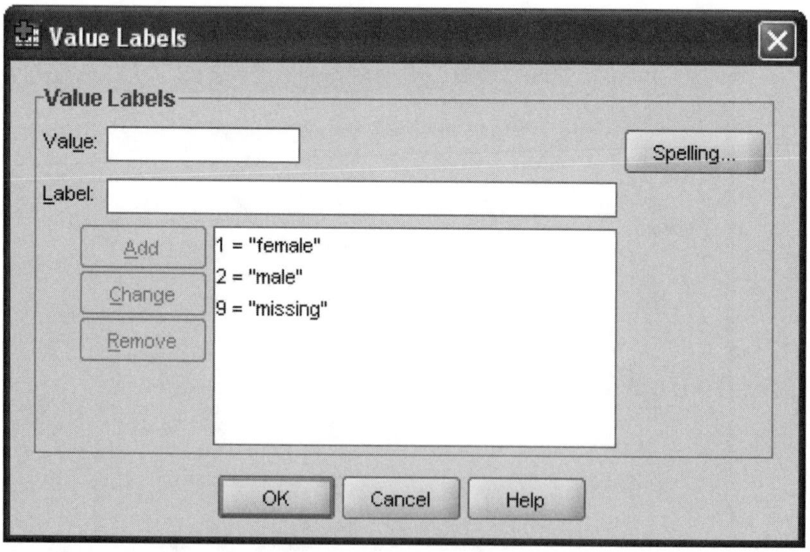

Figure A6.4   Assigning wording to codes: Last step.

SPSS as values that are not ordinarily involved in a computation because they are not a measure of the characteristic. For example, missing values would not be included in computing a mean or a standard deviation. For some analyses where SPSS requires cases to have valid values on all of the variables (e.g., correlation requires cases to have a valid value on each of the two variables being correlated), cases will be excluded from the analysis if that condition is not met.

If a value is not treated as missing, then it is presumed to be a valid value (including typing errors) and will be included in all computations. Hence, it is important to "clean" or "verify" your data prior to data analysis. Although leaving a cell blank will be interpreted by SPSS as a missing data point, it is also possible (and sometimes desirable) to use a specific value or values to indicate that the information on a given case for a given variable was missing. Select a value designating missing that is not in the range of valid values that appear in the data file. The values 9, 99, and so on are popular choices to designate missing provided that they are not valid values. For example, if responses are provided on a five-point scale (1–5), then using 9 for the missing value works because 9 is out of range for valid values. If responses are provided on a ten-point scale (1–10), then using 9 for missing value is not appropriate because 9 is in the range of valid values (you might use 0 for missing value in that circumstance).

Figure A7.1 shows us starting the process of identifying missing values for the variable of **gender**. Clicking in the cell produces a dialog box icon that, when clicked, presents the dialog window shown in Figure A7.2. Choose **Discrete missing values** and type "9" in the first panel as shown in Figure A7.3. Then click **OK** to return to the **Variable View** of the data file. Note in Figure A8 that it shows **gender** with a value of 9 defined as missing. Save the data file with this new information in it.

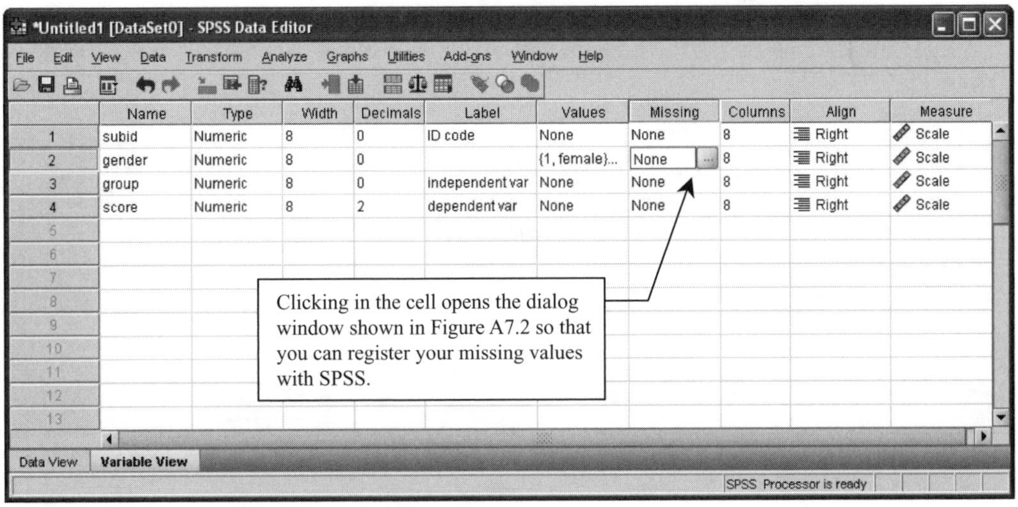

Figure A7.1   Assigning missing values: Step 1.

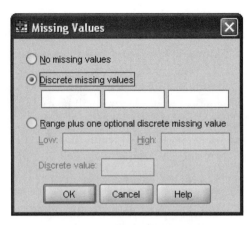

Figure A7.2   Assigning missing values: Step 2.

Figure A7.3   Assigning missing values: Step 3.

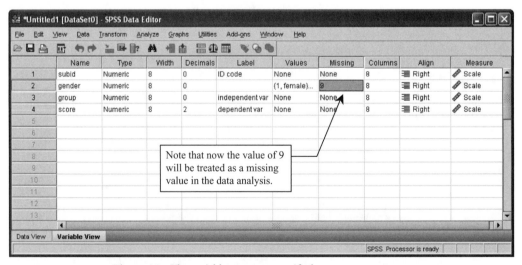

Figure A8   The variables are now specified.

A.8 DATA VIEW OF THE DATA FILE

To display the **Data View** of the data file, click that choice in the bottom left portion of the data file window (see Figure A4). You are now presented with a spreadsheet. The rows are the cases or the participants and the columns (now with names) are the variables. Now you are ready to type your data into the cells. Use the **arrow** keys or the **Tab** key to change the location of the cursor. When you are finished, save the data file.

A.9 READING DATA FROM AN EXCEL WORKSHEET

Let us assume that we have obtained some data from a colleague that is in a Microsoft Excel file named **sample data.xls**. Assume that we want to bring this data set into SPSS. The contents of the Excel file are shown in Figure A9. This very small file contains four cases and four variables for the sake of illustration; the variable names are contained in the first row of the Excel Worksheet.

To import a file from Microsoft Excel (or other spreadsheets) from the main SPSS menu, select **File → Open → Data**. This will open a navigation window to allow you to select the file containing the data. Choose the file format on the drop-down menu for **Excel (*.xls)** files to be read (see Figure A10), navigate to your file, and click it **Open**. Verify your path and, if appropriate, check the box indicating that the variable names are in the first row of the Excel file as shown in Figure A11 (it is a very convenient feature to be able to obtain the variable names from the top row if they are present in the Excel spreadsheet). Click **OK** and the data will now appear in your SPSS data file (see Figure A12).

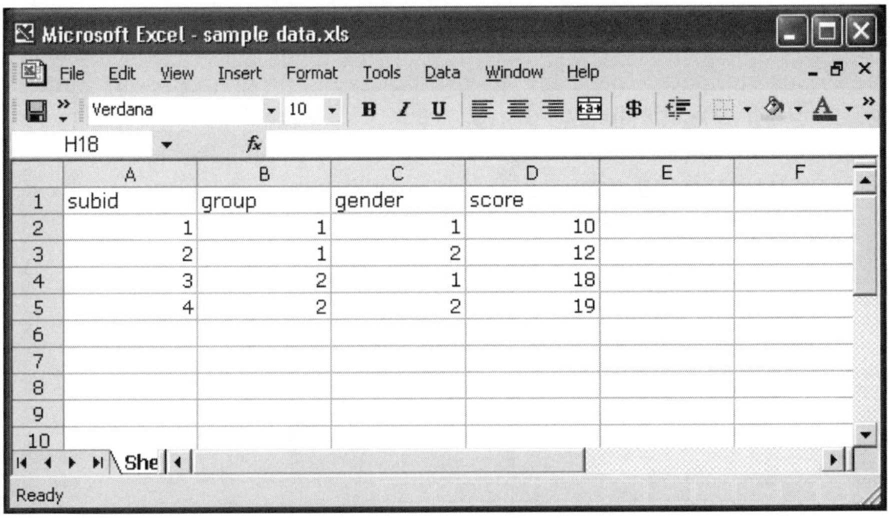

Figure A9   The Excel file that we want to bring into SPSS.

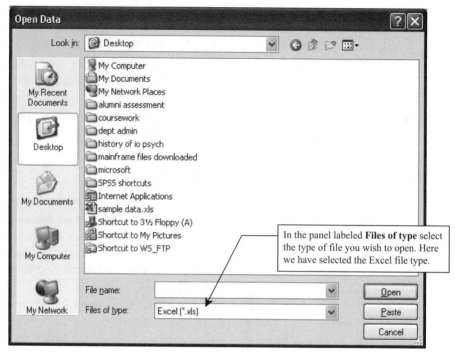

Figure A10 Navigation window where the Excel file format is chosen.

A.10 READING DATA FROM A TEXT FILE

If you are collecting a great deal of data and must hand enter it, typing it directly into an SPSS data file (or an Excel file for that matter) may not be the most efficient method of recording the data. To enter data into different cells in SPSS or Excel requires an extra keystroke (**Tab** or **arrow** key); thus, you make twice the number of keystrokes as data fields. With thousands of pieces of data, you may opt for a more efficient data entry strategy.

Entering the data into a word processing program may be a more attractive option to record large quantities of data. Here, you can have as

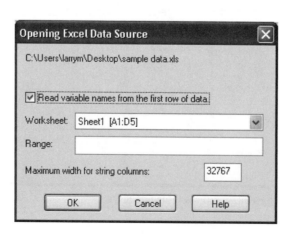

Figure A11 The dialog window for reading an Excel file.

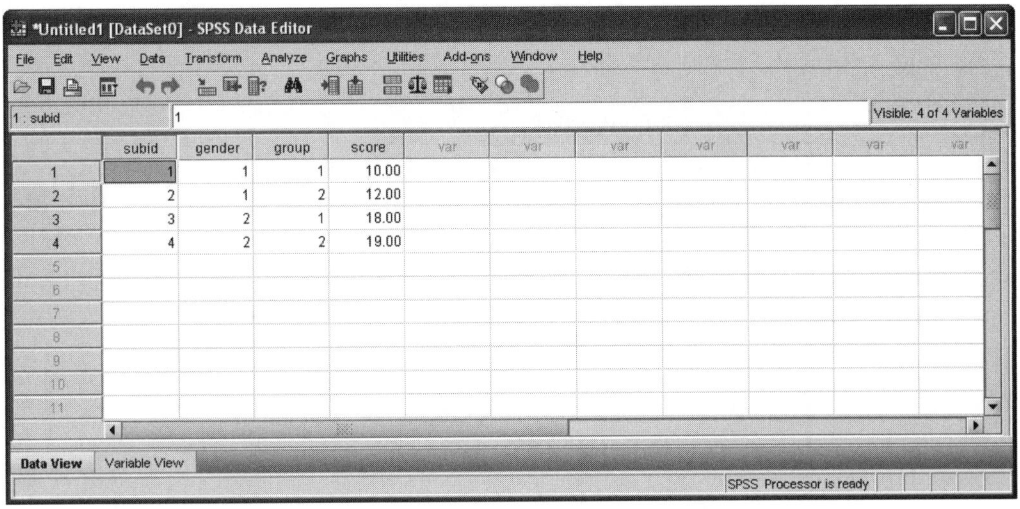

Figure A12   The data in the Excel file are now in an SPSS data file.

many lines of data (they are called *records*) for participants as you wish. The important point to remember is that the document must be saved as a text file (it has the extension **.txt**). A text file is one that contains just ordinary characters (e.g., digits, letters) and some control characters (e.g., tabs, carriage returns) but does not contain formatting information (e.g., bold or italic font, subscripts and superscripts).

To create a text file, you can type your data directly into Notepad (in Windows) or TextEdit (on a Mac) and save the file automatically as a text file; alternatively, you can type your data into a word processing application such as Microsoft Word and use the **Save As** option on its **File** menu to save the document as a text file. We show a portion of a text document in Figure A13. Each participant (case) in this example occupies twenty-seven lines of data.

To bring the contents of a text file into an SPSS data file format, you must open the file through SPSS. Here is one way to do this. From the main SPSS menu select **File ➜ Read Text Data**. This will open a navigation window where you can select which text file you want SPSS to transform to a data file. Once you select a file by highlighting its name and clicking **Open**, you will be brought to the **Text Import Wizard** window, which will take you step by step through the six-step process of bringing the contents of a text file into SPSS. Here is a brief description of the six steps.

A.10.1   STEP 1

Step 1 is contained in the first dialog window you see in **Text Import Wizard**; it is shown in Figure A14.1. You can see the first five rows of text in the bottom panel; this is a quick way to verify that you have retrieved the right file. If you have not previously set up a defined format

Figure A13   Data entered into a word processing program and saved as a text file.

(see Step 6), something relatively new SPSS users will most likely not have done, then answer **No** to the question and click **Next**.

A.10.2   STEP 2

Step 2, shown in Figure A14.2, concerns the arrangement of the variables in the text file. If they are set off by commas, spaces, tabs, or the like, mark **Delimited**; if the same variables are in exactly the same place for every case (this is the easiest structure to type, especially if you do not use delimiters), mark **Fixed width**. *You should plan how you will enter the data with these options in mind rather than have this query come as a surprise to you at this second step.* Then answer the question of whether there are variable names at the top of the file; if you have typed only the data values or if you have multiple lines of data for each case, then the answer is **No**. Click **Next** to continue.

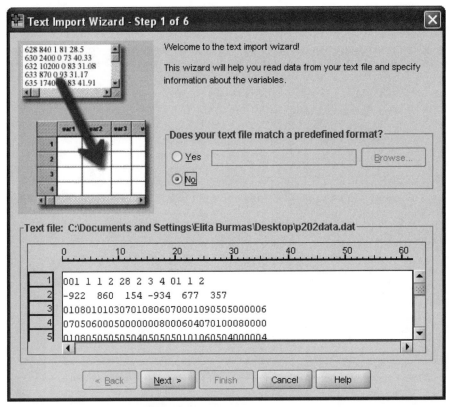

Figure A14.1   Text Import Wizard: Step 1.

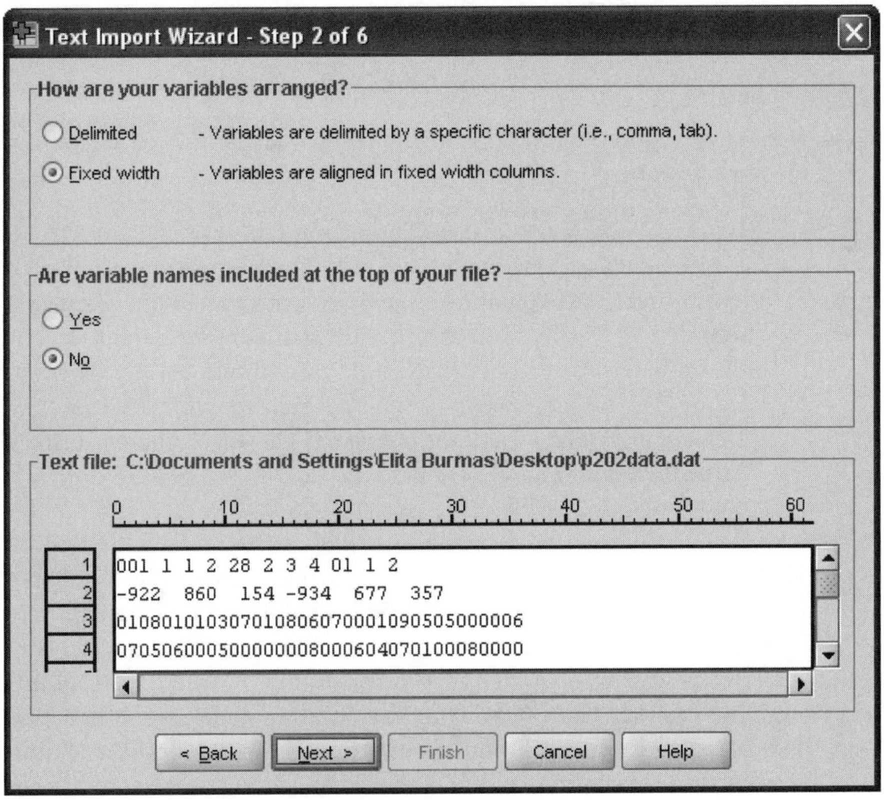

Figure A14.2   Text Import Wizard: Step 2.

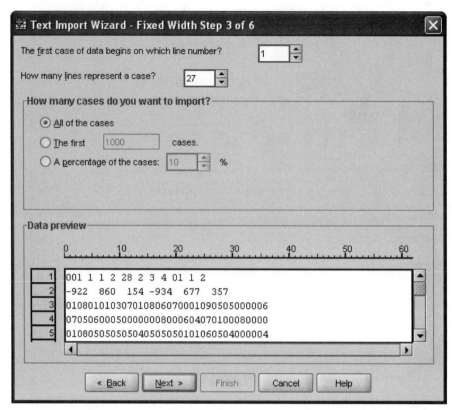

Figure A14.3   Text Import Wizard: Step 3.

A.10.3   STEP 3

Step 3 (see Figure A14.3) asks the following questions. After answering them, click on **Next** to reach the window for Step 4.

- *Where does the first case start?* The answer is very often at line 1.
- *How many lines represent a single case?* You can use any number of lines to record the data from the participants. Each line is called a record. The key here is that you must structure the data in exactly the same way for each participant. In our sample text file shown in Figure A13, we have twenty-seven lines (records) for each case.
- *How many cases do you want to import?* The answer here is usually **All of the cases**.

A.10.4   STEP 4

This step is more complicated to explain than to accomplish. The Step 4 window is shown in Figure A14.4. SPSS now has sufficient information to lay out the numbers in the rows of the data file. All you need to do here is to specify which values comprise variables, that is, where the columns are to be located. Thus, 001 is the identification code for the first case and comprises the first column. Click in the column to produce a column demarcation (a blue line) and drag the line to where you need to have it.

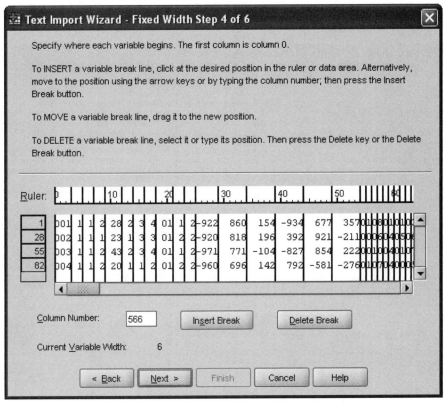

Figure A14.4 **Text Import Wizard:** Step 4.

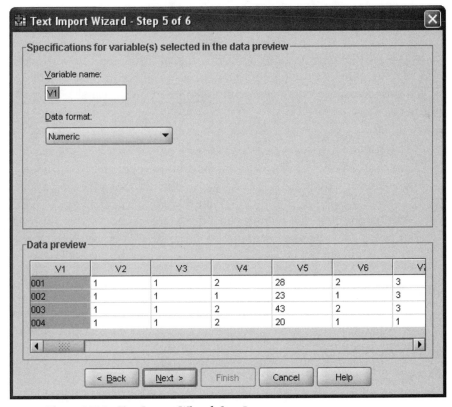

Figure A14.5 **Text Import Wizard:** Step 5.

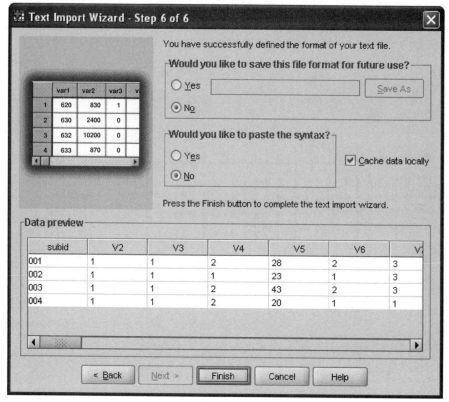

Figure A14.6   Text Import Wizard: Step 6.

The instructions for doing all of this appear at the top of the window. Be sure to place an initial column marker in front of the first variable. In Figure A14.4, we have marked out the columns for the variables.

A.10.5   STEP 5

Step 5 asks you to provide a name for each of the variables in the data file and to specify the data format from the drop-down menu. This window is presented in Figure A14.5. The data format for most or all of the variables you will be analyzing is likely to be **Numeric**. For the naming process, click on the column in the bottom panel that contains the variable you wish to name. Then type the name in the top panel. When you select a new column, the name you just assigned will be seen at the top of the grid.

A.10.6   STEP 6

The sixth and final step in this process addresses housekeeping issues. Its window is shown in Figure A14.6. You need to answer the following three questions after which you click **Finish** to have the new SPSS data file displayed on your desktop. It is advisable to do a **Save As** immediately after the data file is available.

- *Would you like to save this file format for future use?* This refers to the structure of the data file (the first line of data contains these variables in these particular columns, the second line of data contains these other variables in these given columns, and so on); it is what is being referred to in Step 1 when you were asked if the data meet a previously defined format. The vast majority of relatively inexperienced users should answer **No** to this question. However, if you expect to collect exactly the same kind of data in another wave of data collection in the future, then by answering **Yes** you will avoid all the formatting work you just did on the current data set.
- *Would you like to paste the syntax?* There is underlying syntax to everything that you do in SPSS, and many experienced SPSS users can work comfortably with the syntax or even prefer it to point and click. If you answer **Yes**, SPSS will display the syntax in a syntax window (syntax windows use the file extension **.sps**). New users should probably answer **No** to this question until they become more familiar with syntax.
- *Cache data locally?* This places the data in the working memory of the computer and could speed up the data processing done by SPSS. Unless you have an older computer and a large data file, this is probably unnecessary and you need not check the box.

A.11   OPENING SAVED DATA FILES

Once you have saved a data file, you can retrieve it quite easily. From the main SPSS menu select **File → Open → Data** and navigate to the file that you want to open. Click **Open** in the dialog window and the data file will be placed on your desktop.

A.12   THE MAIN SPSS MENU

Figure A15 shows a portion of an existent SPSS data file. At the top of the window you will see the heading on the main SPSS menu (**File**, **Edit**, and so on). You will make use of some of these menus much more frequently than others. When you click on one of these menu items, you will open a secondary menu from which you select what you would like to do. Very briefly, these menu items contain the following:

- **File**: Contains a variety of functions including **Open, Save, Save As, Print**, and **Exit**.
- **Edit**: Allows you to **cut, copy, paste**, and so on.
- **View**: Controls status bars, tool bars, and so on.
- **Data**: Allows you to deal with the data file; among other things, you can select options to **sort** (reorder) the cases, **merge** one data file into another, **restructure** the data file, and **select** a subset of the cases in the data file.

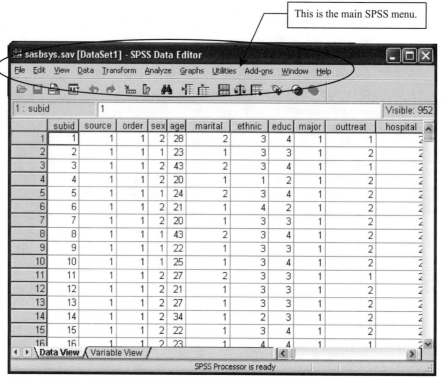

This is the main SPSS menu.

Figure A15   The main SPSS menu.

- **Transform**: Allows you to **compute** new variables (e.g., average item scores together to form a scale score), **recode** variables, and so on.
- **Analyze**: Contains the statistical procedures you use to analyze your data.
- **Graphs**: Contains a variety of preformatted ways to plot your data.
- **Utilities**: Allows you to view information about your variable, such as your variable labels, define sets of variables, and so on.
- **Add-ons**: Contains specialized SPSS modules that can be purchased separately from the primary application.
- **Window**: Allows you to split or minimize your SPSS windows.
- **Help**: Contains documentation explaining how to work with SPSS.

A.13   PERFORMING STATISTICAL PROCEDURES IN SPSS

Figure A16 presents the main dialog window for the **General Linear Model Univariate** procedure, a statistical procedure used throughout the book to perform our ANOVAs. It is therefore convenient to use the main dialog window for this procedure to illustrate the general way in which users interact with SPSS statistical procedures.

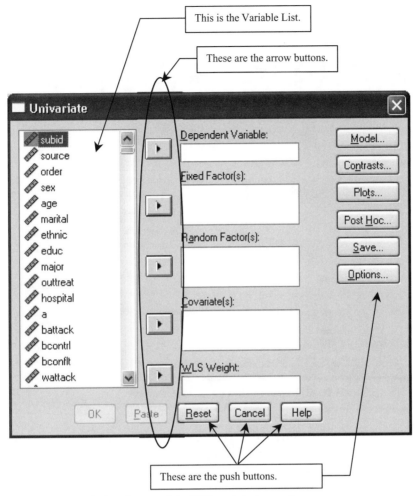

This is the Variable List.

These are the arrow buttons.

These are the push buttons.

Figure A16 A window for one of the SPSS statistical procedures.

A.13.1 THE VARIABLE LIST PANEL

The long panel in the far left portion of the window in Figure A16 lists the variables in the active data file (the data file that is currently open). Variables can be displayed in one of two possible orders (the order they appear in the data file or in alphabetic order) and in one of two possible manners (they are listed by their names or by their variable labels); these alternatives can be specified on the **General Tab** when setting your preferences (see Section A.5.1). In Figure A16, the variables are shown by their names in the order that they appear in the data file.

A.13.2 THE ARROW BUTTONS

To the right of the **Variable List** panel shown in Figure A16, you can see a column of arrow buttons. They show which direction a highlighted

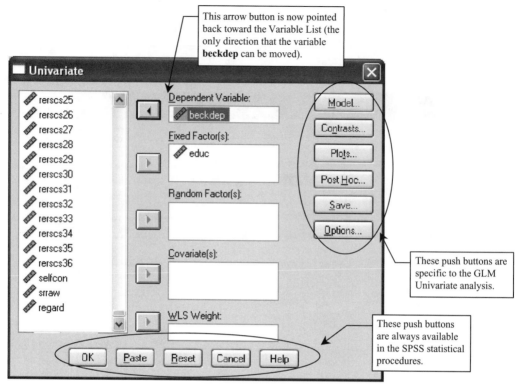

Figure A17   The main window for **GLM Univariate**.

variable will move if the arrow is clicked. Right now they are facing away from the **Variable List** panel and point to the panel to their immediate right. To move a variable from the **Variable List** to a given panel, just highlight the variable and click the arrow button next to it.

If a variable appears in one of the panels to the right of the window, as it does in Figure A17, and you wish to move it back to the **Variable List**, just highlight the variable and click the arrow button. The arrow button will always face in the direction of possible movement once you highlight the variable you want to move. Note that in Figure A17 the arrow button associated with the **Dependent Variable** panel is facing back toward the variable list when the variable (**beckdep** in the panel) is highlighted; that's because going back to the **Variable List** is the only place to which that variable can be moved.

A.13.3   THE PROCEDURE-SPECIFIC PUSHBUTTONS

All of the SPSS statistical procedure dialog windows contain pushbuttons. Some pushbuttons are specific to the procedure that you are using. In Figure A17, the set of pushbuttons in the right portion of the window concern different aspects of the analysis that **GLM Univariate** can perform and are therefore related to this particular statistical procedure.

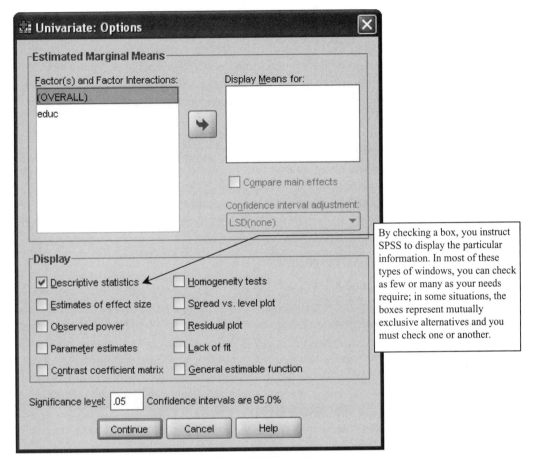

By checking a box, you instruct SPSS to display the particular information. In most of these types of windows, you can check as few or many as your needs require; in some situations, the boxes represent mutually exclusive alternatives and you must check one or another.

Figure A18   The **Options** window from **GLM Univariate**.

For example, clicking the **Options** pushbutton brings you to the dialog window shown in Figure A18.

A.13.4   THE ALWAYS-PRESENT PUSHBUTTONS

The set of pushbuttons that you see at the bottom of the window shown in Figure A17 are contained in virtually all of the SPSS statistical procedures. You will always use the **OK** choice to initiate the statistical analysis you have specified; you may not use the others very often. Nevertheless, here is what each does:

- **OK**: This causes the statistical analysis to be run or it instructs SPSS to accomplish whatever task (e.g., assigning value labels) you have specified. It is active (available to click) only when the analysis is sufficiently specified for it to be performed. Note that in Figure A16 the **OK** and **Paste** pushbuttons are not active because no variables have been specified but are active in Figure A17 because enough variables have been selected for the analysis to be performed.

- **Paste**: This causes the underlying syntax to be placed in a syntax window. This syntax can be edited. Syntax files have the extension **.sps** and can be saved in the same manner as data files. Once saved, they can be opened in the same manner as you would open data files as described in Section A.11. We use this **Paste** function throughout the book to perform simple effects analyses of interactions.
- **Reset**: This causes all of the variables clicked over to the analysis panels to be returned to the **Variable List** and brings the analysis specifications back to its original settings.
- **Cancel**: Closes the dialog box and returns you to the data file.
- **Help** (on Macs it is a "?" button): Brings you to the SPSS help system.

A.14 SAVING OUTPUT FILES

A.14.1 WHAT THE OUTPUT LOOKS LIKE

When you instruct SPSS to perform a statistical analysis, it displays the results in a window in the form of an output file. The results tend to be organized by tables.

We have provided you with an example of some output in Figure A19. Separate tables contain the descriptive statistics and the summary table

Descriptive Statistics

Dependent Variable: BECKDEP Beck depression: bdi

EDUC	Mean	Std. Deviation	N
1 freshman	8.4918	8.17236	61
2 sophomore	7.3333	6.02510	54
3 junior	6.3861	5.70587	158
4 senior	7.4667	7.49880	105
Total	7.1614	6.73586	378

Tests of Between-Subjects Effects

Dependent Variable: BECKDEP Beck depression: bdi

Source	Type III Sum of Squares	df	Mean square	F	Sig.
Corrected Model	214.327 [a]	3	71.442	1.582	.193
Intercept	17350.114	1	17350.114	384.170	.000
EDUC	214.327	3	71.442	1.582	.193
Error	16890.829	374	45.163		
Total	36491.000	378			
Corrected Total	17105.156	377			

a. R Squared = .013 (Adjusted R Squared = .005)

Figure A19 A portion of an output file.

from the standard ANOVA. In the body of this book, we have covered, in detail, how to read all these types of tables associated with the analyses we have performed.

A.14.2   SAVING THE SPSS OUTPUT FILE WITHIN SPSS

Output files have the extension **.spo** and can be saved in the same manner as data files. Once saved, these files can be opened in the same manner as you would open data files as described in Section A.11.

A.14.3   SAVING THE SPSS OUTPUT FILE AS A PDF DOCUMENT

A PDF document is a type of file that is in portable document format. It is a faithful copy of the original, but it is not editable unless you have the full version of Adobe Acrobat or some comparable application. When PDF documents are printed, they mirror what you see on the screen even though your computer may not have the fonts that are used in the document. That's what makes them portable – the portable document format contains within it all the information necessary for the document to be displayed on the screen or to be printed.

Saving the output as a PDF document is extremely valuable for students and those who may not have SPSS loaded on the computer (perhaps their home computer or personal laptop) that they will be using away from the organizational setting so that they can view the results in order to prepare a report. A PDF document can be opened and viewed on any computer. Adobe Acrobat Reader is a free application for both Windows and Mac operating systems that can be downloaded from the Adobe Web site; with it, you can open and view PDF files. Apple also supplies an application called Preview as a standard part of its OS-10 operating system that is functionally equivalent to what Adobe offers.

SPSS will allow you to save a PDF version of the output file that contains your statistical analysis. Since only the SPSS application can open your original output file (and only a version at least as recent as the one that generated the output), it is an ideal way for you to view the *full set* of your results (view a copy of the output file) when you do not have access to SPSS.

To have SPSS save your output file as a PDF document, the output file should be the active window (click its blue bar to make sure that it is active). Then select from the main menu **File → Export**. This brings you to the screen shown in Figure A20.

Select **Portable Document Format** from the **Document Type** drop-down menu. Then select **Browse** to navigate to the location where you want to save the PDF file. When you reach your desired location, name the file. Figure A21 shows the result of this browsing and naming process. Then click **Save**. This will bring you back to the main **Export Output** screen. Click **OK** and wait for the creation process to be finished.

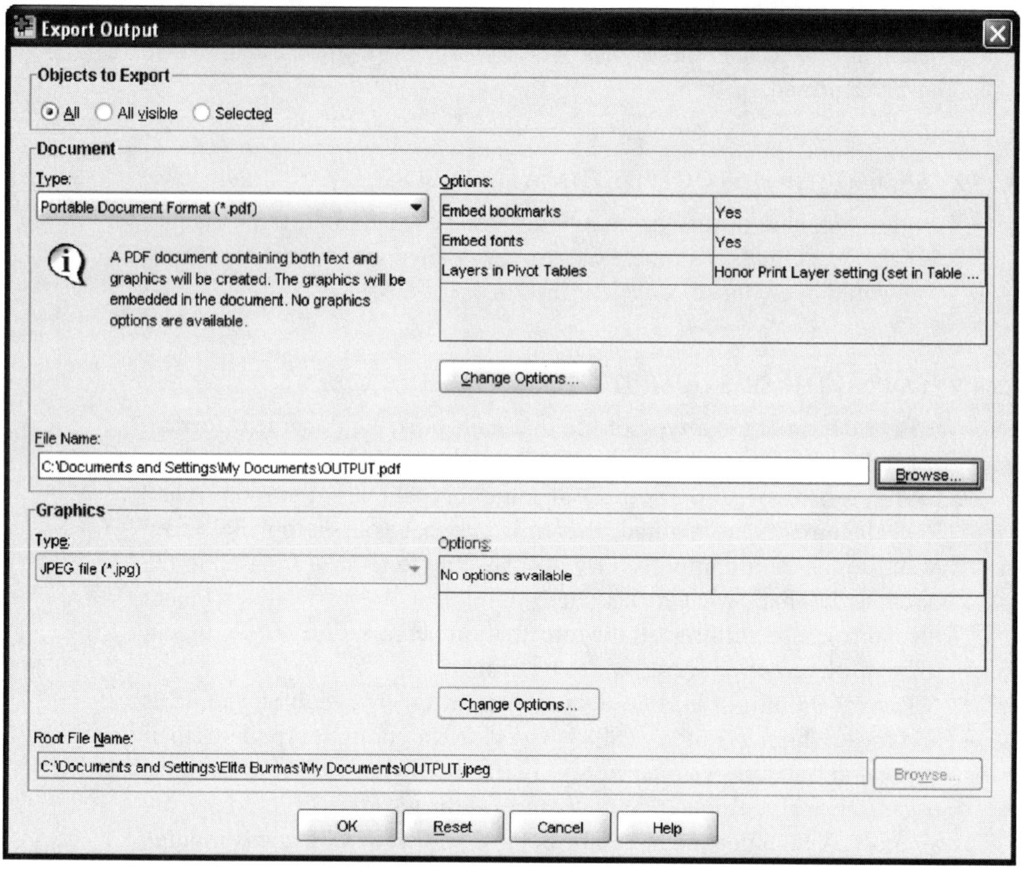

Figure A20   The **Export Output** screen.

A.14.4   SAVING NONEDITABLE SPSS OUTPUT TABLES IN MICROSOFT WORD

As we illustrated in Section A.14.1, the output in SPSS is presented in the form of tables. There are times when you may want to save a particular table in exactly the form that SPSS has produced it, that is, if you want to take a screenshot (picture) of it, and place it a word processing document perhaps during the process of preparing a report. Here are the steps to accomplish to place a screenshot of an SPSS output table in Microsoft Word:

- Make sure that the SPSS output file is the active window and that the table you wish to copy is visible.
- Move the cursor to a position inside the SPSS table that you wish to copy into a word document.
- Click any place in the table. You will see that the table now has a "box" around it.
- From the main menu select **Edit** ➜ **Copy**. The key here is that by treating the table as an object you are "taking a picture" of it.

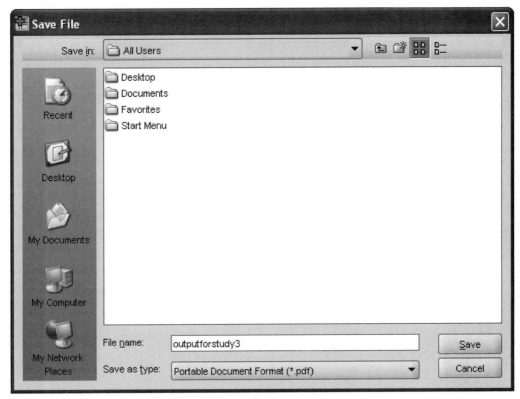

Figure A21 Navigation to the place where the PDF file will be saved.

- Open the Word document into which you want to place the table and set the cursor at the place where you wish the table to appear.
- With the Word document as the active window, select **Edit → Paste**. The picture of the table will now appear in the document.
- Save the Word document to an appropriate location.

You cannot edit the picture of the table, but you will almost always want to resize it by clicking on it and dragging one corner.

A.14.5   SAVING EDITABLE SPSS OUTPUT TABLES IN MICROSOFT WORD

There are times when you might wish to create your own tables in Microsoft Word based on the results you obtained from SPSS but would rather not type in all of the numbers from scratch. It is possible to save the contents of the SPSS output tables in such a way that you can edit them in Word. Here is how to do that:

- Make sure that the SPSS output file is the active window and that the table you wish to copy is visible.
- Move the cursor to a position inside the table that you wish to copy into a word document.
- Click anyplace in the table. You will see that the table now has a "box" around it.

- From the main menu select **Edit → Copy**. Note that by *not* treating the table as an object, you will be able to access the contents of the table.
- Open the Word document into which you want to place the table and set the cursor at the place where you wish the table to appear.
- With the Word document as the active window, select **Edit → Paste**. The table will now appear in the document.
- Save the Word document to an appropriate location.

The output is now in your Word document in the form of a table (tables have their own special format in Word). All of the information from SPSS will be there but it may look a bit different from the way it looked in SPSS (depending on how your system defaults are set up). The good news is that the material can be edited and formatted just as you would edit any Word table (e.g., select a column or row, change the font, widen or narrow the columns).

Primer on SAS

B.1   HISTORICAL OVERVIEW

The SAS Web site provides a comprehensive history of the software and the company. Here is a synopsis of that information. SAS, an acronym for Statistical Analysis Software, is a set of statistical analysis procedures housed together within a large application. The idea for it was conceived by Anthony J. Barr, a graduate student at North Carolina State University, between 1962 and 1964. Barr collaborated with Jim Goodnight in 1968 to integrate regression and ANOVA procedures into the software. The project received a major boost in 1973 from the contribution of John P. Sall. Other participants in the early years included Caroll G. Perkins, Jolayne W. Service, and Jane T. Helwig. The SAS Institute was established in Raleigh, NC in 1976 when the first base SAS software was released. The company moved to its present location, Cary, NC, in 1980.

As is true for SPSS, the procedures it performs are driven by code (SPSS calls it *syntax*) that comprises its own command language. SAS began being used on mainframe computers several decades ago when the only way to instruct the software to perform the statistical analyses was by punching holes on computer cards via a card-reader machine and later by typing in this code on an otherwise blank screen. It should be noted that, unlike SPSS, the vast majority of current SAS users still prefer a code-driven interface.

SAS released its first Windows version in 1993. Windows uses a graphical user interface (abbreviated GUI but thought of by most people as point and click) to make selections from dialog screens. These selections are translated "behind the scenes" to SAS code but the code can be viewed by a click of the mouse. *SAS Enterprise Guide* is the third iteration of SAS's GUI, and runs only in the Windows operating environment. Because *SAS Enterprise Guide* writes code and submits it to SAS as you make selections with the mouse or type text into dialog screens, you also need to be using either a stand-alone computer or one connected to a network on which SAS is installed.

B.2   INSTALLING *ENTERPRISE GUIDE* ON YOUR COMPUTER

SAS Enterprise Guide is ordinarily shipped at no extra charge to the organizations that have ordered SAS. However, possibly because most SAS users write code rather than use the point-and-click user interface, some organizations may not routinely install *Enterprise Guide* when installing SAS. For example, two of the authors who had SAS installed on their university office computers were each not provided with *Enterprise Guide* in the original installation; we had to request the technical support staff at our respective universities to install it. If you have SAS installed on your computer but do not have *Enterprise Guide*, we suggest that you ask for that installation.

B.3   OPENING *SAS ENTERPRISE GUIDE*

We will assume that *SAS Enterprise Guide* 4.0 (or the most current version) is available on your computer and that its icon (a shortcut) is visible on your desktop (if it is not then you can navigate to it in the **Program Files** folder on your internal drive). Open *Enterprise Guide* by double clicking on its icon. This brings you to the window shown in Figure B1.

Everything in *SAS Enterprise Guide* is done within the context of a *project*. The initial screen for *SAS Enterprise Guide* therefore provides choices to you in the context of projects. The top portion of our opening screen lists some of our more recent ones.

You will be entering data into a new project; that is, you will build a data file in *SAS Enterprise Guide*. Select **New Project** from the opening screen. This will bring you to the window shown in Figure B2.

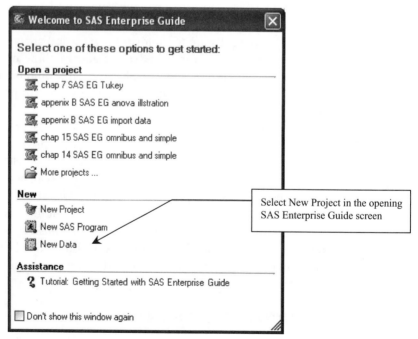

Figure B1   The initial *SAS Enterprise Guide* screen.

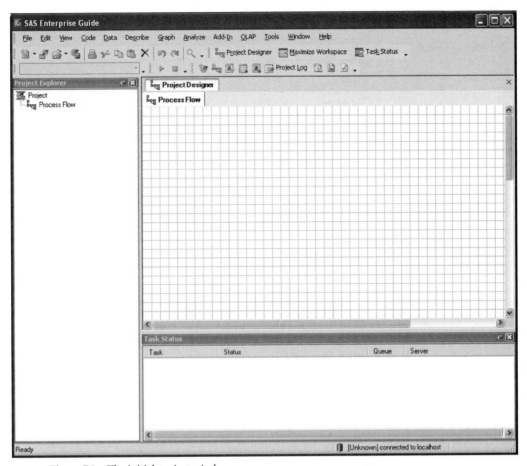

Figure B2   The initial project window.

The initial *SAS Enterprise Guide* screen, as seen in Figure B2, presents the **Project Designer**. It is empty now but at various stages of your work it will contain a data file, the specifications of the analysis, and the results of the data analysis. The window shows a grid that looks like graph paper – this is the background used by the **Process Flow**. Because there is nothing in the project at this time, an empty **Process Flow** window is displayed. You are now ready to enter your data.

B.4   ENTERING DATA DIRECTLY INTO *SAS ENTERPRISE GUIDE*

There are several ways to inform *SAS Enterprise Guide* that a new set of data is to be entered. Many of these are keyed to the various icons on the toolbars that may or may not be visible in your current configuration (you can pass the cursor over the icons to see what they represent). The procedure that is always available to create a new data set, and the one we recommend you use as a new user, is as follows: From the main menu select **File → Open → Data** as shown in Figure B3. This selection brings you to the initial **New Data** screen seen in Figure B4.

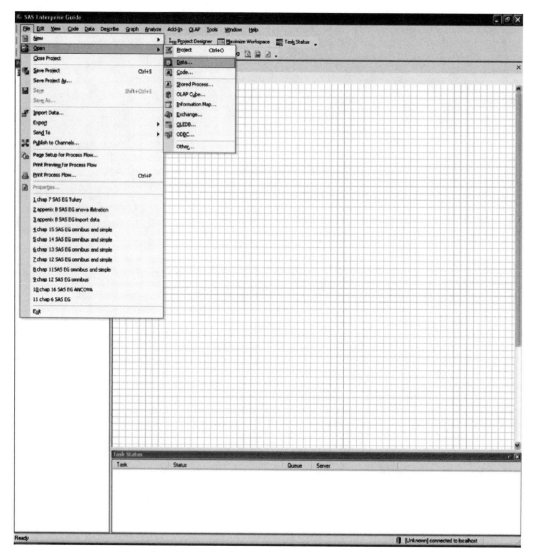

Figure B3   Specifying that a new set of data will be typed in.

The initial **New Data** screen in Figure B4 provides places for you to supply two pieces of information:

- **Name**. This field, highlighted in Figure B4, is used to name the data file that you are about to build. Data file names can be no longer than thirty-two characters, must contain only alphanumeric characters or underscores, and must begin with either a letter or an underscore; no spaces are allowed in the name. Select a name that meaningfully relates to your research project. We will name our file **illustration**.
- **Location**. *SAS Enterprise Guide* will use one of its **Libraries** as the start location. By default, it has selected the **Work Library**. This is fine because once you have entered the data, you will save the project in a location of your choice.

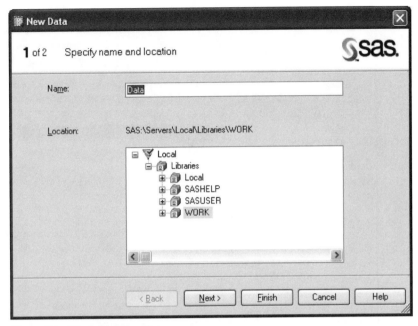

Figure B4   The initial **New Data** screen.

When you have finished with the **New Data** window, click **Next**. This brings you to the second **New Data** screen shown in Figure B5. It is in this window that you can identify the variables and their properties in advance of typing the data. In the left panel are the generic variable names supplied by *SAS Enterprise Guide* (**A**, **B**, **C**, and so on) listed vertically; in

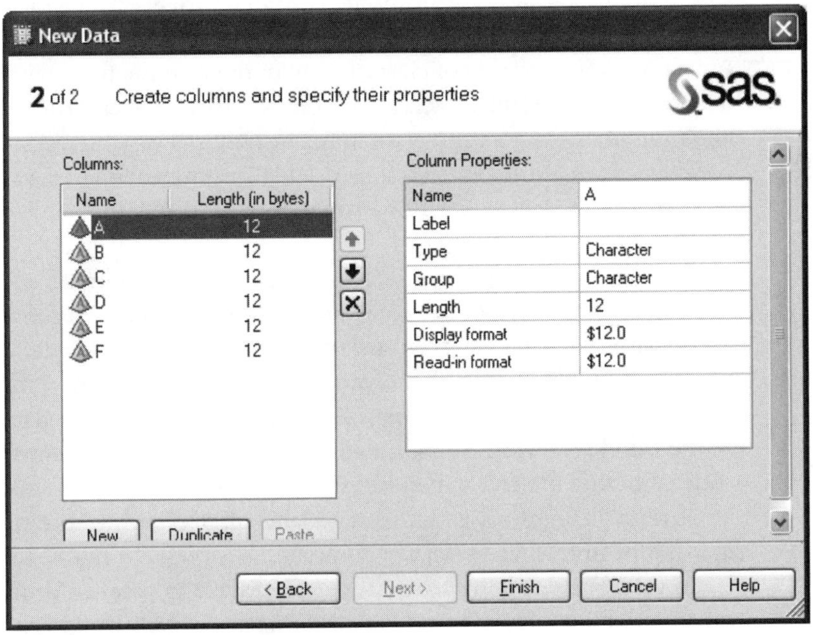

Figure B5   The second **New Data** screen.

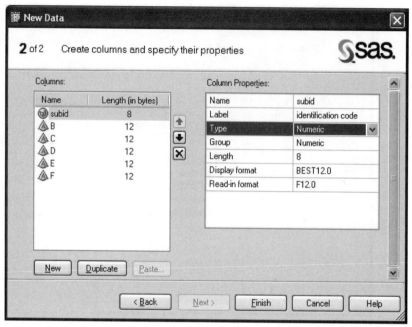

Figure B6   The first variable, **subid**, is now specified.

the right panel are the properties that will be associated with each variable. In the data file, variable **A** will be the first variable and will occupy the first column, **B** will be the second variable and occupy the second column, and so on.

When a variable is highlighted in the **New Data** window, you may specify its properties. For example, consider variable **A**. The icon next to it (a "tent" surrounding an **A**) represents the default of a **Character** (an alphanumeric string of characters with the **A** in the tent standing for *alphanumeric*) variable. Such a variable is a treated as a string of letter and number characters and is a naming or nominal variable. SAS will not perform arithmetic operations (e.g., calculating a mean) on such variables. Note that in the right panel for **Properties**, the **Type** of variable is listed as **Character**.

The first variable we will specify is our case identification variable. Our specifications are shown in Figure B6. Assume that the data set for each participant has been assigned an arbitrary identification code, and we name this variable **subid**. To accomplish this naming, in the **Properties** panel we have highlighted the letter **A** in the **Name** row and typed in **subid**. In the **Label** area, we have indicated that the variable is an **identification code**; although the fact that **subid** represents an identification code may be obvious here, it is a good habit to label all variables whose meaning may not be immediately clear by its name.

In the **Type** panel, we have clicked **Character** to obtain a drop-down menu with the choices **Character** and **Numeric** and have chosen **Numeric**. That selection caused the **Group** choice to switch to **Numeric** as well (the choices are **Numeric**, **Date**, **Time**, **Currency**), which is what we wish. It also caused the icon next to **A** in the left panel to change to a circle containing the numbers 1, 2, and 3 to represent the fact that **subid** has been specified as a numeric variable.

The remaining variables in our illustration data file are all numeric, and we will specify them as well. These other variables are as follows:

- **group** is a variable containing codes indicating if the participant was a member of Group 1 or Group 2.
- **gender** is a variable containing codes to indicate whether the participant was male or female.
- **score** is the value that the participant registered on the dependent variable.

When you have specified these other variables, click **Finish**. This brings you to the empty data grid shown in Figure B7.

The data may be entered as you would do for any type of spreadsheet. Type the value in each cell and use the **Tab** or **Arrow** keys to move from one cell to another. When you have finished, the grid, shown in Figure B8, will

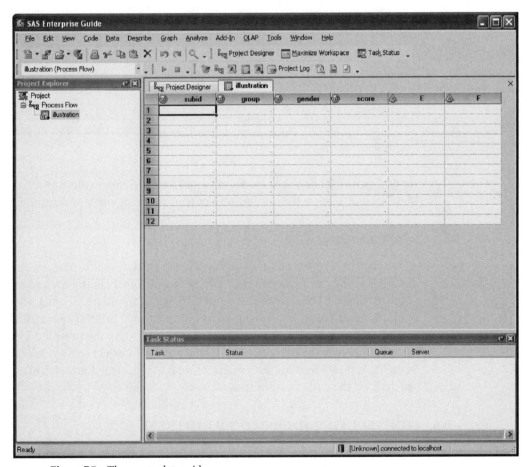

Figure B7   The empty data grid.

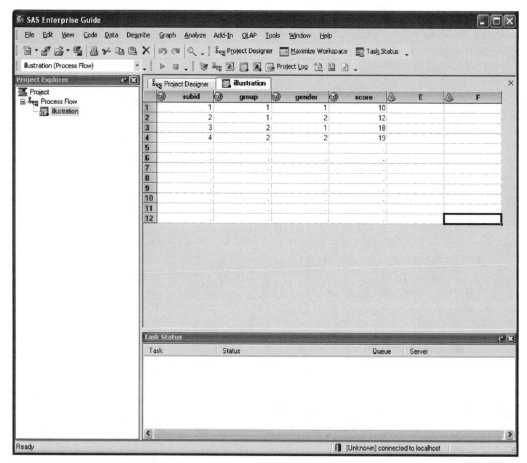

Figure B8   The completed data file.

look very much like the one we completed for SPSS. Save your project as described in Section B.5 to a location of your choosing on your computer.

B.5   SAVING A PROJECT

If we elect to save the project that is currently open, then from the main *SAS Enterprise Guide* menu select **File → Save Project**. This allows us to choose between **Local Computer** and **SAS Servers/Folders**. Select **Local Computer** and navigate to any place on your internal drive or to external media such as a USB flash drive where you want to save the project. Give it a reasonable name to replace the default name of **Project** and **Save**. The data file is now saved in that project. If we had performed any statistical analysis on the data, those output files would also be saved in the project.

B.6   CONSTRUCTING YOUR DATA FILE IN EXCEL

In working with *SAS Enterprise Guide*, we believe that it may be somewhat easier, especially if the data files are relatively large, to enter your data in

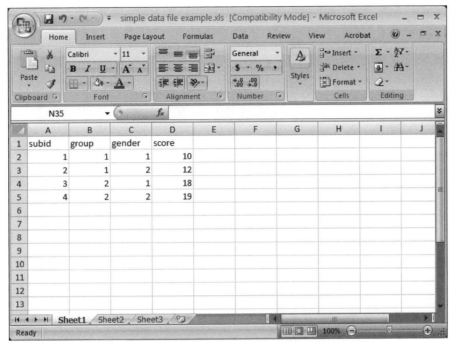

Figure B9   The Excel file that we want to bring into SAS.

Excel rather than entering them directly into SAS. For the purposes of this Appendix, we will assume that you know how to enter data into Excel. We have typed our small set of hypothetical data into an Excel worksheet as shown in Figure B9. After the data file has been constructed, save it to a convenient location on your internal drive or to external media (e.g., a USB flash drive).

B.7   IMPORTING DATA FROM EXCEL

From the main *SAS Enterprise Guide* menu, select **File → Import Data**. This brings you to the screen shown in Figure B10 giving you a choice of opening a project from either **Local Computer** or **SAS Servers/Folder**. We will assume that you are working on a stand-alone computer and that your file is located on your computer or some media (e.g., USB flash drive, CD) that is acknowledged by your computer. Thus, select **Local Computer**.

Figure B10   You need to indicate if you are on a local computer or a network.

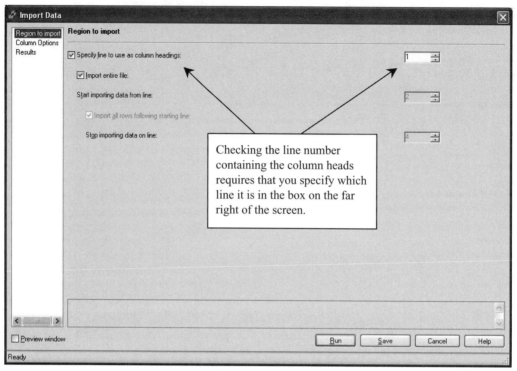

Figure B11   The **Import Data** window.

When you have selected **Local Computer**, you will see the standard Windows **Open File** screen. Navigate to the Excel file containing the data that you have saved. Make sure the **Files of type** panel shows either **All Files** or those with the **.xls** (Microsoft Excel 1997–2003) extension.

Selecting the Excel file results in an **Open Tables** window that asks for the Excel sheet number. You probably used Sheet 1 in the Excel file so double click that. You are then presented with an **Import Data** *SAS Enterprise Guide* window such as the window shown in Figure B11.

Note that in the far left panel are tabs indicating the information about the data file that you might need to address. By default, the active tab is **Region to import** and is the only tab we need to deal with in this example. Be sure that **Specify line to use as column headings** is checked (you should always use headings in your Excel file to name your variables) and that the value of **1** appears in the line number specification box at the far right of the window. Then check the box corresponding to **Import entire file** and click **Run**.

To "run" the **Import Data** routine means that *SAS Enterprise Guide* will transform the data file into the SAS format and will bring it into a project. The screen that appears once the run has been successfully completed is an *SAS Enterprise Guide* window shown in Figure B12.

There are two tabs available in this window, and they are located just above the data file. Note that this window is associated with the

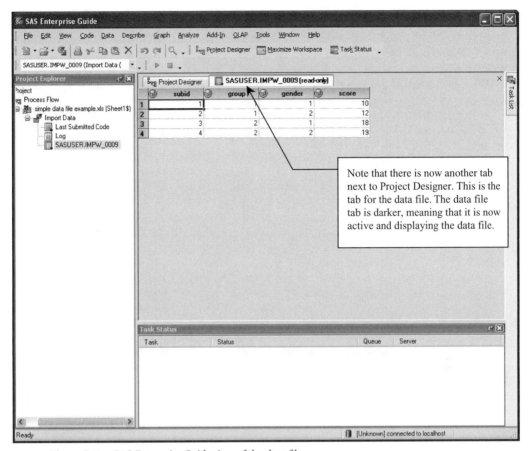

Figure B12   *SAS Enterprise Guide* view of the data file.

tab labeled **SASUSER.IMPW_0009(read_only)**, and that the data file is displayed. The name of the tab indicates that the data are in SAS format (**SASUSER** means "SAS user") and were imported from an outside file. The number 0009 means that it is the ninth project that we worked on since SAS was initially opened in this work session (your number will likely be different).

We mentioned that SAS does its statistical analysis inside a project. To see the context and flow, you can click the tab named **Project Designer** above the grid of the window shown in Figure B13. The **Process Flow** diagram is shown in Figure B13.

The **Process Flow** is a flow chart on a grid background showing the stages you have completed in icon and arrow format. You can see that we started with an Excel spreadsheet named **Sample Data.xls**, which, when we imported it, gave rise to an SAS data file named **SASUSER 1**. Note that in either view (**SASUSER** or **Project Designer**) the main *SAS Enterprise Guide* menu at the very top of the screen is available. It is from this menu that you would select the statistical procedure you want to use.

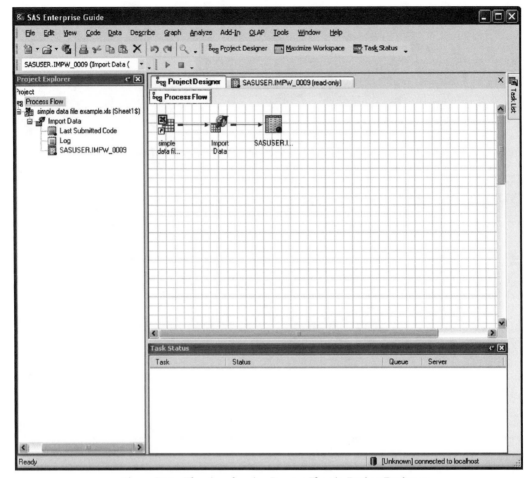

Figure B13   The view showing **Process Flow** in **Project Designer**.

B.8   THE MAIN SAS MENU

Figure B14 shows a portion of an existent *SAS Enterprise Guide* data file. At the top of the window the main *SAS Enterprise Guide* menu (**File, Edit,** and so on) appears. You will make use of some of these menus much more frequently than others. When you click on one of these menu items, you will open a secondary menu from which you select what you would like to do. Very briefly, these menu items contain the following:

- **File**: Contains a variety of functions including **Open, Import Data, Print Preview** (the data file name will appear here), and **Exit**.
- **Edit**: Allows you to **Cut, Copy, Paste, Select All,** and so on.
- **View**: Controls **Toolbars, Task Status,** and so on.
- **Code**: Allows user to run the analysis that has been set up, stop the processing, and deal with macros.

This is the main SAS menu.

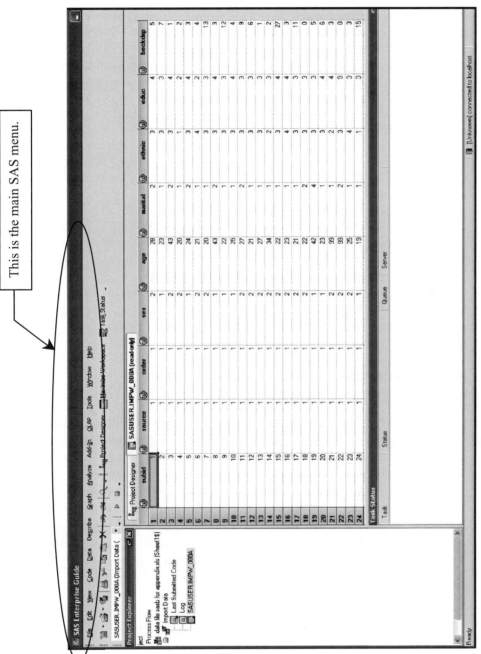

Figure B14 The main *SAS Enterprise Guide* menu.

- **Data**: Allows you to deal with the data file; among other things, you can select options to **sort** (reorder) the cases, **Transpose** the rows and columns, and **standardize** the data.
- **Describe**: Allows you to **List Data** (e.g., identify each case by variables that you designate), acquire **Summary Statistics**, and produce a **Frequency** table on a specified variable.
- **Graph**: Contains a variety of preformatted ways to plot your data.
- **Analyze**: Contains the statistical procedures you use to analyze your data.
- **Add-In**: Gets you to the **Add-In Manager**, which allows you to add, remove, and update commonly used procedures, such as **Standardize Data** and **Summary Statistics**.
- **OLAP**: This acronym stands for online analytical processing. According to the SAS website, the OLAP Server is a multidimensional data store designed to provide quick access to presummarized data generated from vast amounts of detailed data.
- **Tools**: Allows you to access sample data files through **SAS Enterprise Guide Explorer**, place your project in a particular library through **Assign Library**, and produce your statistical output in HTML, PDF, RTF, and so on format through **Options**.
- **Window**: Allows you to reach particular screens.
- **Help**: Contains documentation explaining how to work with SAS.

B.9 PERFORMING STATISTICAL PROCEDURES IN *SAS ENTERPRISE GUIDE*

We will use the data file displayed in Figure B14 to illustrate the process of performing statistical analyses in *SAS Enterprise Guide*. We use a one-way ANOVA procedure to illustrate how to work with *SAS Enterprise Guide* windows.

From the main menu, select **Analyze ➜ ANOVA ➜ One-Way**. This brings us to the main dialog window for the procedure as shown in Figure B15. The navigation panel at the very left of the window will appear in every statistical analysis and allows us to reach different parts of the specifications for the analysis. We typically begin our navigation in the **Task Roles** portion of the procedure. It is here that we select those variables in our data file that will be assigned particular roles in the analysis we have invoked. In this one-way ANOVA procedure, for example, we must specify the dependent and independent variables in the analysis.

B.9.1 THE VARIABLES TO ASSIGN PANEL

The **Variables to assign** panel (next to the navigation panel) in Figure B15 lists the variables in the project data file in the order that they

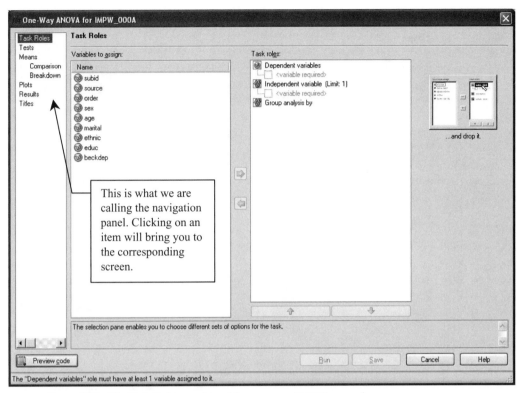

Figure B15   The **Task Roles** portion of the one-way ANOVA procedure.

appear in the data file. To the right of the **Variables to assign** panel is the **Task roles** panel. This panel contains slots to identify the dependent and independent variables in the analysis. To specify which variables are to be used for each role, we highlight and drag variables from the **Variables to assign** panel to the **Task roles** panel. For example, highlight and drag the variable named **beckdep** in the **Variables to assign** panel to the icon for the **Dependent variables**. Then drag the variable **educ** to the icon for **Independent variable**. If you prefer, you can avoid the dragging process by highlighting the variable in the **Variables to assign** panel and clicking the arrow button pointing to the **Task roles** panel to the right of the panel. From the little menu appearing when you click the arrow button, you can select (by clicking the choice) the particular task role to be assigned the variable. When you finish, the window will look like Figure B16.

B.9.2   OTHER CHOICES IN THE NAVIGATION PANEL

Very often, when you select another choice in the navigation panel, you will be presented with a dialog window in which you will mark checkboxes or select from drop-down menus. For example, by selecting the **Means →** **Comparisons** choice in the far right navigation panel, we are presented

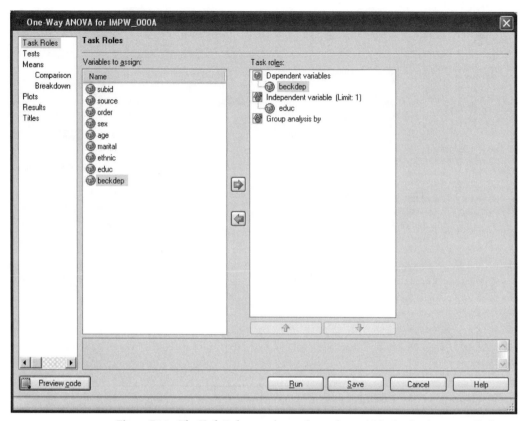

Figure B16   The **Task Roles** panel now shows the variables having been specified.

with the window shown in Figure B17. Next to each of the comparison methods are checkboxes. In Figure B17 we have checked the **Bonferroni** comparison and, by doing so, have made available the drop-down menu for **Confidence level**, for which we have selected 95 percent.

B.10   *SAS ENTERPRISE GUIDE* OUTPUT

When you instruct SAS to perform a statistical analysis, it displays the results in a window in the form that you have specified on the **Tools** menu. We have provided you with an example of some output in PDF form. Figure B18 contains summary table information from the ANOVA procedure we performed. Included in this output is the *F* ratio, its probability of occurrence if the null hypothesis is true, and the R squared (eta squared) index of strength of effect provided by SAS in three separate displays.

Figure B19 presents some descriptive statistics for the groups in the analysis. Recall that the dependent variable in the analysis was **beckdep** and SAS makes this very clear by labeling the columns **Mean of beckdep**, **Std. Dev. of beckdep, and Std. Error of beckdep**. Recall also that the independent variable was **educ**, and SAS displays this information in the very first column of the output.

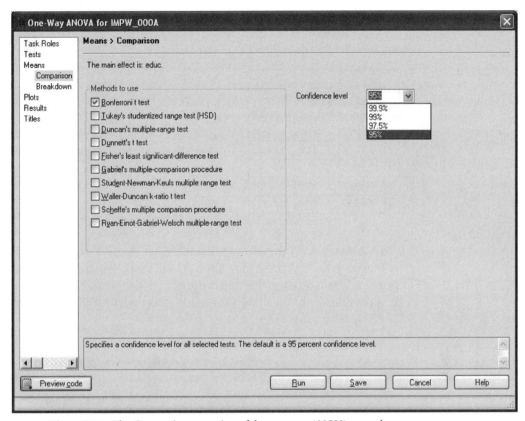

Figure B17 The **Comparisons** portion of the one-way ANOVA procedure.

B.11 SAVING THE SAS OUTPUT FILE AS A PDF DOCUMENT

A PDF document is a type of file that is in portable document format. It is a faithful copy of the original but it is not editable unless you have the full version of Adobe Acrobat or some comparable application. When

Dependent Variable: beckdep beckdep

Source	DF	Sum of Squares	Mean Square	F Value	Pr > F
Model	3	214.32748	71.44249	1.58	0.1933
Error	374	16890.82860	45.16264		
Corrected Total	377	17105.15608			

R-Square	Coeff Var	Root MSE	beckdep Mean
0.012530	93.84113	6.720316	7.161376

Source	DF	Anova SS	Mean Square	F Value	Pr > F
educ	3	214.3274826	71.4424942	1.58	0.1933

Figure B18 Summary table output in PDF format.

One-Way Analysis of Variance
Results
Means and Descriptive Statistics

educ	Mean of beckdep	Std. Dev. of beckdep	Std. Error of beckdep
.	7.16138	6.73586	0.34646
1	8.49180	8.17236	1.04636
2	7.33333	6.02510	0.81991
3	6.38608	5.70587	0.45393
4	7.46667	7.49880	0.73181

Figure B19   Descriptive statistics provided by SAS.

PDF documents are printed, they mirror what you see on the screen even though your computer may not have the fonts that are used in the document. That's what makes them portable – the portable document format contains within it all the information necessary for the document to be displayed on the screen or to be printed.

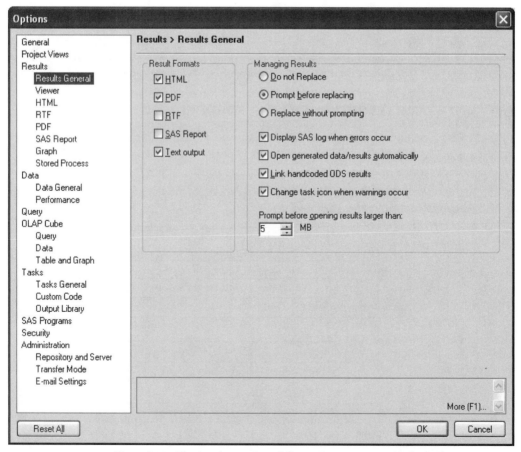

Figure B20   The **Results** portion of the **Options** screen on the **Tools** menu.

Saving the output as a PDF document is extremely valuable for students and those who may not have SAS loaded on the computer (perhaps their home computer or their personal laptop) that they will be using away from the organizational setting so that they can view the results in order to prepare a report. A PDF document can be opened and viewed on any computer. Adobe Acrobat Reader is a free application that can be downloaded from the Adobe Web site; with it, you can open and view PDF files; Mac users can open PDF files with Preview, an application that is bundled with the OS 10 operating system.

SAS will allow you to produce a PDF version of the output to the project. This is a necessary first step toward saving the file on your personal computer. To have SAS display your output as a PDF document, select from the main menu **Tools → Options → Results → Results General**. This brings you to the screen shown in Figure B20. Check the box corresponding to **PDF**. Note that we have also checked the boxes corresponding to **HTML** and **Text output**; thus, we will obtain three versions of the output. Click **OK**. The PDF file is saved in the **Process Flow** diagram, which is on the **Project Designer** tab as shown in Figure B21.

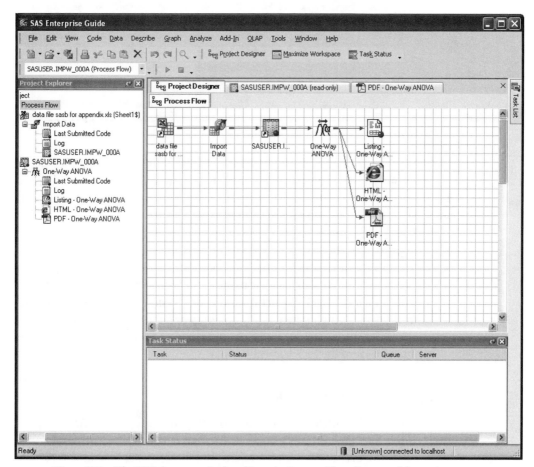

Figure B21 The PDF document is placed into the **Process Flow** diagram of the project.

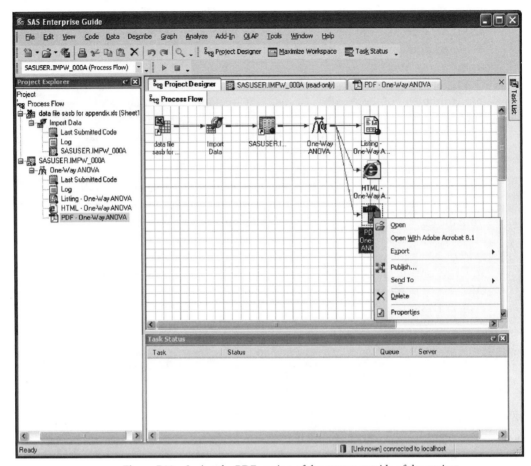

Figure B22 Saving the PDF version of the output outside of the project.

The existence of the PDF file inside of the project is not sufficient for you to access it outside of the project. To now save the PDF to either the internal drive or to an external USB flash drive so that you can email it or transfer it to another computer, right click the icon for the PDF file. The result of this right click is shown in Figure B22. Then select **Export →** **Export PDF → One-Way ANOVA** (this is the name assigned to the file because we performed a one-way analysis; if you performed a different procedure, the file will have the name of that procedure). As an alternative, you can select **File → Export** from the main menu (so long as the icon for the PDF file is highlighted). After making either of these selections, select **Local Computer** from the choices and navigate to the location where you intend to save the file. Then click **Save**.

B.12 ADDITIONAL RESOURCES

Compared to SPSS, there are far fewer resources describing *SAS Enterprise Guide* that are currently available. We suggest consulting Cody and Smith (2006), Davis (2007), Der and Everitt (2007), Slaughter and Delwiche (2006), and SAS Institute (2002) for additional information.

Table of Critical *F* Values

This was obtained from VassarStats, a Web site authored by Professor Emeritus Richard Lowry based in Vassar College. The Web site is dedicated to the free dissemination of knowledge on the world-wide web.

Table C1. VassarStats: Table of critical *F* values (p. 1) [top entry for 0.05 level; bottom entry for 0.01 level]

		df numerator												
	1	**2**	**3**	**4**	**5**	**6**	**7**	**8**	**9**	**10**	**11**	**12**	**13**	**14**
1	161	198	216	225	230	234	237	239	241	242	243	244	245	245
	4,052	4,959	5,404	5,624	5,764	5,859	5,928	5,981	6,022	6,056	6,083	6,107	6,126	6,143
2	1851	19.00	19.16	1925	19.30	19.33	1935	19.37	19.38	19.40	19.40	19.41	19.42	19.42
	99	99	99	99	99	99	99	99	99	99	99	99	99	99
3	10.13	9.55	9.28	9.12	9.01	8.94	8.89	8.85	8.81	8.79	8.76	8.74	8.73	8.71
	34.12	30.82	29.46	28.71	28.24	27.91	27.67	27.49	27.34	27.23	27.13	27.05	26.98	26.92
4	7.71	6.94	6.59	6.39	6.26	6.16	6.09	6.04	6.00	5.96	5.94	5.91	5.89	5.87
	21.20	18.00	16.69	15.98	15.52	15.21	14.98	14.80	14.66	14.55	14.45	14.37	14.31	14.25
5	6.61	5.79	5.41	5.19	5.05	4.95	4.88	4.82	4.77	4.74	4.70	4.68	4.66	4.64
	16.26	13.27	12.06	11.39	10.97	10.67	10.46	10.29	10.16	10.05	9.96	9.89	9.82	9.77
6	5.99	5.14	4.76	4.53	4.39	4.28	4.21	4.15	4.10	4.06	4.03	4.00	3.98	3.96
	13.75	10.92	9.78	9.15	8.75	8.47	8.26	8.10	7.98	7.87	7.79	7.72	7.66	7.60
7	5.59	4.74	4.35	4.12	3.97	3.87	3.79	3.73	3.68	3.64	3.60	3.57	3.55	3.53
	12.25	9.55	8.45	7.85	7.46	7.19	6.99	6.84	6.72	6.62	6.54	6.47	5.41	6.36
8	5.32	4.46	4.07	3.84	3.69	3.58	3.50	3.44	3.39	3.35	3.31	3.28	3.26	3.24
	11.26	8.65	7.59	7.01	6.63	6.37	6.18	6.03	5.91	5.81	5.73	5.67	5.61	5.56
9	5.12	4.26	3.86	3.63	3.48	3.37	3.29	3.23	3.18	3.14	3.10	3.07	3.05	3.03
	10.56	8.02	6.99	6.42	6.06	5.80	5.61	5.47	5.35	5.26	5.18	5.11	5.05	5.01
10	4.96	4.10	3.71	3.48	3.33	3.22	3.14	3.07	3.02	2.98	2.94	2.91	2.89	2.86
	10.04	7.56	6.55	5.99	5.64	5.39	5.20	5.06	4.94	4.85	4.77	4.71	4.65	4.60
11	4.84	3.98	3.59	3.36	3.20	3.09	3.01	2.95	2.90	2.85	2.82	2.79	2.76	2.74
	9.65	7.21	6.22	5.67	5.32	5.07	4.89	4.74	4.63	4.54	4.46	4.40	4.34	4.29
12	4.75	3.89	3.49	3.26	3.11	3.00	2.91	2.85	2.80	2.75	2.72	2.69	2.66	2.64
	9.33	6.93	5.95	5.41	5.06	4.82	4.64	4.50	4.39	4.30	4.22	4.16	4.10	4.05
13	4.67	3.81	3.41	3.18	3.03	2.92	2.83	2.77	2.71	2.67	2.63	2.60	2.58	2.55
	9.07	6.70	5.74	5.21	4.86	4.62	4.44	4.30	4.19	4.10	4.02	3.96	3.91	3.86
14	4.60	3.74	3.34	3.11	2.96	2.85	2.76	2.70	2.65	2.60	2.57	2.53	2.51	2.48
	8.86	6.51	5.56	5.04	4.69	4.46	4.28	4.14	4.03	3.94	3.86	3.80	3.75	3.70
15	4.54	3.68	3.29	3.06	2.90	2.79	2.71	2.64	2.59	2.54	2.51	2.48	2.45	2.42
	8.68	6.36	5.42	4.89	4.56	4.32	4.14	4.00	3.89	3.80	3.73	3.67	3.61	3.56

(*continued*)

Table C1 (*continued*)

							df numerator							
		1	2	3	4	5	6	7	8	9	10	11	12	13	14
	16	4.49	3.63	3.24	3.01	2.85	2.74	2.66	2.59	2.54	2.49	2.46	2.42	2.40	2.37
		8.53	6.23	5.29	4.77	4.44	4.20	4.03	3.89	3.78	3.69	3.62	3.55	3.50	3.45
	17	4.45	3.59	3.20	2.96	2.81	2.70	2.61	2.55	2.49	2.45	2.41	2.38	2.35	2.33
		8.40	6.11	5.19	4.67	4.34	4.10	3.93	3.79	3.68	3.59	3.52	3.46	3.40	3.35
	18	4.41	3.55	3.16	2.93	2.77	2.66	2.58	2.51	2.46	2.41	2.37	2.34	2.31	2.29
		8.29	6.01	5.09	4.58	4.25	4.01	3.84	3.71	3.60	3.51	3.43	3.37	3.32	3.27
	19	4.38	3.52	3.13	2.90	2.74	2.63	2.54	2.48	2.42	2.38	2.34	2.31	2.28	2.26
		8.18	5.93	5.01	4.50	4.17	3.94	3.77	3.63	3.52	3.43	3.36	3.30	3.24	3.19
	20	4.35	3.49	3.10	2.87	2.71	2.60	2.51	2.45	2.39	2.35	2.31	2.28	2.25	2.22
		8.10	5.85	4.94	4.43	4.10	3.87	3.70	3.56	3.46	3.37	3.29	3.23	3.18	3.13
	21	4.32	3.47	3.07	2.84	2.68	2.57	2.49	2.42	2.37	2.32	2.28	2.25	2.22	2.20
		8.02	5.78	4.87	4.37	4.04	3.81	3.64	3.51	3.40	3.31	3.24	3.17	3.12	3.07
df denominator	22	4.30	3.44	3.05	2.82	2.66	2.55	2.46	2.40	2.34	2.30	2.26	2.23	2.20	2.17
		7.95	5.72	4.82	4.31	3.99	3.76	3.59	3.45	3.35	3.26	3.18	3.12	3.07	3.02
	23	4.28	3.42	3.03	2.80	2.64	2.53	2.44	2.37	2.32	2.27	2.24	2.20	2.18	2.15
		7.88	5.66	4.76	4.26	3.94	3.71	3.54	3.41	3.30	3.21	3.14	3.07	3.02	2.97
	24	4.26	3.40	3.01	2.78	2.62	2.51	2.42	2.36	2.30	2.25	2.22	2.18	2.15	2.13
		7.82	5.61	4.72	4.22	3.90	3.67	3.50	3.36	3.26	3.17	3.09	3.03	2.98	2.93
	25	4.24	3.39	2.99	2.76	2.60	2.49	2.40	2.34	2.28	2.24	2.20	2.16	2.14	2.11
		7.77	5.57	4.68	4.18	3.85	3.63	3.46	3.32	3.22	3.13	3.06	2.99	2.94	2.89
	26	4.23	3.37	2.98	2.74	2.59	2.47	2.39	2.32	2.27	2.22	2.18	2.15	2.12	2.09
		7.72	5.53	4.64	4.14	3.82	3.59	3.42	3.29	3.18	3.09	3.02	2.96	2.90	2.86
	27	4.21	3.35	2.96	2.73	2.57	2.46	2.37	2.31	2.25	2.20	2.17	2.13	2.10	2.08
		7.68	5.49	4.60	4.11	3.78	3.56	3.39	3.26	3.15	3.06	2.99	2.93	2.87	2.82
	28	4.20	3.34	2.95	2.71	2.56	2.45	2.36	2.29	2.24	2.19	2.15	2.12	2.09	2.06
		7.64	5.45	4.57	4.07	3.75	3.53	3.36	3.23	3.12	3.03	2.96	2.90	2.84	2.79
	29	4.18	3.33	2.93	2.70	2.55	2.43	2.35	2.28	2.22	2.18	2.14	2.10	2.08	2.05
		7.60	5.42	4.54	4.04	3.73	3.50	3.33	3.20	3.09	3.00	2.93	2.87	2.81	2.77
	30	4.17	3.32	2.92	2.69	2.53	2.42	2.33	2.27	2.21	2.16	2.13	2.09	2.06	2.04
		7.56	5.39	4.51	4.02	3.70	3.47	3.30	3.17	3.07	2.98	2.91	2.84	2.79	2.74

Table C1 (*continued*)

		1	2	3	4	5	6	7	8	9	10	11	12	13	14
							df numerator								
df denominator	31	4.16	3.30	2.91	2.68	2.52	2.41	2.32	2.25	2.20	2.15	2.11	2.08	2.05	2.03
		7.53	5.36	4.48	3.99	3.67	3.45	3.28	3.15	3.04	2.96	2.88	2.82	2.77	2.72
	32	4.15	3.29	2.90	2.67	2.51	2.40	2.31	2.24	2.19	2.14	2.10	2.07	2.04	2.01
		7.50	5.34	4.46	3.97	3.65	3.43	3.26	3.13	3.02	2.93	2.86	2.80	2.74	2.70
	33	4.14	3.28	2.89	2.66	2.50	2.39	2.30	2.23	2.18	2.13	2.09	2.06	2.03	2.00
		7.47	5.31	4.44	3.95	3.63	3.41	3.24	3.11	3.00	2.91	2.84	2.78	2.72	2.68
	34	4.13	3.28	2.88	2.65	2.49	2.38	2.29	2.23	2.17	2.12	2.08	2.05	2.02	1.99
		7.44	5.29	4.42	3.93	3.61	3.39	3.22	3.09	2.98	2.89	2.82	2.76	2.70	2.66
	35	4.12	3.27	2.87	2.64	2.49	2.37	2.29	2.22	2.16	2.11	2.07	2.04	2.01	1.99
		7.42	5.27	4.40	3.91	3.59	3.37	3.20	3.07	2.96	2.88	2.80	2.74	2.69	2.64
	36	4.11	3.26	2.87	2.63	2.48	2.36	2.28	2.21	2.15	2.11	2.07	2.03	2.00	1.98
		7.40	5.25	4.38	3.89	3.57	3.35	3.18	3.05	2.95	2.86	2.79	2.72	2.67	2.62
	37	4.11	3.25	2.86	2.63	2.47	2.36	2.27	2.20	2.14	2.10	2.06	2.02	2.00	1.97
		7.37	5.23	4.36	3.87	3.56	3.33	3.17	3.04	2.93	2.84	2.77	2.71	2.65	2.61
	38	4.10	3.24	2.85	2.62	2.46	2.35	2.26	2.19	2.14	2.09	2.05	2.02	1.99	1.96
		7.35	5.21	4.34	3.86	3.54	3.32	3.15	3.02	2.92	2.83	2.75	2.69	2.64	2.59
	39	4.09	3.24	2.85	2.61	2.46	2.34	2.26	2.19	2.13	2.08	2.04	2.01	1.98	1.95
		7.33	5.19	4.33	3.84	3.53	3.30	3.14	3.01	2.90	2.81	2.74	2.68	2.62	2.58
	40	4.08	3.23	2.84	2.61	2.45	2.34	2.25	2.18	2.12	2.08	2.04	2.00	1.97	1.95
		7.31	5.18	4.31	3.83	3.51	3.29	3.12	2.99	2.89	2.80	2.73	2.66	2.61	2.56
	41	4.08	3.23	2.83	2.60	2.44	2.33	2.24	2.17	2.12	2.07	2.03	2.00	1.97	1.94
		7.30	5.16	4.30	3.81	3.50	3.28	3.11	2.98	2.87	2.79	2.71	2.65	2.60	2.55
	42	4.07	3.22	2.83	2.59	2.44	2.32	2.24	2.17	2.11	2.06	2.03	1.99	1.96	1.94
		7.28	5.15	4.29	3.80	3.49	3.27	3.10	2.97	2.86	2.78	2.70	2.64	2.59	2.54
	43	4.07	3.21	2.82	2.59	2.43	2.32	2.23	2.16	2.11	2.06	2.02	1.99	1.96	1.93
		7.26	5.14	4.27	3.79	3.48	3.25	3.09	2.96	2.85	2.76	2.69	2.63	2.57	2.53
	44	4.06	3.21	2.82	2.58	2.43	2.31	2.23	2.16	2.10	2.05	2.01	1.98	1.95	1.92
		7.25	5.12	4.26	3.78	3.47	3.24	3.08	2.95	2.84	2.75	2.68	2.62	2.56	2.52
	45	4.06	3.20	2.81	2.58	2.42	2.31	2.22	2.15	2.10	2.05	2.01	1.97	1.94	1.92
		7.23	5.11	4.25	3.77	3.45	3.23	3.07	2.94	2.83	2.74	2.67	2.61	2.55	2.51

(*continued*)

Table C1 (continued)

		df numerator												
		1	2	3	4	5	6	7	8	9	10	11	12	13	14
50		4.03	3.18	2.79	2.56	2.40	2.29	2.20	2.13	2.07	2.03	1.99	1.95	1.92	1.89
		7.17	5.06	4.20	3.72	3.41	3.19	3.02	2.89	2.78	2.70	2.63	2.56	2.51	2.46
55		4.02	3.16	2.77	2.54	2.38	2.27	2.18	2.11	2.06	2.01	1.97	1.93	1.90	1.88
		7.12	5.01	4.16	3.68	3.37	3.15	2.98	2.85	2.75	2.66	2.59	2.53	2.47	2.42
60		4.00	3.15	2.76	2.53	2.37	2.25	2.17	2.10	2.04	1.99	1.95	1.92	1.89	1.86
		7.08	4.98	4.13	3.65	3.34	3.12	2.95	2.82	2.72	2.63	2.56	2.50	2.44	2.39
65		3.99	3.14	2.75	2.51	2.36	2.24	2.15	2.08	2.03	1.98	1.94	1.90	1.87	1.85
		7.04	4.95	4.10	3.62	3.31	3.09	2.93	2.80	2.69	2.61	2.53	2.47	2.42	2.37
70		3.98	3.13	2.74	2.50	2.35	2.23	2.14	2.07	2.02	1.97	1.93	1.89	1.86	1.84
		7.01	4.92	4.07	3.60	3.29	3.07	2.91	2.78	2.67	2.59	2.51	2.45	2.40	2.35
80		3.96	3.11	2.72	2.49	2.33	2.21	2.13	2.06	2.00	1.95	1.91	1.88	1.84	1.82
		6.96	4.88	4.04	3.56	3.26	3.04	2.87	2.74	2.64	2.55	2.48	2.42	2.36	2.31
90		3.95	3.10	2.71	2.47	2.32	2.20	2.11	2.04	1.99	1.94	1.90	1.86	1.83	1.80
		6.93	4.85	4.01	3.53	3.23	3.01	2.84	2.72	2.61	2.52	2.45	2.39	2.33	2.29
100		3.94	3.09	2.70	2.46	2.31	2.19	2.10	2.03	1.97	1.93	1.89	1.85	1.82	1.79
		6.90	4.82	3.98	3.51	3.21	2.99	2.82	2.69	2.59	2.50	2.43	2.37	2.31	2.27
110		3.93	3.08	2.69	2.45	2.30	2.18	2.09	2.02	1.97	1.92	1.88	1.84	1.81	1.78
		6.87	4.80	3.96	3.49	3.19	2.97	2.81	2.68	2.57	2.49	2.41	2.35	2.30	2.25
120		3.92	3.07	2.68	2.45	2.29	2.18	2.09	2.02	1.96	1.91	1.87	1.83	1.80	1.78
		6.85	4.79	3.95	3.48	3.17	2.96	2.79	2.66	2.56	2.47	2.40	2.34	2.28	2.23
130		3.91	3.07	2.67	2.44	2.28	2.17	2.08	2.01	1.95	1.90	1.86	1.83	1.80	1.77
		6.83	4.77	3.94	3.47	3.16	2.94	2.78	2.65	2.55	2.46	2.39	2.32	2.27	2.22
140		3.91	3.06	2.67	2.44	2.28	2.16	2.08	2.01	1.95	1.90	1.86	1.82	1.79	1.76
		6.82	4.76	3.92	3.46	3.15	2.93	2.77	2.64	2.54	2.45	2.38	2.31	2.26	2.21
160		3.90	3.05	2.66	2.43	2.27	2.16	2.07	2.00	1.94	1.89	1.85	1.81	1.78	1.75
		6.80	4.74	3.91	3.44	3.13	2.92	2.75	2.62	2.52	2.43	2.36	2.30	2.24	2.20
180		3.89	3.05	2.65	2.42	2.26	2.15	2.06	1.99	1.93	1.88	1.84	1.81	1.77	1.75
		6.78	4.73	3.89	3.43	3.12	2.90	2.74	2.61	2.51	2.42	2.35	2.28	2.23	2.18
200		3.89	3.04	2.65	2.42	2.26	2.14	2.06	1.98	1.93	1.88	1.84	1.80	1.77	1.74
		6.76	4.71	3.88	3.41	3.11	2.89	2.73	2.60	2.50	2.41	2.34	2.27	2.22	2.17

df denominator

Deviational Formula for Sums of Squares

From our discussion in Chapter 3, we noted that the total variability in an experiment or research study consists of between-groups variability (i.e., the effects of the independent variable plus error) and within-groups variability (i.e., the effects of error alone). This variability is a function of how the scores on the dependent variable deviate from the grand mean, or the individual treatment group means, and also how each group mean deviates from the grand mean.

More specifically, we can note that the total variability (total deviation) in a study is composed of how each dependent variable score deviates from the grand mean. It can be expressed as

$$\text{total deviation} = Y_{ij} - \overline{Y}_T. \tag{D.1}$$

As noted above, this total deviation has two component parts, the between-groups and within-groups deviations. The between-groups deviation is a function of each treatment group mean from the grand mean and can be expressed as

$$\text{between-groups deviation} = \overline{Y}_j - \overline{Y}_{} \tag{D.2}$$

The within-group deviation tion of each score from its respective ressed as

$$\text{within-gro} \tag{D.3}$$

Thus, we can su wing manner:

total deviatio

$$Y_{ij} - \overline{Y}_T \tag{D.4}$$

These symbolic relat od within the context of a numer marized in Table D1. These data tigating the effects of study time red in Chapter 6.

The far left column of pendent variable (SAT scores) b dent variable (study time in mont ght-hand

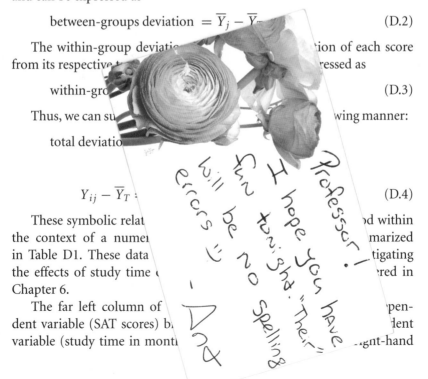

Table D1. Component deviations

		Score Y	Total $(Y - \overline{Y}_T)$	=	Between groups $(\overline{Y}_j - \overline{Y}_T)$	+	Within groups $(Y - \overline{Y}_j)$
					a_1		
		370	(-165.43)	=	(-122.57)	+	(-42.86)
		380	(-155.43)	=	(-122.57)	+	(-32.86)
		420	(-115.43)	=	(-122.57)	+	(7.14)
	$\overline{Y}_1 = 412.86$	420	(-115.43)	=	(-122.57)	+	(7.14)
		430	(-105.43)	=	(-122.57)	+	(17.14)
		430	(-105.43)	=	(-122.57)	+	(17.14)
		440	(-95.73)	=	(-122.57)	+	(27.14)
					a_2		
		410	(-125.43)	=	(-61.14)	+	(-64.29)
		430	(-105.43)	=	(-61.14)	+	(-44.29)
		440	(-95.43)	=	(-61.14)	+	(-34.29)
	$\overline{Y}_2 = 474.29$	480	(-55.43)	=	(-61.14)	+	(5.71)
		510	(-25.43)	=	(-61.14)	+	(35.71)
		520	(-15.43)	=	(-61.14)	+	(45.71)
		530	(-5.43)	=	(-61.14)	+	(55.71)
					a_3		
		530	(-5.43)	=	(17.43)	+	(-22.86)
		530	(-5.43)	=	(17.43)	+	(-22.86)
		540	(4.57)	=	(17.43)	+	(-12.86)
$\overline{Y}_T = 535.43$	$\overline{Y}_3 = 552.86$	540	(4.57)	=	(17.43)	+	(-12.86)
		550	(14.57)	=	(17.43)	+	(-2.86)
		570	(34.57)	=	(17.43)	+	(17.14)
		610	(74.57)	=	(17.43)	+	(57.14)
					a_4		
		560	(24.57)	=	(78.86)	+	(-54.29)
		580	(44.57)	=	(78.86)	+	(-34.29)
		590	(54.57)	=	(78.86)	+	(-24.29)
	$\overline{Y}_4 = 614.29$	620	(84.57)	=	(78.86)	+	(5.71)
		640	(104.57)	=	(78.86)	+	(25.71)
		650	(114.57)	=	(78.86)	+	(35.71)
		660	(124.57)	=	(78.86)	+	(45.71)
					a_5		
		550	(14.57)	=	(88.43)	+	(-73.86)
		610	(74.57)	=	(88.43)	+	(-13.86)
		620	(84.57)	=	(88.43)	+	(-3.86)
	$\overline{Y}_5 = 623.86$	630	(94.57)	=	(88.43)	+	(6.14)
		640	(104.57)	=	(88.43)	+	(16.14)
		650	(114.57)	=	(88.43)	+	(26.14)
		660	(124.57)	=	(88.43)	+	(36.14)

side of Table D1 summarize the individual deviation calculations for the total, between-groups, and within-groups deviations. For example, the first participant's SAT score $Y_{1,1}$ (read "Y one, one") is subtracted from the grand mean $\overline{Y}_T$, yielding $Y_{1,1} - \overline{Y}_T = 370 - 535.43 = -165.43$, the total deviation. Likewise, the between-groups deviation is computed by subtracting the treatment group mean for level a_1 (the zero months of prior study group) $\overline{Y}_1$ from the grand mean, yielding $\overline{Y}_1 - \overline{Y}_T = 412.86 - 535.43 = -122.57$. Lastly, the within-groups deviation is computed by subtracting the first participant's SAT score $Y_{1,1}$ from the treatment group mean, yielding $Y_{1,1} - \overline{Y}_1 = 370 - 412.86 = -42.86$. These summary deviations depicted in the columns of Table D1 labeled "Deviations" provided the ingredients for computing the sums of squares using the deviational formulas.

The deviational formulas for the sums of squares are as follows:

total deviations
$$= \Sigma(Y_{ij} - \overline{Y}_T)^2$$
$$= (-165.43)^2 + (155.43)^2 + \cdots + (114.57)^2 + (124.57)^2$$
$$= 27,367.084 + 24,158.484 + \cdots + 13,126.284 + 15,517.684$$
$$= \boxed{270,268.5}. \tag{D.5}$$

between-groups deviations
$$= \Sigma(\overline{Y}_j - \overline{Y}_T)^2$$
$$= (-122.57)^2 + (-122.57)^2 + \cdots + (88.43)^2 + (88.43)^2$$
$$= 15,023.404 + 15,023.404 + \cdots + 7,819.865 + 7,819.865$$
$$= \boxed{230.496.5}. \tag{D.6}$$

within-groups deviations
$$= \Sigma(Y_{ij} - \overline{Y}_j)^2$$
$$= (-42.86)^2 + (-32.86)^2 + \cdots + (26.14)^2 + (36.14)^2$$
$$= 1,836.980 + 1,079.780 + \cdots + 683.300 + 1,306.010$$
$$= \boxed{39,772}. \tag{D.7}$$

As we note at the top of Table D1, the total deviation (sum of squares total) is equal to the between-groups deviation (sum of squares between groups) plus the within-groups deviation (sum of squares within groups). This relationship can be expressed symbolically as

$$\Sigma(Y_{ij} - \overline{Y}_T)^2 = \Sigma(\overline{Y}_j - \overline{Y}_T)^2 + \Sigma(Y_{ij} - \overline{Y}_j)^2$$
$$270,268.5 = 230,496.5 + 39,772. \tag{D.8}$$

In practice, we will emphasize computational procedures for calculating sums of squares throughout this book. Thus, in the context of a one-way between-subjects design, we will use the following notation to describe the previous sums of squared deviations.

$$SS_T = SS_A + SS_{S/A}$$
$$SS_{\text{Total}} = SS_{\text{Between Groups}} + SS_{\text{Within Groups}}. \tag{D.9}$$

Lastly, remember that while the procedures for calculating sums of squares differ between deviational and computational approaches, the end result is always the same.

Coefficients of Orthogonal Polynomials

Table E1. Coefficients of orthogonal polynomials

# of groups	Polynomial order	Group number in order							
		1	2	3	4	5	6	7	8
3	Linear	−1	0	1					
	Quadratic	1	−2	1					
4	Linear	−3	−1	1	3				
	Quadratic	1	−1	−1	1				
	Cubic	−1	3	−3	1				
5	Linear	−2	−1	0	1	2			
	Quadratic	2	−1	−2	−1	2			
	Cubic	−1	2	0	−2	1			
	Quartic	1	−4	6	−4	1			
6	Linear	−5	−3	−1	1	3	5		
	Quadratic	5	−1	−4	−4	−1	5		
	Cubic	−5	7	4	−4	−7	5		
	Quartic	1	−3	2	2	−3	1		
7	Linear	−3	−2	−1	0	1	2	3	
	Quadratic	5	0	−3	−4	−3	0	5	
	Cubic	−1	1	1	0	−1	−1	1	
	Quartic	3	−7	1	6	1	−7	3	
8	Linear	−7	−5	−3	−1	1	3	5	7
	Quadratic	7	1	−3	−5	−5	−3	1	7
	Cubic	−7	5	7	3	−3	−7	−5	−7
	Quartic	7	−13	−3	9	9	−3	−13	7

Critical Values of the Studentized Range Statistic

df$_{WG}$	α	\multicolumn{9}{c}{Number of groups}
		2	3	4	5	6	7	8	9	10
5	.05	3.64	4.60	5.22	5.67	6.03	6.33	6.58	6.80	6.99
	.01	5.70	6.98	7.80	8.42	8.91	9.32	9.67	9.97	10.24
6	.05	3.46	4.34	4.90	5.30	5.63	5.90	6.12	6.32	6.49
	.01	5.24	6.33	7.03	7.56	7.97	8.32	8.61	8.87	9.10
7	.05	3.34	4.16	4.68	5.06	5.36	5.61	5.82	6.00	6.16
	.01	4.95	5.92	6.54	7.01	7.37	7.68	7.94	8.17	8.37
8	.05	3.26	4.04	4.53	4.89	5.17	5.40	5.60	5.77	5.92
	.01	4.75	5.64	6.20	6.62	6.96	7.24	7.47	7.68	7.86
9	.05	3.20	3.95	4.41	4.76	5.02	5.24	5.43	5.59	5.74
	.01	4.60	5.43	5.96	6.35	6.66	6.91	7.13	7.33	7.49
10	.05	3.15	3.88	4.33	4.65	4.91	5.12	5.30	5.46	5.60
	.01	4.48	5.27	5.77	6.14	6.43	6.67	6.87	7.05	7.21
11	.05	3.11	3.82	4.26	4.57	4.82	5.03	5.20	5.35	5.49
	.01	4.39	5.15	5.62	5.97	6.25	6.48	6.67	6.84	6.99
12	.05	3.08	3.77	4.20	4.51	4.75	4.95	5.12	5.27	5.39
	.01	4.32	5.05	5.50	5.84	6.10	6.32	6.51	6.67	6.81
13	.05	3.06	3.73	4.15	4.45	4.69	4.88	5.05	5.19	5.32
	.01	4.26	4.96	5.40	5.73	5.98	6.19	6.37	6.53	6.67
14	.05	3.03	3.70	4.11	4.41	4.64	4.83	4.99	5.13	5.25
	.01	4.21	4.89	5.32	5.63	5.88	6.08	6.26	6.41	6.54
15	.05	3.01	3.67	4.08	4.37	4.59	4.78	4.94	5.08	5.20
	.01	4.17	4.84	5.25	5.56	5.80	5.99	6.16	6.31	6.44
16	.05	3.00	3.65	4.05	4.33	4.56	4.74	4.90	5.03	5.15
	.01	4.13	4.79	5.19	5.49	5.72	5.92	6.08	6.22	6.35
17	.05	2.98	3.63	4.02	4.30	4.52	4.70	4.86	4.99	5.11
	.01	4.10	4.74	5.14	5.43	5.66	5.85	6.01	6.15	6.27
18	.05	2.97	3.61	4.00	4.28	4.49	4.67	4.82	4.96	5.07
	.01	4.07	4.70	5.09	5.38	5.60	5.79	5.94	6.08	6.20
19	.05	2.96	3.59	3.98	4.25	4.47	4.65	4.79	4.92	5.04
	.01	4.05	4.67	5.05	5.33	5.55	5.73	5.89	6.02	6.14
20	.05	2.95	3.58	3.96	4.23	4.45	4.62	4.77	4.90	5.01
	.01	4.02	4.64	5.02	5.29	5.51	5.69	5.84	5.97	6.09
24	.05	2.92	3.53	3.90	4.17	4.37	4.54	4.68	4.81	4.92
	.01	3.96	4.55	4.91	5.17	5.37	5.54	5.69	5.81	5.92
30	.05	2.89	3.49	3.85	4.10	4.30	4.46	4.60	4.72	4.82
	.01	3.89	4.45	4.80	5.05	5.24	5.40	5.54	5.65	5.76
40	.05	2.86	3.44	3.79	4.04	4.23	4.39	4.52	4.63	4.73
	.01	3.82	4.37	4.70	4.93	5.11	5.26	5.39	5.50	5.60
60	.05	2.83	3.40	3.74	3.98	4.16	4.31	4.44	4.55	4.65
	.01	3.76	4.28	4.59	4.82	4.99	5.13	5.25	5.36	5.45
120	.05	2.80	3.36	3.68	3.92	4.10	4.24	4.36	4.47	4.56
	.01	3.70	4.20	4.50	4.71	4.87	5.01	5.12	5.21	5.30

Note. This table is abridged from Table 29 in E.S. Pearson and H.O. Hartle (Eds.), *Biomerrika tables for statisticians* (3rd ed., Vol 1), Cambridge University Press, 1970.

References

Allison, P. D. (2002). *Missing data.* Thousand Oaks, CA: Sage.

American Psychological Association (APA). (2001). *Publication manual of the American Psychological Association* (5th ed.). Washington, DC: Author.

Anderson, N. N. (2001). *Empirical direction in design and analysis.* Mahwah, NJ: Lawrence Erlbaum Associates.

Baxter, B. (1940). The application of factorial design to a psychological problem. *Psychological Review, 47,* 494–500.

Baxter, B. (1941). Problems in the planning of psychological experiments. *American Journal of Psychology, 54,* 270–280.

Boik, R. J. (1979). Interactions, partial interactions, and interaction contrasts in the analysis of variance. *Psychological Bulletin, 86,* 1084–1089.

Bonate, P. L. (2000). *Analysis of pretest-posttest designs.* London: Chapman and Hall.

Bruning, J. L., & Kintz, B. L. (1968). *Computational handbook of statistics.* Glenview, IL: Scott Foresman.

Brown, M. B., & Forsythe, A. B. (1974). Robust tests for the equality of variances. *Journal of the American Statistical Association, 69,* 364–367.

Campbell, D. T. (1957). Factors relevant to the validity of experiments in social settings. *Psychological Bulletin, 54,* 297–312.

Campbell, D. T., & Stanley, J. C. (1963). *Experimental and quasi-experimental designs for research.* Chicago: Rand McNally.

Campbell, D. T., & Stanley, J. C. (1966). *Experimental and quasi-experimental designs for research.* Chicago: Rand McNally.

Cardinal, R. N., & Aitken, M. R. F. (2006). *ANOVA for the behavioural sciences researcher.* Mahwah, NJ: Lawrence Erlbaum Associates.

Carmer, S. G., & Swanson, M. R. (1973). An evaluation of ten pairwise multiple comparison procedures by Monte Carlo methods. *Journal of the American Statistical Association, 68,* 66–74.

Clark, H. H. (1973). The language-as-fixed-effect fallacy: A critique of language statistics in psychological research. *Journal of Verbal Learning and Verbal Behavior, 12,* 335–359.

Clark-Carter, D. (1997). The account taken of statistical power in research published in the *British Journal of Psychology. British Journal of Psychology, 88,* 71–83.

Cochran, W. G., & Cox, G. M. (1957). *Experimental designs* (2nd ed.). New York: Wiley.

Cody, R. P., & Smith, J. K. (2006). *Applied statistics and the SAS programming language* (5th ed.). Upper Saddle River, NJ: Pearson Prentice Hall.

Cohen, B. H. (1996). *Explaining psychological statistics.* Pacific Grove, CA: Brooks/Cole.

Cohen, J. (1962). The statistical power of abnormal-social psychological research: A review. *Journal of Abnormal and Social Psychology, 65,* 145–153.

Cohen, J. (1969). *Statistical power analysis for the behavioral sciences.* New York: Academic Press.

Cohen, J. (1977). *Statistical power analysis for the behavioral sciences* (revised ed.). New York: Academic Press.

Cohen, J. (1988). *Statistical power analysis for the behavioral sciences* (2nd ed.). Hillsdale, NJ: Lawrence Erlbaum Associates.

Cohen, J., Cohen, P., West, S. G., & Aiken, L. S. (2003). *Applied multiple analysis for the behaviorial sciences* (3rd ed.). Mahwah, NJ: Lawrence Erlbaum Associates.

Coleman, E. B. (1964). Generalizing to a language population. *Psychological Reports, 14,* 219–226.

Collyer, C. E., & Enns, J. T. (1987). *Analysis of variance: The basic designs.* Chicago: Nelsen-Hall Publishers.

Cook, T. D., & Campbell, D. T. (1979). *Quasi-experimentation: Design and analysis issues for field settings.* Boston: Houghton Mifflin.

Crutchfield, R. S. (1938). Efficient factorial design and analysis of variance illustrated in psychological experimentation. *Journal of Psychology, 5,* 339–346.

Davis, J. B. (2007). *Statistics using SAS Enterprise Guide.* Cary, NC: SAS Institute.

Der, G., & Everitt, B. S. (2007). *Basic statistics using Enterprise Guide: A primer.* Cary, NC: SAS Institute.

Duncan, D. B. (1955). Multiple range and multiple F tests. *Biometrics, 11,* 1–42.

Dunnett, C. W. (1955). A multiple comparisons procedure for comparing several treatments with a control. *Journal of the American Statistical Association, 50,* 1096–1121.

Durbin, J., & Watson, G. S. (1950). Testing for serial correlation in least squares regression I. *Biometrika, 32,* 409–438.

Durbin, J., & Watson, G. S. (1951). Testing for serial correlation in least squares regression II. *Biometrika, 38,* 159–178.

Durbin, J., & Watson, G. S. (1971). Testing for serial correlation in least squares regression III. *Biometrika, 58,* 1–19.

Einot, I., & Gabriel, K. R. (1975). A study of the powers of several methods of multiple comparisons. *Journal of the American Statistical Association, 70,* 574–583.

Everitt, B. S. (2001). *Statistics for psychologists: An intermediate course.* Mahwah, NJ: Lawrence Erlbaum Associates.

Field, A. (2005). *Discovering statistics using SPSS* (2nd ed.). London: Sage.

Fisher, R. A. (1921a). Some remarks on the methods formulated in a recent article on the qualitative analysis of plant growth. *Annals of Applied Biology, 7,* 367–372.

Fisher, R. A. (1921b). Studies in crop variation. I. An examination of the yield of dressed grain from Broadbalk. *Journal of Agricultural Science, 11,* 107–135.

Fisher, R. A. (1925). *Statistical methods for research workers.* Edinburgh, England: Oliver & Boyd.

Fisher, R. A. (1935). *The design of experiments.* Edinburgh, England: Oliver & Boyd.

Fisher, R. A., & Eden, T. (1927). Studies in crop variation. IV. The experimental determination of the value of top dressings with cereals. *Journal of Agricultural Science, 17,* 548–562.

Fisher, R. A., & Mackenzie, W. A. (1923). Studies in crop variation. II. The manorial responses of different potato varieties. *Journal of Agricultural Science, 13,* 311–320.

Garrett, H. E., & Zubin, J. (1943). The analysis of variance in psychological research. *Psychological Review, 40,* 233–267.

Glass, G. V., Peckham, P. D., & Sanders, J. R. (1972). Consequences of failure to meet assumptions underlying the analysis of variance and covariance. *Review of Educational Research, 42,* 237–288.

Guilford, J. P., & Fruchter, B. (1978). *Fundamental statistics in psychology and education* (6th ed.). New York: McGraw-Hill.

Hartley, H. O. (1950). The maximum F-ratio as a short-cut test for heterogeneity of variance. *Biometrika, 37,* 308–312.

Hays, W. L. (1981). *Statistics* (3rd ed.). New York: Holt, Rinehart and Winston.

Honeck, R. P., Kibler, C. T., & Sugar, J. (1983). *Experimental design and analysis: A systematic approach.* Lanham, MD: University Press of America.

Howell, D. C. (1997). *Statistical methods for psychology* (4th ed.). Belmont, CA: Duxbury.

Huberty, C. J., & Morris, J. D. (1989). Multivariate analysis versus univariate analyses. *Psychological Bulletin, 105,* 302–308.

Hummel, T. J., & Sligo, J. (1971). Empirical comparison of univariate and multivariate analysis of variance procedures. *Psychological Bulletin, 76,* 49–57.

Jaccard, J., Becker, M. A., & Wood, G. (1984). Pairwise multiple comparison procedures: A review. *Psychological Bulletin, 54,* 589–596.

Keppel, G. (1973). *Design and analysis: A researcher's handbook.* Englewood Cliffs, NJ: Prentice Hall.

Keppel, G. (1982). *Design and analysis: A researcher's handbook* (2nd ed.). Englewood Cliffs, NJ: Prentice Hall.

Keppel, G. (1991). *Design and analysis: A researcher's handbook* (3rd ed.). Englewood Cliffs, NJ: Prentice Hall.

Keppel, G., & Saufley, W. H. (1980). Introduction to design and analysis: A student's handbook (1st ed.). San Francisco: W. H. Freeman.

Keppel, G., Saufley, W. H., & Tokunaga, H. (1992). *Introduction to design and analysis: A student's handbook* (2nd ed.). New York: W. H. Freeman.

Keppel, G., & Wickens, T. D. (2004). *Design and analysis: A researcher's handbook* (4th ed.). Upper Saddle River, NJ: Pearson Prentice Hall.

Keppel, G., & Zedeck, S. (1989). *Data analysis for research designs.* New York: Freeman.

Keuls, M. (1952). The use of the Studentized range in connection with an analysis of variance. *Euphytica, 1,* 112–122.

Kinnear, P. R., & Gray, C. D. (2006). *SPSS 14 made simple.* Hove, East Essex, England: Psychology Press.

Kirk, R. E. (1995). *Experimental design: Procedures for the behavioral sciences* (3rd ed.). Pacific Grove, CA: Brooks/Cole.

Kirk, R. E. (1996). Practical significance: A concept whose time has come. *Educational and Psychological Measurement, 56,* 746–759.

Kline, R. B. (2004). *Beyond significance testing.* Washington, DC: American Psychological Association.

Klockars, A. J., & Sax, G. (1986). *Multiple comparisons.* Newbury Park, CA: Sage.

Levene, H. (1960). Robust tests for equality of variances. In I. Olkin (Ed.), *Contributions to probability and statistics.* Palo Alto, CA: Stanford University Press.

Likert, R. (1932). A technique for the measurement of attitudes. *Archives of Psychology, 140,* 5–53.

Likert, R., Roslow, S., & Murphy, G. (1934). A simple and reliable method of scoring the Thurstone attitude scales. *Journal of Social Psychology, 5,* 228–238.

Lindquist, E. F. (1940). *Statistical analysis in educational research.* Boston: Houghton Mifflin.

Lindquist, E. F. (1953). *Design and analysis of experiments in psychology and education.* Boston: Houghton Mifflin.

Mauchly, J. W. (1940). Significance test for sphericity of *n*-variate normal populations. *Annals of Mathematical Statistics, 11,* 37–53.

Maxwell, S. E., & Delaney, H. D. (2000). *Designing experiments and analyzing data: A model comparison perspective.* Mahwah, NJ: Lawrence Erlbaum Associates.

Maxwell, S. E., Delaney, H. D., & O'Challaghan, M. F. (1993). Analysis of covariance. In L. K. Edwards (Ed.), *Applied analysis of variance in behavioral science.* Taipei city, Taiwan, Province of China: CRC Press.

McKnight, P. E., McKnight, K. M., Sidani, S., & Figuerdo, A. J. (2007). *Missing data: A gentle introduction.* New York: Guilford Press.

Meyers, L. S., Gamst, G., & Guarino, A. J. (2006). *Applied multivariate research: Design and interpretation.* Thousand Oaks, CA: Sage.

Murphy, G., & Likert, R. (1937). *Public opinion and the individual.* New York: Harper.

Myers, J. L., & Well, A. D. (1991). *Research design and statistical analysis.* New York: HarperCollins.

Newman, D. (1939). The distribution of the range in samples from a normal population, expressed in terms of an independent estimate of the standard deviation. *Nutrition Research, 10,* 525–533.

Oehlert, G. W. (2000). *A first course in design and analysis of experiments.* New York: W. H. Freeman.

Page, M. C., Braver, S. L., & MacKinnon, D. P. (2003). *Levine's guide to SPSS for analysis of variance* (2nd ed.). Mahwah, NJ: Lawrence Erlbaum Associates.

Pawar, M. (2004). *Data collecting methods and experiences: A guide for social researchers.* Chicago: New Dawn Press.

Pearson, E. S., & Harley, H. O. (1970). *Biometrika tables for statisticians* (Vol. 1, 3rd ed.). London: Cambridge University Press.

Pearson, E. S., & Harley, H. O. (1972). *Biometrika tables for statisticians* (Vol. 2, 3rd ed.). London: Cambridge University Press.

Pedhazur, E. J., & Schmelkin, L. P. (1991). *Measurement, design, and analysis: an integrated approach.* Hillsdale, NJ: Lawrence Erlbaum Associates.

Rubin, D. B. (1987). *Multiple imputation for nonresponse in surveys.* New York: Wiley.

Rubin, D. B. (1996). Multiple imputation after 18+ years. *Journal of the American Statistical Association, 91,* 473–489.

Ryan, T. A. (1960). Significance tests for multiple comparison of proportions, variances and other statistics. *Psychological Bulletin, 57,* 318–328.

Salsburg, D. (2001). *The lady tasting tea: How statistics revolutionized Science in the twentieth century.* New York: W. H. Freeman.

SAS Institute. (2002). *Getting started with SAS Enterprise Guide* (2nd ed.). Cary, NC: SAS Institute.

Scheffé, H. (1953). A method for judging all contrasts in the analysis of variance. *Biometrika, 40,* 87–104.

Scheffé, H. A. (1959). *The analysis of variance.* New York: Wiley.

Shadish, W. R., Cook, T. D., & Campbell, D. T. (2002). *Experimental and quasi-experimental designs.* Boston: Houghton Mifflin.

Sidák, Z. (1967). Rectangular confidence regions for the means of multivariate normal distributions. *Journal of the American Statistical Association, 62,* 626–633.

Slaughter, S. J., & Delwiche, L. D. (2006). *The little SAS book for Enterprise Guide 4.1.* Cary, NC: SAS Institute.

Snedecor, G. W. (1934). *Analysis of variance and covariance.* Ames, IA: Collegiate Press.

Snedecor, G. W. (1946). *Statistical methods* (4th ed.). Ames, IA: The Iowa State College Press.

Stevens, J. (2002). *Applied multivariate statistics for the social sciences* (4th ed.). Mahwah, NJ: Erlbaum.

Stevens, S. S. (1946). On the theory of scales of measurement. *Science, 103,* 677–680.

Stevens, S. S. (1951). Mathematics, measurement, and psychophysics. In S. S. Stevens (ed.), *Handbook of experimental psychology* (pp. 1–49). New York: Wiley.

Tabachnick, B. G., & Fidell, L. S. (2001). *Computer-assisted research design and analysis.* Boston: Allyn and Bacon.

Tabachnick, B. G., & Fidell, L. S. (2007). *Experimental designs using ANOVA.* Belmont, CA: Thomson Brooks/Cole.

Thompson, B. (2002). "Statistical," "practical," and "clinical": How many kinds of significance do counselors need to consider? *Journal of Counseling and Development, 80,* 64–71.

Thompson, B. (2006). *Foundations of behavioral statistics: An insight-based approach.* New York: Guilford Press.

Thurstone, L. L. (1927). A law of comparative judgment. *Psychological Review, 34,* 273–286.

Thurstone, L. L. (1928). Attitudes can be measured. *American Journal of Sociology, 33,* 529–554.

Thurstone, L. L. (1929). Theory of attitude measurement. *Psychological Review, 36,* 222–241.

Thurstone, L. L., & Chave, E. J. (1929). *The measurement of attitude.* Chicago: University of Chicago Press.

Toothaker, L. E. (1993). *Multiple comparison procedures.* Newbury Park, CA: Sage.

Tukey, J. W. (1952). Allowances for various types of error rates. Unpublished IMS address (cited in Oehlert, 2000).

Weinfurt, K. P. (1995). Multivariate analysis of variance. In L. G. Grimm and P. R. Yarnold (Eds.), *Reading and understanding multivariate statistics* (pp. 245–276). Washington, DC: American Psychological Association.

Welsch, R. E. (1977). Stepwise multiple comparisons procedures. *Journal of the American Statistical Association, 72,* 566–575.

Wilcox, R. R. (1987). New designs in analysis of variance. *Annual Review of Psychiatry, 32,* 29–60.

Wilkinson, L., & the Task Force on Statistical Inference. (1999). Statistical methods in psychology journals: Guidelines and explanations. *American Psychologist, 54,* 594–604.

Winer, B. J. (1962). *Statistical principles in experimental design.* New York: McGraw-Hill.

Winer, B. J. (1971). *Statistical principles in experimental design* (2nd ed.). New York: McGraw-Hill.

Winer, B. J. Brown, D. R., & Michels, K. M. (1991). *Statistical principles in experimental design* (3rd ed.). New York: McGraw-Hill.

Yates, F. C. (1937). *The design and analysis of factorial experiments.* Harpenden, England: Imperial Bureau of Soil Science.

Author Index

Subject Index